A Science Career Against all Odds

Bernhard Wunderlich

A Science Career Against all Odds

A Life of Survival, Study, Teaching and Travel in the 20th Century

Springer

Prof. Dr. Bernhard Wunderlich
200 Baltusrol Road
Knoxville TN 37934-3707
USA
Wunderlich@CharterTN.net

Additional material to this book can be downloaded from http://extras.springer.com

ISBN 978-3-642-42608-7 ISBN 978-3-642-11196-9 (eBook)
DOI 10.1007/978-3-642-11196-9
Springer Heidelberg Dordrecht London New York

Softcover re-print of the Hardcover 1st edition 2010

Cover design: WMXDesign GmbH, Heidelberg

Printed on acid-free paper

Springer is part of Springer Science+Business Media (www.springer.com)

Preface

Today is Sunday, June 17, 2007. Father's Day. Naturally, the obligatory, carefully selected cards, phone calls, and small gifts arrived from the children and grandchildren. Best wishes for Father's Day were also the first words in the morning from Heidel, my wife of 54 years, although for many years I had made the comment: "I am not your father." But, in the frame of my life's experiences in the 20th century, as I intend to summarize them over the next few years, the 17th of June has much deeper significance. This was the day in 1953 when we finally fled from our life of oppression which had lasted 20 years. Two successive dictatorships, one of Hitler and the other of Stalin, caused the most horrific slaughter of civilians and soldiers, eclipsing all prior history. During these first years of my life, I was plainly lucky to survive. After this day, I had a much better chance to experience the freedom needed to lead a life of creativity, satisfaction, and ultimately prosperity, all directed largely by our own decisions.

The 17th of June 1953 was a Wednesday. I stayed in the apartment of my parents in my hometown of Brandenburg, in the German Democratic Republic (GDR), the former Russian occupied zone of Germany. The summer vacation of the Humboldt University in East Berlin, some 40 mi further east, had just started. But, I was alone with my father, "Vati." My mother, "Mutti," and two little brothers, Joachim and Manfred,[1] had left several weeks earlier to the safety of West Berlin for an indeterminate stay with my Tante Minna (Mutti's sister). We were to follow in a week or two after completing the surreptitious closing of Vati's roofing business. Despite voluntary 'downsizings' since the end of the war in 1945 to less than 10%, it was leaked to us by friends in the SED (the "Sozialistische Einheitspartei Deutschlands," the political party controlled by the Soviet Union) that the label 'capitalistic enterprise' was unavoidable and that take-over by the state was to follow. Somewhat on purpose, the week before, I had failed an examination in'Marxism-Leninism,' the required lectures for students of science. I had never attended these lectures. Heidel, at that time my fiancée and law student at the same university, had to flee several weeks earlier to West Berlin because of conflicts of conscience in working on exercises which included indictments of political detainees which might have been used later for the actual trials, ending commonly with inhuman punishments. My older brother, Frank, who worked close to Berlin in government employ, also had just fled the GDR with his wife, Dagmar, mother in law, and baby son, Eberhard.[1] Eventually, we all were recognized as political refugees, attained the rights of citizens of the Federal Republic of Germany (FRG), and received financial support to get on our feet.

On this Wednesday morning, I only wanted to go to town and purchase a new Zeiss microscope. I had convinced the sales lady in the state-owned department store ("HO") that I urgently needed it as a student of chemistry. The cost was to be 400 DM-Ost, a price equivalent to my whole summer's stipend. Since I did not intend to return to the university, it would have been safe to resell the microscope in the FRG. It had to be registered in my name and then would have been forbidden to be exported. After being safely in the FRG, however, I could not be held

---

[1] A brief genealogy to identify the various relatives is given on pages 11-1–4.

responsible for it. The price of such desirable objects in the FRG was the same in DM-West and could have been our support for the first months of freedom. Otherwise, conversion of DM-Ost to DM-West was at that time 10:1 in our disfavor.

Joseph Stalin had died about three months before, and the firm grip of the Soviet Union on the GDR seemed to loosen and discontent with the government was beginning to be voiced more openly. People were hoping for a change. In East Berlin, a strike was called on Tuesday to protest against the increased production quotas. When the regime failed to respond immediately, the workers took to the streets and demanded also a change in government. The rebellion, then, quickly spread throughout the GDR. When I walked up the "Steinstraße" in Brandenburg, at about 10 a.m., a surprising scene showed in front of the Court House. A noisy crowd had gathered as shown in Figure 1 (taken an hour later by a news photographer). Shouts were heard to release political prisoners and many court documents were scattered in the street. I would have loved to join the crowd, but was aware of the danger. Our plan to unite the family in West Berlin would have been put in jeopardy. Soon, the borders would be sealed. I hastened to the department store in the "Hauptstraße," but to my dismay, it showed a hurriedly hand-written sign 'Closed Today.' Plan B had to be enacted. I walked quickly the 20 min to Vati's business office in the Werder Straße 28. Vati and I decided, we must leave immediately. All unfinished business had to be left undone. With only our briefcases, we got railway tickets to Potsdam. To Berlin, none were sold anymore, an ominous sign. This was reinforced by the conductor who said, today, you will not make it to West Berlin anymore.

Figure 1: *In front of the Court House on June 17, 1953.*

Indeed, from Potsdam all the "S-Bahn" trains to East Berlin which went through West Berlin were cancelled. Rumors circulated that Russian tanks were rolling toward Berlin. The only thing we could find, was a slow, local train which bypassed West Berlin in a circle, to ultimately enter East Berlin from the south, taking more than an hour. We took it, but after half an hour, it also stopped on the tracks near Teltow. This was one of the first suburbs straddling West Berlin's southern border, and many people were walking this afternoon along a foot trail leading through the border fence into and out of West Berlin. The trail, patrolled by the East German police, reinforced by the army, was used by West Berliners who worked in the East and vice versa. There were barely any guards on the side of West Berlin. The city was not fully divided at this time. "Die Mauer" (the Wall) was erected only in 1961. There even was an official money-exchange system in place. Anyone with a job in the 'other' Germany got his or her wages in the currency of the home address, and naturally had papers to freely cross the border (as long as no contraband was detected by the border guards). With only briefcases as luggage, we left the train, crossed the tracks, and blended into this crowd of workers on their way home. With beating hearts we approached the re-enforced and armed border guards. I went first. As a student in East Berlin, I had an identification card stating in German, Russian, French, and English that in any emergency, I was to be given every help to reach the university (see on pages 4-12–13). In this case, it was best to travel through West Berlin. The East German guard looked at the card and let me pass. I immediately took off, not looking back to Vati. At least one of us made it across! This was wisdom, gained from the weekly border crossings when traveling between Brandenburg and the university. From a safe distance, I could then see that the guard had confiscated Vati's passport and held it in his hand with several others. Vati stood behind the fence, a few steps aside, among the other persons forbidden to cross. What was I to do? As usual, I knew that no clever argument would help. So, I played as calm, relaxed, and disinterested as possible. Waiting a few minutes, so that some 20 or 30 others had crossed the border and hopefully dimmed the memory of the guard, I walked back the few steps and asked Vati through the fence, loud enough so that the guard could hear: "Why didn't you come with me?" He answered: "The guard took my passport, I have to wait." Next, I turned to the guard and asked indignantly: "Why didn't you let my father pass? We both belong together!" My face must have had the right degree of innocence so that he said: "This is your father? I didn't know. He can go." Nothing else was said until we were safely on the West Berlin side. This was a close shave and both of us were fully aware that we just had won big in the lottery of life. If my Vati would have been detained, it would quickly have been found out that he was fleeing East Germany and multi-year incarceration, if not worse, would have been the likely consequence.

Befehl

des Militärkommandanten
des Stadt- und Landkreises Brandenburg
vom 17. Juni 1953

1. Im Stadt- und Landkreis Brandenburg wird der Ausnahmezustand verhängt.
Die ganze Macht geht in die Hände der Militärkommandantur über.
2. Jegliche Versammlungen, Demonstrationen im Stadt- und Landkreis werden strengstens verboten.
3. Gastwirtschaften, Kaffeehäuser, Kinos, Theater und andere öffentliche Gebäude werden bis auf weiteres geschlossen.
4. Der Aufenthalt auf den Straßen der Stadt Brandenburg und im Landkreis Brandenburg von 20 Uhr bis 6 Uhr ohne besondere Genehmigung ist verboten. Transportverkehr in der Stadt und im Landkreis ist außer der Volkspolizei nur mit Bescheinigung gestattet.
5. In volkseigenen und anderen Betrieben sowie in der HO und anderen Geschäften wird weitergearbeitet.
6. Personen, die gegen diesen Befehl verstoßen, werden dem Militärtribunal übergeben.

Militärkommandant
für den Stadt- und Landkreis Brandenburg
Oberstleutnant Ismodenob

Figure 2: *Martial law declaration in Brandenburg. Issued June 17, 1953. Photograph of an original.*

The uprising was soon quelled throughout the GDR. Soviet troops and tanks intervened. At least 21 people were killed, and hundreds of others wounded. In the wave of retribution that followed, some 1,300 were sentenced to prison for taking part in the uprising, which the East German government portrayed as a plot by West Germany and the United States. A copy of the declaration of martial law in Brandenburg is reproduced in Figure 2. It was posted later in the day, after we had left. Interestingly, it was not proclaimed by the civilian government of the city, but by the Soviet Military Command who supposedly had relinquished control to the civilian government. We all knew, they had not! The hesitation of the Soviets had passed. Revolution and free elections could certainly not be permitted, as was proven later in the brutal suppression of the uprisings in Hungary and Czechoslovakia. The six points made in the martial law declaration of Figure 2 translate as follow:

1. Martial Law is declared. All control is in the hand of the Soviet Commandant.
2. No demonstrations and meetings are permitted.
3. All public buildings are closed.
4. From 8 p.m. to 6 a.m. there exists a curfew.
5. State-owned business continues to operate.
6. Persons who violate this order will be turned over to the military tribunal.

At the Teltow border, however, we had not yet made it to our family in West Berlin (Rudow, a suburb). A march of about 15 mi was necessary. There was a street-car terminal after about one-half mile. But we had no western currency! In the spirit of the Berliner who rises to any emergency, the conductor told us: "Today there is a revolution in the East. We hope they will get rid of the communists. You do not have to pay today." He gave us a ticket to anywhere we wanted to go. Happily we arrived about an hour later at our destination. It was about 6:45 p.m. by then. The family had anxiously gathered around the radio and listened to the reports about freeing of political prisoners in Brandenburg, looting of the court house and other government buildings, and ultimately, the actions of the Soviet troops, leading to the closing of the border, arrests, and shootings.

For us, the uprising was over. Fortunately, it ended without harm to our family. A new life could begin in the second half of the 20th century. This new start was again from level zero, quite similar to the end of the war in 1945. Figure 3 catches one of the many serious discussions needed to plan the future between Vati and Onkel Paul. The only way to get out of West Berlin and reach the Federal Republic of Germany (FRG) was now to be flown out by plane.

Figure 3: *Vati and Onkel Paul a few days after the 17th of June discussing our future. The garden of Onkel Paul was close to the border and one can see the watch-tower with armed guards in the distance, keeping us in West Berlin.*

To fully understand the significance of this new beginning, additional episodes require to be recounted and other questions need to be tackled: How did our whole family survive the two dictatorships? How did I, eventually, come to study chemistry? How could we achieve so much when starting with so little? How were childhood and education affected by the environment we grew up in? How was our further life affected by war and oppression? How did the teaching change toward the end of the century and how could it be improved? How did travel fit into the professional and private life? These and many other questions will be tackled as they arise out of the overall description of the customs of society which, in my opinion, are not always positive. Since I like photography, pictures out of my collection will be used to document the narrative and thus save 'thousands of words.'

The later life was naturally heavily intertwined with my effort to research, understand, and teach the chosen profession, polymer science. The three appendices available with the on-line version of the book display the details. Appendix A is a catalogue of all scientific publications and a list of my students and coworkers. Appendix B contains a newly updated manuscript of a personal review: 'Thermal Analysis of Macromolecules,' the central field for all my work. Appendix C, finally, is a listing of the various places we lived at and of our travels throughout the world for science sake (with lecture titles), for pleasure, and for what I like to call the 'need to see.' The Epilogue, Chapter 10, contains a brief conclusion and an outlook into the beginning of the 21st century. On pages 11-01–12, there is a brief, illustrated genealogy. It should be consulted to make the connection between the family members. The Table of Contents, below, is designed so that its quick 'reading' points to the general aspects of life, profession, politics, and travel. The Indices, finally offer access to people that crossed my path in person or by action, and serve as an extensive listing of the subjects, objects, places, and topics mentioned in more than passing in the text.

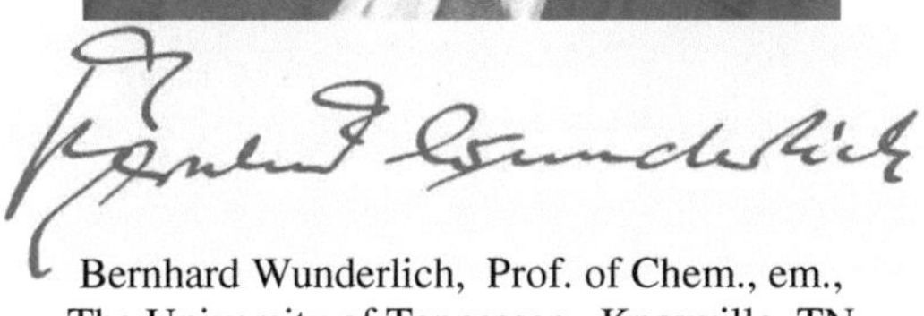

Bernhard Wunderlich, Prof. of Chem., em.,
The University of Tennessee, Knoxville, TN
and Rensselaer Polytechnic Institute, Troy, NY

6/17/2007

Acknowledgments

First and foremost, I would like to acknowledge the more than 55 years of unwavering support by my wife, Heidel (Heidi). Without her help, many of the projects simply could not have been completed. I hope she always meant it only jokingly when she said on occasion: "I am only wife number two. Wife number one is 'Mrs. Chemistry.' " Similarly, our children had to grow up knowing that often research, teaching, and travel had to take precedence over what other people may call a 'normal' family life.

Next, it must be acknowledged that without coworkers, little could have been accomplished in quality and quantity of research. The 144 postdoctoral fellows, graduate students, and technicians contributed 340 person-years to the work listed in Appendix A. In addition, many friends and colleagues contributed with constructive discussion and criticism, and sometimes also by participation in joint publications (see Appendix A). Much progress is based on such interaction and exchange of ideas. Perhaps, it should be mentioned that there also were the not so friendly comments from colleagues who were sharply negative. For me, these also were valuable. They were incentives to do better in the future and to offer more precise explanations, so that the different approaches which I liked to take, would be understood by everyone.

The limited space does not permit to give a full appreciation of the importance of the ideas drawn from the literature. More about them is summarized in Part I of Appendix B, where my publications are listed which detail the thousands of literature sources. Much of my knowledge about the link to industrial applications grew out of the connection I had for over 25 years as an exclusive consultant for E. I. du Pont de Nemours and Company (see Appendix C, page C-4).

Support for instrumentation in thermal analysis was received from the three major equipment makers: TA Instruments, Inc., Mettler-Toledo, Inc., and the Perkin-Elmer Corp. Without this help we would not always have been able to be at the forefront of experimentation.

Financial support for research came from outside the universities. Over the 50 years, about five million dollars were spent by government agencies, companies, and private foundations to support coworkers, myself, research equipment, expenses, and travel. This is little, when compared to expenditures needed to support 'research centers' or to build 'megamachines.' It amounts to about $10,000 per publication, a modest amount when compared to the knowledge and ideas gained.

Sustained funding came from the National Science Foundation (Division of Materials Research, Polymers Program, where particular thanks must go to the long-time program directors Drs. N. M. Bikales and A. J. Lovinger). During 1988–2001, additional support was connected to my position as Distinguished Scientist at UTK and ORNL, as described in the Footnote to page 9-6. On my arrival in Knoxville in the year 1988, ORNL was managed for the U.S. Department of Energy by Martin Marietta Energy Systems, Inc. In 1995, the management was taken over by Lockheed Martin Energy Research, Inc. Since the year 2000, ORNL is managed and operated by UT-Battelle, LLC. The entire list of the supporting agencies and companies is given in Chapter 7 in the Footnote to page 7-5.

This autobiography, naturally, did not only rely on my memory. Many relatives had to contribute data. A major help were Heidel's recollections of names, dates, facts, and places. Finally, both Heidel and Brent were of great support in proofreading the various versions as they were generated during the concentrated writing stage.

Frequent use to reconstruct the past was also made of the accumulated large number of memorabilia. For our personal life, there were the many photographs in the 95 albums assembled by Heidel which contained not only pictures, but also other items linking to events of importance that happened over the years. These included the following: Occasional airline tickets (with itineraries dates, and prices), event tickets, hotel and restaurant bills and sometimes also the menus, announcements (like Figures 16 of Chapter 5 and 11 in Chapter 10), newspaper articles (like seen on page 8-122 and as Figure 6 in Chapter 6), dried flowers, cards (received on special occasions), letter excerpts, etc. From my years of study, there were report cards (as shown in Figure 16 of Chapter 1 and Figure 9 of Chapter 3), high school and college papers, study books (as shown in Figure 6 in Chapter 4), "Übungsscheine," graduation certificates (as reproduced on page 6-23), etc. Also kept were the basic science and mathematic books and notes (as discussed on pages 4-17–18). For the later professional life there existed, from 1963 on, a full line of 'annual reports of activities.' These reports were continued, even when not required, until the end of 2009 and allowed to recall the happenings of every year. They were more valuable to me now, than for the people who originally requested them. The feedback when reporting special events and recognitions (which I rarely advertised otherwise) was minimal. This makes me believe that hardly anyone ever read those reports, particularly after 1964, when my tenure had been decided.

The lists of publication, lectures, and coworkers were also started in 1963 and proved sufficiently valuable to be attached as Appendices A and C. The personal review "Thermal Analysis of Macromolecules" found its start in 1996 and was regularly updated. Its final form is found as Appendix B, completed in 2009. Until I gave up my office at UTK in 2009, a full collection of all correspondence was available as backup of my memory. It was stored in four standard filing cabinets and has resisted the occasional onslaughts of efficiency efforts by my secretaries. (One secretary even tore all files that had no entry in the prior five years and threw them into a large waste container for disposal—fortunately there was no shredder. When I discovered the 'clean up' in progress, it took several weeks to resurrect the files and tape the torn letters.)

It was fortunate that the bombing of our house in 1945 involved no fire (see Figure 9 of Chapter 2) and personal papers and books were not of value to looters. The two periods as refugees were such that, with some effort, many of the memorabilia could be retained. The major move to the US contained the large amount of baggage documented in Figure 21 of Chapter 11. A good part of this was the memorabilia and my basic books of study. The last big move in 1988 filled the moving van which is shown in Figure 114 of Chapter 8 with no restriction on weight.

Helpful to verify facts and in the search for historical information were: The internet, chronics (like "Chronik der Menschheit," Chronik Verlag Harenberg, Dortmund, 1988), encyclopedias (like "Encyclopædia Britannica," printed editions of 1771, 1911, 1973, DVD edition 2007, and "Brockhaus' Konversations Lexikon," Brockhaus, Leipzig, 1895), and various dictionaries helped with the proper wording. To find the most suitable German–English and reverse translations, the ≈1500 pages of "Langenscheidts Handwörterbuch" (Langenscheidt, Berlin, 1981) helped enormously. Useful internet sites existed about the political developments in Germany from 1945 to 1990. The comments about most of the over 50 sites on the UNESCO World Heritage List we visited over the years were checked with: http://whc.unesco.org/en/list. The official web sites of the universities I was connected with, offered succinct details about the founding of the schools and their goals, often not known to me in such detail before. Finally, the days of the week and of the holidays mentioned could easily be confirmed at the site http://www.timeanddate.com/calendar.

Contents

Front Material

CHAPTER-04 *"Humboldt University in Berlin, 1949–53"*

Pages:

CHAPTER-07 ***"Cornell University, 1958–63"***

Pages:

CHAPTER-10 "*Epilogue*"

Pages:

thick-walled ice-house. The ice was cut from the nearby lakes during the winter and sold in the summer. My Opa (Chapter 11, Figure 3) told us stories about the roofers who had little work during winter due to the snow and would find additional income from the lakes by cutting ice for summer delivery and also thatch as roofing material. For us, the yard was a playground for fun and mischief. Particularly attractive was the big bin with sand, needed for making the mortar. It took the place of a beach for building sand castles. A large shed was filled with old and dusty wagons, not in use for many years. We would climb up to the driver's seat and pretend to handle the horses. Soon, we also learned how to sneak into the decrepit, dark ice-house via a back door. It had huge blocks of stone strewn all over the floor. What fun to play hide and seek in the dark.

Figure 8 *The first automobile in the Wunderlich Family, named Archibald, a 1932 Opel P4, shown during one of the first motorized family outings.*

Two more pictures are shown on this page. Figure 8 illustrates <u>*our first car*</u>. Its purchase in 1935 was a major event. Vati had to take driving lessons which involved not only learning the traffic rules and how to steer and shift the car, but also studying the internal workings of the automobile. We children wanted to learn with him. Since this time, I know the mechanism of a four-cylinder engine and what the gear shift is for. I was also interested to learn about the later installed heater. It made use of the heat from the engine block (not the radiator). Heaters were not standard equipment, but the flower vases between the side windows were, as were the directional signals mounted on the left and right of the windshield. They were lifted by magnetic coils and glowed with a red light. To start the engine, one had to set a number of controls and crank the engine by hand. This was too difficult for me, but driving on the lap of Vati on deserted roads was sometimes permitted. Mutti, in winter (with sufficient coaching and for special fun), was once our driver and she pulled the three male Wunderlichs on a sled behind the car until the cord broke and we all tumbled into the snow. More common than the motor car excursions, however, were Sunday afternoon walks, children in front (holding hands) as seen in Figure 9. Thus, the early period until 1936, I would consider as my 'normal,' early childhood.

Figure 9 *Easter 1936. The family in their Sunday's best, walking to a lake-side Café.*

A few comments about the first part of the early childhood concern some recollections, the family structure, and the position we children had in the family. When Frank started school *in 1934, life changed for me*. From then on, I was alone in the mornings, and pretty soon, he also had to do homework in the afternoons. So, I often had to find things to do by myself. There were no children of my age in the neighborhood, the early 1930s were years of low birthrates.

My entertainment was to play in the back yard, surrounded on three sides by wings of the apartment houses. On the fourth, separated by a tall brick wall was a rare one-family home. This unique house, surrounded by apartment houses, we called the "Villa." It faced the Kleist Straße and still stands today. A drive curved around the house and led to a garage in back. A garden area sported an artificial grotto. When climbing a trash-can, we could marvel at this 'palace.' When all seemed deserted, Frank and I could even sneak across and explore the garden and grotto.

In front of our house, there was a small flower garden. In one corner, I had a place to play with my toy animals and cars. It was even fun to squeeze between two fences separating the gardens of adjacent houses and pretend to be in a car or on the bridge of a ship and navigate the street. The iron rods of the fences were bent at the proper height to serve as imaginary steering wheels.

Mutti had not only to do the daily housekeeping and shopping, but also was in charge of the bookkeeping for the business and had to make frequent trips to town. Since I could not always accompany her, I remember the dejected feeling when she was away. As seen from Figures 7 and 9, Frank's duty was to be my babysitter, something he did not cherish and I did not appreciate, but being left alone was worse.

The routine at home went as follows: After our common breakfast, Frank went to school. Vati often was already working out of the house. Particularly in the summer, work would start at sunrise to take advantage of the cool morning hours. All joined for lunch at about 1:00 p.m. for the main meal of the day (" das Mittagessen"). Depending on the season, Vati might stay home for an hour or more to do some paperwork and to rest. Not later than 3:00 p.m. he would take off again to supervise and actively participate in the work until about 6:00 p.m. The supper we had together again. The noon and evening meals were always eaten in the living room.

The work week of 48 h (with longer hours for the boss, Vati) allowed a weekend of half a Saturday and a free Sunday. Frequently one or both of these days saw visits to or from relatives or friends and excursions to the surrounding meadows, woods, and lakes (see Figure 9). During these early years, frequent visits were made to the house of Onkel Richard (II, Vati's brother-in-law), a 15 min walk, or to Oma and Opa Siedlung's apartment and large garden (Mutti's parents), a ¾ h excursion by street car and walking (≈1.5 mi).

Figure 10 *Vacation in the Harz Mountains, 1937–39.*

After the integration of Archibald, our car, into the family routine (see Figure 8), *the range of exploration was expanded*. The Harz mountains at Lautenthal, seen in Figure 10, were our occasional destination for week-long vacations when school was not in session. In the summer, Vati would

deliver the family on a weekend, and pick us up again at the following weekend. In winter, when there was less work to be done, he would stay with us. Of special fun were the Christmas travels to Lautenthal. We left on Christmas morning. The gifts were given, as customary in Germany, on Christmas Eve, and some of the new toys were taken along to play.

Another noteworthy trip of about 125 mi went to Landsberg on the Warthe river, to visit some more distant relatives of Vati [see page 11-2, now Gorzow Wielkopolski (Poland)]. The interesting discoveries on this trip were the 'new' relatives and the fact that they lived next to the fire station. This allowed us to inspect the fire engines and the tall tower for drying of the fire hoses. Other distant relatives from Vati's side visited Brandenburg in 1934 or 1935. They traveled with a large merry-go-round to fairs. It was called "die schwankende Karoline" and not only rotated, but the "Karoline" also moved up and down. I did not see the fun in it and felt only the nausea from the unexpected motion, a feeling which I had again on the Atlantic Ocean, some 20 years later.

Other relatives living not far from us offered a feeling of belonging to a larger family. There was _Vati's father (Opa)_. He was visiting occasionally and later joined Vati's business as a supervisor (after he gave up his own, not very successful enterprise). He had separated from Oma, and it took me quite some time to figure out his relationship to "Tante" Lorchen (see Chapter 11, Figure 3) with whom he lived in the St.-Annen-Straße 32 of Figure 11 [near (1) in Figure 6].

On rare out-of-town stays of Vati and Mutti, we boys were separated. Frank stayed with Oma Siedlung, and I with Tante Lorchen. We rather would have switched baby sitters. Oma Siedlung lived in the "Saarburger Straße" [Figure 6, (10)], not far from her "Schrebergarten" (an allotment garden, championed in the 19th century by M. Schreber, MD). There were more possibilities to play at the "Siedlung" (a development of apartments and one-family houses for workers of the "Mitteldeutsche Stahlwerke," where Opa Siedlung worked). Tante Lorchen, in contrast, was very lively and talkative (which earned her the nickname "Tante Hatata"). Perhaps, I should not have complained, she arranged steamboat trips on the lakes around Brandenburg, and once, I remember being taken to the "Panoptikum" a small curio cabinet in the basement of the city hall ["Rathaus," (1) in Figure 6]. There, I could see my first moving pictures. A large wheel had a stack of thousands of pictures mounted on its rim, showing stages of motion of cowboys and Indians on horseback. On turning it in front of a viewer, a short movie was produced. It was so simple that I could understand the workings of these 'movies.' When I was later taken by Mutti to the "Capitol," a real movie theater, for an afternoon performance of a fairy tale ('Snow White and the Seven Dwarfs,' ≈1937), I looked for and could recognize the analogous sequentially projected pictures by looking back into the projector.

Figure 11 *In the back of Opa's Apartment, ≈1936.*

Oma, Vati's mother lived about ½ mi further toward the city limits in the back wing of the Wilhelmsdorfer Straße 46. She spent much time with Tante Else (Vati's sister) who lived close to her at the end of the short Koppehlstraße. My first recollection of Oma dates back to a walk we took together, probably in 1935, to her apartment. Later in the day, there would then be a family get-

together at Tante Else's house. Oma was holding my hand when walking along the "Wilhelmsdorfer Straße." She was not saying much, but I was telling her long stories, and she was seemingly listening attentively. By now, I think she must have considered problems she had discussed with Mutti before we left. When we passed the side street named Grüne Aue, I told her that I owned the property next to the corner as 'my car park' and that I stored all my automobiles there. She did not react as I expected by telling me that this could not be true. She seemed to accept my imaginary possessions. In reality, this place has, indeed, a large number of private garages which I may have seen earlier. This reaction of Oma bothered me for many years. It ultimately led to my never telling anyone my day dreams, in fear they would be accepted as reality.

At another occasion, Oma displayed her quick handling of an emergency. Frank and I were playing close to her apartment on the ice of the "Büttelhandfaßgraben" (named for the brook where earlier the executioner washed his hands). The creek served as runoff from the often flooded meadows from the city limits to the Havel river. At the point where we played, it widened to about 100 feet and had a reasonable depth. The older boys were making a "Bebbie" (a small quake) by running in step with linked arms over the not too thick ice. After a few passes, the ice would develop many small cracks and produce several-inch-high waves under their feet. For us, this was fun to watch, but should have been a signal that it was not safe to further explore the ice. Next, *Frank broke through the ice* and fell into the water up to his waist. With a little help from me (since we were required to hold hands—see Figures 7 and 9) he made it quickly to the shore. Our closest refuge was Oma. She quickly recognized the situation the dripping Frank was in, undressed him, and put him to bed. Next she told me (!) "watch Frank, I run to the store which has a telephone and call Mutti to get some dry clothes." Since Vati had at that time already a telephone for the business, this was fast action. Indeed, Mutti arrived quickly and all was resolved. Our shock and a long sermon about safety on the ice and the special dangers at the edge of the water was considered sufficient punishment for us when Vati was told in the evening of our misadventure.

Tante Grete, Tante Minna, Tante Frieda, and Tante Lieschen were the four aunts we did not see too often. Tante Grete, Vati's older sister lived in Berlin, about 40 mi from Brandenburg. For a picture of her, see Chapter 11, Figure 8. She worked at the famous Wertheim department store and was their expert in sales of fine china. Much of Mutti's good porcelain came through her connection. She visited occasionally and also was the source of theater, concert, and opera tickets for Vati and Mutti who would on these days drive by car to Berlin in the afternoon and return at about midnight. The two of us stayed home with little supervision. I remember that on one occasion there was a strong thundershower. It was pouring rain as we had not seen before. The drains in the street backed up and pushed more water to the lower-lying houses. The lowest was Vati's yard at the Wilhelmsdorfer Straße 73. It was completely flooded by a running torrent from the street. As soon as the rain stopped, Frank and I went outside and enjoyed jumping barefoot through the water. When we got to the yard, we had difficulties to keep our short pants dry (see Figure 7).

Tante Minna, Mutti's oldest sister (see Figure 5 of Chapter 11), also lived in Berlin. We had little contact with her, until we left Brandenburg in 1953. Tante Frieda (also in Figure 5 of Chapter 11) lived in the same development as Oma Siedlung. She was Mutti's sister-in-law. Her husband, Onkel Hermann (see Figure 7 of Chapter 11), worked as security officer and fireman at the Opel truck plant. We rarely saw them. None of the family got along with Tante Frieda, and ultimately even her marriage broke up. Finally, Tante Lieschen (see Figure 3 of Chapter 11), Mutti's older sister, lived already in the United States of America, the USA. She had emigrated in 1922 to join

Figure 12 *Vati, Mutti, Tante Lieschen, Oma and Opa Siedlung, and in the front row Bernhard, Jim, and Frank. Picture taken in Oma's Schrebergarten.*

her husband, Onkel Richard. We knew her from her letters and effort to convince Vati to also go to the USA in the late 1930s.

In 1937, Tante Lieschen visited us in Brandenburg with her youngest son, my cousin Jim, as seen in Figure 12. It was a time to get all relatives together. Frank and I had to share our toys with Jim and he, his with us. At the end of the visit, each of us picked the best of the other's toys to keep. We, unfortunately did not think too much of the American toys. I remember having kept two of Jim's race cars, they looked racy, but were not true to scale. For us, this was a fatal flaw. We got to know Tante Lieschen much better after moving to the US in 1954 (Chapter 6, Figure 2, and also Chapter 11, Figures 4 and 6). We also fondly remember the CARE packages she sent after WW II (page 3-18).

The Position of the children in the family was much different from what one sees today in the US, and, naturally, customs have also changed in Germany. Considerable emphasis was placed on discipline when we grew up. Still, we felt loved and knew we belonged to the family, but we also recognized that we were at the bottom of the family order and had to learn much and behave to advance. At the top was Vati, then came Mutti, the two of us were a separate group, distinguished by our age. Corporal punishment was accepted, but in our family, it was rare and never caused any injury. It only was a last resort when lectures were not successful and extreme displeasure of the parents needed to be expressed. Mutti did hand out occasional slaps, but usually only assessed restrictions or promised review of the misdeed with Vati, who then had to do the punishment. Whenever I did not agree with the penalty, I consoled myself by recollecting all the things that went undetected or unpunished.

A typical example of the handling of misconducts occurred in 1935. I played in front of the house that had Vati's yard in the back. I felt I had a right to be there. I was hanging upside-down, my legs on the iron railing of the stairs leading into the first floor of the house. I just wanted to see the world from a different angle. Suddenly, a well-dressed middle-aged, somewhat heavy-set gentleman came up to me and said angrily: 'Look, you are not permitted to do this!' I slowly righted myself, got to my feet and answered defiantly: 'You can't tell me anything, I live here.' He turned red, moved toward me, and wanted to grab me (or hit me?). I was faster and ran into the yard. He followed. I was surprised how fast he could move. When the distance between us got too small, I dove under one of the old wagons stored in the dusty shed. He also got on the floor, disregarding his being well dressed. Next, I jumped into the wagon. This was a mistake, since he now had me cornered. He grabbed me, I cried, but to my great surprise he knew where we lived and marched me home. Mutti opened the door and immediately understood the situation, took my hand and pulled me behind her with an assuring word, and angrily spoke, as I had never heard her before. It sounded somewhat like: 'Herr Kleiber, you are not permitted to punish my child. If he deserves punishment, my husband or I will do it. I ask you to never touch my child again!' I felt really bad,

since I realized that Mutti knew his name. Now I remembered, this was the man who leased the yard to Vati. He argued little, but told a story of the incident which did not quite jibe with my memory. He said something about my 'trying to fight with him' and 'a good chance that I could have broken his railing had I continued,' and 'you should have better supervised and taught your son not to do such things.' When he had left, Mutti turned to me and asked what was that all about? I told her, and she answered 'we will see what Vati has to say about it' and 'try to stay out of the way of this fellow.' When Vati got home, to my surprise he had a good laugh about my supposedly trying to hit Herrn Kleiber. For a long time it was a story told at the dinner table to friends. To me, Vati reiterated my need to be more courteous to grown-ups and that I should not aggravate the fellow any further since he had to get along with him. To my surprise there was no mention of any punishment. Frank and I had our revenge half a year later when we saw Kleiber's son, about one year younger than I, in the yard. We had a penny, so we asked him: 'Do you want this penny?' He did. Frank then dropped it into a barrel with tar and I told to him: 'Get it!' He tried and stuck his arm as far as it would go into the tar. But he did not reach the penny. Afterwards, my mother had to tell Frau Kleiber the secret roofer's wives know about removal of tar from clothing: 'Use margarine to dissolve the tar, and then remove the mess with soap and water or dry-cleaning fluid.'

This leaves the question *how did we slowly rise to be a full member in the family* and society? There were some well-established stages. As soon as we could be trusted to safely cross the street and not to throw balls when grown-ups were about, or disturb passers-by, we could play in front of the house without supervision. We knew by then that children always were first assumed to be at fault in any incident with grown-ups. Then, as soon as we could eat properly with fork and knife, we were taken along into restaurants (a not too frequent occasion). Next, when we could tell time, Oma Siedlung would give us a wrist watch, to be able to be on time. As soon as these and similar steps were overcome, we were so much a part of the family that when guests were coming, we would be introduced, were expected to take part in the conversation if we could contribute, and certainly, always would take part in the meal (with fork and knife). After dinner, we could get excused, but never could leave without bidding a proper goodbye or good night to the guests.

It seems to me as if today's custom to forgive children's misbehavior keeps them in a separate society. I even think that the overly large effort to protect children from harm is of limited use. At some point, this special protection must be removed. It certainly is harder to learn the needed discipline at age 18 than at age three. A good compromise seems still missing. Many years later, we had a scary incident supporting this point. Our son Brent was brought home by the school bus and was dropped off, as usual, on the opposite side of the street. A speeding car overtook the bus at the moment when Brent stepped around the front of the bus and with certainty would have hit him. But he had been taught by us never to cross a street without first stopping and looking, even in the presence of a school bus. It saved him, and probably also the future of the driver.

How about the street gangs? Yes, roots of such behavior existed in the 1930s and probably much earlier. There were not many children of similar age in our neighborhood, as I mentioned above. Frank, being three years older, had more luck. There was a group of his 'friends' with whom we sometimes met to do mischief or to play ball. I was the tolerated appendix because otherwise, Frank was not allowed to play with them either. I was amazed at what forbidden mischief was considered when discussing what to do. It included petty theft, breaking windows, and hitting other children not belonging to the neighborhood. By having the majority of us constantly reminded by our parents not to misbehave, however, I know of none of these misdeeds that was actually carried

out. But, for safety's sake, we were careful to look up and down the street to avoid meeting children from adjacent neighborhoods, they might have considered similar deeds.

Within the family, we played with cousin Bertie (son of Onkel Richard II) and his friends. When trying to get to Onkel Richard's house in the "Koppehlstraße," we had a problem. There lived a family with a dozen children of a wide spread of ages. They were 'enemies' of Bertie, and by association, of us. Perhaps Bertie made it worse than it was, since I do not remember ever being in a fight with them, but I do remember running a few times to avoid a confrontation. Playing with Bertie involved either exploring the furniture workshop and storage shed (where an old, inoperative car was kept—great for playing), or walking to the nearby railroad tracks and crawling under the bridges to reach the meadows beyond. This was our 'prairie.' But, I hated playing cowboys and Indians. As the youngest, invariably, I was tied to the stake and 'tortured'—but ultimately always survived unscathed. Frank knew, if I would complain to Mutti, the 'fun' would be forbidden. More interesting was to find a penny, putting it on the railroad track, and having it flattened by the train.

Playing ball on a vacant lot close to home was also not without problems. We usually played with my ball. Since as the youngest, I had the best chance to get it back when it sailed over the fence. Perhaps, I could make the saddest face (see Figure 7, above). Once, however, I did not get the ball back and was told to come back with my father. We decided, the punishment from Vati would be less than the sorrow over losing the ball. Things turned out different. When Vati asked for the ball, he was shown a broken window. What happened then surprised us. Vati just paid for the window. The owner agreed that 'boys have such mishaps,' and I got my ball back. The only punishment was that we all were admonished to be more careful in the future. We were!

As one would expect from a six-year-old, I had no idea about the *economic and political developments in the country*. The economy had improved somewhat over the years. This was caused by the enormous rearmament effort and extensive government projects such as improvements of highways which led to the building of the "Autobahn" as we know it today. It was the first, nationwide four-lane interstate road system in the world. In addition, the support of construction of apartment houses and setting up of price-controls of agrarian products and fixing wages helped to stabilize the economy. Other efforts were to help the propaganda by producing economical, good-quality, basic radios ("Volksempfänger") and to develop the social program of the Nazi party (see page 3-37) by founding the "Kraft durch Freude" organization (KdF, which translates to Strength Through Joy). The KdF provided affordable vacations and entertainment to two million people in 1934. This increased to 25 million by 1939. Opa and Tante Lorchen, for example, went to the Norwegian Fjords with KdF. Ships were refitted or built for this purpose. One was the ill-fated MS Wilhelm Gustloff, launched in 1937 (it sank in 1945, drowning 9,343, see page 2-25). She was the first cruise ship ever built. Not fully completed projects included the 'Colossus of Prora,' a mile-long spa on the beach of the island Rügen in the Baltic Sea. There 20,000 guests could have vacationed, all in rooms with ocean view. Also, the "Volkswagen" (VW Beetle) was a KdF project. The car was designed in 1933 and unveiled in 1938 with a top speed of 62 mph, going 33 mpg, all at a cost of 990 Reichsmark, RM (US $236). The average weekly wage of a worker was 32 RM (5 Mark-a-week was the sales slogan—after 750 RM were saved, the car was to be delivered).

In 1936, Vati led the whole family to the roof of our apartment house. We could observe the travel of the Zeppelin Hindenburg, the largest airship ever built, on its maiden voyage to the Olympics in Berlin. Its route touched Brandenburg, and the time was well publicized. It was exciting to be permitted on the roof, and the airship was admired by us as it floated past.

Political situations were discussed with increasingly more subdued voices. The opinions of my father, as a craftsman who was traditionally a Social Democrat, were kept from the children for fear they may get out of the house. After the burning of the Reichstag Building in Berlin (1933), the Parliament passed the 'Enabling Bill' ("Ermächtigungsgesetz"), permitting the suspending of most fundamental rights guaranteed by the Weimar Republic. Less than three months later, all non-Nazi parties, organizations and labor unions were forbidden. Concentration camps were created, initially for opposing communist and social democrat leaders. After the death of President Hindenburg in 1934, Hitler assumed the twin title of "Führer" *and* chancellor and abolished the office of President. He also took over the leadership of the army. In January 1935, a plebiscite in the Saarland returned this territory to Germany from international control, set up after WW I. On this occasion, Hitler introduced the general conscription and increased the size of the "Reichswehr" (army). In 1936, the demilitarized Rhineland was reoccupied by Germany, the Rome–Berlin–Tokyo axis was proclaimed, and the volunteer "Legion Condor" dispatched to fight in the Spanish Civil War. Still, the highly successful 1936 Olympics in Berlin and Garmisch-Partenkirchen and the constant reiteration that the goal of the new government was the continuation of the foreign policies of the Weimar republic convinced most western democracies of the peaceful intentions of Hitler.

The beginning of schooling in 1937 was the start of a new segment of my life. My school was the "Jahn Schule" for boys (named after the "Turnvater Friedrich Ludwig Jahn," 1778–1852, who introduced physical education into elementary schools). Figure 13 shows the school, not with me in front, but with my brother Manfred on his first day of school in 1952. The left wing was the Jahn Schule. In the back it has large, tree-shaded grounds for recess and a fully equipped sports arena with a soccer field (see Figure 5). The indoor gym with its six large windows is in the center of the building complex. The right wing, a mirror image of the left, was the "Hermann Löns Schule" for girls (named after a German poet, 1866–1914). A short time later, Heidel, my future wife, went there. Both schools were separated in the back by high fences and had fixed, separate hours for the use of the sports facilities.

Figure 13 *Center and right wing of my elementary school. Now this is the von Saldern Gymnasium (high school). Manfred's first day. The left entrance leads to the wing of the Jahn Schule.*

My first day of school, I remember clearly. Mutti walked the three blocks with me to school. After a short wait in front of the gate, we were directed to my classroom (8a). There were three parallel 8th grades in 1937 (8a to c). I still have the distinct smell of the oiled floor in my nose. It was the first time that I saw such floor construction. Its dark-brown floor boards were periodically oiled and then sprinkled with sawdust which was swept up and took along all other dust. Outside of the class rooms were long rows of hooks for the coats and hats. When all mothers had finished the formalities, our teacher, Herr Menzel, assigned seats. For the 36 students, the room had 18 double benches, arranged in three rows. Each bench had an attached table top with ink wells (still empty for the first few months) and a shelf, underneath, for books. On each side was a hook for

CHAPTER-11 ***"A Brief Genealogy"***

INDICES

APPENDICES

(These Appendices are only available in the On-line book as an Electronic Supplement)

APPENDIX-A ***"List of Publications and Research Associates"***

This Appendix contains in Parts A–D a complete list of all scientific publications written between 1955 and 2008. Listed are the sequential numbers according to the date when considering writing the manuscript. In Part A these are furthermore sorted by the year of appearance. In Part E, a list of the research associates is given, separated in level of appointment. The number of years in my laboratory is given, together with a brief characterization of the research topic and later employment.

APPENDIX-B ***"Thermal Analysis of Macromolecules"***

In Appendix B a scientific publication is presented. It reviews the development of my research over the years. Earlier versions were written for the NATAS Symposia on Thermal Analysis and Applications in 1996 and 2006. (See pages C-49 and C-63–64 in Appendix C, and publications407 and 575 listed in Appendix A).

APPENDIX-C ***"Permanent Homes, Travel, and Lectures of Bernhard Wunderlich"***

This Appendix C is a listing of the travels to give lectures (titles, location, occasion, and dates) outside the normal teaching duties. Since this travel was often extended to inspect 'need to see' sites, the list includes also travel for 'need to see' only, and occasional vacations. The entries are subdivided in addition by dates in sequence of places of residence, so that one can identify my whereabouts in the various years.

CHAPTER-01

Early Childhood, 1931–1939

Nobody remembers when one was born. Parents have to tell about the birthday, or the birth certificate must supply the information. In the "Familien-Stammbuch" (official family record) of Vati and Mutti[1] I read: *Bernhard Otto Klaus Wunderlich, born May 28, 1931.* The citizenship of Vati was given as "preußisch" (Prussian). Since I want to rely in this autobiography as much as possible on my memory and notes, not on stories circulating in the family, I have little to tell about the first three years of my life. I just don't remember much (anything?). Even my first picture which was in Vati and Mutti's collection, I do not remember. It is shown as Figure 1.

Something I do remember, though dimly, was my somewhat delayed *christening in the church of St. Katharinen* in Brandenburg. It occurred in December 1933 and was held, along with that of my three-years older brother Frank, on the insistence of Oma Siedlung.[1] Vati had a dispute with our Protestant minister about the advantages of cremation. Since Vati was not permitted his opinion, he did not feel obliged to support the church for a while. Personally, I was not happy about the christening procedure, but was surprised to have my Oma sing so beautifully and loudly in the church with its strong echo. Just recently, when we revisited Brandenburg in 2004 with our whole family, as a treat on the occasion of our golden wedding anniversary,[2] I recognized the corner where the christening was held and could describe the occasion to our grandchildren.[1]

Figure 1 *I, probably 1933.*

From **later studies of history** and discussions within the family, I learned that in 1931 Germany was a difficult place to enter the world. *The 'Good Old Times'* my grandparents talked so much about, referred to the second, united "Kaiserreich" of Germany which had ended with the lost World War I, WW I, in 1918.[3]

---

[1] For a brief genealogy to identify the relatives mentioned, consult the pages 11-2–4.

[2] For information on travel, see Appendix C, for this trip, also the top of page 10-11.

[3] Second Reich: 1871–1918. [The "First Reich," was linked to the 'Holy Roman Empire,' HRE and included lands in western and central Europe. First, it was ruled by Frankish (Carolingian) kings. In 919, the central part of the HRE was split off and ruled by German kings (Henry the Fowler, Ottonian Dynasty). The HRE lasted from Charlemagne's coronation in 800 ("Karl der Große," who called himself 'Augustus and Emperor') until renunciation of the imperial title, 1806.]

Figure 2 *A 100 RM bill from the 'Good Old Time Money;' 4.20 Reichsmark = US $1.00.*

The Second Reich was followed by *the Weimar Republic*. It was founded in 1919 and, with the abdication of Kaiser Wilhelm II, calmed the revolutionary efforts at the end of WW I. This first republic, however, suffered from the impossible financial burden of the reparations asked of Germany as result of the peace treaty of Versailles (1919). After long discussions, the reparations were finally fixed at the Paris conference of 1921 at 226×10^9 "Goldmark," payable over 42 years.[1] An immense inflation followed the next few years, decreasing the value of the German Reichsmark (RM) from 4.2 to the US Dollar to 4.2×10^{12} to the US Dollar.[1]

When I was five and able to count, Oma Siedlung (Chapter 11, Figures 3, 5, and 7) brought me about 30 bills of 100 RM of the 'Good Old Time Money' to play with. It was her life's savings. By that time, she had given up hope that this money would ever be revalued. The fun her grandson had with it was a small compensation for her tremendous loss. The bills made such an impression on me, that 68 years later, I bought the 100 RM bill shown in Figure 2 for the equivalent of about 20 DM (see page 3-28) to keep with my coin collection. By waiting such a long time, the money had recovered some value in numismatics. In contrast to the pre inflation 100 RM note, which in its time was a small fortune, coins with much higher face values than the 100 RM notes were issued during the inflation. The provincial emergency coin of Figure 3 was made of gold-plated brass to appear valuable, but one could not purchase a single slice of bread with it (and its actual use as coin is not certain). Now, it is valued at about US $50 by coin collectors. A 1924 coin of the same design, with a face value of 10^{12} RM (one billion) served as souvenir of the inflation. In 2008, the gold-plated version was valued at US $1,000!

Figure 3 *Inflation-coin of the year 1923. Despite the large denomination of 10,000 Mark, this coin would buy almost nothing; it was issued too late, the inflation had outpaced it.*

A new currency (the "Rentenmark") was introduced in *1924, when the inflation was stopped*. The overall economy picked up slowly, and the political unrest calmed to some degree. This positive development, however, was of short duration. By the time 1931 rolled around, the '1929 Stock Market Crash' had occurred in the US and led to the 10-year economic slump affecting all western, industrialized countries. In Germany, the unemployment reached six million by 1932 (over 30% of the work force) and the industrial output had decreased by 50%. This economic stress was the major reason that in 1933 Hitler was appointed Chancellor

[1] In the counting, customary in the US, 1×10^{12} is one trillion. In most other countries, 1×10^{12} is one billion (the US billion is equal to the German milliarde = 1×10^9 or 1,000 million).

by Hindenburg, the President. Hitler was the leader of the strongest elected party. His standing in the polls had decreased during the economic improvement, but increased with the unemployment. Once chancellor, he quickly developed into a dictator, proclaiming the ill-fated 'Third Reich.'

Besides the economical catastrophe of the nation, our family had to overcome its private crisis. Vati, who was a master roofer and tin-smith, had started his own business some years earlier, independent of Opa, Vati's father, who also was a roofer, as was his father. The crisis arose when Vati developed a severe case of asthma which disabled him sufficiently to make for an uncertain future. Fortunately, because of the national health insurance system in Germany dating back to the efforts of Chancellor Bismarck in the 19th century, Vati could spend time in many different sanatoria, searching for a cure. In the last of them, his asthma quickly disappeared, but returned only a few days after coming home. Through his own observation he alerted the physician that it seemed to be some allergy to dust from the mattress and pillow filling. As soon as Vati purged all feathers and kapok[1] and had it replaced with horse hair, and gave up his hobby to breed and race carrier pigeons, he was well (as long as he could avoid the allergens). Personally, I was not aware of any of these problems. I was part of a well functioning family of which a picture is given in Figure 4.

Figure 4 *Our family on the 4th Advent 1934, Bernhard, Mutti, Frank, and Vati in our apartment (Wilhelmsdorfer Straße 70).*

In "Brandenburg an der Havel" (our hometown), we lived at **Wilhelmsdorfer Straße 70**, a four-storey apartment house built about 1900 at the corner to the short Kleist Straße (see Figures 5 and 6, for the location). It was destroyed during the final days of World War II, WW II, in 1945.

At the ground floor, there was a shop for writing materials, paper products, tobacco wares, and fishing supplies. First, we lived on the top floor where the apartments were somewhat smaller and, in keeping with the custom of the time, the toilet was between two floors and had to be shared. In the basement, there was a large communal coal-fired copper kettle in a room called the "Waschküche" (laundry room). This room could be scheduled a week or more in advance for any major washing by any of the families. Besides washing of clothes, *major cooking events could also be scheduled*. For example, sterilizing the annually prepared preserves, and in our family, the annual slaughter of a fattened pig.

The pig was killed at the rather new slaughterhouse at the edge of the city and then moved to the laundry room where it was processed with help of a butcher who brought his tools for the job. The special cuts and sausages were cooked in the big kettle. The salami meat was stuffed in natural

[1] The seed pods of the kapok tree contain a fluffy, yellowish fiber that is a mix of lignin and cellulose. It is used extensively for filling of mattresses, pillows, and upholstery.

Street Map of the Neighborhood Where I Grew up, 1931–1949

(My hometown Brandenburg an der Havel, the inner city is in the NE, about a one mile walk, as can be seen from the smaller scale map of Figure 6 on the next page.)

Figure 5 *Present day satellite photo (Google Earth) of the area of the streets were I first lived. Wilhelmsdorfer Straße 70 was at the left corner with the Kleiststraße, the missing house above the small "Villa." Towards the NE, Wilhelmdorfer Staße 73, Vati's business sheds were behind the second, somewhat smaller house, butting against the houses of the Kleiststraße. The two trees mark the fence of the garden to the restaurant "Gambrinusquelle." The Jahn Schule, for my first four years of school, is on the upper left corner (see also Figure 13 and note the sport field in back). On the south it faces the Jahnstraße on the east, the Franz-Ziegler-Straße. (At that time, the Franz-Ziegler-Straße was named Kaiser-Friedrich-Straße.) Later, I went to the Theodor-Körner-Schule (see Chapter 2, Figure 4). In 1947, the school was renamed Johann Wolfgang von Goethe High School from which I graduated in 1949 (see Chapter 3). It is noticeable by its bright roof and faces the Jacobstraße with its small side via its two parallel entrance streets (Wredowplatz). The double row of trees toward NW frames the Jakobsgraben which connects to the Havel. The second, smaller building on the lower entrance streets was the "Wredow'sche Zeichenschule," where Frank obtained part of his civil engineering training. On the west side of the Franz-Ziegler-Straße in the lowest portion of the set-back is number 35, where we lived starting in 1946 (Figure 5, Chapter 3).*

Figure 6 *"Google" view of Brandenburg (2005). The lower river is the Havel, above, the "Silo Kanal" (city bypass). The circles mark: (1) The former "Rathaus" in the "Neustadt" (new city). (2) The "Altstädtische Rathaus" (old city) with the "Roland." (3) The "Dom" island, the oldest part of the city. (4) The Marienberg. (5) Wilhelmsdorfer Sraße 70. (6) The Jahnschule. (7) The old army garrisons. (8) Werderstraße 28. (9) The Goetheschule. (10) Oma Siedlung's address Saarburger Straße 72. (11) The Mitteldeutsche Stahlwerke. (12) The Hohenzollernstraße 11. (13) The Haydnstraße 20, the apartment of Heidel's parents. Overall, the E–W distance ≈3.5 mi.*

casings, meats salted, and the lard rendered. Smoking of the sausages was done at the butcher's. Some of the freshly cooked meat ("Wellfleisch") and the pork tartare were special delicacies and eaten right in the basement kitchen. When now traveling in Germany, we still try to find a restaurant which features an attached butcher to enjoy the "Schlachteplatte" (the fresh meats and sausages, preferably served on the day of butchering).

Two other common areas in the house were the rest of the basement and the attic. In each of these, every family had a separate walled-in area for storage (in the attic, unused furniture, appliances, old clothes, etc., and in the basement, potatoes, coal, canned goods, wines, beer, etc.). Other provisions such as sugar, flower, sausages, eggs, etc. were kept in a pantry, off the kitchen, the "Speisekammer." Perhaps, it should still be mentioned that the cooking area in the kitchen consisted of a coal-fired baking oven, on top this oven was a "Grude," a second oven, run with coarsely ground coal which glowed for many hours to keep water and slow-cooking meals hot all day. In addition, there were a wood-fired hot plate and two gas burners.

All rooms, except the kitchen, were heated with tall, tiled ovens which were heated with soft coal and stayed hot for many hours after the initial burning. As soon as my brother, Frank, and I could manage, our daily chores were to fetch the coal, potatoes, and any other items from the cellar.

The renters were also required (according to a strict schedule) to periodically clean the hallway, stairs, and toilets. In addition to the kitchen, the apartment consisted of an entry hall, a living room, bedroom, and a children's room. Of the upstairs apartment, I have little recollection, except that across from us lived a nice old lady, "Tante Kemmer," who was my 'best friend.' A most precious gift she gave me was a Bible, beautifully illustrated and leather-bound, which I studied many hours (before I knew how to read).

In 1934, we moved to the ground floor in the same building. Now we had our own bathroom with a coal-fired boiler for the hot bath water. What progress! The living room is seen in Figure 4. In addition to the cleaning of the hallway, the ground-floor renter also had to clean the sidewalk in front of the house. Of special fun was the snow removal which had to be done fast, so that walking in front of the house was not impeded. Everything was well organized and we children were part of the house community. (Not only as the nuisance we sometimes must have been—fortunately, I forgot).

Figure 7 *First school day for Frank in 1934.*

Figure 7 reveals the picture taken on *the first day of school of Frank*. I remember fussing because I wanted to be in the picture, despite Mutti's objections that I was not properly dressed and washed. The picture was taken at the entrance to the storage sheds for the equipment and supplies for Vati's business in the back of the Wilhelmsdorfer Straße 73 where he had rented the space which earlier had served as a yard for a horse-drawn delivery service for ice and was built around a large,

the regulation school bag, such as shown in Figures 7, 13, and 15. Mutti directed me to the last center bench, one of the somewhat larger benches. Not chosen because of the '**W**' in our last name, which in later years often banished me to the last row. Then, Mutti could say good by and wait for me outside for perhaps half an hour.

Figure 14 *Mutti reading to us in 1937. Does it not look like Frank can follow the writing, and I cannot?*

Teacher Menzel gave us the first lesson, instructing us how to behave and laid down the rules of behavior. He was my "Klassenlehrer" for the next four years. I was looking forward to read books such as the ones Mutti read to us regularly. Figure 14 is a typical picture, taken by Vati.

But, I cannot say that I liked my teacher–he was accepted and respected. As I remember, he only once made use of his right to corporal punishment. Otherwise, I must have been a well-behaved boy, as seen my grades in behavior ("Betragen" = 1, always! See Figure 16 on the next page). The punishment arose two years later when a somewhat smaller boy of another class was bullying me. Since I did not respond to his taunting for several weeks, he thought I was a perfect outlet for his aggressions and slowly others joined, in what he thought was fun. He did not notice that I was deeply hurt, so on that day, enough was enough. On the way home, after the next remark, he was on the ground with a bloody nose. We were both brought back to school by passing teachers. Without much investigation we both got the same punishment, but I was never bullied again.

The first day at school, as is customary in Germany, had a splendid ending as is illustrated in Figure 15. The "Schultüte" (literally 'school cone') was given to me, filled with sweets and fruits. As can be seen, it is of rather large volume, so that the treats lasted a week, sweetening the beginning of the new segment of my life. The same type of "Schultüte" can also be seen in Figure 13. Even in the US, we made sure that Caryn and Brent got theirs on their first day at school. The small leather pouch hanging on my left side in Figure 15 had a metal insert and contained the mid-morning snack (see also Figure 7). By the time for the noon meal, school was over. Only toward the end of the war and thereafter, when food was scarce, was first milk, and later also a nutritious snack served in school.

Figure 15 *Back home after my first school day with "Schultüte."*

Everybody had to walk to school. If the distances got too long, the street cars were available. The middle and high schools, however, were not neighborhood schools and there, bicycles were a common sight. From surrounding villages, a train ride might also have been necessary to get to the special schools. My walk to school involved the crossing of two streets and passing of three stores located on the ground floors of apartment houses. One was a textile shop, dealing also in perfumes and soaps. It was of no interest to me (but see the incident in 1938, described on page 1-31). The next, was a store for school supplies and also toys. It was of greater interest. I stopped there often on my way home to window-shop. The third was a bakery, but not 'our' bakery, which I knew well and where I often fetched the freshly baked breakfast rolls ("Brötchen"). This job I even had before going to school. I do remember the heavenly smell, and that the counter was so high above the glass-enclosed goods that I barely could hand over the coins. The bakery on the way from school, however, had a special attraction: It advertised leftovers for only five pennies: "Altes für nur einen Sechser[1]"—I could read by then!). Since I liked to eat, I worked on Mutti for a long time until she finally gave in and handed me the money. I was proud to get a huge paper sack of end pieces and other cake leftovers. Unfortunately, when I got home, it did not taste very good. At the expense of five pennies, I was cured. It may be, that I learned in this way early in life to be critical of advertising slogans. Whatever sounds too good to be true, does not taste good.

What I learned in school can be discussed best with my grade reports, displayed in Figure 16. Note, that for class 7 and later, the grading was changed. Between 1 (very good) and

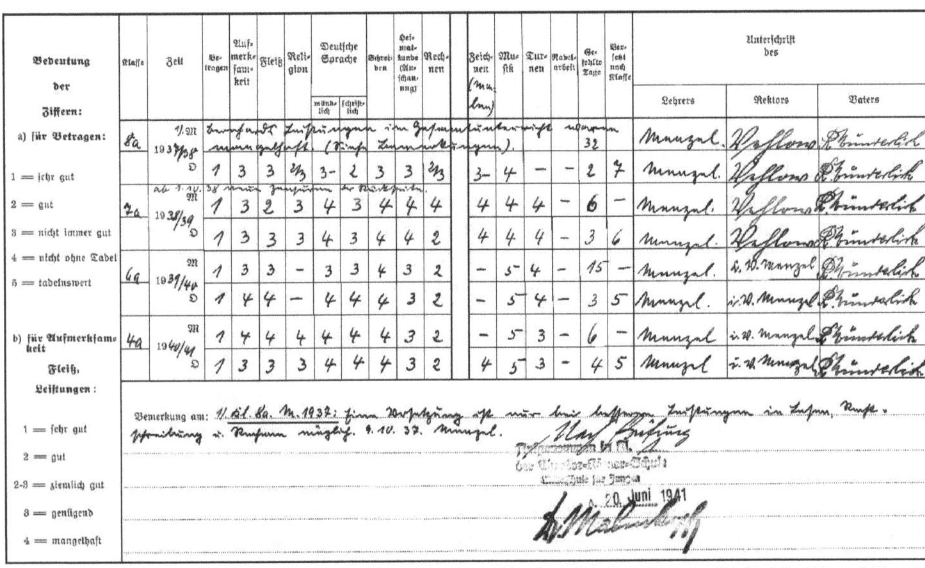

Bedeutung der Ziffern:

a) für Betragen:
1 = sehr gut
2 = gut
3 = nicht immer gut
4 = nicht ohne Tadel
5 = tadelnswert

b) für Aufmerksamkeit, Fleiß, Leistungen:
1 = sehr gut
2 = gut
2-3 = ziemlich gut
3 = genügend
4 = mangelhaft

Klasse	Zeit	Betragen	Aufmerksamkeit	Fleiß	Religion	Deutsche Sprache mündlich	Deutsche Sprache schriftlich	Schreiben	Heimatkunde (Anschauung)	Rechnen	Zeichnen	Musik	Turnen	Nadelarbeit	Gefehlte Tage	Versetzt nach Klasse	Unterschrift des Lehrers	Unterschrift des Rektors	Unterschrift des Vaters
8a	1. M 1937/38	[illegible]													32		[illegible]	[illegible]	[illegible]
	O	1	3	3	2/3	3-	2	3	3	2/3	3-	4	–	–	2	7	[illegible]	[illegible]	[illegible]
7a	M 1938/39	1	3	2	3	4	3	4	4	4	4	4	4	–	6	–	[illegible]	[illegible]	[illegible]
	O	1	3	3	3	4	3	4	4	2	4	4	4	–	3	6	[illegible]	[illegible]	[illegible]
6a	M 1939/40	1	3	3	–	3	3	4	3	2	–	5	4	–	15	–	[illegible]	[illegible]	[illegible]
	O	1	4	4	–	4	4	4	3	2	–	5	4	–	3	5	[illegible]	[illegible]	[illegible]
4a	M 1940/41	1	4	4	4	4	4	4	3	2	–	5	3	–	6	–	[illegible]	[illegible]	[illegible]
	O	1	3	3	3	4	4	4	3	2	4	5	3	–	4	5	[illegible]	[illegible]	[illegible]

Bemerkung am: [illegible]

20. Juni 1941

Figure 16 *Report card from 1937 mid-year (M, early October) to Easter 1941 (O, March/April).*

---

[1] If you wonder why five pennies, a fiver, is called a "Sechser," (sixer) you can join me. It took me a good number of years to learn that the old Thaler (= 1/30th pound of fine silver), the base coin before German unification in 1871, was divided into 360 pennies. Thereafter, one Thaler was three RM, with 1 RM = 100 "Pfennige" (pennies). The old small coins stayed in circulation for some time. Five pennies (new) = 1/20 RM, thus, was equal to 1/60 Thaler, or six pennies (old)!

2 (good), a 'new 2' (substantially above average) was inserted, replacing the old ⅔, and increasing all larger numerals by one. The grading scale closed now with 6 (fully unacceptable) instead of 5 (objectionable) as failing grade. The 'new 3' was called 'satisfactory without reservation' and the average was placed between the old 3 and 4. Overall, my performance based on this scale was average, with highlights in behavior ("Betragen") and arithmetic ("Rechnen"), and disasters in music and days missed ("Gefehlte Tage"). Did this performance permit an extrapolation to the future?

The missing days point to my catching all childhood diseases. The first, in 1937, was scarlet fever, which almost made me repeat the 8th grade (see note at the bottom of Figure 16), but by Easter 1938, I had caught up. In these days, scarlet fever was not a hospital disease, but Frank had to leave the house and stay with Oma Siedlung for several weeks. The whooping cough, chicken pox, a diagnosed weak case of diphtheria, various colds, and ear troubles often moved Frank again out of the house and gave me my poor school attendance record.

In addition to these illnesses, there were the 'little' accidents. In my early school years, when playing at the loading dock of the old ice house near Vati's storage shed, I used a big wooden pole with an iron tip to push the imaginary boat into the middle of the imaginary river. Unfortunately, the iron tip slid on the cobble stones, and unfortunately, the river was solid. My right arm hurt like never before. With a sad face I went home. Only with difficulty could I clean my hands to appear for dinner. It was impossible to get the spoon with the right hand into my mouth. Vati soon recognized my dilemma and stated: 'You broke you arm, look, it is not straight. Let's get to the doctor.' The dinner was interrupted and he drove me to our doctor. When I was home again, my right arm was solidly packed in plaster. It did not hurt anymore, and I was proud of my 'badge of honor.' Now I could take the spoon officially into the 'wrong' hand and everyone found out quickly that *I had no difficulty using the left hand*. Just when writing, I had to watch that the left-handed product was not the mirror image.

Seeing the effort made today to make children choose whichever hand feels better for writing and other activities, I think I would have had a 50% chance to become left-handed. I am glad I did not. The world is a dictatorship of a right-handed majority. The little effort (enforcement) to distinguish the 'good' and 'bad' hand had great benefit. Another connection to my tending to use the left is that I still am sometime confused about right and left. This got largely fixed, but not by saying: 'Right is where the thumb is left.' Once, I hurt my right ring finger and developed a scar. Ever since, and even now, when the scar is long gone, when in doubt, I can quickly rub my pinkies against the ring fingers. Where I 'feel' (or imagine) the scar, that is where *right* is.

The excuse for poor grades in music may have to do with my notoriously bad memory for melodies and tones. Despite this, I enjoy classical music and abhor most modern music.

The good grades in arithmetic, '2' which in the new scale meant 'far above average' was not recognized as anything special, but may have led to the ease of passing the examination for admittance to the middle school in 1941 (see bottom stamp in Figure 16).

The remaining grades are all 3 or 4, i.e., average. I paid only average attention ("Aufmerksamkeit"), did not work more than most ("Fleiß"), religion was slightly better, perhaps because of my 'study' of the Bible from "Tante Kemmer" (see page 1-6). German ("Deutsche Sprache"), oral and written ("mündlich und schiftlich"), were also not too good. Here I think the missing days came into play. My interest in word origins and literature originated much later, after I finished school. Could my poor penmanship have anything to do with a slight preference of the left hand? Later courses in English, Latin, and fragmentary Russian were worse than German.

In local history and sport ("Heimatkunde" and "Turnen") I could not see reasons for making a major effort. Most exploration of German history and geography came from later coin collecting and travel. Sport, as it was done in school held little interest (despite the name Jahn Schule). Out-of-doors, it consisted mainly of track and field, and indoors, of gymnastics exercises. Soccer and other competitive group sports were not played in my neighborhood, and I knew of no one in the family I could take as a role model. The occasional games during the one or two hours per week in school did not raise any competitive spirit, one had hardly time to get to know the team mates.

In 1936, there was quite some excitement about the Olympics in Berlin and the world championship in boxing was won by the German Max Schmeling in the USA, but I could see no connection to my effort in school. Later, in high school, I did well in rowing, a natural because of the many lakes around Brandenburg. Finally, I even attempted to master down-hill skiing on the only hill in Brandenburg, the "Marienberg" (elevation 190' above the level of the Havel, see 4 in Figure 6). I gave up in 1941 when at the end of a perfect downhill attempt one ski went left of a tree, and the other right. I barely made it home on my own power. If swimming had been on the school program, I might have done better. But, I could not really stay very long in the water. After half an hour, I turned blue and shivered sufficiently, to quickly have to leave the water. Ice skating also had a problem. Although I did well on the enormous areas of ice on the miles-wide flooded meadows around our town, I had difficulties in figure skating. Since my shoe size did not fall into the range of mass production, I always had to wear oversize boots. Later, my better walking shoes were hand-made by special order and at great expense. Good store-bought shoes I could find only after arriving in the US (best fit is a size of 8½–9 EEEE). Clamping ice skates to poorly fitting boots did not steady the feet, so that I could not do acceptable figures, despite my interest in figure skating. Nothing but long-distance skating and simple forward and reverse circle attempts were left, no figures of eight ever met in the center. Competitions in speed skating were unknown to me.

What about art ("Zeichnen und Malen")? Also a not so good subject of mine This flaw found an explanation when it was discovered about 10 years later that I was red-green color blind—not so good for painting. Teaching of technical drawing in high school was more to my liking.

But enough of the re-analysis of my report card with perfect hindsight. Most important for me was that I accepted the discipline to go to school, found what I liked and what not, and I pleased teachers and parents.

Figure 17 *The first book I read.*

The 'school' did its job to teach me the three **R**s. Reading at home I started promptly after progressing sufficiently in school. Figure 17 illustrates how happy I was when first reading my own book. For the second **R**, the writing, I document the strong impression writing on the slate tablet made by the fact that I still remember the jingle which reminds me how to write the letter **i**: "auf, ab, auf—Pünktchen drauf" (translated: 'up, down, up—dot on top.' Note for the first 2 years we still learned the "Sütterlin Schrift" for writing by hand. This German writing was introduced by the graphic artist and teacher Ludwig Sütterlin (1865–1917). It was taught in most German schools between 1915 and 1941. Similarly, the printing of most German books was done in "Frakturschrift," (also known as the Gothic type, or more precisely 'black letters,' of which my computers has the font Blackletter 686 BT installed—see also the book title shown in Figure 17). It was in common use from the 16th century to about 1940. An example of the "Sütterlin Schrift" and "Fratur" printing is given by 'birthday' "Geburtstag" and the printing of parts of the alphabet on the right. After 1940, we had to relearn how to read and right in what was called the "lateinische Schrift," (Latin writing), readable worldwide.

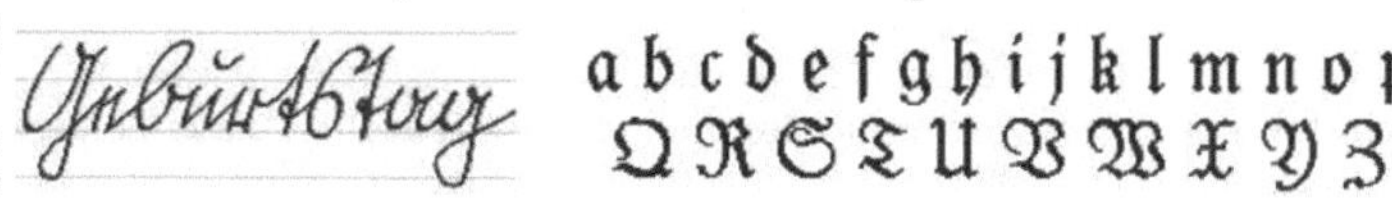

Finally, the third **R**, arithmetic was always fun for me. When we were taught to make chain calculations, the teacher would slowly call out the operations and when he was finished, he wrote the answer on the back of the blackboard. For example, he would give the problem:

$$5 + 7 - 3 \div 3 \times 7 + 9 \ \{= 30\}.$$

Then he would turn to the class and asked who has the answer? I usually already raised my hand before he asked the question. Not necessarily because I knew the answer, but I knew how much additional time I needed to complete the calculation. In case he called on me for the result, I could get up sufficiently slowly to finish calculating and give the proper answer (we were never permitted to shout out the answer without first getting up!). Naturally, the others thought I could calculate as fast, or faster than the teacher. After a while, the teacher did not call on me very often because he wanted to find the students who needed extra help. When later introduced to the number for 'π' (= 3.1415926) I found it fun to produce, when asked, π = 3.14159265358979323 which I had found at home in a book. I do not know why 17 places, but they could easily be grouped.

Vati and Mutti also seemed satisfied with the school. They felt they had turned me over to professionals to do a job. They interfered as little with the teacher as possible. Parent-teacher conferences were rare for the average student. Vati would say: 'I don't want to have 'Teacher Menzel' tell me how to fix a roof, so I can't tell him how to teach you.' Vati and Mutti made sure, however, that I followed instructions given in school and always finished my homework.

Observing the schooling of our children and grand children, and noting the accomplishments of freshman college students, *I feel happy not to have to start school today*. For example, the gender difference makes a handicap for boys in the first school years (and kindergarten). Boys do not have the early verbal skills and dexterity of girls, as documented often by their ranking in class. If I look at the decoration in typical class rooms, there is little that supports education of boys. The majority of teachers in the first grades seem to be women and may have difficulties serving as role models for boys. The differences lessen in the intermediate grades, but the last years of high school, again, are troublesome in coeducational schools. The problems seem to lie, as before, in the different rates of development of teenagers to adults. Today's efforts to eliminate the differences by offering identical teaching to all students all the time are probably futile. To stay with my experience, I gave help in mathematics and science in the last two years of high school for the parallel girls' grades. This job came about because of my involvement in the evening high school ("Volkshochschule"), run by a teacher of the girl's school (see page 3-20). There, I noted the differences in the approach to learning. Naturally, all subjects should be taught to all, but one should first study the diversity between students, and then adjust the teaching to recognize the different and changing abilities. It is more efficient and successful to teach groups of students in similar stages of development.

For me, **the beginning explosion of technology must be placed in this time period**. It was the time when the remnants of the horse-and-buggy era disappeared and were replaced by the motor car, truck, and bus, challenging the monopoly of the railroads. The radio became available and provided an instant source of news and a continuous stream of classical and popular music. Unfortunately, the conveyed messages also included subtle and overt political indoctrination. The 78-rpm record was invented and permitted favorite music to be available on demand. Airplanes slowly became a means of transportation and electrical gadgets made their entry into the households. My toys mirrored this change in technology. Horse and wagon changed to automobiles and race cars. Action toys became motorized, first with spring-driven motors, then battery-powered, finally via low-voltage transformers. As mentioned above, a central ice manufacturing facility was built within the new slaughterhouse in Brandenburg, replacing the old ice houses.

The local industry needed some time to recover from the downturn in the early 1930s. The well known, highly modern "Brennabor Werke," which had the largest passenger car production in Germany in 1925, were forced to stop production in 1933 and kept only their baby carriage and bicycle manufacturing, before adding armament production. Similarly, other companies recovered slowly, until being heavily involved in modernizing production and turning to armaments for WW II. Ultimately trucks, modern tanks, aircraft (including early jets and rocket planes), anti-aircraft guns, etc. were developed, tested, and produced in Brandenburg.

I still remember the *horse-drawn beer wagons and garbage collections* as remnants of earlier times. The (mostly) docile heavy horses were an unending source of wonder. At present, in the US, only the 'Budweiser teams' can recreate this view. The use of horses was coupled with the need for special street cleaning crews who had to remove the horse manure. Even the street car maintenance was affected by the horses. Workmen with special tools to scrape the groove in the rails would walk along the tracks ("Ritzenschieber," groove cleaners, considered one of the lowliest jobs).

The motorized garbage collection was especially well designed. Every house had several uniform, about four-feet-high, metal garbage cans, slightly tapered down toward the bottom. When filled, these were too heavy for one person to lift. The garbage collectors, clad in leather aprons, would sport hooks at proper height from a harness to connect to handles on the sides of the cans. Two men would steady themselves by linking arms above a can and walk it to the large motor van. Arrived on its side, they would hook the can onto a pair of rails and with a lever start a compressed-air-driven mechanism that whisked the can to the top of the collection tank, turned it upside down, shook it a few times to empty all, and transported it back to the grounds. All done in a few seconds, a marvelous sight. Empty, the can could be twirled on an angle, guided by one hand on a solid knob on the lid, and propelled by the other hand. From time to time the same compressed air, generated by the truck engine, would drive a big piston to compress the collected garbage, again, a noisy affair, worth watching. All this was of great interest to me and certainly was more advanced than the garbage collection we have at our present house, some 70 years later. Here the garbage is still heaved onto the truck by hand. Only the compression is automated.

My first experience with plane rides was a disaster. In school, money was collected for a round trip above the city with a "Lufthansa" plane on the occasion of the "Flugtag der Arado Flugzeugwerke." It was a beautiful summer day (8/27/1938). Perhaps a dozen school-classes marched to the company airport and waited at the edge of the grassy field. When after a long time no plane appeared, the message was, the plane has engine trouble, no flights today. My first flight, from Berlin to Frankfurt, had to wait until 1953. Technology does not always work.

Naturally *the business of Vati also was slowly modernized*. First, he invested in the telephone, mentioned above already, then in a car (see Figure 8) to pull a two-wheel trailer to haul supplies to the working crews so that they did not have to waste time pulling the customary four-wheeled carts to the job site. This was a considerable saving in time and effort. Later, engine-driven hoists driven by diesel engines were added, and even motorized carts for jobs on large industrial roofs. Tar kettles were fitted with oil heaters (instead of wood and coal fires) and connected to pumps to spray the hot tar instead of brushing it on. Even camouflage paints were sprayed to cover large roof surfaces and even the tarmac of airfields of the Arado and Briest aircraft facilities.

After entering school, the common questions I was asked by visiting friends of the family and relatives turned to: 'How are things in school?' And: 'What do you want to be when you grow up?' These questions were taken quite seriously, and reasonable answers were expected. Yes, it was possible to say: I don't like my teacher, or music is not my field of interest, but it surely was to follow with a short sermon, pointing out the problems arising for the future when not liking the teacher and then not doing well in a subject. If the dislike covered one of the three **R**s, a serious discussion was likely. All reinforced my understanding that school was not 'child's-play.'

Being frequently confronted at age 6–8 with questions about the choice of an occupation made me think: *What do I really want to be*? For my older brother, it was self-understood that he would learn the trade of the family and later, take over the business. Whenever Vati could take one of us to a job site, it was Frank. This was not only because he was older and of more use, but also, it was his later job and 'it is never too early to learn.' On these occasions, I had to stay home with Mutti. She understood that I would rather have liked to go along with Vati. As a treat, she lightened my work at home and 'sweetened' my stay. She knew my liking for special foods and sweets.

So, what did I want to be? A role model for any child could be the chimney sweep. He showed up practically every month, walking on the roof from chimney to chimney and inspecting the fire grates in the apartment. Choosing this profession promptly got the comment: 'Take a look at the dirt he has to wash off every day!' No, this was not right for me, washing was a chore. How about a garbage collector? Answer: 'Have you noticed the smell he has to deal with?' Yes, I had. By trial and error, I finally thought I like animals, I like the countryside, I liked the forests around Brandenburg. Why not become a farmer? Vati had farmers among his customers and when he visited them, I was occasionally taken along and delighted in inspecting the stables, barns, and machinery. When I proclaimed officially, 'a farmer is what I want to be,' I found no objection after the initial surprise. The grown-ups pointed to the hard work, the non-existent vacations and the relative isolation in the villages. I found this not so bad. I thought such life was fun and not hard work, and living near the woods and fields was a pleasure. Gradually, my new toys on birthdays and at Christmas were toy animals and later whole toy farms, as shown in Figure 18. This picture I took

Figure 18 *Barnyard of Lineol animals, many of which I owned as a child; some I still have (Lineol is the trademark of a toy manufacturer in Brandenburg).*

during a visit in Brandenburg's "Heimatmuseum" in 2006. The famous Lineol process started with sawdust and a polymerizable mixture with animal glue. The molded raw animals and soldiers were cleaned up ("Kamele kratzen" scraping camels), and then hand-painted by scores of home workers. Frank, in contrast to my animals, could boast a considerably larger army of Lineol soldiers than mine and a castle with medieval knights. The final nudge to my choice of occupation came in 1939, when one of Vati's farmer friends proposed that I should, after completing school, apprentice on his farm and study the trade. His suggestion was, since he had no children, that after proving I could run the farm successfully, I would inherit the farm with the usual proviso that he would live for the rest of his life with my support in the "Altenteil," the part of the farm designed to serve as living quarters for the (more or less) retired elders. This agreement was a serious offer and broke down only after the farmer was disowned during the Russian-inspired communist "Bodenreform" in 1945 (dispossession of farms, see page 3-9). At that time, I had to rethink what to do with my life.

This way of finding a job, can be compared to the present, when children are to 'find themselves,' often even past college age. From my experience, much valuable knowledge is attained as a teenager and it is of greater use if one has made a career choice by then. Yes, one should reexamine the early choices and change them, if needed. Often, adjusting the goals can be to a related direction without losing the benefits of learning about the earlier choice. Caryn also first studied chemistry and Brent chemical engineering, related, but not their final careers, and both of our grandchildren seem to make similar changes.

Toys played an important role in shaping interests. They were largely given at Christmas. Children's birthday parties were not as common as today. Usually relatives and friends with children were invited for coffee or dinner. Meals and cakes particularly liked by the birthday-boy could be wished for, but gifts were frequently clothing and other practical (uninteresting) things.

Figure 19 *A Christmas tree, as has long been traditional in our family.*

Christmas was a different story. The first Sunday of Advent saw a wreath with some special cookies and "Honigkuchen" (honey cake), but serious Christmas shopping was done only in the last two weeks when the stores stayed open on weekends. Our typical Advent wreath is shown in Figure 4, above. **The 'real' Christmas for us was Christmas Eve**, the day when presents were exchanged. It was a day when work stopped at noon. The children were full of anticipation. Songs helped to increase the excitement, such as: "Morgen (Heute) Kinder wird's was geben ... " [children, tomorrow (today) gifts will be given]. After the noon meal, a bath was taken, and we were put to bed. Surprising noises could be heard in the living room. Santa Claus (or Vati) was at work. Frank and I made the connection much earlier than Vati and Mutti thought, but who argues with the one who brings presents? It was, and still is, our custom not to erect the Christmas tree before Christmas Eve. The decoration was always the same as it is today in our house and shown in Figure 19. The tree is decorated with green, shiny glass-bulb ornaments in sets of three and bunched icicles, but little else. Chocolate treats are occasionally hidden on the branches and picked off during the holidays. The lights were initially real candles, but since about 1937 they are electric. This is safer, and the tree could then be lit more often and longer.

After getting up from our nap and dressing in our Sunday's best, late afternoon cake was served. Finally, with the traditional Christmas music from the radio, and later the phonograph, we were led into the glittering living room. The gifts were not wrapped, but decoratively displayed. Only glances at the gifts were permitted at that moment! First, each of us had to recite a Christmas poem, memorized days or weeks before. The length of the poem increased with age from a simple four-liner, to many verses. Only then could the inspection and initial tryout of the gifts proceed. From earlier years, I remember a wooden sales-stand dominating the gifts. It was built by Vati, white in color with a large table, chair, a cash register, a multi-colored, striped cloth-cover as roof. It was filled with all sorts of sweets, the goods for sale. Particularly suitable were items of marzipan in shape of fruits, vegetables, and meats, ready to be sold by one of us to the other. Care had to be taken too, so that the goods would last a few days and not cause overeating.

After the first few hours of play, dinner was served. It was always eel and carp from the Havel river, boiled in a brown-beer gravy (nonalcoholic "Malzbier"). I loved the fish, but hated the gravy, so I had to scrape the gravy from the fish and build potato dams on my plate to keep the gravy from soaking into the rest of the potatoes and the vegetables. Because it was Christmas, I could get away with leaving the dam and gravy on the plate. After dinner, playtime resumed until we were too tired to go on. In the earlier years, Oma and Opa Siedlung came to celebrate together with us.

On Christmas Day, we sometimes walked to the house of Onkel Richard (II) to join our cousin Bertie. On Christmas Day, the noon meal consisted of goose, gravy, potatoes, and kale as the vegetable. In the later years, we got up early on Christmas Day and drove to Lautenthal in the Harz mountains for a few days vacation in the snow with sledding and skating. The religious aspects of Christmas were in the background. Naturally, the traditional Christmas stories were read, told, and also taught in school, but reinforcements through visits to church were not on the schedule. The 26th of December was a second holiday, free of work for Vati, and with closed stores.

Figure 20 *SS Bremen, of the North German Lloyd, built 1929, 51,656 GRT, attained the 'Blue Riband' on both her first Atlantic crossings (speed 28 knots).*

Over the years my toys changed from playthings ("Spielzeug"), as shown in Figures 18, above, to hobbies, illustrated by a famous ocean liner in Figure 20. The other interest that was furthered, was the building of things. Frank was annually increasing his metal erector set (made by the Märklin company), while I was expanding my wooden set, the "Matador Baukasten." Many different shapes such as cylinders, blocks, plates, laths, and wheels were available. All had equidistant holes penetrating the pieces so that they could be connected with wooden pegs or posts. They also could be made to turn when using special wooden axles. My proudest achievements ranged from houses, stables, towers, and cranes to aircraft carriers and periscopes. The latter let Frank and I sit underneath the table and observe the fleet of our "Wiking Modelle," moving across the blue ocean (see also

Figure 22, below). This picture was taken in 2005 in the "Stadt und Spielzeugmuseum" in Traunstein, Bavaria. There are only few museums with a collection of "Wiking Modelle.[1]" By now, these are rare since most were of war ships, not deemed worth keeping after the lost WW II.

Once Vati surprised me two weeks before Christmas by showing me a beautiful model of a big cannon, fitting in scale to our Lineol toy soldiers. He said, he wanted my opinion. 'Would Frank like it as a Christmas gift?' The cannon was the latest in realism. The breech could be loaded with special brass grenades. The hammer on the outside would ignite a paper percussion cap, this, in turn, would ignite a charge of two or three caps inside the breach and propel the grenade some 5 to 10 feet. The grenades, in turn, were hollow and into their head one could place another cap and a loose brass weight, to explode on impact. Around this last percussion cap there were holes in the grenade, so that in the dark, a credible explosion was seen, as well as heard. My judgement was an enthusiastic: 'Frank would love it!' You can imagine my surprise when on Christmas Eve, this cannon was *my* present. The early showing was seemingly only a trial run to see if I would sufficiently appreciate such an expensive toy. Frank, in turn, got a howitzer, less sophisticated, I thought. Now, our playing with soldiers was more realistic and I, with my much smaller army, had an occasional chance at winning, due to the superior artillery.

The most sophisticated toy was a modern H0-scale Märklin train-set. The initial parts were given in 1938 and extended in 1939. This gift was difficult to install since our electric supply was still 220 V DC, common in part of Brandenburg until after 1945. This direct current voltage could not be reduced by a simple transformer. Vati had to buy a 220 V DC motor, coupled to a 12 V AC generator. It hummed away in the corner when the trains were running and was almost as interesting as the trains. By that time, the premier toys had definitely become a hobby which lasted a lifetime. I still have a nice collection of classical H0 trains, including one identical to my first engine with Prussian compartment cars, shown in Figure 21. Such equipment was in use for local passenger trains in the 1920–50s. By now, I also keep Z-scale Märklin trains running in my office, the world's smallest, fully operable electric trains, 6.5 mm track width, to a scale 1:220.

Figure 21 *A typical 1938 Märklin model train.*

After the war, I repaired our first, bomb-damaged H0 trains for my brothers Joachim and Manfred. When fleeing to West Germany, the trains had to be left behind. My present collection was begun on my first sabbatic in Germany in 1967/68 for Caryn and Brent (and me) to play with.

After 1939, only war-related toys were manufactured. This meant, we could collect only war ship "Wiking Modelle." The British fleet is shown in Figure 22, again, photographed in the museum in Traunstein. I owned the battleship "Hood," the fourth from the right, I believe.

[1] Model ships of the Wiking Co., founded by F. K. Peltzer (1903–1981). They were to scale 1:1250, made of die-cast metal and later plastics. By now, the old models are rare collectors' items.

It remains to be mentioned that by 1938 *books were also appreciated as gifts*, although no big libraries were amassed. Memorable were several travel-adventure books, descriptions of WW I heroics, and books about life in the various German colonies. My specially prized books dealt with the experiences of a tramp, working his way as stoker on a ship to the USA and his adventures; the experiences of an 18-year-old boy on his very own small motor cycle through Sweden; the fate of a whaling company marooned at the turn of the century for years on an island (ultimately being rescued after they discover gold); and the exploits of the light German cruiser "SMS Emden" stationed at the German naval base at "Tsingtao" (Qingdao), China, at the outbreak of WW I.[1]

Figure 22 *The British war fleet in "Wiking Modelle."*

Special fun were books by Karl May.[2] Frank received two for Christmas "Winnetou," about American Indians, and "Am Stillen Ozean" (Pacific), about the South Sea and China. I had to catch up when these books became available again after the war. Of specific interest for today are descriptions of the never-ending struggles between the extremist Shiite and Sunni Moslems. The intolerance and the violent attacks between these and with any other 'infidels' described by Karl May could have been taken from today's newspapers. Politicians should read them in order to understand that their problems cannot be solved by interference of infidels from across the ocean.

Vati liked travel logs of polar expeditions, and explorations through central Asia. His favorite author was Sven Hedin.[3] We borrowed his books and found them highly interesting.

This discussion of Christmas gifts should point out *how important the proper gifts may be* to shape ones interests. Naturally, I have mentioned only the important gifts I remember in this connection. There were additional gifts, but their number was more limited than one is used to see for today's middle-class children. Carefully thought-out gifts, supporting or developing a special interest, are invaluable. Also, the number of toys given should be sufficiently small, so that full attention can be given by the recipient to explore their full potential. Often problems are hidden in today's toys not yet widely recognized. Let me give only one example which troubles me.

[1] The Emden captured and sank dozens of British ships in the Indian Ocean and finally was scuttled on Coco Island. One of the most romanticized incidents of WW I.

[2] 1842–1912. A famous author of adventures of a fictitious German traveler visiting the American Indians, the Balkan and the Middle-east, as well as China and the South Pacific Islands. These stories were written without the author having seen the locations, but with copious quotations in the pertinent languages (which the fictitious traveler always spoke fluently).

[3] World-famous Swedish explorer and geographer, 1865–1952. He mounted many expeditions to Asia, resulting in fist maps of the Pamir, Taklamakan Desert, Tibet, and the ancient Silk Road. He was the geographer to first recognize the Himalayas as a single mountain-range.

Probably, I would have enjoyed today's computer games as much as anyone, although I usually remove all games included with the operating systems to gain more memory for other applications. The problem I noticed was in simulations of driving and racing of automobiles. To win, these games teach to take risks that go beyond ones skills and not necessarily how to proceed safely. If there is a crash, one is not dead, but one can push the reset button and start over! Is here a direct link of such teachings and the skyrocketing cost of automobile insurance for teenagers?

The years immediately before the war brought increased prosperity to many German families. Vati was initially refused to be considered for government jobs, but when it was obvious that he would do a better job than many others, he was forgiven his opposition to the Nazi Party and his not becoming a member.

Figure 23 *Front of an Opel Super 6, 1938 as Vati's.*

With the growth of his business came the opportunity *to buy a new car in 1938*. This became the largest travel of our family up to this time. We all took the train to Rüsselsheim near Frankfurt/Main (a trip of about 300 mi). There we went to the plant of the Opel AG, a leading German automobile manufacturer (by that time already owned by GM). At the appointed time, from the end of the assembly line came a shiny, dark-blue Super-6 like the one shown in Figure 23. The car of Figure 23 has a different color and was photographed recently at the German Automobile Museum in Amerang. This was an up to date car with first-rate technology (six cylinders, 55 hp). We all piled in. Vati got some instructions, and the next morning, we started our way back with a detour along the Rhein. It was a beautiful trip. There were plenty of stops to admire the river, mountains, vineyards, and quaint cities. While Vati and Mutti could taste the wines, we were happy with grape and apple juices, carefully chilled with ice, something we were not familiar with. This apple juice is still my standard to judge today's juices. Not long after getting the new car, I got a hold of a large-size (coffee-table) book with glossy, first-rate color illustrations, discussing the technical advances that went into making the new 1938 cars by Opel. Besides showing shots of the up to date assembly line, details were given about the front grille, prominently shown in Figure 23. This was described as a marvel of new technology. It was of one-piece, injection-molded ("Spritzguß") aluminum and represented the cutting-edge technology in 1938. I was impressed by all. After visiting automobile museums in the US, I recognize that this was the beginning of the globalization of technology. Practically identical grills were at the same time in high style in the USA.

With the new car, came more trips into the surroundings of Brandenburg. For example, one weekend we went to the Park in Wörlitz (60 mi south of Brandenburg). The claim to fame of this park for me was a long and exceedingly flexible suspension bridge over a gorge ("Kettenbrücke"). Mutti and I got our shake-up of the year, but finally made it over the undulating bridge. The lunch at the restaurant calmed us down. For the first time, I had a pork-tartare roll ("Hackepeter Brötchen") with a sliced pickle on top. This was even better than the sandwiches mentioned on page 1-6. Again, *something to be remembered to judge later culinary treats*.

Continuing with the remembering of special culinary treats during this early time of my life, I cannot omit the "Eintopfsonntag" (One-pot Sunday meal). In 1938/39 the meat and butter production in Germany fell behind the demand, and efforts were made to restrict consumption. One of these was the widely advertised "Eintopfsonntag." Specific sundays were declared as "Eintopfsonntage" and the population was urged to give up on this Sunday the typical meat, potatoes and gravy dinner. To help the effort along, the army posted their field-kitchens, the "Gulaschkanonen," to central positions in the city and distributed for a nominal price the standard military pea soup. They did a marvelous job. I doubt that the soldiers got such a high-quality meal. This "Eintopf" is my standard to judge pea soups by. Unfortunately, today's pea soups usually suffer (for dietary reasons) from a limited quantity of smoked bacon needed for top quality.

The "Eintopfsonntag" reminds me further of the annual *open house staged by the large military garrisons in Brandenburg*. A special day, on which Frank and I once visited one of the main garrisons along the Magdeburger Staße [see Figure 6, at (7), covering about ¾ mile of the north and south sides of the street starting from the Zanderstraße [marked as the newly redirected, yellow Highway 1]. By that time, I had forgiven the army for an event that occurred several years earlier. When a military formation moved past our house with their marching band, I walked along them, as five-year-olds might do. The commanding officer was on the sidewalk. He saw me in the group of children as I was proudly displaying an old, original military belt buckle. He barked: "Take it off!" It still displayed the motto: "Gott mit uns." Opa Siedlung had given it to me. Naturally, I was upset and never again found it interesting to walk with a column of soldiers. This may have been a first contributing factor to my never serving in anybody's army.

At the open house Frank and I had fun. We inspected special army trucks, tanks, and guns. On the parade ground, a mortar battery was set up and was shooting blanks with big noise. The visitors could buy these charges and shoot them off. When we got closer, we noticed that they sold about 5" large cubes of cardboard which had a strip on one surface which, when rubbed, would after a delay of five seconds explode the cube. Igniting one cube, dropping it into the mortar, and following it three seconds later with a second one would cause the second cube to fly about 50 feet. It exploded then close to the time it landed. This was great fun! We immediately saw the potential of these cubes and bought one. But, we did not shoot it off, rather we took it along to the Kutschers, friends of Vati and Mutti. They lived close by, at the foot of the "Marienberg" [Figure 6, (4)]. At their house we met with Vati and Mutti after the open house, as was planned before.

The Kutschers had a son, Helmut, who was slightly younger than I, and an even younger daughter Charlotte, of little interest to us. When Frank and I were together, we did not necessarily get along. But, *in the presence of other children, Frank and I were the best of friends*. Near the Kutschers' house was the park area where we frequently played, running up and down the hill, and exploring the fauna and flora. A place of particular interest was a neglected mound where we had discovered a number of rabbit holes. We talked Helmut into believing that this was a fox den with several exits. Then we showed our explosive device and posted him to watch several of the holes and urged him to catch the fox when he was scared out of the den. As we expected, we only scared Helmut. As soon as he heard the muffled explosion of the ignited box, he took off. There were no rabbits, no fox, and no Helmut, just our fun of having pulled off a prank.

When looking at the map of the city in Figure 6, one can see that the inner city is easily visited on foot. An electric street car, the "Straßenbahn," existed. It was called "die Elektrische" and is still running today (in contrast to the horse-drawn one, given up at the turn of the century).

It started from the railway station (near 8) and went to the northern part suburbs, to Plaue, the neighboring city in the west, to the old city (2), and to the "Stahlwerk" (11). In addition, there was a cross connection from the "Rathaus" (1) to the outskirts beyond (5) through the Wilhelmsdorfer Landstraße. Figure 24 shows a collection of old photos as displayed in the street-car museum in Brandenburg. The service was further extended by buses to outlying areas. In addition, we had access to bicycles to explore the lakes, beaches, and woods.

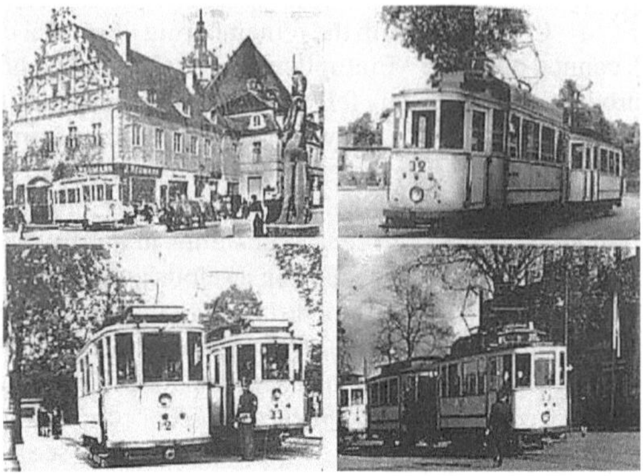

Figure 24 *Electric street cars in Brandenburg in the 1930s. Upper left in the city center (Neustadt). Upper right probably Wilhelmsdorfer Landstraße, "Endstation." Bottom in front of the railway station.*

This meant <u>our areas of play covered the whole city</u>. By walking and using the street car we could inspect all stores of interest and visit the parks with Mutti or alone. The Marienberg was most attractive in winter. A professional sled run was built with a drop of, perhaps, 120 of the 190' elevation of the "Berg" (mountain or better, hill). Several properly banked curves gave the run special appeal. We tried it first with Vati. By 1939 Frank and I together (but not I alone) had gained enough experience to have the courage to make it down the whole stretch. On the back side of the sled facility was a ski run. I have described already my unfortunate experience on this course (see page 1-18).

Key locations of our roaming outside the city were the many beaches. The Grillendamm was the city's premier, fully-supervised swimming beach. It faced the bottom section of the Beetzsee [somewhat north of (3) in Figure 6]. The Beetzsee stretches about 4 mi N and then 7 mi NW, and is up to one mile wide. In addition to the white sand, the Grillendamm beach had a tower for jumping from 1, 3, and 10 m diving boards. We slowly graduated from 1 m to 3 m, but avoided the 10 m. It was even uncomfortable for me to just standing at the top. After both Frank and I were sufficiently good swimmers, we were given the small entrance fee and could go to the Grillendamm by ourselves. With my friends from high school (male and female), it later was a common excursion goal for the summer weekends. Another group of lakes near Brandenburg are the "Plauer Seen." They are more compact than the Beetzsee. They are suitable for swimming and boating and contain several islands for camping overnight. The Havel and the Silo Kanal flow into them. This group of lakes starts about ½ mi WSW from (5) on the map of Figure 6 and is about 4 mi in diameter.

A near beach was at the Buhnenhaus Restaurant, close to the mouth of the Havel. For us it was best to use the bicycle to go there. In 1937, when my reading was not so solid, Frank and I got enough money from Mutti to have cake and something to drink at the restaurant. We went first to the end of the Wilhelmsdorfer Landstraße to use a cut-off path along the "Plane," a creek which empties into the Havel near the restaurant. When we turned into this path, Frank said: "We probably should not go this way because this note on the signpost says 'road impassable!' " I didn't agree and convinced Frank to go on, mainly because I had no idea what 'impassable' meant (in German "unpassierbar"). I was soon to learn the meaning. The "Plane" had risen over its banks. First, the path got muddy, then, it was a few inches under water. Next, we had to push our bicycles. Finally, we were up to our short pants in water and only could make-out the creek by the bushes on its side. We ultimately made it to the restaurant, just in time to weather a strong thundershower. By the time cake and soft drinks arrived, we were dried out and largely recovered. I had learned a new word. But we were not in the mood for swimming, and we chose the safe, longer way back home.

Learning a new word was always of interest. Certainly, at that time the most puzzling word for me was "Särge." German is a largely phonetic language which means in a very short time you learn the alphabet, and then you can read everything. The spelling is the same as the pronunciation. This does not mean that you understand what you are reading, but if you miss a word or two in a sentence, you usually are able to make sense of it from the rest. In contrast, I noted, when American children encounter a never seen word, they commonly stop reading. We could later help Caryn and Brent by telling them: "Sound out the word (if need be, in German pronunciation) and continue. You may figure out the word from the rest of the sentence." But this word, "Särge," I found in several locations in the city as a single word on signs, mounted on the wall of houses, indicating something, but what? I could have asked, but I thought I should be able to figure it out. One sign was in the Jahnstraße, one in the Koppehlstraße, next to the house of Onkel Richard (II), and there were a few others. Surprisingly, all these signs were of identical size and appearance. About 3×2 feet, with wavy and shiny silver lettering on a black background. Mysterious! I think it must have taken a year until, all at once, I saw in Onkel Richard's storage shed a whole stack of coffins. I knew the German word for coffin, "Sarg," and naturally I could derive the plural: "Särge." I just had not made the connection that one would write this word all by itself as an advertisement and mount it on the wall of a house. If the sign would have said: 'We have "Särge" for sale,' I would have had no problem in the first place. As soon as I was enlightened, I immediately knew that all the places where I saw the signs were on houses of cabinet makers. One problem was solved.

Economy and Politics in 1937–1939 accelerated in the directions already foreshadowed by the descriptions on pages 1-13–14. Now, as an eight-year-old, I noted that *our personal economic situation was improving*. The amount of Vati's work was increasing, and modernization abounded. From discussions at home I knew that Vati's first big job for industry was at the "Mitteldeutsche Stahlwerke" (steel mill), involved in armament production, as shown in a poster, probably from about 1937. This was the company which had brought Opa Siedlung to Brandenburg in its early years. The poster indicates the pride in their annual steel production figures (1914, 40 kt and 1934, 280 kt) and the quality of their tank production. [The word "fertig" in Sütterlin writing (see pages 1-18–19) means completed!] When the order to make the roof for a new facility came to Vati, I learned with interest how it was possible to earn more by proper planning and manipulating of the finances than from the actual work.

Figure 25 *Propaganda from Brandenburg's steel mill. From a poster in the "Industriemuseum."*

The contract was signed after the financing was guaranteed by a bank. This done, Vati went to the supplier of the roofing felt, tar, and other materials in Berlin and managed to arrange quick filling of the order and economical shipping by barge to the terminal of the "Stahlwerk" on the Silo Kanal [see Figure 6 (11)]. As soon as the material had arrived, it was moved to the construction site. The building had a roof with the biggest area in Germany (Europe?) at that time. The new, motorized equipment needed little time to move the felt and other materials to the storage sheds and onto the roof. Shortly thereafter, Vati saw the director of construction and asked for an "Abschlagzahlung" (partial payment), guaranteed by the value of the material and the fulfilled condition that work had started. When he was granted the substantial sum, the supplier was paid. He had given a thirty-day period before his bill was due, and no loan was needed from the bank, but a substantial profit had already been made. This type of operation was repeated many times for the other expanding industries in Brandenburg. It changed Vati's business from a small, local roofer to a trustworthy firm known to nationwide industries. Figure 26 revealed Vati's pride in his profession. It represents his trademark used on all his business documents. It was designed by him personally in 1940, and I still have his golden signet ring, which was ordered by Mutti (probably for Christmas 1941).

Overall, the employment in industry and craft in Germany had increased 30% by 1936 beyond the level of 1933, but had not reached the 1925 level. In 1939, the increase over 1933 was 95%, exceeding the 1925 level (not counting the added countries of Czechoslovakia and Austria). The deficit spending, however, increased substantially.

The newly created state projects were already summarized on page 1-13. Early forerunners of the "Autobahn[1]" had opened already.[2] But now, *the first segment of the nationwide "Autobahn" was opened* in May 1935 between Frankfurt and Darmstadt. By 1941, when Autobahn construction was halted due to WW II, 2,250 mi were completed. In 2006, the Autobahn accounted for 7,500 much improved miles, about 23% of the total net of "Bundesfernstraßen."

Figure 26 *Vati's business seal, carved in form of a plate (translated: Value the Crafts). Now displayed in my brother Joachim's house in Hastings, NE.*

[1] The first use of the term "Autobahn" was in 1928 by the HaFraBa (an association for the planning of an automobile toll road from the Hansatown Frankfurt to Basel).

[2] 1921: Completion of the intersection-free Avus (Automobil-Verkehrs- und Uebungs-Straße GmbH, Berlin, founded in 1909), with construction of the road beginning in 1912, length: 12 mi, also used as race track with a banked turnabout and visitor stands, now part of A115 of the "Autobahn"); 1932: Opening of the 12-mi section of the Köln–Bonn, four-lane limited-access highway, planned in 1926, now incorporated into the A555. (Inaugurated by Konrad Adenauer, then Mayor of "Köln" (Cologne), later 1949–63 Chancellor of the Bundesrepublik Deutschland, FRG).

The dictatorship of Hitler was strengthened further during this time period (see also page 1-14). Although I was not aware of the dramatic changes in the government, it became obvious that certain things were not permitted to be talked about in public anymore. For safety's sake, these subjects were not discussed in front of Frank and me. Even between the friends of the house differences were made. Herr Kutscher, mentioned on page 1-27, became a member of the Nazi Party (see page 3-37). This did not mean he did not stay a friend, but the discussions in his presence were limited. It was noteworthy that after the war, he was incarcerated for several years for his actions as a member of the Nazi party before and during the war. After his release from prison, he became a member of the SED, the "Sozialistisch Einheitspartei Deutschlands"), then the leading communist party. Now he was 'convinced' of his mission, and we had no further contact with him.

A very disturbing incident occurred on November 12, 1938, known in history as the "Kristallnacht." When I left school on this day, the textile store in front of our school (see page 1-16) had a larger number of people milling around. We students needed to investigate. I saw that the store windows were broken, and some grown-ups and school children were picking out small items. Store windows were broken occasionally, but one did not loot the contents! In particular, why was there no police? I quickly went home to tell Mutti. She could not explain this either. When Vati came home to eat, he reported that more store windows were broken in town and that the synagogue was burning. He said this was directed towards Jews and forbade Mutti to go with us children to watch the fire and commotion in town with the comment: "It is dangerous to be seen." Fires were rare in Brandenburg. Under other circumstances, it would have been an attraction worth witnessing. I did not know at that time what a synagogue was, and also had no contact with any children known to me to be Jewish. The incident was left unexplained. Only slowly did I become aware of the persecution of Jews, as well as the existence and main purpose of the concentration camps. The full tragedy was learned only after the war, as is detailed in Chapter 3.

The aggressive foreign politics quickly broadened beyond what one might call the inner-German affairs, already mentioned on page 1-14. In March 1938, Austria was annexed and proclaimed to be a part of Germany. In September, Great Britain, France, Italy, and Germany signed the 'Munich Agreement' (Appeasement) which forced Czechoslovakia to permit Germany to annex the "Sudetenland." This was an area of Czechoslovakia where 2.8 million Germans and 800,000 Czechs lived. At the end of WW I, on the break-up of the Austro-Hungarian Empire, the "Sudetenland" was made part of an independent Czechoslovakia, created in 1918.

Was this the end of German expansions? No, in March 1939, Germany supported, or better 'forced,' the independence proclamation of Slovakia from Czechoslovakia. Promptly thereafter, German troops marched into the remaining Czechoslovakia which was made the German Protectorate of Böhmen and Mähren. Furthermore, also in March, an area north of the river Memel, the "Memelland" was reunited with Germany after it had been separated from East Prussia in the Versailles Treaty. This precipitated territorial guaranties for Poland by France and Great Britain. The countermeasure was a nonaggression pact between Germany and the Union of Soviet Socialist Republics, the USSR, in August. With this pact, Hitler foiled planned guaranties between Great Britain and the USSR. The nonaggression pact included a secret protocol that made part of Poland a sphere of interest to the USSR. Today this 'sphere of interest' remains under Russian control (see page 2-2). On August 25, the territorial guaranties were followed by a formal alliance between Great Britain, France, and Poland. This was to forestall an attack by Hitler on Poland, probably known to them to be planned for August 26. This alliance delayed the invasion by but six days.

At the end of this chapter dealing with **the importance of the first eight years** in my development, it may be of interest to see if *one could have deduced at that time what my future might bring*. In retrospect, this is always easier. In my thoughts, these eight years in Germany represented a close to normal childhood. Despite practically no memory of the first three years, it seems that many traits that developed in the future were already noticeable at age three. At age eight, I felt as an independent person, who was in need to be part of the family, but I also was happy to be by myself with my own thoughts. Already at this age, I would try to stick to my decisions once they were made, and was not to be bullied by my brother who always was my keeper. A strict discipline (as we thought) was forced upon us, but since it was consistent, applied equally to both of us in our different roles, and strictly enforced, it was (begrudgingly) accepted. In hindsight, I now see the early teaching of discipline as a valuable asset for the rest of my life.

After writing down the memories of the life at the Wilhelmsdorfer Straße 70, our first car, and later occurrences, I find that my interest in technical details reached back to these early times. No one seemed to have noticed. Who would have thought, that I could recall after 70 years many of the details of the early machinery I had contact with, and even knew their functioning reasonably well. At the age of eight, I certainly was well aware of the role of improvement of technology in our life, an observational effort that continued for the rest of my life. Also, I was much more interested in business details than Frank, as documented by my descriptions of Vati's business dealings. Again, nobody, least of all I, paid attention.

The role of relatives outside the immediate family remained limited. The reason seemed to have been Vati's strict ideas of what was right and what was not. Once he was crossed, it took a long time until things were back to normal, if ever. According to stories told later, Opa, for example, was at odds with Vati. This reached so far that Vati left home and, without Opa's support, finished his professional education and started his own business. Slowly, the problems between them were overcome, and Vati later employed Opa, but Vati never again trusted him fully.

The description of the first years in school was sufficiently detailed to give a picture which subjects were to my liking and which were not. Certainly, the discipline at home contributed to the stellar grades in behavior (see Figure 16). Again, the consistent grades and my liking of arithmetic were noted, but never acted upon. The question what I really wanted to be, was resolved by trial and error, and one might ask the question: What would have happened if the ability to see technical details and the good grades in arithmetic would have been taken into account when choosing a possible profession? Farming would probably not have been the first choice, although modern farming needs much technology. Putting myself with this much hindsight in Vati's position, I would have tended to suggest builder, civil engineer, or architect. This would have been far enough from Frank being groomed to take over Vati's business, and it could, perhaps, even suggest a possible partnership in general construction without undue overlap.

The story of the role of toys in education was discussed in detail and coupled to our Christmas customs. It stresses the importance of *proper* gifts. They must not only satisfy the child's wishes for possible fun, but also help developing his mind. You notice, that my favorite toy, besides the farm animals, was the extensive, annually growing erector-set. The interest in technology was further strengthened by model cars, trains, and ships. My 1930s Mercedes model race car even had a working differential drive, and was steered via a realistic rack-and-pinion mechanism. Both were open for inspection and were well understood by me. I can still demonstrate this car today. The same make of toy (Schuco) was bought again by me in 2002 in the Amerang Automobile Museum in Bavaria. Only toys true to reality were appreciated. Compared to the volume of toys frequently seen in today's children's rooms, ours were few in number. The enforced discipline to clean up after play meant no disorder in our room by the time we went to bed. In the morning, we had a clean beginning except for incomplete creations with the erector sets or the trains which took longer to set up. Had I chosen one of the above alternative job directions, little would have changed the learning from toys since there was sufficient stimulation in addition to the farm animal collection.

The role of Mutti during these eight years needs a few additional comments. As mentioned in the ranking of the family (see page 1-11), she came second, but from the view of Frank and me she came first. Vati was the role model since we were boys, but she was our support to get though the early life's adversities. All major decisions were outwardly done by Vati, but he did nothing, as far as we ever knew, that Mutti did not agree with (and vice versa). There was no friction in the family and at that age Frank and I were not permitted to create any. Did we love Mutti? Yes, certainly. Did we say it daily? No, it was not necessary. Was Mutti 'only' a housewife? No, in the critical years of the start-up of Vati's business, she was the complete office staff, dealing with the customers by phone and in person, handling the bookkeeping, taxes, and finances, representing the company for the day-to-day business at the bank, and handling the wages of the workers. This was done in addition to running the household, teaching herself to become a gourmet cook to satisfy her husband and children who all were developing quickly into appreciative, discriminating eaters. In addition, she was a gracious host when family, friends, or business acquaintances took part in celebrations or dinners.

Figure 27 *Mutti, picture taken in 1938.*

The world changed drastically in the next years. The 20th century represented the darkest period for Germany. Without the strong presence of Mutti and Vati, we probably would not have survived unscathed. How they managed through this period to increase our family with Joachim and Manfred to a four-son unit, is simply heroic. Some 20 years later, our family was in the USA, strong and united (see Figure 1 of Chapter 11). A new successful life with freedom from oppression was possible. It led to a comfortable retirement of Vati at 66 years of age and started with what he said were the best five years of his life. He spent this time traveling through the US, Mexico, and Canada with the Wally Byam Caravan Club in his Airstream travel trailer. He died when he was 82. Mutti survived him without financial worries for 14 years. Her wish, when her memory failed, to be able to read notes of earlier times, as well as my own regrets not having asked all the questions I should have when Vati and Mutti were still around, brought me to the decision to put together this autobiography, and naturally, I also had to add my growing involvement in Chemistry.

The historical, economical, and political developments on in this first and the later Chapters were updated, supplemented, and checked using the sources listed in the Acknowledgments (placed after the Preface). Particularly in Chapters 1–3 my memory needed augmentation for the reader to understand the life in Germany. Perhaps, in this way the book is of value not only to people I had contact with, but also to others who like to know about the life of a scientist in the 20th century.

CHAPTER-02

Years of War, 1939–1945

The war started when I was eight years old. *The center of my hometown, Brandenburg, looked as it had for 250 years*. The "Altstadt" ('old city,' first recorded in 1170) and the "Neustadt" ('new city,' first mentioned in 1196) were combined in 1715 on order of King Friedrich Wilhelm I of Prussia (see pages 1-5 and 3-36). Figure 1 is a photograph of the distinctive "Rathaus" (city hall) with the "Roland" in front, the sign of the market-right of the city. On the left is the ornate "Kurfürstenhaus," the house of the "Kurfürst," (an elector of the first German Reich, see page 1-1, Footnote 3). The electorate was established in the 15th century and the "Kurfürst" stayed in this building when in town. He never had his residence in the house, but the old and new towns were the biggest, most important cities of the "Mark" (Province) Brandenburg. City and "Mark" were named after an older castle on the Havel river. When the Kurfürst made Berlin his residence, Brandenburg lost the importance it held in the 16th century.

Figure 1 *"Rathaus, Kurfürstenhaus," and the "Roland" (on the right), the three "Wahrzeichen" (landmarks) of Brandenburg. (See also the upper left photo of Figure 24 in Chapter 1.)*

The street running towards the "Rathaus" is the "Steinstraße," the first cobbled street of the New City. The "Hauptstraße" branches to the left, the "St.-Annen-Straße" and the market place go to the right. By the end of WW II, none of these buildings remained, and the "St.-Annen-Straße" was fully destroyed when the Soviet forces entered the town (see Figures 17–20 and 24). The Roland survived, having been stored outside the city. It is now standing in front of the "Rathaus" of the Old City on the other side of the Havel (see Chapter 3, Figure 26, and Chapter 1, Figure 6, point 2).

September 1, 1939 was a Friday. For our family, the day started very early in the morning. I woke up from loud talking, walking, and banging on doors. It started at the top floor and moved slowly down to us on the ground floor of the Wilhelmsdorfer Straße 70.[1] *Two soldiers knocked at the door and delivered Vati's[2] "Stellungsbefehl"* (draft notice). He had to be, with his car, at 7 a.m. at a sports facility outside the city ("Turnerheim"). When Frank and I got up

---

[1] For residences and travel, see Appendix C, for locations, Chapter 1, Figures 5–6.

[2] For names of family members and their relationships see Chapter 11, pages 2–4.

a few hours later, Vati had left already and Mutti was fully in charge. She gave us the news. She also told us that Vati had reached her by phone and we should go to the "Turnerheim" with our bicycles later in the day and bring things he had forgotten in this hurry, and also something to eat.

When we arrived in the afternoon at the sports field, there was an enormous crowd of men. Several tables were set up where the processing was done. After a little while, we found Vati and his car. His preliminary papers were already complete. He was attached to an officer as driver of his own car and had to await transfer to the barracks when all others were properly processed. We delivered our goods, and this was the last we saw of 'private Vati'—until later in the evening of the same day when he came home on indefinite leave. The officer found a more experienced, soldierly driver. Vati had made a convincing case that because of his asthma it was a health-risk for him not to sleep on 'safe' mattresses (see page 1-3), and it was more important to get on with repairing and building of roofs over the barracks and armament factories than to join the army.

Once home, he worked the connections he had with the various industries which were pressed to meet the manufacturing quotas. Within a few weeks, Vati received a proper deferment from the army. Even in 1945 he was exempt when everyone between 16 and 60 years of age was drafted to the "Volkssturm," the last-ditch effort to win the war. A few months later, Vati even got permission to extract his car from the army, and we were looking forward to seeing his 'new' car, painted in army camouflage. But, we received the message, officer, driver, and car were lost in Poland. What would have happened if Vati would have remained as driver? Since many men were drafted, it did not take long for Vati to find a second-hand car for his business, an "Adler Trumpf".

Of greater significance was the other happening of September 1, 1939. What was to become *World War II had started at 4:45 a.m.* with the well-planned German attack on Poland. The story given over the radio in the morning was that this move was in retaliation to a Polish attack on a German radio tower and on border posts. Even when we first heard it, it was not credible.

Within two days, Great Britain and France declared war on Germany, pursuant to their week-old alliance with Poland created to forestall the fate that had befallen Czechoslovakia (see page 1-31). In five days, the Polish resistance collapsed, and on September 27 the "Blitzkrieg" ended with the fall of Warsaw after major bombing attacks on the city. As in Czechoslovakia, German territories lost after WW I were reincorporated into Germany, and the rest of Poland remained occupied and was given a German civilian government with the capital in Kraków (see page 2-21).

On September 17, the Soviet army also marched into Poland. The eastern provinces were annexed. This action was in accord with the non-aggression pact of the USSR with Germany. The USSR keeping this part of Poland after WW II violated the promise given by Great Britain before and during the war when Polish exiles joined the British army. Overall, the territory lost to the USSR in 1939 and the territory gained from Germany in 1945 moved Poland west by ≈100 mi.

On October 6, 1939, Hitler sued for peace under the condition of a status quo, but this was rejected by Great Britain and France. Would an end to the war at that time have prevented WW II and its horrors? Definitely not. Hitler's planning included further expansion to the east. He would, however, have loved to gain a year or two to complete his rearmament.

Let me now summarize *the first phase of World War II* by following the major operations from 1939 to 1940. This period can be called the European phase of the war. In November 1939, the USSR bombed the Finnish capital of Helsinki and attacked its borders. By March 1940, the Finns were forced to sign the peace agreement of Moscow and lost considerable territory. Furthermore, the USSR annexed the Baltic Republics (Latvia, Lithuania, and Estonia) and parts of

Romania. Between April and June 1940, Germany occupied the neutral countries of Denmark and Norway. In May, German troops marched into Belgium, Holland, Luxembourg, and continued into France. On May 14, Rotterdam was bombed in support of the ground offensive. By the end of June, an armistice was reached with France, leaving the southern half of the country under the government of Marshal Pétain in the city of Vichy, while in London, Général DeGaulle[1] called for active resistance. The northern half of France was kept occupied by Germany, with Paris as its capital.

Finally, when it appeared safe, Italy joined the war effort of Germany. Only five months later, Italy attacked Greece. The invasion of England by Germany, planned for September 1940, was postponed in favor of the air war. Finally, with Italy's attack of Egypt across the Libyan border, the war spilled over into Africa and became a World War. Also, DeGaulle's exile troops conquered Libreville (Gabon, French Equatorial Africa) in an effort to weaken the Vichy government, but failed in a similar attempt in Dakar (Senegal, French West Africa).

What did I learn about the war as an eight-year-old boy? The details of the war came over the radio in form of special announcements of the victorious progress of the war ("Sondermeldungen"). Otherwise, life continued with few changes. The rationing in goods and food-supply were instituted only slowly. Mutti, taking care of two boys and still helping in Vati's business, did not have to work for the war effort. The excursions by car, however, were curtailed. The last vacation was at Christmas 1939 to our beloved Lautenthal, and a short visit to Stralsund occurred in 1940, to see Tante Grete and my ailing Oma (see Chapter 11, Figures 8 and 4).

Tante Grete had moved earlier to Stralsund (≈1938) to open the porcelain department of a branch of the large retail store Wertheim she was working for in Berlin. Oma, who was not too happy in Brandenburg had moved also to Stralsund to stay with her daughter. In Brandenburg, she was unhappy to be in the middle of the frequent quarrels within the family of Onkel Richard, and, I think, she also felt still too close to Opa, who lived together with Tante Lorchen. As workers became scarce due to the war effort, Opa helped more and more in Vati's business.

During the visit in Stralsund, I was particularly taken by a visit of the harbor. Two motor boats with closed cabins caught my interest. One had large windows, space for about 8–10 people, and a captains-bridge. The other was a cabin cruiser. For a long time thereafter, I took imaginary trips on these boats and was wondering how far these boats could ultimately carry me. Could I go beyond the Baltic Sea? This was the first body of water I had seen to reach beyond the horizon. Finally, we met Inge, Tante Grete's friend, who spoke the much different local dialect. I loved to listen to her, and her voice is still in my ears. Oma, though, did not get well. She died in 1942.

This brings me to discuss *the decrease of our family from 1938 to 1942*. First, Opa Siedlung died in 1938. His and Oma Siedlung's grave are shown in Figure 2. We had visited Opa Siedlung more frequently when he was sick, but Vati thought, it was best not to take us along to the funeral so that we could remember him the way he was in the past, as is shown in Figure 3. Oma Siedlung had difficulties adjusting to life alone and moved to Tante Frieda, a choice Vati had predicted to be a mistake. Oma Siedlung quickly got involved in the not-so-smooth family-life of Onkel Hermann. Also, she missed her garden and could not work in Tante Frieda's garden as she had thought she would. We boys were not supposed to know all the details, but overheard many of the bits and

[1] After the liberation of France, he was head of the French Provisional Government from 1944 to 1946, and later, 1958–1969, he was elected to head the 'Fifth Republic of France.'

pieces of the problems when she was visiting us in the Wilhelmsdorfer Straße to lighten her heart. After two years, she finally moved into a home for the aged (the "Gödenstift"), not far from us. She visited now more frequently and Mutti also brought meals to her since the food in the home was not always to Oma's liking. Again, after she fell ill, we did not see her again, and the funeral was held at the "Krematorium" without us boys being present.

Figure 2 *Path and grave of Opa and Oma Siedlung at the "Krematorium" in Brandenburg, photographed in 1972 on the first visit after emigration to the USA.*

We missed her very much. She always watched over our performance in school and gave little presents for good grades. When I went along to the market with her, I could count on getting something good to eat. After a few experiments with different foods, I found, best was the hot dog of horse meat. Not only was it a bigger hot dog, but it also tasted better. I could verify this by trying one of these on a much later visit to Brandenburg when horse meat became fashionable again during the mad-cow-disease scare ("bovine spongiform encephalopathy," BSE, first seen in Great Britain in 1985). With three of the grandparents dead, the connections with Onkel Hermann's and Tante Else's families strained, and with Tante Grete, Tante Lieschen, and Tante Minna out of town, we had practically shrunk to our nuclear family.

The school went on without much noticeable change, as can be seen in Figure 16 of Chapter 1. Perhaps, the final half-year showed some bigger effort, since I wanted to make the transfer into the middle school for boys in 1941 ["Theodor-Körner-Schule," named after the patriotic poet (1791–1813) who died in Lützow's volunteer corps, fighting against Napoleon in 1813]. Since in the past, nobody of our family had any education beyond the required eight-years of elementary school, it was a big event for Frank and me to go to the middle school (also called "Bürgerschule"). Figure 4 shows a picture of the school taken on one of our later visits to Brandenburg. (For its location, see Figures 5 and 6 of Chapter 1.)

Figure 3 *Oma and Opa Siedlung in the garden, 1937.*

The German school system was well structured. Required was an eight-year schooling, the elementary school. Everyone started in this elementary school. Anyone

Figure 4 *Our family on a nostalgic visit in 2004 to Brandenburg in front of the Theodor-Körner-Schule (with was renamed in 1947 to Goethe Schule, and is now the Fontane Schule, all names of German poets).*

who had to repeat a grade more than twice, was moved to a school providing special education, known by us students derogatorily as the school for dummies. Later, when I was in high school, I met one of their teachers, the father of a girl friend, and found that he was better educated and more caring than many of our teachers. I spent enjoyable hours with him, discussing high-school-level questions and beyond. After four years of elementary school, a decision had to be made either to continue with elementary school, a choice most students made, to go to the middle school (the "Bürger" school) for six years, as I did, or, as top choice, to go to high school (the "Gymnasium") for eight years, as I finally did. For the higher educations, a minimal fee had to be paid.

Any change in school did not have to be a final decision. Depending on the performance, changes in either direction were possible. This accommodated different rates of development and changes in effort by the students. Staying in elementary school meant you would most likely become an apprentice in a craft or trade and still have several years of formal schooling in your chosen profession (usually one day a week). From middle school, the same path was open, but one could continue formal education in a trade school to reach a technical certificate ("Handelsschule," perhaps similar to a community college in the US). On good performance, one could even enter a technical university with this certificate. A high school graduate usually went on to a university. The cost of the university also was minimal, except for the living expenses. In my going to middle school, I certainly was more than qualified to become a farmer. Frank graduated from middle school in 1944, and pursued a trade-school education in civil engineering after the war (completed in 1951).

Corporal punishment was still acceptable in the early grades of school, but rapidly became obsolete. I remember one time I barely avoided the teacher's wrath. It was in 1939 or 1940. We had to learn a poem. When Herr Menzel called the first student to recite it, he could not. He received three whacks with the cane across the left hand (the right one was needed to write!). The second student did not do better. Same treatment. On it went. 'W,' was last, so, under the table, I could open my book and learn quickly. It took long enough to go through the 35 students for me to know the poem sufficiently well to get by. Herr Menzel was probably also tired by then.

At about the same time, I was not quite as lucky at home. I had built a powerful bow and arrow out of the spokes of an old umbrella. The bow was beautifully elastic and in our room I was pulling it for practice. All at once I (honestly accidentally) let go, and the arrow hit the ink bottle on the table. The bottle hit the wall, the cork flew out, hit the ceiling, and ink ran in a broad stream from the ceiling to the floor. I never figured out how this was possible, especially since the bottle was not broken! This deed was punished with an earned whipping. The last one I ever got from Vati, and the only one I remember today. It turned out that the damage to the wall was not as bad as it looked. Vati had a spare roll of the wall paper, and quickly covered the disaster.

Thinking, while lying on the sofa in the children's room (which in the evening was made into my bed) was developing into one of my most treasured form of solitary entertainment. Originally, it concerned my imaginary deeds, travels, possessions, and stories. Later, it concerned the unraveling of the functioning of things I had observed. Even later in life, it dealt with the scientific problems I did not understand. Possible solutions could be worked out in this way. I found, it was valuable time spent. It took some time until I could convince Mutti of this. Initially, she only observed that sometimes (often?) I ultimately fell asleep, but still, the problem or story was many times completed when I woke up, although I could not fully remember the detailed thought processes. In writing these Chapters, I use the time in bed before falling asleep or when waking up too early, to recall what has occurred at the time under consideration.

It may be, that this method of solving problems is the explanation why, when I am not fully sure about a solution of a problem, I am much more often right than not. Not too long ago, I happened to mention in discussing a scientific problem with a well-known colleague: "You know, we both are not sure who is right, but from experience, I know, in such situations I am usually right." He became quite upset by this remark and did not think this was good scientific arguing. I did not think so either, but a year later, it was proven by others that I was right in this particular case.

My quarrels with Frank became more physical as we grew older. Due to his being always responsible for me when we were alone, he felt that I should always do what he decided. I certainly did not think so. Analyzing the problem, I knew that as I grew older, the difference in physical strength would diminish and, perhaps, one day we would be equals. The flaw in this reasoning, I also knew. As a grown-up, I would not fight anymore. In the meantime, when a major confrontation arose, I hit Frank with all my might, then ran to the bathroom, locked the door, and waited for Vati and Mutti to come home and to resolve the reason for the fight.

All this was brought to an end when years later Frank had an argument with Vati. In the heated discussion it became clear to me that Frank truly thought I was always preferred by Vati and Mutti. I was exactly of the opposite opinion, he, as the older one, was always preferred. This made it clear to me that Vati and Mutti tried very hard to treat both of us fairly, but not necessarily equal. The age difference made things appear different to us. Ever since then, we got along better. In fact, later, some situations were reversed from what they had been before. When earlier we were only the best of pals when we were with other children we could argue with or play jokes on, as soon as Frank got interested in the opposite sex, the two of us were the best of pals when he was between girlfriends. Then, the two of us went together to movies, concerts, or dances and had a great time.

The first signs of war showed in 1940. Food ration cards covered all items, and the specialty foods Vati was so fond of, bought at 'Mehnert's' in the Jacobstraße, disappeared from the store and could only be obtained by 'good customers.' Later, they were not available at all.

When the air raid alarm was sounded, all lights had to be covered. All windows were to be darkened with special paper, so that no light could be seen from the street. Similarly, the automobiles had to have their headlights covered, so that only a minimal amount of light would show where one was going. Sand and water buckets were distributed in the attic to fight possible fires from incendiaries. Mutti was asked to take a first-aid course in case of injury due to enemy action. Vati thought, with two active boys around the house, this was a good idea anyway. Next, we were issued gas masks. One could never trust the enemy! The masks were very ugly, the whole head was surrounded with a rubber cover, except for the glasses for the eyes and the screw-connection for the heavy air filter. During the testing, one quickly got hot in the mask and it was difficult to breathe through the filter. Fortunately, they were never needed.

Next, the basement of the apartment house was rearranged to create an air raid shelter with reinforced ceiling, steel door, and a special ventilation shaft to the outside. Later, a passage in the basement corridor was broken through the thick, bearing walls to the neighboring houses, closed by a thin ¼ brick closure resting on a bottom row of loose bricks. In case of a fire, this closure could be opened by pulling out the loose bricks and offered an escape route. Little did we know that these connections were not only life-savers during the later fire-bombing of the German cities, but ultimately, they were also of value in defending the cities against enemy ground troops.

Finally, Vati also had to learn how to close burst water pipes made of lead by fusing them with a well-aimed blow of a heavy hammer. This should prevent any drowning from broken water mains. The hammer was stored in the shelter together with first-aid supplies and the gas masks. Today, one may be surprised to know that at that time water pipes were made of lead. Similarly, many of the paints used were based on lead oxides. It was widely known, that lead compounds are poisonous. We were even taught in school that cooking in lead containers may have contributed to the downfall of the Roman Empire by deteriorating the mental and physical capabilities of the population. Also, when a new house was built, it was self-understood that for the first few months no water was to be drunk from the faucet or used for cooking. After that time, the internal surfaces of the pipes were sufficiently oxidized that the amount of lead in the drinking water was thought to be minimal. And, naturally, it was unthinkable that lead paint would ever be ingested by peeling off the paint and eating it! Today, perhaps due to more sensitive analytical methods for detecting lead ions, one does not accept the old limits of safety.

Was all this war preparation necessary? '*Yes*,' must be the clear answer. On August 26, 1940, 50 British bombers flew for the first time to Berlin and passed over Brandenburg. The sirens went off at 12:45 a.m. and the all-clear was given about three hours later. All crowded into the air raid shelter. For us school children, it was later arranged that in case of long air-raid alarms, school would be delayed. On this first serious event, bad weather hindered the planes, and no bombs actually exploded in darkened Berlin. We heard the planes overhead and the shooting of the "Flak" ("Fliegerabwehrkanone," anti-aircraft gun). Near Brandenburg some explosives missed all targets (if there were any) and some incendiaries burned down the local theater in the Blumenstraße. The "Flak" was not very effective, but we heard the shrapnel raining on the city roofs. This gave Vati additional work in replacing the broken tiles and mending holes. Later, shells were only aimed to explode above the city when it was absolutely necessary. For us children, it was a new excitement to collect these often bizarre-looking shell fragments. I remember trading one of my most valued toys for a particularly big piece, about 3" long. On this and many other of the early air raids, Berlin was always the primary target, Brandenburg was left largely unscathed.

During the war, Brandenburg experienced more than 400 air-raid alarms (64 in 1940, 33 in 1941, 10 in 1942, 69 in 1943, 120 in 1944, and110 additional ones in 1945). As the terror bombings became heavier, a large part of the civilian population was protected by huge, above-ground concrete bomb-proof bunkers, built in strategic locations (see Figure 28, below). Without these precautions,casualties would have been much larger than the estimated 600,000, including 70,000 children. This compares to 55,000 flyers lost by the British Bomber Command (44% of the total).

Clearly, World War II did not only pit armed forces against each other, but also against civilians.[1] The first air attack of 1940 was much more effective than the minimal material losses. Hermann Göring, the "Reichsmarschall" of the German air force, had promised that his name would be "Meier" if British planes would ever be seen above Berlin. He, indeed, was called "Meier" from this day on when it was safe to do so. Two weeks later, the German bombing of London started and continued for 57 nights with 100–200 bombers each night. Since German planes could start from bases in France, they even reached the city of Coventry and caused extensive damage. By December 1940, this German air campaign was halted. The civilian casualties were 23,000.

A more detailed answer to the question of the prior paragraph can now be given. Yes, the precautions to protect civilians were necessary, and no, the bombings of the cities were not necessary. On both sides, civilian morale was not undermined, in fact, the will to endure was strengthened. The war in Europe had finally to be determined by armed forces on the ground. To subdue the Japanese war effort, it took dropping the much more horrible atomic bombs to avoid the ground invasion of Japan by the US forces. Brandenburg, although of strategic value due to the armament industry, was for a long time only a secondary target for bombers who did not make it to Berlin. Only in 1944/45 did Brandenburg become a primary target. In practically the last raid on April 20, 1945, our house in the "Werderstaße 28" was destroyed, as will be described below.

My move to the "Mittelschule" in 1941 went without problems ("Bürgerschule," "Theodor-Körner-Schule," see Figure 4 and page 2-4; and for its location, see Chapter 1, Figures 5 and 6). I kept none of my old classmates, which meant also, that I had to go to and from school alone or with Frank, in case our schedules matched.

The new school was much different. Every subject was taught by a different teacher. The walk to school was the same length as before, but there were more things to see (see Chapter 1, Figure 5). First, there were five or six stores. I could stop at Vati's business location at the Wilhelmsdorfer Straße 73, and had to pass the Brennabor factory where grenade casings and gun carriages were made. Finally I had to cross the "Jakobsgraben," an early canal, connecting the lower and upper Havel to maintain a water-level difference for the flour mills. The "verrückte Kapelle" (Jakobskapelle) was just beyond the bridge. The attribute "verrückt" could have two meanings. The first was 'moved,' and, indeed, the chapel had been moved by some 30 feet when the modern bridge over the Jakobsgraben was built. The other meaning of the word "verrückt," however, is 'crazy,' a much more interesting attribute used by all the students who went past it. Figure 5 shows a recent picture of the newly renovated, quaint little church. Even when we moved to the "Werderstraße" a few months later, my walk to school was of similar length (see Chapter 1, Figure 6).

The teaching was at a somewhat higher level and looking at my grades, in retrospect, they showed a distinct improvement. All grades that were '4' in the elementary school (see Chapter 1, Figure 16), were now better by one ('3'). The good grades in math were kept with the new teacher.

---

[1] Jörg Friedrich, "The Fire, The Bombing of Germany, 1940–1945." Columbia University Press, New York, 2006. (Translated from "Der Brand," Ullstein, Munich, 2002). A scholarly book, bringing together original sources from all parties involved, written sufficiently long after the war to be impartial (see its Chapter 2, Strategy). Dates and numbers given are correlated with this book.

Figure 5 *The Jakobskapelle near the Jakobsgraben.*

Additional 'good' grades were received in history and biology, but the poor grade in music remained. A new subject on the report cards was "Sammeltätigkeit" which was to judge the effort I made to collect materials (nonferrous metal, clothing, etc.) for the war effort and money for various political causes. These collections were to be made by going from door to door, as well as in the street with a collection box (often in return for little trinkets like pins or candles). Since I never liked ringing doors of strangers or accosting people in the street, I did not do well.

This last activity was supported by *the Nazi youth organizations*. All boys of age 10–14 were drafted into the "Jungvolk (Pimpfe)" and of age 14–18 the "Hitler-jugend." Analogous organizations existed for the girls. The selection was based on school enrollment records. Shortly after I entered the middle school, I was ordered to report on Wednesdays and Saturdays in the afternoons to the Rochow Schule, an elementary school, not far from our home ("Kleine Gartenstraße," Figure 6). The school provided class rooms for the meetings and the school yard was an ideal exercise ground. The leaders of these groups were specially educated and had to teach political orientation, physical education, arrange for war games in the surrounding countryside, and inform us about the basic glories of soldiering and its thrills, drills, and skills. Terrible! I did not like to follow unreasonable orders or sing in public, and usually liked to follow my own thoughts. Only excursions into the countryside and outings in the evening with gatherings around an open fire interested me. Overall, it looked like I would not make a good soldier.

In my effort to avoid this new 'duty,' I made an important discovery. When I did not feel like going to school, there was no chance that Mutti would write an excuse. But, for the "Jungvolk," the slightest complaint made Mutti immediately comply. She did not like to see me go there either. It was not long that it was known that one could not bank on my presence. When, a year later, I came down with diphtheria and had to miss months of school, to be commented on below, I finally had a permanent excuse and never went to these meetings again. I actually was forgotten.

The Jungvolk uniform was freely available while other clothing was scarce. Mutti liked us to wear the short pants of the uniform. They were made of very durable, black corduroy. The brown shirts, in contrast, she did not like to see on us. The uniform was completed by a black scarf, and a leather knot to hold the scarf in place. This dislike of the "Jungvolk" did not mean, that I was not playing with my toy-soldier army and fight navy battles with Frank, or observing our "Wiking Modelle" as from an "Unterseeboot" (U-boat) from below the table, using self-built periscopes (see Chapter 1, Figures 20 and 22). There was enough propaganda about the seemingly unending victories of the German army, air force, and navy to make such play relevant. But, I never forgot my animal collection (Chapter 1, Figure 21), which included exotics, like elephants and lions.

The move to a new apartment and business location in the "Werderstraße 28" occurred in 1941. Not only did we have a larger, more comfortable, and modern apartment, Vati's business could also be located on the same property. Figure 6 shows an air photo (see also Chapter 1, Figure 6). The history of the house, located in the center of the ellipse, goes back to the beginning of the 20th century. As an orphanage, it was owned and operated by the "Brandenburger Verein für Waisenpflege." In the 1920s, it was changed into an apartment house. By that time, it was owned by the "Corona Fahrradwerke," shown as the buildings lining the bottom portion of the right-angle bend of the Werderstraße and the left side of the beginning of the "Große Gartenstraße." When Vati rented one of the two second-floor apartments and all the surrounding property within the red outline, it was handled by the "Bau und Sparverein," a communal building and credit union. Vati's friend Herr Kutscher (see pages 1-27 and 1-31) was their financial officer and was instrumental in arranging the lease. He also was helping Vati with his much expanded bookkeeping for the growing business.

Figure 6 *Brandenburg, the railroad station and Werderstraße, the left of the three streets radiating upward. The other two, are the Große and Kleine Gartenstraße. A British surveillance photo of 1945.*

Above the house, a rather substantial garden is seen. The garden was separated from the street by a 10-foot-high brick wall. The bottom part of the property shows Vati's two storage sheds and the garage and office building. By then, Vati had two secretaries, relieving Mutti from her work in the business. The senior secretary was Frau Ladwig, a former shop owner (for chocolates and sweets) who had to close her shop because her husband was drafted, and chocolates were in short in supply. The junior secretary, was Fräulein Sonnenberg. A new, modern bookkeeping system was introduced and much admired by me. With the new, modern machines to calculate and automatically distribute the entries into the different office records, the bookkeeping became a model of efficiency.

The wing for the office and garage was newly built after we moved in. It was located on the right in Figure 6 and connected to the bottom shed. I spent as much time as possible watching the carpenters work the beams with chisels and wooden mallets. The whole building was assembled

Figure 7 *Business property of Vati, photographed in 1947. Top photo: Vati's offices window (left) and the secretary's window (right). In the background, the burned-out "Corona Fahrradwerke." Bottom photo: View in the opposite direction, showing on the left the corner of the remaining part of our cleaned, bombed-out house. In the back are the houses facing the "Große Gartenstraße" (compare to Figure 6).*

in the old-fashioned way, without any iron nails. The head-carpenter made it his business to explain all the details and advantages to me. I followed his words very attentively. I am sure, he would have gladly taken me as his apprentice.

Being close to the railway had the advantage that the materials could now be quickly brought to the yard for storage. The disadvantage was that frequent whistles from the switching yard could be heard all night. Surprisingly, it did not take long and it did not disturb our sleep anymore. Figure 7 contains two photographs of Vati's business. It was not much affected by the bombing, perhaps, because of the special construction which would give, when hit by a blast.

On the first day when I got home from school to the new apartment, I was in for a surprise. When I opened the large and tall front door to the house, I faced 10 steps of a triple-width staircase with a very high ceiling. A characteristic smell hit my nose. I can still remember it today. Later I identified it as resulting from the highly polished linoleum floor. To the right of the door was an interesting contraption for scraping and brushing ones shoes which I had never seen before. A glazed double door and wall were at the top of the wide stairs and gave access to the actual stairwell of normal width, leading to the second floor. This construction remained from the use of the building as an orphanage when it served to receive and screen visitors. In our second-floor apartment, the living room was on the right from the entrance corridor, fitted with all-new furniture. A beautiful credenza displayed Mutti's porcelain sculptures. In the center between two high windows which faced the street, stood a new, large "Telefunken Radio" with record player, and in the middle was a large, elliptical table with six chairs, expandable to seat 10. Above the table was a heavy, eight-arm, oak lighting-fixture with colored shades of crackled glass covering the light bulbs. A wrought-iron support reached to the ceiling. A new oriental-style carpet covered the floor. To the right was Vati's home-office and "Herrenzimmer" with a large desk in front of the window, a sofa along one wall with a low table in front of it. Mutti's old china-cabinet was on the other wall. The apartment had central heating, the kitchen was modern, fully equipped with gas-oven and burners, as well as hot and cold water, connected to the automatic gas heater for the water in the bathroom. All was new, clean, and fancy. Two bedrooms where at the end of the corridor, Vati and Mutti's facing the back of the house, Frank's and mine, the front. Our room was also larger than before. Most impressive were the high ceilings all throughout the apartment (≈12').

True to my predestination as a farmer, *I was put in charge of the about 20 chickens* made up of Leghorn, Sussex, and Rhodeländer, a good international mix. In addition, four geese, and later even a big tom-turkey, were added to the menagerie. Vati had fenced-in an area for the poultry along the whole length of the garden. He had also built a modern chicken coop, attached to the wall (see Figure 6). It had a roosting area, enough space for indoor-running, a big window, and eight cubicles with self-closing doors for the laying of eggs. Three times a day I went there, to provide food, release chickens from the cubicles, and to keep records of egg production and keep all clean. A good chicken would lay up to 200 eggs per year, more than needed to be surrendered to the collection agency since eggs were rationed. In return, we received feed and minimal compensation. For fun, Vati also had some miniature chickens. A black Paduan with feather crest and a few Gypsy chickens including a specially colorful cock made up the flock which usually kept apart from the rest. I practically knew every chicken by sight. A gander by the name of Phillip was so well trained, that I could call him and, without restraint, walk with him through the garden.

If you ask me the question, I have heard directed to 4-H prizewinners: "How do you feel that the animal you cared for so lovingly is going to be butchered?" I can answer: "The chickens I raised, tasted better than any bought from the store." For me, it always was a fact of life that livestock are raised as food. This does not mean that I did not get attached to the chickens, geese, or the turkey, or that I did not treat them lovingly. The chickens after slowing laying eggs, made chicken soup that today cannot be duplicated since only fryers are on the market. I am sorry that many of today's laying chickens ultimately end-up as dog-food, and that the fryers have a too short lifetime and are too restricted in running space so that they do not make a good soup.

Josef and Stanislaus were two Polish laborers forced to work in Germany. They were employed by Vati starting in 1941. Both spoke some German from earlier stays as temporary workers. Josef was rather quiet and thoughtful, Stanislaus, eager, aggressive, and talkative. Both were patriotic Poles. They had to wear a yellow 'P' on their jacket. We children got along splendidly with both of them. But, what employee would risk not getting along with the boss's children? Vati arranged their release from the camp for forced laborers and Josef moved into a room in the raised basement of our house. He became the supervisor of the home-base operation and took care of many of Frank's and my chores as well. As time went on, Stanislaus approached Vati to help him to apply for German citizenship. When Vati asked Josef, would he want to do the same? He said no, I am a Pole and like to remain it. Vati respected the wishes of both, but the application of Stanislaus was later denied. To finish the story, toward the end of the war, when the German loss was obvious, Josef approached Vati and said: "I think, I do not want to return to Poland after the war. Life in Germany is much better than what I had before. After the war, I would like to stay." But, after the war we did not see him again. We heard that all Poles were repatriated by force.

Workers for projects at the Arado Aircraft facility and the military airport in Briest were also recruited from among the prisoners of war (POWs) and even the "Zuchthaus" (penitentiary). Among the latter was a trained roofer. From all I could ascertain, it was Erich Honecker, although his name would have meant nothing to us at the time (see Footnotes on pages 3-22 and 9-24).

Some of the more "normal" experiences involved joint mischievous behavior with Frank. We were old enough (?) that we could be left alone evenings. One of these times we inadvertently chained the front door. When Vati and Mutti came home, they could not wake us. When I went out early to feed the chickens, I was surprised to see a long ladder leaning against the bedroom window. The answer was given at breakfast with a stern warning. Don't chain the door and sleep so deeply that you do not hear our ringing the bell. Fortunately, Vati noticed that the bedroom window was open a crack. He got a ladder and climbed through the window. We were amused when thinking about Vati climbing the ladder in his best suit, and neighbors possibly taking him for a cat burglar.

Another time, the punishment was automatic. Vati, because of his asthma, did not smoke but he always had cigarettes and cigars available for guests. He used to say: "*I certainly am glad not to have the habit of smoking*. The only exception is when I am in serious business discussions and have to make quick decisions. Then smokers have an advantage over me. They can always interrupt the proceedings to light a cigarette, pick up their cigar, or work their pipe. While I have to come up with an answer instantly, they can take a minute or two to think it over without anyone noticing their indecision."

One evening, Frank and I decided to test Vati's cigars. We thought it more dignified than smoking a cigarette. We bet: "You can't finish it." Then, went to the central furnace room. We knew, otherwise Mutti would smell the smoke. After guests smoked, she always aired the apartment for hours. Frank picked a black cigarillo, I, a standard cigar, somewhat bigger, but more friendly looking. After not too many puffs, and plenty of coughing, I did not feel too well, and threw the cigar into the fire. Frank tried harder in order to win the bet, but finally, he also gave up and both of us hurried to the bathroom where we had to throw up. When Vati and Mutti came home, they saw our pale faces and knew something was wrong. Their concern was quickly dispelled when they smelled the smoke on Frank. They knew then that we were not sick, and we had to confess.

From that time on, I was opposed to smoking. To test Vati's comment that smoking might help to bridge difficult moments in life, I tried it once more when I was 16 or 17. Cigarettes were still scarce at that time, but in social get-togethers, I found some of my friends would smoke and, may be, it helped them. So I tried it at home. I made it half through one cigarette. This was terrible. My conclusion was that something that causes such an irritation in your throat and tastes so horrible must be bad for you, i.e., I quit smoking on that day more than 60 years ago after having consumed a total of $^1/_{10}$ of one cigar and $^1/_2$ of one cigarette. Frank, unfortunately got addicted when he was for a short while in the army. He did not manage to give it up after coming home.

The winter of 1941 came early and brought the best ice skating of my life. The meadows south of the city had flooded and, without snow, were frozen solid with a layer of "Spiegeleis" (reflecting ice, or black ice). Vati joined us on a sunny, cold day. We went beyond the railroad tracks where we found the clean, smooth ice and could mount our skates to the boots. We were almost afraid to skate on the ice because one could clearly see the grass one to three feet under the surface, but the ice held. Only in few places did willow bushes interrupt the immense plane. We went for miles without stopping, almost reaching the next village (Schmerzke). It was a most exhilarating experience. This was the most effortless skating possible. It taught me also that it is wrong what many people thought, namely that a thin layer of water lets you glide on ice, but not on polished stone. It was a very cold that day and there could not have been a water layer on the ice, but skating was marvelous. When the weather warms, a layer of water is discernable, but skating is difficult. Many years later, I read in a scientific journal that Canadian scientists had solved the problem I noticed. On the surface of solid ice, the H_2O-molecules can swivel in place, making it possible to glide over the ice. Only at very low temperature is this property lost. A water layer keeps the skates from contacting the mobile ice molecules and, thus, hinders skating. In fact, this subject of high mobility in certain crystals (mesophase crystals) or on surfaces was later an important research topic on mine (for details, see Appendix B, Sections D to G).

A more serious ice-skating experience occurred later in the year after the thick ice had melted and refrozen several times to more normal conditions. Frank and I wanted to go skating on the Jakobsgraben, just around the corner. At that time, this was an open area, with good access for skating. When we arrived, there were a number of people standing at the edge of the ice, trying to help a small boy, perhaps a little younger than I, to get out of the ice he had broken into. It did not work. The older ones, who might have been able to help, were too heavy and broke through the ice long before reaching him. All poles or ropes that were available were too short. The boy splashed in the water and broke more ice, but it was not enough to reach a stable surface. The rescue-squad had been called, and finally, we could hear their siren. By then, the boy had stopped moving and by the time the squad launched their rubber dinghy, only his hat swam on the water. They broke through the ice step by step until they reached the hat, and with a grappling hook, pulled the boy out of the water. The efforts to resuscitate him were not successful, as we heard later. We had observed the whole tragedy and did not feel like skating on this day anymore. All the warnings we had heard many times before were repeated again when we told Mutti why we did not feel like skating. The Jakobsgraben was a particularly dangerous waterway. It drained the water of the "Breite Bruch" (wide marsh) and currents of water of different temperature circulated in the creek. Every frequent skater in Brandenburg knew this. Only after long periods of cold weather could one trust the Jakobsgraben. We also knew it before, and now, we stayed off the Jakobsgraben even longer.

In spring, Brandenburg was visited by *a small flotilla of model war-ships*. A battleship (the "Bismark"), a panzerkreuzer (a pocket battleship, the "Graf Spee"), and two smaller cruisers. The steel models were marvelous. They were sufficiently large so that one could sit inside the 6–12' long ships and steer the electrically driven vessels. Mutti went there with me and I inspected the details. My next fun was to mentally construct not only war ships, but also a "Zeppelin" which could be stored on a trailer for transport. I pondered for quite some time how much weight it might be able lo lift?—Enough for one person? At that time, I did not have the knowledge to figure it out.

Increasingly the *rationing of the food restricted our diet*. The first shortages actually showed up before the beginning of the war (see page 1-27). Not that we did not get enough to eat, but an increasing number of specialties were not available as the war continued. Early in 1942, Vati finally got the message, the normal shopping channels Mutti had patronized were drying up. The reserve of ≈30 chocolate bars she had amassed early in 1941 had shrunk to few, and the delicatessen from 'Mehnert's' were not available anymore. Vati did not like to frequently buy on the black market, it was too dangerous. A better way was to barter with the farmers, fishers, bakers, and butchers among his customers for part of their bills in kind. It worked. In this way, a whole support-network was set up. It turned out to be even more important after the war.

One time, I woke up at night. When I passed by the kitchen, I saw a piglet on the table being carved up. Vati and Mutti were taken aback. I was not to know. If I would have talked, it could have led to dire consequences. My thoughts, I remember even now: This is illegal. They should not have gotten it. But then, I certainly liked to eat. If I keep quiet, it will be all right. Indeed, I did not ever mention it, and Mutti thought, I was not fully awake to see or understand what I had seen.

Other help with the food supply was my 'poultry farm' and the garden products. It supported the family with eggs, meat, vegetables, and fruit and supplemented the rations. The flower and grass section of the garden was restricted to a rock garden, covering a drop in the ground level, and a sitting area in front of a newly built gazebo with a branch-line from the radio in the apartment.

A near disaster was caused by this second loudspeaker in the gazebo. One day, late in the evening, a neighbor from the other side of the Werderstraße, who was unknown to us, rang the door bell and said: "Sorry to disturb you, but *the speaker in the garden transmits the opening chords of Beethoven's 5th* for all of us to hear. You may want to turn it off." Without waiting for an answer he turned and left. This was not a complaint about loud music, but a warning that we had forgotten to turn off the loudspeaker in the garden and everyone could hear that Vati was listening to the BBC, the British Broadcasting Corporation's short wave station, a capital offense. The BBC opened its program for Germany with the four chords of Beethoven's 5th Symphony. Everybody in the neighborhood had heard it, but nobody turned us in. The speaker was permanently disconnected from then on. Luck was a large part of survival if you were at odds with the dictatorship.

The second phase of World War II started in February 1941 with the "Deutsche Afrika Korps" under General Rommel landing in Libya to support the faltering Italians. In April, Greece and Yugoslavia were overrun by the German army. These fronts were opened since Greece had resisted successfully the Italian attack, and Yugoslavia had a change in government and political alliance in March. Most important, however, was the surprise attack on the USSR by German, Romanian, and Italian troops on June 26. By December, the three Baltic Nations were freed (?), Leningrad and Moscow were reached. But, the fast progress was stopped by the especially strong winter of 1941/42 and the first counter-offensive of the Red Army.

On December 7, 1941, the Japanese attacked Pearl Harbor and started the war with the USA. Five days later, Germany declared war on the USA, based on the German-Italian-Japanese alliance of 1940 which had strengthened the original Axis (see page 1-14) and was arranged to keep the USA out of the European war. At the end of the six-year WW II, 51 countries had declared war against Germany and four of its allies had changed their allegiances (Italy in 1943, and Bulgaria, Romania, and Hungary in 1944).

For me, the year 1942 brought a severe illness. A number of years back, I had come down with what was probably wrongly diagnosed as a mild case of diphtheria. I recovered quickly, and Frank could return home after about one week of quarantine at Oma Siedlung. When I had similar symptoms one Friday in the spring of 1942, our pediatrician (Dr. Milatz) thought this could only be an inflammation of the tonsils since one should not get diphtheria twice. Mutti was not so convinced, particularly, since it got much worse over the weekend. A return visit of the doctor on Monday, quickly confirmed Mutti's fears and I was whisked to the isolation station of the hospital. Of the first week in the hospital I remember very little. After the first day, I remember only what seemed to be a continuous state of sleep with occasional moments of bright consciousness, usually brought about by a routine at the children's ward, the evening songs of the nurses out in the hallway.

When I slowly recovered, I could talk to Mutti every day for a while through the closed window. No visitors were allowed in the room. Gifts were delivered by the nurses. Vati would usually stop by on his way to different job sites. But nothing seemed to be normal with this illness. After a while, Mutti found out that she had difficulty understanding me through the window. Next, I noticed that when drinking and when eating soup, the liquid would run up my nose and drip out. Just pinch your nose shut, I thought, and indeed, it worked. Then, I noted that when I wanted to read, everything appeared double. My recipe to fix this, was to hold my hand over one eye. This did not fool Mutti too long. When she brought me a particularly nice book, she asked: "Don't you like the book? You wanted it, and now you have not even taken it from the table to look at it." I obediently took it and pretended to look it over. Unfortunately, Mutti discovered that I held the book upside down. This ended my pretense, I had to tell her all the problems of the last few days.

Her reaction was instantaneous: "How could the doctor and nurses not have noticed this? Wait for a few minutes, I will call Vati." He was in his office and in a short time he was at the hospital. I heard him argue with the doctor, and just half an hour later he had found out that I was not infectious anymore, and he had gotten my release. I was driven home, to be taken care of by our own pediatrician. He found out quickly that in addition to the paralysis in the throat and eyes, I could not walk either. Nowadays, this neglect in the hospital would have been good reason for a law suit which could have left me rich for the rest of my life. Not then. A few harsh words were all that Vati could get off, but I must say, support was given afterward by the hospital to affect my recovery.

The convalescence took quite some time. The throat and nose problems disappeared first. I could eat and talk again. Being home with better food helped also. Next, I could coordinate my eyes to read for longer and longer periods of time. The walking was more difficult, perhaps worsened because of the long bed rest. Vati would, on sunny days, carry me down to the garden and place me in a lawn chair. Then he would fetch Phillip, the gander. Phillip would come running, and as if he understood my problem, would sit next to my chair and when I dropped my arm, nibble on my fingers. The chair was placed under our old apricot tree which in that year had a specially plentiful harvest of large and sweet fruit. A delight to eat.

With the help of the hospital administration, Vati could arrange first class train tickets for two from Brandenburg to "Bad Mergentheim," a spa, a 380 mi trip to the south. Mutti and I should stay there for a three-week convalescence. Since Mutti had often problems with her gall bladder, this was a choice that could serve two purposes. The hospital arranged for a nurse who got me and Mutti into the crowded train and arranged for similar preferential treatment at the stations where we had to change trains. This was a special privilege in war-time.

Once we arrived in Bad Mergentheim, I was introduced to the spa routine, daily special pine-scented bubble baths (good), massages (tolerable), and for Mutti, the mineral spring waters (terrible). Much more important was that on the first morning, I could make it out of bed by myself and could walk small distances. The length of the excursions increased every day. Vati could hardly believe it and made the trip by car on the next weekend to see the miracle. The following weeks I was outrunning Mutti and had my own special places to visit. The best I remember was a shooting range in the hills were I learned how to shoot respectably, helped by the instructor and the allowance from Mutti to pay for bullets. The key suggestion of the instructor was as follows: "You can't hold your rifle sufficiently steady when pointing it continuously at the target. You must move it slowly across the target and smoothly press the trigger at the proper moment when you cross the mark." It worked. Vati was impressed. Mutti's health also improved, although I thought this foul-tasting mineral water could not have been the cause. One had to drink it, while walking around the nice park. Next to all plants was a sign saying don't pour the spring water onto the plants! If the plants could not tolerate it, and if it doesn't taste good, how can it be good for you?

A final, great impression for me was a visit to the "Pfarrkirche Stuppach," a few miles from Bad Mergentheim. There I saw the famous Madonna, painted by Matthias Grünewald. A small reproduction is shown in Figure 8. Perhaps it fitted the spirit of the miracle we just experienced.

Frank spent this time at the Baltic Sea as part of the of the broadly arranged evacuation of children to avoid the bombing threats ("Kinderlandverschickung, KLV"). As mentioned already, by the time I was ready to return to school, I did not report back to the "Jungvolk," and to my pleasure, they had forgotten that I existed. Returning to school presented one difficulty. The instruction in English was advanced sufficiently, so that I could not follow. Extra help was arranged for me in the afternoons with an English teacher of a parallel school, who later for a while was also my English teacher in high school ("Frau" Hirth). The extra help succeeded to some degree, but not enough so that I ever excelled in the subject.

Figure 8 *The Grünewald Madonna of Stuppach, painted 1518. From a postcard.*

The next noteworthy excitement came in winter 1942. Vati was able to leave for a two-week skiing trip to the Alps with business friends, as he had done the year before. We admired his photographs of the little town of Obergurgl in Austria, completely covered with snow and the 12,372' "Wildspitze" in the background.

Then, one of these days on going home from school, I discovered a new smell. A smell that was later only too common, typical of burning wood, paper, textiles, and tar, doused with water to prevent full combustion. It is in the air whenever a building burns, and the fire is being extinguished with water. Soon *I saw a fire engine in front of our house*. Running home, I found Mutti, Josef, and the secretaries huddled together while the firemen put out the last of the fire in the office wing. In a storage compartment behind the office, the fire had started in an electrical installation and both offices and the garage had major fire, smoke, and water damage. Vati's car was pushed out of the garage by Josef and survived unscathed. Everyone was finally happy that only property damage was done, but Mutti was hard to console. The fire occurred while she was in charge.

What would Vati say on his return, scheduled for the next day? Opa also returned from his job site, but his comment that Vati certainly would not blame her, she did not accept. I could not help her either, although I knew Vati sufficiently well to know that he would assess the problem and immediately go to work to arrange for fixing the damage and not to find blame where there was none. This is what happened. I could see Mutti's relief when Vati had returned and all was under control. The material losses of much greater magnitude which occurred later, did not worry Mutti to the same degree. They happened when all of us were together and none of us could be blamed.

The third phase of World War II started in 1942. The initiative of the war slowly changed from Germany to the Allies (Great Britain, the USA, and the USSR). Large-scale bombardments of German cities began. First was the old town of Lübeck which was largely destroyed (3/28/42, the close-by Dornier aircraft manufacturing facility, however, remained intact). Next, a first attack with over 1,000 bombers occurred at the end of May on Cologne ("Köln"). The vision of such 1,000-bomber terror bombardments of German cities as a military goal to destroy the morale of a country was first proposed by Winston Churchill in 1919(!) (see Footnote on page 2-8, pg. 51 ff. of the cited reference). These bombings found their peak in February 1945 with the destruction of Dresden, an undefended city of little industrial or strategic importance. Some 60,000 civilians were killed in the inferno created by the fire ignited by scientifically determined mixture of incendiaries and explosives fitted to the nature of the dwellings. Some estimates of casualties are much higher. Many refugees were not registered in the city and many victims were cremated without counting.

Later in 1942, the German "Panzerarmee" was halted in their advance in Africa and had to capitulate in May 1943. An initially successful final major German U-boat operation in the Atlantic started in August 1942, but came to a precipitous end in May 1943. The USSR enclosed the 6th German Army in Stalingrad at the end of 1942 and forced their capitulation early in 1943. Resistance groups in the occupied countries and also in Germany became increasingly active.

In July 1943, Allied forces landed in Sicily to protect the interests of Great Britain in the Mediterranean from those of the USSR (!). This postponed the Allied landing in France. Mussolini, Italian's dictator, was deposed by King Victor Emanuel III in 1943 (see also page 2-23).

The reaction of Germany to these setbacks was the declaration of the 'total war' by Goebbels in February 1943 in the "Sportpalast" in Berlin with the (in)famous slogan "Wollt Ihr den totalen Krieg?" For most of us, this was taken as the final sign that the war was lost.

At the end of 1943, the first conference of the three Allied Leaders (Stalin, Roosevelt, and Churchill) took place in Teheran to coordinate the last phase of the war in Europe with major offensives by the USSR and an Allied landing in France. Also, the division of Germany after the war was discussed. It was to be subdivided into several independent states, their make-up to be discussed later. The westward movement of Poland was settled (see page 2-2). Furthermore, it was agreed that Germany's East Prussia, and the Baltic Republics were to be annexed by the USSR.

Vati and many of his friends of equal persuasion had already feared the loss of the war when Hitler attacked the USSR in 1941, but this early into the war, *we also feared for Germany if it would win the war*. Although most of the atrocities committed in the name of the 'Third Reich' were not known to the public until after the war, everyone, even we children, knew that certain things were not to be expressed openly. Heavy penalties could be counted on for wrong political opinions. How much personal freedom would be left in case of a victorious Germany? And even more so, how would life be in the occupied rest of Europe? I remember in 1941, thinking: 'How come, I was born a German? It must have been purely by accident. Could I not just as well have been a small boy in one of the 'enemy' countries? Would I feel different then? Most certainly!' I was still too small at that time to come to an answer, but such thoughts slowly changed my outlook on life. I gradually learned that Vati was not in favor of the German government. But what could he do? Any open rebellion would certainly destroy him and his family. The only solution was to offer as much passive resistance as possible, keep your own opinions hidden, and, as soon as you can trust the children, make sure they see what is right and what is wrong. Finally, he made a major effort to assure that we children were not taken-in by the continuous Nazi propaganda.

One tool that was developed was to formulate jokes, ridiculing the actions of the Nazis. The weak safety net was then to say 'it is only a joke.' A special situation arose one day when I needed to speak to Vati and saw him in the back of the house, talking to a well-dressed visitor. I knew, there was nothing wrong with my interrupting the conversation by greeting a stranger with the greeting, "guten Tag" (good day). Vati introduced me as his son, and the stranger said: "This is not how we greet each other today, tell me, how it is done properly!" I immediately recognized that he wanted me to say "Heil Hitler!" But I did that only when it absolutely could not be avoided. Still not getting the point that he might be a Nazi official, I answered him with: "I am sorry, I thought you are a friend of my fathers." It took a lot of quick thinking by Vati to mollify the one who was in reality not a friend, and to convince him that I had made a bad joke.

Planning for the end of the war was necessary. The most important goals had to be to survive the bombing and to stay out of the armed forces. For the first, Vati reinforced our air raid shelter against the will of the other tenants. They thought that the huge posts and support-ceiling Vati had erected took away too much breathing space. Later-on, he earned their gratitude when it proved that this improvement most likely kept them from harm. Nobody lost their life when bombing destroyed our house on April 20, 1945. Figure 9 illustrates the damage.

In 1943, it became also a distinct possibility that the war might drag on, so that Frank, and finally I, might be drafted into the army. Fortunately, we all survived, and even Vati kept his deferment, even when the "Volkssturm" was formed. The "Volksturm" provided the last-ditch army which included all ages that could hold a rifle and would certainly have included me, had I been more visible in the "Jungvolk."

Figure 9 *Our house, Werderstraße 28 in 1945 after preliminary clean-up after the end of the war. Our apartment was the one at the top floor on the right, most of the contents fell into the apartment below.*

Another concern was the survival of our livelihood. Vati's company had grown over the last years by the work with large industry. It, most likely, had to go back to providing roofs and repairs for private housing, farms, and small industry (it did). The degree of automation would not be sustainable, so that a supply of presently not needed, simple tools were bought and stored, such as pulleys for hand-hauling of materials up to the roof, hand-drawn carts to replace trucks, and wood-burning kettles to replace the automated ones (unfortunately much of this equipment was looted at the end of the war). The destruction of the cities by bombings guaranteed a large amount of work after the war. As the industry moved eastwards to smaller cities and into the woods to escape the bombings, Vati's business had to follow the jobs. He had opened self-contained branch businesses. Hopefully, one or more might remain (they did not). Since his automobile might not survive (as it did not), a motorcycle and bicycles were bought and hidden away for the uncertain future (but were stolen).

Being brought up to oppose all aspects of war did not mean that at that age it was not fun to see and hear about its tools and strategies. *On a school outing we went to Potsdam* to visit historic buildings, but we were also shown a large, realistic diorama of trench warfare of World War I, complete with artillery and planes. The soldiers were of the same type as my toy soldiers (Lineol, made in Brandenburg). The sand box must have been 40 feet long and invited detailed inspection. This was much better than what we had at home and I studied it with interest. Such illustrations of war were sufficiently impersonal that one did not imagine the suffering of the wounded in the depicted field hospital, or the tragedies caused by the death of soldiers. Presentations have changed somewhat today, but many books describing past and present wars are still glorifying the deeds in battle when there is nothing glorious about it. It is particularly upsetting to me that frequently one want to soften the pain of the surviving families by declaring all casualties in war as heroes.

In the summer 1943 I was also eligible to join the "Kinderlandverschickung", the evacuation of school children to the countryside, "KLV." Our school-classes with their main teachers were sent to occupied Poland, far away from the reach of bombers. The small town I was sent to was Rabka, a health spa south of Kraków in the foothills of the Carpathian Mountains which stretch from the Czech Republic to Poland and Slovakia, the Ukraine, and finally to Romania.

The good-bye in the evening at the train station in Brandenburg was rather subdued. It was my first travel alone from home. Mutti was very sad to see me go because she thought I would be very home-sick, and I was. To some degree the many new impressions and the presence of most of my school mates helped to overcome the difficulties. New money, new food, and the beautiful, hilly countryside took my attention. The hotel we were quartered in was changed to bunk-beds, eight boys to one room. A large dining room, several rooms for schooling, and the kitchen made up the rest of the house ("Villa"). The trip to Rabka, however, was not without problems. In Berlin we came into a night air raid. The train stopped and we all had to run to take shelter. This was the only time I was inside a monstrous grey, concrete, bunker. We were placed in a long corridor, and had to sit for an hour or two on the floor until the all-clear sounded and we could march back to the train. Dead-tired, we arrived in Rabka and were distributed to the various hotels.

The school routine in Poland was as it was at home, just that a limited number of teachers had to cover all subjects. Every morning was school, every afternoon, special leaders from the "Hitlerjugend," about 16–20 years old, would try to indoctrinate and train us in soldering skills and Nazi ideology. Fortunately, they were not very good at it. They were full of hate toward Jews and East Europeans, and of ideas of Germanic superiority. Their effort was to convince us that this gave them the right to colonizing the eastern lands. Their message was not persuasive to me. I liked the school periods much better in which the teachers cautiously tried to counter this propaganda.

On Sundays, we could explore the village on our own. I quickly discovered a nice park in the center of the village where I could spend some of my "zlotys" and "groszys" on such delicacies as the freely available fresh, sour milk. It only tastes good from unpasteurized and unhomogenized, milk, which today is hard to get anywhere. We did not need to learn much Polish. As a tourist town and spa, most of the people around us knew some German. We also felt no animosities toward us.

My major complaint was that the quality of the food served at the house was not the best. Particularly the milk soups and puddings were regularly burned at the bottom of the big pots. Even if I was served from the top, I could taste the burn which was to me (and still is today) a most repugnant taste. I'd rather not eat at all (for a while). Once I was caught pouring the soup down the drain and got a strong reprimand. When we wrote home, our mail was censored and graded by our teacher. All errors were corrected, and certain disallowed messages were blacked out. When I complained about the food and asked for some goodies to be sent, Mutti showed me the letter after my return home. All references to the specific food that did not taste good were blocked out, and a remark in red was added: "Your son gets enough to eat, please do not send any food to him." Still, Mutti knew her son, and regular emergency rations arrived.

My dislike of the military drill came to a head when we expected a visit from the German Governor of occupied Poland (the "Generalgouvernement Poland"). We had to march to the train station in formation and a few hundred boys from the various houses were lined up in the sun. The train was late and it got hot. After perhaps one hour, I had enough. I sat down. Immediately our leader appeared. Get up! My answer: "I can't, I feel so bad, I will throw up and might faint." He came up with a, for me, brilliant solution. My neighbors left and right were ordered to accompany me home and see that I was taken care of by the house staff. With hanging head I stumbled away until we were out of sight. Then we all three ran home and had an afternoon off. My recovery was miraculous. The Governor never came. The others marched home, late for dinner.

The many war-games I also did not appreciate either. On one of those occasions, the boys from a neighboring house had stolen our flag. All were sad and felt a big loss. Except I, I tried to console my friends by saying: "Look at the bright side, we now don't have to raise and lower the flag with the daily boring ceremonies." I never understood why this comment was not appreciated. Something about 'honor,' I was told. All this convinced me, if it was possible, never to join the military or similar other organizations. This does not mean that I dislike legitimate armed forces or the people choosing to serve in it as their career, it just is not for me.

Much nicer were the two major excursions we made. One was to Kraków, where we could visit the world-famous Wavel Cathedral and Castle, the 15th century St. Mary's Church with the largest Gothic altar carved by Veit Stoss, and the 13th century Main Square (now, since 1978, all a United Nations Educational, Scientific and Cultural Organization, UNESCO, World Heritage Site). On our visit in 1972, I could refresh my memory of Kraków and take the photos of Figure 10.

A second excursion brought us to Zakopane at the foot of the Alpine-like Tatra Mountains (which are part of the Carpathian Mountains, near the Slovak border). With a cable car we made it up into the (now) Tatra National Park with the highest peak in Poland (Mount Rysy, 7,500'). For me this was the first time to see really high, bare, rocky mountains.

Two other memories were a gypsy camp which was indescribably dirty. The gypsies were poor and seemed to have no idea about proper hygiene. I think we were brought there only to support the claim of their racial inferiority. The other memory was the frequent gunfire in the hills from the resistance groups. Vati had heard about the Polish partisans and was concerned about the weakening of the eastern front.

A delay of our return home was caused by difficulties to arrange transports due to the bombings and an outbreak of scarlet fever which kept us in quarantine. With all other childhood diseases, I also had scarlet fever before, and thus, was not affected. By that time, the stay away from home was way over the planned three months and the mood was rather down among us boys. After my return home, Vati told me that he had saved two jerrycans of gasoline, so that in an emergency, or if the stay would have lasted much longer, he could have

Figure 10 *Kraków, 1972.* **(A)** *Top picture, the Wawel.* **(B)** *Bottom picture, the St. Mary's Church.*

driven the 800 mi round trip to fetch me. Fortunately, this was not necessary. I arrived home and was glad to be able to resume normal schooling and to tend my animals. Also, I finally could again successfully avoid the "Jungvolkdienst," as I had done before, described on pages 2-9 and 17.

On getting home, I had to get used to the bombing again. Once, I was rudely awakened in the middle of the night by our window being shattered and the shards peppering my bed. Vati was standing in the door, urging us in no uncertain terms to quickly get into the bomb shelter (in pajamas and barefoot). By that time, the information for the civilian population about bombing dangers was much improved. By using a wire transmission ("Drahtfunk") to a small radio we had continuous information where and how many planes were over Germany. An (assumed) target-city for their raid also was constantly updated. In this case, it was supposedly the usual bombing of Berlin and Vati thought we needed the sleep more than the safety. He was wrong. One bomber seemingly was in distress and unloaded its bombs on Brandenburg, with one bomb exploding not far away away from us. Such accidents were always liable to happen, one was never safe as a civilian. Later in the war, it was even customary for fighter pilots to step up the terror and shoot at civilian trains, farmers in the field, workers going to the factory, and even students on their way to school. The air raid alarm had to be taken seriously, all the time.

In 1943 there also were increasing daylight bombings by the US air force. On clear days we could observe the American formations of some 20 or more aircraft flying at great height (>30,000'). On such days at home, we did not take the air alarm too seriously, we thought we could see the danger in time. At least once, this did not work too well. I came from the chicken coop with a number of eggs in the feed pail when I heard the screeching sound of falling bombs. At the entrance to the shelter, I collided with Vati, Josef, and the secretaries. All five of us had started running at the same distance. We made it through the door simultaneously, although the door, at most, was wide enough for two. The bombs detonated far enough away to cause no damage, except I now had scrambled eggs in my pail. Fortunately we could laugh about the incident.

During this time *we heard infernal noises from the aircraft manufacturing* and testing facilities in Brandenburg and Briest. Through Vati, who was involved in keeping up the roofs and could look what happened below, we found out that these noises came from the testing of jet engines for planes (Me 262). Imagine, planes without propellers! At the same time, rocket engines were tested as well, with even louder noises (Messerschmitt, Me 163 and Heinkel, He 176). Since 1942, the German industry had production-ready jet aircraft. Their deployment was, however, delayed.

On one of the days in 1944 with good view, we observed a single jet aircraft shoot what looked like almost straight into the sky from the airbase Briest, near Brandenburg. Vati told us, he had seen an hour before one of the most famous German air aces, Dolfo Galland, come from the rail road station. Was he the one who tested the jet? Looking up his biography, I could, indeed, find out that he had shot-down seven planes with the new jet, Me 262, and was finally wounded in 1945 when crash-landing one of these. In short time, the jet caught up with the American formation and before it left our field of view, the jet started to shoot down one or more planes out of the formation. Naturally, seeing such demonstration of a superior aircraft made us hope that the bombings could be stopped soon. But this was not to be.

Despite the bombing, German industrial armament production reaching its maximum in mid 1944. The bombings were still directed largely toward civilian targets. In 1944 Germany built 38,122 aircraft, compared to 12,234 in 1941, but it could neither compete with the US industry, nor could it train sufficient personnel. During its war effort, the USA built about 300,000 aircraft.

How did the children react to the bombing was a question Heidel and I were asked in the 1970s by our neighbor, Bill Geddis, in Troy NY. When we talked about the war, he told us that he was a bombardier in the Pacific theater. Only later did he admit to us that he had also flown missions over Germany. He told us that the idea that he was bombing children bothered him greatly to this day. We could put his mind at ease. Yes, it was a terrible time, but children are sufficiently resilient that they suffer greatly only when they are actually wounded or lose members of their family. They do not know better and accept the environment they grow up in as normal. Adults are much more affected by the anticipation and fright of being bombed and the possible consequences.

The last phase of World War II began with the landing of the Allied Forces in the Normandie on June 6, 1944. By then, in addition to the destruction of the cities of Germany, the armament industry and the facilities of synthetic fuel production became a major goal for the bombing. Heidel and I could find out much later when touring the Normandie in 2001 that the cities of France, Belgium, and the Netherlands which were in the path of the Allied Forces, were also heavily bombed in support of the advancing troops. These cities, however, were not prepared for the bombing and civilians had fewer chances of surviving the destruction of their houses.

The highly propagandized *expectations in Germany that technology might turn the fortunes of war* ("Wunderwaffen") were not realized. The V1 introduced in mid-1944 and later the V2 rockets were marvels of technology. Their creators served after the war (more or less voluntarily) for many years in support of the space programs of the US (Wernher von Braun) and USSR (Helmut Gröttrup). The impact of these rockets on the war, however, was insignificant. The only sufferers were the Londoners, hit by over 3,000 (from a total of ≈5,000) of these rockets, with some 4,000 casualties. This compares to the 50,000 who died in Hamburg with 'Operation Gomorrah' by 3,000 British and American bombers (July 26 to August 3, 1943). A single raid by 729 aircraft within this operation, started a firestorm that, alone, killed 40,000 civilians. The weather condition and the specially chosen mix of explosives and incendiaries were just right to start this firestorm with devastating effect. Most of the casualties suffocated or were burned to death.

At the second *conference of the Allied Heads of State in Yalta* between Stalin, Roosevelt, and Churchill on the Crimea in the USSR on February 4–11, 1945, last military plans were made and the post-war politics was set along the decisions of 1943 in Teheran (see the pages 2-18 and 26). During the war, it was decided to replace the ineffective League of Nations founded after WW I. Delegations from the USA, USSR, Great Britain, France, and the Republic of China planned a *United Nations (UN)* in 1944. The founding members signed its charter in 1945.

Ultimately, Mussolini is shot and Hitler commits suicide. Mussolini died on April 28, 1945. He was deposed in 1943 (see page 2-18), and the shooting was on orders of the Liberation Committee of Italy (see Footnote on page 9-20). Hitler died on April 30, 1945 in his bunker in Berlin. This eliminated two of the main dictators of the 20th century, leaving only Stalin. But the Allied leadership changed also. Roosevelt dies on April 12, 1945, and Churchill leaves office on July 27, 1945 after the Labour Party wins the parliamentary election. The third meeting of the heads of state, which occurs after the capitulation of Germany, sees Truman, Stalin, and Atlee signing the *Communiqué of the Potsdam Conference* on August 2, 1945. Little did they know, that this is now considered the beginning of the Cold War which lasted until 1990, almost to the end of the 20th century. Only after the end of the Cold War was the last scourge of Europe removed. The Potsdam Conference affected how the two Germanies developed, but also strongly influenced my choice of occupation. All this is described in much more detail in Chapters 3 to 5.

Figure 11 *Joachim in my arms in 1944 on the garden swing of the Werderstraße.*

More pleasant things have still to be reported from our family, despite the bleak presence and even worse outlook. My brother *Joachim was born on January 12, 1944*. For a little while, this increased our family from 4 to 5 and he was received with great joy. Mutti got help from Gerda, a 19-year old girl from the neighborhood, who was quickly treated as one of the family. Figure 11 shows one of my efforts to make Joachim feel welcome.

The next change in the Werderstraße was the expansion of our apartment by one room into the space of our neighbors across the hall, the Marocke's. It was planned that our room, which was next to Vati and Mutti's bedroom, would be Joachim's nursery and we 'old' boys would move into the new room. It did not quite develop that way, as will be seen below.

Frank finished his middle-school education in 1944 at Easter. As planned, he started working as an apprentice in Vati's business and once a week went to formal schooling for the craft. He expected to get his 'bachelors degree in roofing' in three years, and his 'masters degree' 2–5 years later. After this, he would have been qualified to lead the business. But, not much after his 16th birthday, he was drafted to the "Arbeitsdienst" ('work commando,' usually required for teenagers before being drafted into the army). He left home and was stationed in northern Germany. Besides working on building projects, he was trained to man anti-aircraft cannons. By this time, the "Arbeitsdienst" was used more and more to give a pre-military education. By 18 years of age he should have been taken over by one of the branches of the military. Again, things did not progress as expected in these dreadful last two years of the war.

In 1944 increasing numbers of *refugees from the eastern front started arriving* in Brandenburg and readjustments had to be made in the housing arrangements. The sizes of apartments were matched to the number of people sleeping there, and we ultimately were required to offer one room to a lady-refugee. After Frank had left the house, Joachim was sufficiently old to be moved into the planned nursery, and I moved into Vati's "Herrenzimmer" (which was not used as home office anymore). This move was much to my liking since I could do my homework on Vati's desk and the sofa was equipped with a folding arm rest, so it was a comfortable bed. The room was separated from the living room by a curtain only, but by 1944 only few guests were entertained. Everyone wanted to be home in time for the practically daily air raids aimed at Berlin. My walks to and from school were now always alone, since Frank had left. Occasionally, I went to one of my two closer friends in the city and old town (Bessen and Neudeck) to discuss homework and play with different sets of toys. The first of these friends had moved to Brandenburg from the middle of Germany, the other was the son of the owner of a shop for business supplies and toys.

In February 1945 *the Theodor-Körner-Schule was closed* and refurbished as a hospital. All students had to carry the biology collection and other easily movable displays from the middle to the high school, the Saldria, short for "von Saldern Schule." Treasured was moving hawks and other birds of prey. The Saldria was founded in 1589 and located at the edge of the Havel since 1867, half-way between (1) and (4) of Figure 6, Chapter 1. My walk to and from school was now more than one mile. A street car from the train station could be used in bad weather. The school hours were staggered between the different student bodies. Only daytime air-raid alarms caused problems. They meant either a long dash home, or using one of the nearby air raid shelters. No street cars ran during air-raid alarms. Overall, schooling became somewhat haphazard.

During these days *Frank came home on furlough from the "Arbeitsdienst*." At the same time, with 16½ years of age, he was drafted into the army and ordered to report at the south-eastern front. Vati had lengthy discussions with him, urging him to disobey the order and hide in Brandenburg. Frank, however, disagreed. He was already inducted, and if discovered by the frequent SS-patrols ("Schutzstaffel," uniformed, paramilitary Nazi elite army and police) meant immediate execution. He chose to leave. On arrival, his unit was in full retreat and his lieutenant made sure that the 'kid' was safe. The goal was to surrender to the Americans, rather than the Russians. They succeeded. He ended up in one of the huge POW camps under open skies on the Rhine Meadows. Food was scarce and it was cold and wet. He knew some English and could act as a much needed translator.

When it became clear that (according to the agreement in Yalta) many of POWs of the Americans were to be shipped to the USSR, the Americans discharged 'children.' On foot, taking occasional jobs on the march home to get some food, he made it by fall 1945. I remember, as soon as he appeared in the office in the Werderstraße, Vati called home and I had to bring new clothes. Soviet soldiers would still pick up any German partially in uniform and make him a POW, even if he had proper papers declaring him to be a civilian.[1]

Finally, the end of the war closed-in. The unprecedented and unprepared *flight of the population out of the eastern area of Germany began*. They had to travel in the deepest of winter the up to 600 mi on foot, with horse-drawn wagons, or by ship across the Baltic sea. The full magnitude of the mass migration and the enormous numbers killed by hunger, illness, torpedoing of the refugee ships, and aircraft strafing is only now being assessed. Since it was little reported in the German press and radio, we didn't know the worst of it.[2] For example, the Soviet submarine S-13 sank within two weeks the 26,000 ton passenger ship MS Wilhelm Gustloff (KDF ship, see page 1-13) and the 14,600 ton liner SS General von Steuben. Both were filled with refugees and wounded soldiers of which 13,000 drowned.[3] These were more than 50% of all casualties during the evacuation across the Baltic Sea by the German Navy from January to May. Despite this greatest sea disaster of all time, this evacuation was amazingly successful. About 2–3 million people were evacuated by 800 naval, cargo, and passenger ships.

---

[1] From the over 3 million German POWs in the USSR, over 1 million died. The last surviving POWs being released only in 1955, after Stalin had died. Frank was extremely lucky.

[2] Transcribed from: Wikipedia, 2007: "Flight and Expulsion of Germans During and after WW II." It refers to the flight, forced migration, and ethnic cleansing of Germans and ethnic Germans ("Volksdeutsche") from various European states and territories 1943–1945 and after the war, 1946–48. Exact numbers are difficult to come by, but best estimates are 13.5–16.5 million people. The German government's official estimate of deaths due to expulsion was 2.2 million. In 1950, one could find 1.6 million (12%) of the pre-war ethnic Germans still living in eastern Europe.

[3] In 1990 (!), Mikhail Gorbachev posthumously rewarded Marinesko, the captain of the S-13, with the title 'Hero of the Soviet Union.'

In late March we could hear a continuous rumble in the west, just like a continuous thunderstorm. From day to day it became louder. We could hear the artillery, marking the front line. The US Army was approaching the Elbe river. On April 11, they reached the Elbe. This meant they were only 60 mi from Berlin, and half that distance from Brandenburg. But by then the front seemed to become stationary. We did not know that the fate was already settled at the Teheran and Jalta conferences (see pages 2-18 and 2-23). Not much later, we could hear the same rumble from the east. The Soviet troops began with their offensive on Berlin on April 16. The encircling of Berlin was completed on April 25. One day before, on April 24 at 8:30 a.m. the unending slow wailing of the sirens, and later ringing of the church bells signaled "Feindalarm" (enemy alarm). If not before, by then, our hopes were gone that the US Army may reach Brandenburg first. This we thought would save us from the Russian occupation of which the German press told such horrible stories to strengthen the will to fight.

The *first direct air attacks on Brandenburg occurred in 1944*. On each, April 18 and August 6 about 100 planes attacked the city. The American bombers tried in these cases to destroy mainly the industrial facilities. But many of the bombs went awry. The city records[1] show 31 and 64 civilians fatalities, respectively. The city had by that time 10 high-rise concrete bunkers with a wall thickness of about 6 feet. By the end of the war as many as 50,000 may have been seeking safety in these bunkers. After the war, all bunkers were demolished (see Figures 12 and 28).

The *last two bombings occurred March 31 and April 20, 1945* and were of much bigger magnitude. On March 31, Easter Saturday, between 9:30 and 10:00 a.m., 265 planes dropped 719 tons of bombs, causing 552 casualties. Half were prisoners of war and foreign laborers who did not have as good air raid shelters. About 100 houses were fully destroyed in the "Altstadt." The high school, where we would have had classes on any weekday, received a direct hit (see Figure 21).

On Easter Sunday, Vati decided: "*We have to get out of here*. The most likely next target for the bombers will be our train station." He was correct. By April 10, a careful survey was made by British reconnaissance planes. A copy of one of the pictures is shown as Figure 12. At this time, all was still whole in our part of the city. With a magnifying glass, one can make out every single railroad car, as well as the street cars in front of the station. During the next week, our family, together with the families Richter and Bergen, business friends of Vati's from the Arado aircraft company, moved into a "Baracke" (wooden shack) with three large rooms, located in the woods between the villages of Krahne and Reckahn, six miles out of town. This shack was built on an unpaved road. It was erected as living quarters for the workers of a planned assembly facility for aircraft, out of view of bombers. The building of the plant had not begun. It suited us fine.

The children of these three families consisted of Karl-Heinz Bergen and Lutz Richter, two about 8-year old boys (the first very quiet, the other very rambunctious), Monika Richter (a rather withdrawn 14–15-year old girl), and Joachim and myself (Joachim was about 14 months old, and I, his 14-year old babysitter). Joachim and I got along well, and I managed to teach him how to walk. Soon he soon could maneuver quickly along the soft forest floor. Otherwise, I used much of the spare time to explore the surrounding countryside.

---

[1] The exact dates and quantitative information is taken in case of Brandenburg from the summary: "Das Jahr 1945 in der Stadt Brandenburg." A paperback book published by the Kulturbund Brandenburg, written by the working group: "Stadtgeschichte," published in 1997.

After work, all three men returned daily from Brandenburg in Vati's car and supplied us with any missing items from home. For me, the school was suspended until a replacement building could be found, an unlikely event at that time.

After a few days, it was decided that we were safely out of town, but we had no protection against direct enemy action or 'friendly' fire at the location. It was decided <u>*to build a small bunker to safely house all 11 of us*</u>. Figure 13 shows the leftover ground formation from this bunker after 46 years of undisturbed growth of vegetation. Vati's workers dug the L-shaped trench, still discernable in the photograph, taken in 1991. Railroad ties were used to reinforce the sides, and a roof was fashioned out of two layers of these ties. All was covered with the sand dug from the trench. It looked solid enough to protect us from trigger-happy soldiers who might shoot straight into the opening. We could all fit in the right-angled second leg of the L, and thus would have a second chance. Herr Bergen fashioned a white flag to be fixed above the entrance. We hoped that this would encourage soldiers to wait until we peacefully emerged from the opening. We had not drawn lots to decide who would go out first.

Die Brandenburger Bahnhofsvorstadt auf einem britischen Luftbild, 10 Tage vor dem Angriff (aufgenommen am 10. April 1945)

Figure 12 *The area around the train station (see also Figure 6.) The "Bahnhofsbunker" is marked by* O.

On Friday, April 20th came the anticipated, second air raid by 200 American bombers. It was about 11:00 a.m. when I heard the sirens. I scrambled up the slope of the woods to reach a wheat field from where I had a clear view in the direction of Brandenburg. It was too far away to make out buildings, but soon I could hear a series of explosions which were to destroy the railway connection between the two fronts, but largely missed.

<u>*I never heard such loud explosions again*</u>. Soon, a thick black cloud arose from the direction where I knew our home to be. In the evening, Vati reported that our house was a total loss due to a large bomb that fell into the middle of the street, causing the front wall to collapse (see Figure 9, above). All people in the strong air raid shelter of the basement survived and could crawl out of the ruin. Vati brought along some of the personal belongings strewn all over the rubble. Now we could not go back into the city again. The bombing had occurred so few days before the fight for the city began, so that no detailed damage reports are available. One estimates about 1,100 casualties. And certainly, this part of the city never again looked as it does in Figure 12.

Figure 13 *Trench left by our bunker.*

Increasing numbers of civilians walked past our shack, and from their own experiences told that for once, the Nazi press told the truth when describing the terror caused by the victorious Soviet soldiers. In "Das Jahr 1945" I read that in part this behavior was incited by the order of Soviet Marshal Zhukov of April 16: "Soldiers take revenge, behave such that this victory of our army is not only remembered by today's Germans, but also by the generations to come. Soviet soldiers, show no mercy!" (Ref. Footnote above, transcribed from the German on pg. 26 of the cited book.)

Our three families decided: _We must try to escape the Russian Front_. This was easier said then done. A German order did not permit fleeing from the front, and any able male caught not defending the remaining cities could be shot.

Early in the morning a few days later, I heard a big commotion. "Herr Richter" had seriously burned his foot. According to the story told to us, he wanted to bathe his foot and his wife, Thea, had inadvertently poured boiling water over his foot. Thea was a nurse, and quickly bandaged his foot expertly. But now, he could not walk for weeks. Only many years later, when reading more about the heinous behavior of the Soviet troops toward civilians in Berlin and Brandenburg, the many suicides this prompted, and the large number of able-bodied men who were shot by the SS troopers when caught not having reported to the "Volkssturm," did I get the idea, that, maybe, "Herr Richter" planned this 'accident' as a margin of safety. This would explain why Vati and "Herr Bergen," the remaining able-bodied grown-up men, were so upset with the Richters and showed less than the expected sympathy. Unfortunately, when this occurred to me, I could not ask Vati anymore.

Late in the afternoon, we took off. Vati drove the car with a trailer carrying belongings and food. "Herr Richter" needed the right front seat. The three wives, each with their youngest child on the lap, were squished into the back. This left "Herrn Bergen," Monika Richter, and myself. The three of us followed the car on bicycles. No major road was traveled, not to run into German or Soviet troops. Fortunately, Vati knew the area from serving customers and his occasional hunting trips. Late in the evening (naturally without lights) we arrived in a small village and found refuge with a farmer Vati was familiar with. We settled in the barn and Vati and "Herr Bergen" went to the tavern to reconnoiter (and possibly find a beer). They found the beer. No Russians had been sighted. We were still on the German side of the front. After a short while, two SS officers arrived and announced to the assembled group: "We declare this village a citadel which will be defended to the last man. You and your neighbors must report by sunup at the church with spades to dig the fortifications. By noon you will get guns and ammunition and you will be joined by some SS units." Then they left, probably to prepare the next village for the defense line. We quickly loaded up the car again and left with directions from the farmer, using the back door moving between the fields. There was a sufficient moon, so that we did not get lost.

By next morning, _we arrived in Lindau_, a somewhat larger village. The road into the village is shown in Figure 14. We had traveled about 30 mi from Brandenburg.

Figure 14 *Lindau, photo taken in 2005. Except for the new road, it is not much different from May 1945.*

Figure 15 *The forester's house (on top) and the barn (bottom). It has changed little since 1945. The picture was taken on a nostalgic visit to Lindau, 2005.*

In another 9 mi we could reach the Elbe, where we knew the Americans to be. The front had become rather silent by then. We found refuge in the house of the forester, outside the village. The house is still standing today, but nobody lives in it anymore. Today, being a forester in this area is not a full-time job. These days, he lives in the village and drives to the forest by car. Figure 15 shows a picture of the main house. The barn, shown at the bottom, was our domicile and housed a number of other refugees. The stable (with the smaller door) housed a pair of horses and two cows.

For the next few days we rested. Vati, "Herr Bergen" and the forester discussed the situation. He still had a radio and let us know that Hitler was dead by then, and the war should end in days. Berlin had just capitulated. He was still hoping that the Americans would come to town and keep out the Russians. He was concerned about this possibility. In his forest there was a shot-down Amercan fighter plane with the pilot still at the controls. Alone he could not get the pilot out. He thought, since he now had help, that the three men should go and bury the pilot, not to incense the Americans, should they come. This chore was done the next day. Actually, I was not to know about it, but small boys have an excellent hearing, even if they behave as if they are not listening. A day later, I explored the plane wreck and all the surrounding forest.

The following day, a long column of *American trucks with soldiers seated in the back entered the area*. They did not take the trouble to make it to the forester's house which was about one mile off the main road shown in Figure 14. I made it to the village and saw that there must have been still some German soldiers. The creek was littered with German guns, helmets, and ammunition. The Americans attempted little contact with the civilians.

A few more quiet days went by, when suddenly someone cried: "*The Russians are coming*." The Americans had slipped out of the village without us noticing it. What a difference the Soviet entry made. A long column of "Panjewagen" (horse-drawn, four-wheeled carts) with soldiers, some trucks which looked like built in 1920, and many soldiers on foot. They were on the same road, but now a large group took the turn to the house we stayed in. We all withdrew in a close group into the straw. The first thing we noted, was that Vati's car was being confiscated. Obviously by soldiers who did not know how to drive. They took off, but next morning we found the car in the ditch, not 300 feet away, awaiting help to be pulled out. Next, a group of soldiers entered the barn, looking for watches and jewelry with the request we later heard so often: "Uri, uri." They were satisfied as soon as they had all watches, chains, bracelets, and rings. Immediately, Mutti told me, here is our jewelry box, throw it away so that they do not find out we have not given all. The Bergens thought it best to drop their jewelry box into the toilet (hopeful that a farm worker would find the treasure later). I hurled ours onto the second floor of the barn, as far away as I could.

Next, there was a soldier who looked for a woman ("Frau komm"). He was, however, scared off when he saw that there were several able men in our group. We heard later that he went into the forester's house and raped a woman there. Next, we were all herded into the main house and told to live there. Not a generous offer, dictated by the communist ideology. The barn and stable were needed for their horses. The troops slept in the open and started the next day to cut trees for huts in the forest. The officers also occupied a portion of the main house and protected us from further harm. Vati's trailer was still there, it was searched thoroughly, but most of the clothes were left.

The next day was a most important day, one of the Soviet officers informed us in broken German: "Krieg aus, Deutschland kaputt." The unconditional surrender of Germany occurred on:

May 8, 1945.

The 'Third Reich' which was to endure 1,000 years, existed only for 12 terrible years. It was to last longer than the 887 years of the 'First Reich,' the Holy Roman Empire of the German Nations (see page 1-1). It did not even reach the 47 years of the 'Second Reich' which collapsed at the end of WW I, or even the 14 years of the ill-fated Weimar Republic.

For Germany's future, *this unconditional surrender was the zero hour* [it took effect at 23:01 on May 8, (i.e., 17:01 EST, and 2:01 on May 9, Moscow time)]. The worst was over, although the relief of having gotten rid of the dictator was short lived. For Japan, the worst was still to come. The atomic bombs dropped in August 1945 over Hiroshima and Nagasaki killed 140,000 and 80,000 people, respectively, and led to the surrender of Japan. During a later visit of Japan, when we were visiting the Hiroshima memorial (see page 8-99), our Japanese host pointed out the brutality of dropping the bomb. After some discussion, I finally asked him: "Considering the situation of Japan at that time, what do you think the 'Empire of Greater Japan' would have done if it had the atomic bomb?" Without hesitation, he answered: "Bombed the US." The same would probably have been true for the European war theater. When the atomic bomb was developed in Oak Ridge, not far from my later laboratory at the National Laboratory, it was not yet decided where to drop the bomb.

The war being over, one could set up statistics.[1] The estimated loss of life worldwide due to WW II was estimated to be 72 million. Of these, 47 million were civilians, including 20 million deaths due to war-related famine and disease. The military toll was about 25 million, including ≈5 milliom prisoners of war in captivity. The countries that lost more than 10% of their populations were: Germany, Latvia, Lithuania, Poland, and the Soviet Union (alphabetically). These figures include the 5.75 million deaths of the Jewish holocaust, over half of which were Polish citizens. Of the almost two billion (2×10^9) people involved in the conflict, 3.7% died as result of the war. Never before had the world seen such a slaughter and loss of property.

---

[1] The figures about the worldwide casualties were summarized from the internet: http://en.wikipedia.org/wiki/World_War_II_casualties, as listed in 2007.

The terrible progressions of the war and my personal experiences illustrate the increasing understanding of cause, action, and consequences while growing up. In the frame of the rise and fall of the dictator, my personal life seemed inconsequential, but in this Chapter it could be shown how far a 'normal' childhood was possible. Highlighted are the enormous amounts of luck our nuclear family of five had during these turbulent times. The suffering was easier to bear since everyone was in the same situation. The experience gained by growing up in this first (homegrown) dictatorship was invaluable for the survival in the second, Soviet dictatorship. Comparing it to today's experiences of children, I do not feel deprived. Definitely, the childhood was over in 1945. One had to grow up fast. There was no time to go through the 'terrible teens,' as they are called by today's frustrated parents. There also was no time to 'find yourself.' I was part of the grown-up world with 14 years of age. Figure 16 shows that we were soon to be securely located in the Soviet Occupied Zone, which in 1949 became the GDR (German Democratic Republic). Not German because of the Soviet domination, no democracy, but a dictatorship, and certainly not a republic. Figure 16 shows the large loss of area of Germany and its division into four occupation zones (see also Footnote 1 of page 3-1). The capital cities of Germany and Austria (which was separated again from Germany) were divided into Sectors for the different Allies. Not shown is the large loss of land of Poland to the USSR, i.e., Poland was certainly not treated as one of the Allies.

The casualties and destruction in Brandenburg, which had grown during the war to a population of almost 100,000, were 5–7,000 soldiers (on the various fronts), ≈3,500 civilians (during air raids and the fierce fighting which ended on May 2), and over 600 Soviet soldiers (in the fight for the city). About 2,000 apartments were destroyed, 70% of the industry was inoperable, and 39 bridges were blown up. In Figures 17–28 the destruction is summarized. Recovery, started in the following years, but approached completion only in the 21st century (see Figure 26 of Chapter 9).

Figure 16 *The divided Germany (from: http://www.zum.de/whkmla/histatlas/germany/194548.gif).*

Figure 17 *The "Kurfürstenhaus", symbol of the city in May 1945. Compare to Figure 1. The building should have been restored!*

Figure 18 *The ruin of the "Rathaus," see Figure 1. The Roland was safely stored during the war and is now in front of the surviving "Rathaus" of the "Altstadt," and displayed in Figure 26 of Chapter 3.*

Figure 19 *The "Reichspost" (main post office). The building was razed and the empty lot was used to build a shopping mall, opened in 2009.*

Figure 20 *The St. Annenstraße, the main street to enter the city. In this area of the city, the major door-to-door fighting took place in 1945.*

Figure 21 *The "von Saldern Schule," high school after a direct hit on 3/31/1945. The "St. Johannis Kirche," in the background, was also hit, and two ships sank in the Havel (foreground).*

Figure 22 *"Jahrtausendbrücke," demolished to hinder the advancing the Russian Army. The temporary walkway was built later in 1945. The bridge was reopened for traffic on 10/4/1947.*

Figure 23 *The St. Pauli Kloster" in back of the St. Annenstraße. Rebuilding, completed in 2008, after initial stabilization of the walls and tower.*

Figure 24 *The "Neustädtische Markt" with a view of the backside of the "Rathaus," which was not saved as a historic building, but razed.*

Figure 25 *View of the Hauptstraße" toward the "Rathaus" showing the lost businesses.*

Figure 26 *View of foot traffic with hand-carts across the demolished "Jahrtausend Havel Brücke" with partially destroyed shipyard on the left, photographed in 1946, see also Figure 22.*

Figure 27 *Temporary bridge across the "Kanal" ("Steintorbrücke"), needed to move traffic into the city including the street car service which before was only shuttle service between the destroyed bridges (39 of the 63).*

Figure 28 *"Mühlentorbunker," above-ground air raid shelter. Demolished after the war with damage to the houses, shown after a failed first attempt at blasting (note the political slogan).*

CHAPTER-03

Final Years in Brandenburg, 1945–1949

Starting from Zero Hour on May 8, Germany ceased to exist as a state.[1] We were still refugees in the house of the forester of Lindau, 30 mi from Brandenburg, where our house was bombed on April 20, 1945 (see Chapter 2, Figure 9). A few days after the bombing, we had fled together with the families Bergen and Richter from Brandenburg to avoid the advancing Soviet Army. We ended up in Lindau, as described on pages 2-28–30. Now, we had to survive the next few days after having been overtaken by the Russians. As a not quite 14-year-old boy, I could still roam around rather safely and freely. There was no danger that I would be mistaken for a soldier and imprisoned, and I had no valuables that could be taken from me. In the woods, I inspected the huts built by the soldiers out of small trees and branches as their living quarters. Sometimes they let me know that I should stay away (прочь!). I could not search for Mutti's[2] jewelry which I had thrown to the upper floor of the barn in a quick decision to hide it. The barn was now filled with the soldiers' horses and off-limits to me. There also was nothing edible to be searched for that early in the year.

A few days later, a Soviet officer was looking for some people who could work in the village (рабóтать!). Vati and I were willing to go. There was a chance that some food was to be found to replenish our dwindling reserves. *We worked for the farmer of the first house* in Lindau (see Figure 14 of Chapter 2, property on the left side of the road). We had to load dried chips of sugar beets onto Russian trucks. The chips were last-year's harvest and were to be brought to the sugar refinery. Wether the farmer, in return, got any sugar, we did not know. The chips were not great eating, but

---

[1] The Allies agreed on partitioning Germany after WW II into three zones of occupation—a large Soviet Zone in the east, a British zone in the north-west, and an American Zone in the south-west. Austria was to be separated from Germany. The Saarland was to be placed under French administration, as after WW I. Germany's territories east of Oder and Neiße were given to Poland in compensation for its eastern territories which remained part of the USSR; the northern half of East Prussia was annexed by the USSR. The German population of these territories, as well as the German population located within the borders of restored eastern European states such as Czechoslovakia, Hungary, and Yugoslavia, was expelled. The refugees moving into what remained of Germany exceeded 10 million. Breslau, one of Germany's largest cities, was renamed Wroclaw; Danzig, the city of Schopenhauer, Gdańsk; Königsberg, the city of Kant, Kaliningrad. A 4th zone of occupation was established by the recognition of France as a victorious power. This zone was located in the south-west. Berlin was treated separately, partitioned into 4 Sectors. (For a map, see Figure 16, Chapter 2.) Summarized from: *http://www.zum.de/whkmla/region/germany/ger4548west.html.*

[2] For a family member's name, relationship, and picture, see Chapter 11, pages 2–12.

they tasted sweet and certainly must have had some calories. While working, we did not get paid, but did get something to eat, and a small amount of food to take back home.

At that time, the money we had was of little use. There were no goods to buy, especially not in this small village. *One had to make use of the known channels of barter for food*, but these we had only closer to Brandenburg, i.e., we had to try to go home as soon as possible. The problem was to find a safe route where we would not meet any uncontrolled soldiers. When working on the upper floor of the barn, we were close to the roof, known to any roofer as a good place to hide things. Indeed, in one corner we found an unmarked bottle of booze. Not something we needed to raise our spirits, but an item valuable to trade with. It did not take long for Vati to find out that one of our supervising Russians was in charge of transportation, not only for the chips, but also area-wide. With some haggling back and forth, he offered, in return for the booze, train transport for our three families to Belzig, a town 18 mi from Brandenburg, closer to the customers of Vati's. From Lindau to Belzig, we had about 22 mi to travel by train. By the next day, our trailer had been fitted as a hand cart. Fritz Richter had traded with the forester for a small four-wheeled handcart he could sit in, so that he, with his burned leg, could be pulled by his wife and daughter.

Figure 1 *The railway station in Lindau (in 2005).*

In the early morning, *off, we went to the railroad station* of Lindau, shown in Figure 1, as I saw it again in 2005. The rail line still existed in 2005, but the freight yard (in the bottom photo) had fallen into disrepair. By then, the goods produced and used in Lindau were moved by truck. Still, not much had changed in this area during the 60 years that had passed. On this morning in May 1945, there was a locomotive standing under steam with a single freight car attached. We found the officer, gave him the bottle, and after Vati had tasted a sip, to make sure it was no poison, the deal was made. With the help of some of the soldiers present, a second freight car was attached to the train. They also helped us to lift the carts into the boxcar, closed the door, and the train took off, rattling at breakneck speed toward Belzig. On arrival, we were helped out of the freight car and the Bergens and we marched in one direction, the Richters, who had another goal, went into another.

We pushed our cart about two miles along the main road towards Brandenburg and stopped in a small village by the name of Lütte. Here, Vati had worked on a roof for the butcher, and we were welcome. During the march we did not meet any other soldiers or marauding forced laborers. The butcher, who also farmed on the side, had a large garden and in it, a garden house, which we occupied for the next six to eight weeks. I was immediately picked out by the butcher. He said: "I can use him. He can start learning to be a butcher and live in the main house." There I was, two problems solved. I was to become a butcher, and, what was more important, we had enough to eat. Figure 2 shows the butcher shop in 1992 as I photographed it on a first travel back into the village of Lütte after the communist dictatorship collapsed (see pages 9-23–26). Obviously, the butcher had prospered. The store was remodeled, and one can see an addition that used to be the old barn. By that time, I could not meet anyone who was around when I was working there. The garden behind the house was still there, and the road going to the fields was still the same sandy path.

Figure 2 *The butcher shop in Lütte as seen in 1992.*

Next morning, I, the butcher apprentice went to work. To my surprise, animals were still slaughtered in this village. Life seemed normal. I learned how a pig is killed (by being shot with a special slug), and how a cow is first stunned by a well-aimed blow with a heavy hammer, and then it has the artery opened at the neck with the blood collected and stirred in a pail. The carcass of the pig has its hair scraped-off in a hot water bath, while the cow is hung from a ladder, and skinned. Then, I had to clean the intestines, to be filled later with the sausage meat. The butcher prepared the cuts of meat. Everything was used, either as a cut of fresh meat, for smoking (loin of pork and the hams), as fresh ground pork ("Mett") or beef ("Schabefleisch," beef tartare), or to be made into fresh or smoked sausages. Most delicious in taste and dreadful in content was the liver sausage. Every little scrap of fat and meat deemed edible was trimmed off the cuts and scraped off bones, skin, and innards to be added to the sausage filling. Liver was the smallest amount. The secret of good liver sausage lies in its spices, texture, balance of fat and meat, low level of filler, and its faint taste of liver. For many years thereafter, I was very hesitant to eat liver sausage. The salami, blood sausages, headcheese, and bologna were much better in this respect. In these you can still recognize most of the ingredients.

Not all the work in Lütte was butchering. The animal stables needed to be cleaned out and many things were to be carried from one place to another. Most memorable was a delivery of feed grains. Many neatly stacked bags were lying on a horse-drawn wagon, and I was told: "Come, bring the sacks to the top floor of the barn." I looked at the butcher in disbelief. The sacks were of "Doppelzentner" size (220 pounds!). I had never even attempted to lift anything close in weight. He said: "You are strong. I will put them properly on your back and you will see, it is easy. Everybody has to learn it." In no time, I had one sack sitting across my shoulders and I thought my legs would buckle under me. But, they didn't. With considerable effort, I even made it step by step to the upper floor, and with enormous relief, dropped one sack after the other onto its proper place. I don't remember, but I must not have looked too happy and fit after a while, since I was relieved of the duty long before all was done. Ever since, I blame my not too infrequent back pains on this experience. But, I also learned that you can move much more than you think, if you put your mind to it—or are forced to do it.

This experience came to mind when we had bought a property in the Adirondack State Park, close to Troy, NY in 1977 and had set up a gazebo as a place to enjoy the countryside (see Chapter 8, Figure 88A). To secure the screen, I had brought some 80 bricks. Our car was parked at the foot of the hill, about 800 feet away. After we had brought up our picnic accessories, I asked Brent, our then 17-year-old son, to bring up the bricks. He looked somewhat surprised since he thought there were too many to handle by him alone. So I said: "Take as many as you can carry. I will help you with the last batch." After a little while, he came, all exhausted, carrying five bricks stacked between his hands in front of him, a most uncomfortable position. My answer to his effort: "I guess you need some training in carrying things. Come back down. I bet you, I can carry ten times as many and you can carry the rest." His answer: "Impossible!" By taking a strong tarpaulin and positioning the stacks of bricks sufficiently high off the ground so that they could be placed comfortably on our backs, we managed to move them easily (?).

The return to Brandenburg had to wait until the end of July 1945. Two weeks before, Vati and Herr Bergen had made the trip to Brandenburg by bicycle to see if it was safe, and where we could possibly live on arrival. After we waited for three anxious days, they returned. The verdict, when moving as a group, we could chance the trip. The apartment of the Bergen family on the Büttelhandfaßgraben 5 was still intact and not ransacked. We could all stay there for a few days. By then, the "Bau und Sparverein," (see page 2-10), the owner of our bombed-out house in the Werderstraße, would have one of their apartments in the Hohenzollernstraße 11 ready for us to move into (Chapter 1, Figure 6, marked as point #12). The business was still in the same state as Vati had left it at the end of April. Opa also had lost his apartment in the St.-Annen-straße 32 and had moved into an apartment in the Große Gartenstraße, in back of Vati's business yard (see Chapter 2, Figures 6 and 7). He had taken charge of the business and kept the looting to a lower level than otherwise. It took quite some time to sort out the personal property of the two families who occupied the right side of the house, shown in the photograph of Chapter 2, Figure 9.

When a load of goods was to be delivered from Lütte to Brandenburg, *we joined the trek by horse-drawn wagon and on foot*. The trip went north, along what is now called the "Deutsche Alleenstraße." Although occasionally there were signs of WW II in some of the small villages, the impression was surprisingly normal. Work in the field went on. Figure 3 shows the approach to Brandenburg with the characteristic skyline in the distance. When we reached the city, this changed abruptly. This was the route the fighting had taken into Brandenburg on April 24, 1945. In the Potsdamer Straße and the branching St-Annen-Straße to the center of the city, all houses and factories that were not reduced to rubble, where burnt-out shells (see Figure 20, Chapter 2). The St. Annen bridge into town was demolished. We had to turn south to reach the Büttelhandfaßgraben.

Our earlier apartment house in the Wilhelmsdorfer Straße 70 was demolished, and the attached house, was a burned-out shell. We unloaded at Bergens' apartment.

Figure 3 *Approach to Brandenburg as seen on the road from Belzig, taken in 1945 (photo of 1972).*

Next, Vati and I walked to the Werderstraße, the location of the business yard. A similar picture offered itself. The streets had barely been cleaned of rubble. Many houses were burned. At the bridge across the Jakobsgraben, one of the few remaining passable bridges, stood a burned-out Russian tank. This south-east quarter of the city had suffered most due to the fight for Brandenburg which lasted until May 2 and cost many Soviet soldiers' lives (see page 2-31). The German casualties were never assessed. All over the city, ammunition depots and abandoned small-caliber anti-aircraft cannons were left. Most depressing was the awful smell of the city. I never forgot it throughout my life. Still, it was accepted as a fact. More important for us was to get back to some type of living. *This certainly was our personal zero hour*.

Less than a week later, we moved into our own apartment in the Hohenzollern Straße 11, a part of the city which was not as badly destroyed as the city center and the area around the railway station. Here, we put up whatever furniture was recovered and still useable. There was enough left to furnish our smaller apartment, although nothing fitted together anymore. At least the bedroom furniture was reasonably intact. Within a few days, I had again the comfortable feeling of being 'at home.' This same change of feeling from being lost while moving from one place to another to that of being safe at home has never left me, even when moving under normal circumstances. But in 1945, it was particularly satisfying, it meant we were not refugees without a home anymore. *Little did we know that periods of being refugees would happen twice more* in the next 10 years.

Life was not fully back to normal that quickly. The city utilities returned only slowly. The occupying troops did not feel safe and were afraid of hidden soldiers and militant Nazis (Werewolf terrorists). Frequently, and always in the middle of the night, *Russian soldiers would undertake* "*Razzias*" [(military) police raids]. Every apartment was searched for anyone not properly registered and for possible weapons. On these occasions, items of interest to the soldiers frequently were confiscated. There were no watches left by that time, but soup spoons and little ornaments were highly valued. Only later, were these raids reinforced by German police and became less threatening. By that time, however, the goal was, in addition, to discover illegal food reserves. As one learned from these raids, the hiding places in the attic and basement had frequently to be changed. Luck and some foresight played a big role that we lost little on these occasions.

Soon thereafter, provisional *school started again after a break of about four months*. The middle school was changed back from being a hospital (see page 2-24). It was now called (erroneously) the Wredowschule[1] because of its location at the Wredowplatz. The name Theodor-Körner-Schule was not acceptable anymore. The main building was intact, except for its windows. They were closed first with cardboard, later with translucent plastic sheets (PVC). It took years until full glazing was available. Since there were insufficient buildings, boys and girls had to share one middle school. Teaching was done in shifts, switching every two weeks. Boys in the morning and girls in the afternoon, and vice versa. Most of the teachers were still the same and the plan of study was not readjusted, so that much of the missed subjects were never taught, i.e., there are a few subjects that I did not learn, such as old history (German, as well as Roman and Greek) and even some chapters of German grammar were omitted. Modern German history was avoided as much as possible. Not enough reliable information was available, and most teachers may have been uncomfortable, telling us about the horrible stories which slowly came to light. As students, we

---

[1] The legitimate "Wredow'sche Zeichenschule," founded by the sculptor Wredow (born 1805 in Brandenburg), was next door to our school on the Wredowplatz, see Figure 5, Chapter 1.

found it very funny when one of our old teachers on a Monday morning, got in front of the class, raised his right arm and said "Heil Hit... ." In the middle of the word he recognized that times had changed and he finished with "Guten Morgen."

New books, paper, pens and other supplies were non-existent. Most teaching was done by lecturing. The learning, by studying notes and doing dictated homework. I could meet my need for writing paper by using Vatis left-over business stationary from the 1942 fire (see page 2-17). Around the edges, the sheets were blackened for an inch, but otherwise they were of good quality.

Since with 14 years I did not only have a good appetite, but in short order, I also ran out of fitting clothes and shoes. Fortunately, food and leftover cloth from the German production of uniforms in Brandenburg could be traded-for by fixing leaky roofs. The milk for 1½ year old Joachim, I had to pick up from a friend who had a goat in the area across the Silo Kanal (see Chapter 2, Figure 6, the top waterway in the picture). This was a two-mile round-trip, including an unsafe crossing of the "Silo Kanal" by a crowded, often half-flooded, make-shift, hand-hauled ferry. Mutti often had to be the tailor for our clothes, or Vati had to find double portions of material to 'pay' the tailor. This barter business went on for the next four to five years. The last items of value included leather jackets and coats, and my first pair of hand-made shoes. Because of my wide instep, all ready-made shoes had to be several sizes too big and were hard to maneuver in (see page 1-18). The just mentioned leather coats were good enough still to be worn in 1954 when we arrived in the US (see Vati in Figure 21, Chapter 11). As of today, my leather jacket still hangs in our coat closet. It served me well during the days of snow-shoveling (see Chapter 8, Figure 76). Vati's effort to supply enough food beyond the rather skimpy ration cards saved us from any real, sustained hunger.

Initially we had to set our few remaining clocks to the inconvenient Russian time (−3 hours), it was changed back to the better suited central European time only in September. Newly erected street signs were written with Cyrillic letters, for example, Тра́уе́рбе́рги for "Trauerberg," a place on my way to school. The signs replaced those lost in the war, gave information to the Soviet soldiers, and were needed to indicate street names changed to erase the Nazi memory. The main activity in the streets was the dragging of big wooden crates on skids to the railroad. They were filled with dismantled *remaining machinery for shipment to the USSR as reparation* for WW II. We were wondering if the machines were ever of use in the Soviet Union since many small items continually fell out of the boxes, and little care was taken during packing and transport.

Shortly after the electric power in the city had reached a reasonable number of hours per day, it had again to be limited. Insufficient generating power was left after transport of too many generators to the USSR as reparations. The frequent "Stromsperren" (power shut-downs) were unpredictable, and in the late afternoons, this meant, school had to be suspended. Even in these hard times, a cancelled school hour was something to enjoy. In time, I could not understand this student mentality anymore. It seems to be worldwide and cover all levels of education—and I must admit I also enjoyed it then, unless it affected one of my special subjects. All throughout my later teaching career in the US, although by then the students were very much aware of the cost of education, a cancelled lecture was always a happy occasion. Students never asked for a make-up (or money-back), and when a make-up lecture was arranged, it was rarely welcome. Obviously, *something is wrong with education if the consumer does not value the goods*! (Is there a solution on page 8-75?)

During the first year, when *walking alone through the city, one was not always safe* from Soviet soldiers. It was possible to be robbed of property, such a valuable bicycle, jewelry, and if still present, a watch. Also, there was the chance of being picked up for a work detail or, even worse, to be held for minor or nonexisting reasons under the pretext to search for former army officers, members of the Nazi party, or just 'enemies of the state.' More positive was the singing of the columns of Russian soldiers when they marched through town. The beautifully deep solo voice supported by the full choir was something which I still enjoy. Surprisingly, these performances easily overcame the hate amassed due to the early terror, perpetrated by the soldiers.

To help against randomly being picked up in the street, all people legitimately employed had a card to be able to prove their need to walk the street. The same card also entitled one to increased food ration cards if the job demanded it. As a student, I did not have such identification. But since I was sufficiently tall, it happened that I was picked up by Russian patrols on my long way to school, and trucked to the bombed-out Arado aircraft facility to work. The collected detail consisted of 20–30 men who could not prove their need to be at that time at that point in the city. I escaped, when waiting to collect the tools for our job which was to clean-up the rubble. It took me more than two hours walking to arrive back home since I had to choose a circuitous detour, not to be caught again. In the evening after this unscheduled excursion, Vati said: "You better learn how to become a roofer with me. Then you have the proper papers and also get a better food ration."

In this way, within three months, *I started a second apprenticeship. This time, as a roofer*. I was properly 'initiated' by the other workers on the first job, although everything was held in friendly limits. If for no other reason, I was the boss' son. I remember, we were to fix a tile roof near the Marienberg. We could not reach the roof from within the house. A long fire ladder on wheels was gotten to enter the roof from the outside. Weighing the least, I had to climb to the top of the extended, wobbly ladder which reached a height of two stories. Looking down through the last rungs, I did not feel too well. Then the ladder was rolled the remaining 10–20 feet on its wheels to touch the roof. I really did not have to do anything since the ladder was then secured by supports on the ground. Since nobody seemed to have recognized my fear of height, I had passed the first test. To fix the roof without additional tiles, the original "Doppeldach" (doubly covered roof) was changed to a singly covered one. The doubly covered tile roof was quite typical for our area. The lattice work of wooden laths was covered starting at the gutter by hanging a single layer of tiles on the first lath and cover it by a second layer on the same lath, overlapping the joint between neighboring tiles. If one of the tiles would break, the roof would still remain water tight. The war, however, had left bigger holes. The lattice work was repaired, and then, because there were no new tiles, the double layers were replaced by single layers. Out of a half-destroyed roof, we could make a water-tight whole one. A temporary solution, but many of these 'fixed' roofs had to wait years for final repair. Another type of tile roof is called the German roof. It has a lath-spacing half as wide as the doubly covered one, giving a similar overall density of tiles, but is sturdier and nicer looking. The German roofs were fixed by leaving every second lath empty.

The next initiation test came when stacks of tiles needed to be moved on the roof. One takes a heavy pile on the shoulder, but then needs both hands to keep the pile steady. To walk without stretching the arms for balancing was the problem. I passed the test and learned that it took me not long to lose my fear of height when I had to. In fact, after a week's time, the fear changed into recklessness and I had to be lectured sternly to never work without a safety belt in critical locations. I survived all, and learned a lot. Again, my employment lasted only six to eight weeks.

The 'high point' of my career as a roofer came when *I was permitted to work on the tower of the "Dom"* (Cathedral), shown in Figure 4, a picture from a later visit. The "Dom" was built in 1165 and had considerable damage on the slate-covered tower. To fix it, a partial scaffolding was

Figure 4 *The Dom of St. Peter and Paul in Brandenburg, recent photo.*

erected out of the top windows of the tower and we could get to work, suspended on a harness. The main tool was the slate hammer, shown in the logo of Vati's business in Figure 26 of Chapter 1. The pointed tip was used to cut a hole into the slate blank to nail it to the underlying wooden planking. The blunt end, opposite to the pointed one, was the hammer, to drive the nail. The left side of the hammer had a knife's edge above the (leather) handle to cut the slate to the proper shape using well-aimed chops, something left to the skilled roofer. Many of these special techniques I still had to learn, but never did.

In fall 1945 we moved into a bigger apartment in the Franz-Ziegler-Straße 35 shown in Figure 5. This location was more convenient to Vati's business (see Chapter 1, Figures 5 and 6). The apartment had three bedrooms, in anticipation of Manfred's arrival (born in May 1946) and, optimistically, Mutti also expected Frank to return soon from his excursion to the German army (see page 2-25), having left in February 1945. The apartment was on the top floor (the center four windows). Vati had even managed to have a telephone installed, so that he could be in touch with the office from home, as well as his bigger customers. Not long after our move, indeed, Frank appeared at the office and was quickly brought home. He then resumed his part in being trained for the business.

Next, I changed my occupation again. In October, my teacher visited our home and asked Vati to have me finish middle school. On October 1, 1945, all schools in the Soviet Zone had officially opened. To my surprise, the teacher expressed that I was a good student and would do much better later in life with a completed middle school education, due to be finished at Easter 1947, in only 1½ year. By then, it was safer to go to school. Also, my distance to walk was much shorter, no more than it was before we left Brandenburg. After some discussion, I was back in the old middle school.

By this time, the *Soviet Military Administration had permitted anti-fascistic political parties* and unions to be created. The new parties were based on the ones of the Weimar Republic. The first two parties were the SPD (Social Democratic Party of Germany) and the KPD (Communist Party of Germany). Participation was cautious and much of the leadership was taken by former party officials who were persecuted during the Nazi time. The national leadership of the KPD consisted of returning émigrés from the USSR who had already been trained there for their future job. With the two other parties founded in fall, the CDU (Christian Democratic

Figure 5 *Franz-Ziegler-Straße 35. Picture of 1972, with GDR slogan (The might and unity of the socialist union of nations may grow!).*

Union), and LDP(D) (Liberal Democratic Party of Germany), the block of anti-fascistic political parties was constituted. An early cooperation between KPD and SPD was brought about by the dire need for survival of the city. By the end of 1945, the stance of the KPD, however, became threatening to the democratic development, and the cooperation with the SPD deteriorated. In April 1946, finally, the KPD and SPD were united to the SED (Socialistic Unity Party of Germany). This unification was by no means supported by the majority of the members of the SPD. No vote was ever permitted, and the leaders were pressured at the "Vereinigungsparteitag" in Berlin into agreement. For us, the overall political changes were initially of little interest. Of overriding importance was to survive and to begin a new life. Yes, we had newspapers and radio to follow the developments, but only the immediate effect on our situation seemed to matter. The bombing of Tokyo, Japan, and even the dropping of the atomic bombs created little discussion. We were numbed by our own horror. We heard about the end of the war in Asia, but this was far away.

In August 1945 two items caught our attention, *the Potsdam Conference and the formation of the "Alliierte Kontrollrat*" (Allied Control Council), established in Berlin to govern all German occupation zones (see page 2-23, 31, and 3-1). It consisted of Marshal Georgy Zhukov for the Soviet Union, Field Marshal Bernard Montgomery for the United Kingdom, General Dwight Eisenhower for the United States, and General Jean de Lattre de Tassigny for France.

We agreed heartily with the main items of the Communiqué of the Potsdam Conference: "All laws which were the basis of the Hitler regime were to be abolished. All war criminals and Nazi officials are to be incarcerated and tried. All members of the Nazi party who were more than nominally involved and all other persons who oppose the Allied goals were to be removed from employment in the public sector and leading positions in major industrial firms. They were to be replaced by capable persons able to contribute to the development of democratic institutions." The articles, which detailed the restriction of German industry were published later. They had a large influence on my final education and choice of occupation.

In November, in accord with the Communiqué, the process against the main war criminals began at the international court created in Nürnberg. We hoped this would be the beginning of accountability for criminal behavior during wars, and that similar trials under the auspices of the newly constituted UN would be arranged for every future war. For us, the trial succeeded in exposing the unbelievable crimes perpetrated by the Nazi government which were much more horrific than any of us knew. But all this had to wait until the second half of 1946 (see page 3-39).

One topic of importance to me was the decision by the Soviet Military Administration of September 3, 1945 to initiate a land reform ("Bodenreform"). All four of the newly established political parties endorsed this plan. *Any farm larger than about 250 acres was to be dispossessed* without compensation, divided into small parcels and turned over to farmhands, refugees, expellees, or small farmers. There went my dream of becoming a farmer. I did not want to become a farmhand and had to search for a new goal in life. From 1952 on, the small farmers 'voluntarily' united into agricultural production cooperatives, LPGs ("Landwirtschaftliche Productionsgenossenschaften").

In the meantime, *the daily bread and a safe and dry and warm place to live* was the most important question. Thinking about the immediate future, one noted that coal and potatoes, generally stored by every family in the fall for the winter, were in short supply. These shortages forebode problems for the upcoming winter. The housing situation had worsened because blocks of apartment houses were requisitioned for the Russian occupation forces and as prisons for enemies, perceived and true. The food rations were insufficient to survive on, although outright and widespread hunger, as seen in other parts of Germany, did not appear in Brandenburg because of good organization and cooperation from the surrounding agricultural areas. At least most of the time, the rations were delivered, although long queues were common in front of the stores. From August to October 1945 the weekly mean rations consisted of approximately 4 lb potatoes, 2½ lb bread, ¼ lb jam, 1¾ oz pudding powder, and additionally monthly, 1 lb sugar; ½ lb vegetables, ½ lb onions, ¼ lb meat (or double horse meat) and small amounts of fat (or none), ¼ lb of salt, and 2¼ oz of ersatz coffee. (The source of data is referenced in the Footnote to page 2-26). There were six different categories of ration cards. They were worst for the non-working population, with a factor of, perhaps, two better for the best. Naturally, there was a black market, but the prices were outrageous. I remember that ½ lb of butter cost 220 RM, (more than one month's salary). Indeed, fat was a major bottleneck. A common "Ersatz" (substitute) which was surprisingly good in fooling ones taste buds, was cream of wheat cooked with water, spiced with thyme and some fried onions. On a slice of bread, it passed for lard (perhaps because we did not have had any for so long).

Vati had to use all his connections to keep us from harm, but also help out his workers and friends. Buying on the black market was kept to a minimum, it was not only expensive, but also a criminal offense, and one could easily get caught. One of Vati's customers and friend, Herr Asmus owned a fish store. He could partially 'pay' for his roof repairs with cod liver in oil. This came in 2 lb cans with a black market value of 60 RM, a bargain, compared to butter! We found out that it also tasted good. Today we still use it on sandwiches (available in ≈6 oz cans), but it is hard to find. When we discover it in a delicatessen store, or more frequently, on our trips to Europe, we always bring it home. For me, when spread on buttered sourdough bread, it tastes better than French liver pâté. The excess oil was used for frying, and, as well known, was also of great benefit to our health due to its content of vitamin D. Herr Eilert, another customer and friend of Vati's, owned a coal depot. He was of importance to keep us warm through the winter. Also, he owned a truck, refitted for gas, generated from wood (chips and saw dust). About half an hour before leaving, he started the gas generator and then he could travel for perhaps 50–100 mi by keeping the fire smoldering.

The Autobahn passed close to Brandenburg. There it was possible to meet truck drivers at the rest stops (accidentally?). They were delivering goods to Berlin. Vati managed on one of these occasions to obtain a barrel of 900(!) salt herrings. These were stored in the basement and lasted us for a whole year. After watering to reduce the salt, they were eaten as such. Typically they were served with boiled potatoes, mixed with diced, cooked potatoes and other ingredients as "Heringssalat," eaten smoked as kippers, spread chopped with pickles and onions on slices of bread as "Häckerle," or served rolled up with pickles and onions as "Rollmops." It was a most versatile food and a delicacy (if not served daily). Again, herring is occasionally still on our menu today.

Once (in 1946), I could even go with Vati in Eilert's "Holzgas" truck to Berlin to shop for roofing material. This was quite an exciting trip. What a shock it was to see how wide-spread the destruction was in Berlin. Through many of the streets in the center of the city, temporary tracks for narrow-gauge, steam-powered construction trains were laid to help clearing the enormous mountains of rubble. It took many more years until all of this was done.

Since at that time Vati had no car or truck, he also would borrow from Herrn Asmus a three-wheeler with a small motorcycle engine and seats for two. A covered box in back made it into a delivery van. With its help, he could supply his farmer customers with roofing material. Since all roads in and out of Brandenburg had check points to avoid illegal transport of food, he had to use special tricks to get the bartered food home. Most successful was to fill the box with three 55 gallon steel barrels. The one in back was specially prepared with a cut-out bottom and covered with tar on the outside only. The other two were true, dirty, tar-filled barrels. On the road, it was impossible to check the contents of the barrel in back without getting fully 'tarred.' Vati made it always home with his contraband of potatoes, vegetables, fruit, and sometimes even butter and meat. The main problem with this transportation was that the weak engine needed often both, passenger and driver, to help to push the vehicle up-hill.

The Soviet Administration helped also in our quest to survive. They needed the houses and barracks under their command serviced and had the necessary clout to requisition the needed materials, usually in larger quantity and better quality than needed. This also benefitted Vati's civilian customers. Sometimes, we would also be supported directly. Once, when the grain mills in Brandenburg needed service, Vati found a sympathetic soul and brought a 50 lb box of butter-cookies home. This was something we had not seen for years. There it was, in the middle of the kitchen, smelling heavenly and open to overeat on. Not just calories, but also delicious.

Since most stores had little to sell, some owners made use of the situation and opened trade shops. You could bring anything of value, get some credit points, and you were then able to pick something else of equal or lesser value. One felt as being transported to the stone age before the invention of money. In 1948, an effort was made by the state to benefit from the flourishing black market. A state-controlled "Handelsorganisation, HO" was founded to officially sell goods that were otherwise not available, charging many times the customary price. I remember going through the HO on opening day to see for the first time the precious goods. Their merchandise ranged from textiles to shoes, to photographic equipment and accessories, to food items. This was the beginning of state-sponsored inequality. The HO was later expanded to special stores, hotels, and spas where you could pay with western currency. Even worse, stores and vacation hotels were set up where only preferred classes of the political leadership were served.

Most productive for me was to read books that fell into my hands. *The new books helped me to find a new occupation*. They were recovered from the rubble of our apartment and belonged to the lady in the downstairs apartment. Her husband was an engineer who did not return from the war. She thought, I would be interested in them. Some books were of popular science, others were professional. One from 1915 was by Alfred Wegener (1880–1930), and described his theory of plate tectonics and continental drift. I found it fascinating and was surprised later that scientists had not accepted his thesis until after 1960. Another book dealt with the shape of the earth. It suggested life did not exist on a sphere, but on its inside ("Hohlwelttheorie"). It was instructive to see how difficult it is to distinguish fact from fiction. A third book described experiments that the author thought proved an increase in mineral mass on growing seeds in water and air, i.e., the 'creation' of matter. It looked scientific and I could find nothing wrong. Likely, the author had not assessed his errors properly! A fourth book was a compendium of baffling everyday observations which could be explained scientifically and was a treasure trove of valuable information and knowledge.

Two other authors of interest were Max Eyth (1836–1909), a German engineer who, in England, constructed with the industrialist J. Fowler early steam-plows, and installed them all over the world. An example is shown in Figure 6. Eyth was an outstanding story teller, the first who described modern technology as created by the engineers. It was his goal to show on examples and observations out of his life the long path from the creative ideas of inventions, through extensive and tedious calculations, and actual model constructions, to the ultimate success (or failure). The other author was K. A. Hofmann. His book was the 2nd edition of "Inorganic Chemistry," printed in 1919. The book was an enormous source of interesting facts and historical footnotes! It was college-level chemistry and I could soon use it in high school, as well as later at the university.

Figure 6 *Two-locomobile setup for plowing with steam power, as developed in the 19th century.*

The reading about automation got me to copy technical drawings of all types of machinery, preferentially steam engines and diesel motors in cross section, so that one could make out the principle of operation. Vati helped in this effort. Since he needed drawings of roof constructions, drafting tools were available in his office. He gave me an old drafting board, a singed compass set (from the 1942 fire) and the needed T-square and triangles. With his help, I could find India ink, drafting paper, and even a shop that could make blue prints from my 2 × 3' 'masterpieces.'

Anything that had to do with learning and useful work had full support of Vati. He liked to let his many friends in town know that, after having lost my chance at farming, I wanted to look into engineering. The interest in steam engines went from arrangements as shown in Figure 6 which still linked my interest to farming, to steam engines of railway locomotives, ships, and finally electric generators. Figure 7 illustrates the type of picture I liked. It was a pity that the 19th century had past and machinery got increasingly more complex and harder to understand.

A most-cherished, modern drawing I made was a cross section of a two-stroke diesel engine for aircrafts! (Junkers, Jumo 1932–45). It was used as an engine in air ships and high-flying reconnaissance planes. When writing this page, I was surprised when checking 'diesel engine for aircraft,' Google gave a very enlightening article not only of the old Jumo engines and their fate, but also of the modern developments. Diesel engines for aircraft are not dead, as I had thought to document with my literature search.

The interest in diesel engines, naturally, made me also a fan of the 1942 UFA biographic movie "Diesel" with Willy Birgel (Rudolf Diesel, 1880–1930) and Hilde Weissner (Martha Diesel). Over the years I owned two diesel-powered cars and can attest to their fuel-efficiency, simplicity of operation, and interesting smell. The cars are a link to these early avocations (see pages 8-40 and 9-10).

Figure 7 A *steam-driven generator.*

A "Schulreform" (reorganization of the school system) was initiated in 1946 for the school-year starting in April. The organization into elementary, middle, and high school as separate branches of the education system (see pages 2-4–5) was abolished in favor of an "Einheitsschule" (unified school) with grades from 1–12. Since I was in my 9th year of schooling, I had the choice of either remaining for one more year with the old curriculum (to complete an equivalent of the intended middle school), or I could transfer to the 10th grade of the new school and continue to the 12th grade (teacher, grades, and parents permitting). After long discussions, considering that I also had to decide on a new career, all concerned, including myself, concluded that it would be best to try transferring to the high school and finish after 12 years with the "Abitur," a state examination. Note, that counting all the absences from school due to illness, war, and turbulent times thereafter, I would have then only had barely 11 years of schooling.

The school building was again the same as before, just that now the lower and upper grades were added. Figure 8 shows a picture of the front of the same school which was now called the "Oberschule für Jungen" (as long as I knew the building, it changed its name several times, see pages 2-4, 8 and 3-5, and Figure 4 of Chapter 2).

Figure 8 *Goethe Schule in 2002. See also the Figure 4 of Chapter 2.*

Some of the new teachers had a PhD and the level of instruction was much higher. The high school class of ≈30 students was constituted from former students of the "Saldria" (see pages 2-24 and 32), us middle-school students, and several students from a teacher seminary. Our high school was later renamed the "Johann Wolfgang von Goethe Schule," after the famous German poet (1749–1832). It was considered a "Realgymnasium," a high school for mathematics and natural sciences, not a classical humanistic high school, as was its forerunner, the "Saldria." The main subjects for the last three years were, weekly, mathematics (6 h), chemistry (3 h), physics (3 h), and biology (2 h). The foreign language was English (3 h), a second, Latin (1–2 h). Russian was supposed to be taught, but the school did not have a qualified teacher.

I immediately liked the school. My grade average jumped by one numeral and kept an average of '2,' meaning it was substantially above the average for the last three years (compare also to Chapter 1, Figure 16). Five of the students from the earlier "Saldria" were from my original elementary school class in the "Jahn Schule" (see pages 1-14).

Our mathematics teacher was Professor Dr. Josef Geißler (called by us "der Lange," the long one). Previously he was a professor at the Technical University of Vienna. It was said that he could not continue there because of his political leanings. He was great. He challenged the elite of 3–5 students in all the different branches of mathematics, including projective geometry and differential equations. To, perhaps, half of the rest he taught a lot. Everyone else was discouraged, so that in the 12th grade there were only 19 students graduating. Professor Geißler did all his geometry free hand on the blackboard, and we tried during the breaks to reach his perfection. We managed quite exact circles and intersecting straight lines. Important was to mark a center, and then make sure that the joint in the shoulder was placed above this center. A circular motion of the arm would then, with some practice, produce the perfect circles.

<u>*Next in the list of my preferred teachers was Dr. Karl Geißler*</u>, our physics teacher (called by us "der Kurze," the short one—compare Figures 12 and 18). Before, he was a research physicist at Arado, the aircraft factory, and chose teaching as his second profession. We found his absolute honesty heart warming. He had no problem to say: "Sorry, I don't know, let us figure it out together," or "I will look it up and let you know next time." Sometimes after such remarks, a general debate ensued with us 'experts' and led to everyone learning a lot. One incident remained in my memory. I think it was when he was discussing the 'voltage divider,' mounted on the wall in back of the physics and chemistry class room, fitted for demonstrations (see Figure 14, below). In front, it had large volt meters, mounted on a big marble plate, hiding the circuitry and rheostats. Horst Rückert, one of the experts, exploded after a remark of Dr. Geißler: "You are wrong, ha, ha, ha—absolutely wrong. You cannot save energy with a voltage divider!" We all thought this was somewhat rude and we could see that "der Kurze" turned red. He answered: "I will bet you, *I* am right! If *you* are right, you will get a grade of '1,' if not, you will only get a '2'." Horst replied: defiantly: "Thank you for the '1.' " Indeed, Horst was right. This was important for all of us, because during the year if we had not fully reached the (ambitious) goals set by the curriculum, nobody was permitted a grade of '1.' The curriculum was based on uninterrupted instruction in the lower grades, something we did not have. This bet broke our grade barrier for physics.

The next lucky assignment of a teacher became obvious in the first chemistry class. Again "der Kurze" entered the room. He said: "No, I am not in the wrong room, despite that I am in no way qualified to teach chemistry. We must learn it together!" I noticed, that the material was not very new to me, I had read much of it in my chemistry book (see page 3-11). Dr. Geißler usually was one day ahead of us with each lesson. Soon, I was two days ahead, and later, even a whole week. Quickly my book, or I was the source for new information in chemistry. My final report card, the ("Reifezeugnis"), Figure 9, showed a "1 ≡ sehr gut" in chemistry, as well as in mathematics and physics, biology and geography.

<u>*On the list of my admired teachers was also our director, Herr Albert Büstrin*</u>. He was activated from retirement because of an acute shortage of new teachers who were cleared not to have a Nazi past. He did his job well for a good number of years. When five years later, Heidel and I announced our wedding also in Brandenburg's local

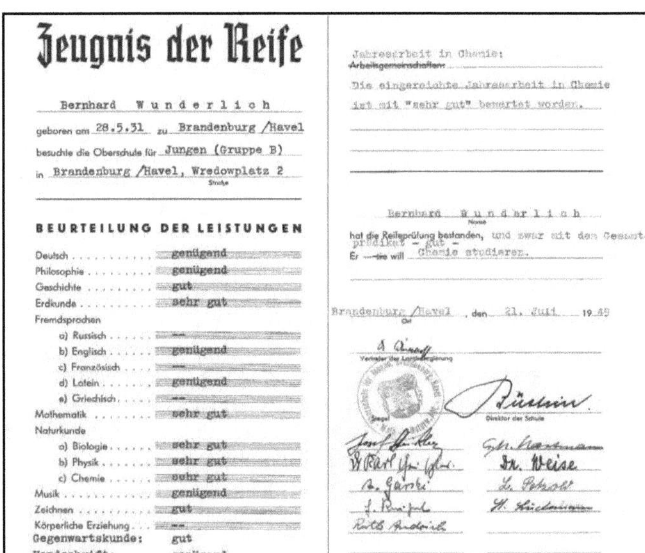

Zeugnis der Reife

Bernhard Wunderlich

geboren am 28.5.31 zu Brandenburg /Havel

besuchte die Oberschule für Jungen (Gruppe B)

in Brandenburg /Havel, Wredowplatz 2

Straße

BEURTEILUNG DER LEISTUNGEN

Deutsch	genügend
Philosophie	genügend
Geschichte	gut
Erdkunde	sehr gut
Fremdsprachen	
a) Russisch	--
b) Englisch	genügend
c) Französisch	--
d) Latein	genügend
e) Griechisch	--
Mathematik	sehr gut
Naturkunde	
a) Biologie	sehr gut
b) Physik	sehr gut
c) Chemie	sehr gut
Musik	genügend
Zeichnen	gut
Körperliche Erziehung	--
Gegenwartskunde:	gut
Handschrift:	genügend

Jahresarbeit in Chemie:
~~Arbeitsgemeinschaften~~

Die eingereichte Jahresarbeit in Chemie ist mit "sehr gut" bewertet worden.

Bernhard Wunderlich

Name

hat die Reifeprüfung bestanden, und zwar mit dem Gesamtprädikat - gut -

Er ~~sie~~ will Chemie studieren.

Brandenburg /Havel, den 21. Juli 1949

Ort

Vertreter der Landesregierung

Siegel

Direktor der Schule

Figure 9 *The final grades in 1949 for the "Abitur."*

newspaper with our West German address, he had the courage to sent a nice congratulation letter. Besides his work in administration, he also did a good job in teaching history.

As "Klassensprecher" (speaker of the class) for the last years, I got to know him well. I was the liaison to the students and often had to deal with their transgressions. Usually I was advised of these first (and often was involved myself) and tried to achieve good terms for both sides. For example, the last class of the day was biology with Fräulein Hesselbarth. Once, she was late, so we decided not to leave, but to do as if we had. We quickly hid across the hall. Fräulein Hesselbarth appeared, opened the door, and saw the empty class room. She set down her briefcase, checked the schedule, and quickly went down to "Herrn Büstrin" to complain. We got back into the class and sat absolutely still and heard her arguing in the hall what evil deed we had done, leaving without waiting the required 15 min. She also expressed her discontent that we always were unruly in her class (correct). The door opened, absolute quiet on all sides for a long minute! Then Herr Büstrin asked me: "What does this mean?" My answer: "I am sorry, Herr Büstrin, but Fräulein Hesselbarth was late and we just went briefly across the hall to check on our drafting assignment, she must have left to see you before we got back." Big relief on all sides and a weak reprimand to Fräulein Hesselbarth: "Don't always jump to conclusions, I only have the best experience with class 11!" Note, in high school, the classes were now counted forward from 1–12, earlier in elementary school, backwards from 8–1 (see the page 1-14 and the report card of Figure 16 of Chapter 1).

A final outstanding teacher was Herr Studienrat G. M. Hartmann. I have to admit this attribute, despite of a mutual dislike. "Herr Hartmann" taught German, Philosophy and also Latin. His idol was the poet Goethe. His stature, hair style, and general behavior was close to what we thought Goethe might have had. He thought, Goethe was such a genius that his writings in geology, botany, anatomy, physics, and history of science were also significant. I begged to differ and thought Goethe's science was outdated, not prophetic, and of little use today. In addition, my performance in German, Latin and Philosophy was not earth-shaking. I did not make more than satisfactory progress in his subjects.

A good example of our mutual dis-admiration is his handling of a particular exam in Latin. At this time I was giving help in mathematics in the parallel girls school, where Latin was a more prominent subject. I traded math-help for a translation of the exercises in our Latin text. Both schools used the same text, just that we were one year behind the girls. One day, Herr Hartmann said: "Today I want to see what you really know. Go ahead for two chapters in your text. There is an exercise for you to translate into German. The new words, you can look up at the end of the book. You have one hour." I opened my book, and found the translation attached to the page. I quickly copied it, not changing a single word. Then, I gave the text to my neighbor on the left, and then, right. Both were better students in Latin than I, but they appreciated help and distributed it further up and down. The title of the article was something like 'A Miraculous Rescue.' How fitting! When Herr Hartmann returned the papers, he commented that he must not have emphasized one special point in translating Latin texts. Too many of us had made the identical mistake. Everyone who copied my text got a '2.' Without this mistake, he said, he would have given a '1.' But I, having verbatim the same text, only got a '3.' How come? Probably in his opinion, I could do no better. I must say, I enjoyed reading the poetry by Goethe only many years after I left school.

An opposite experience was also demonstrated to me. Professor Geißler, as a rule, never checked homework. But I always had so much fun doing it, so that I took great care and had always most precise drawings. To prove the rule, once he checked. Going through the rows, he looked at everyone's effort. Almost nobody had done it. When he came to my desk, his face lit up and he took up the homework and showed it around. "This is how a perfect homework must look!" I was somewhat embarrassed. If it would not have been such fun to do, I would probably not done the homework either. I always was telling my students later: "Look at the homework examples in your book. If you see you can do it, don't do it, it is a waste of time. If you see that you cannot do it, you must do it. Take as long as necessary. You must find out what you missed in the lecture." This story about Professor Geißler, however, had an amazing ending. Eberhard Witte, one of the poorer math students and a good friend of mine sat behind me. He quickly pulled my homework across and put it in his book. Wouldn't Professor Geißler pick out this paper also and showing it and said: "Look at this now, see how sloppy it is." He never found out the cause of the unrest in the class. Everyone had seen the swap of homework.

The other grades shown in Figure 9, I felt, were generous. The teachers in these subjects were kindhearted not to be too harsh with the star pupil in math and science for 1949. Nothing worse than a '3' was given, not even in music. Drawing, history, and social studies made it at the end to a '2.' *Better overall grades were not given at graduation in this year*.

Gradually a new dictatorship was instituted despite all promises of democracy. The years 1946 to 1947 saw an ominous beginning, which was finished by the founding of the German Democratic Republic, GDR ("Deutsche Democratische Republik, DDR ") on October 7, 1949 with Wilhelm Pieck as President (see Figure 23 and pages 2-31 and 3-29). Just like the dictatorship of Hitler, it developed out of the weakness of an insecure democracy. In both cases, there was no mandate from the electorate. The economic difficulties were in both cases of considerable help in the development. Hitler was home-grown only after the annexation of Austria, Pieck, born in Germany, belonged to the SPD and later became the leader of the KPD in exile. From 1935 he lived in the USSR and returned, properly indoctrinated, in 1945 to lead the development of the SED into a Stalinist dictatorship, based on centralization, totalitarianism, and the pursuit of communism.

Figure 10 *Mutti with her four sons in 1947. Frank (19), Berhard (16), Joachim (3), Manfred (1).*

The year *1946 brought also the increase of our family* through the birth of Manfred on May 12. Our family was now complete and a picture from 1947 is shown as Figure 10. (For later pictures from 1949 to 1958, see Chapter 11, Figures 1, 2, and 18.) Despite the difficult time, all went well, and I could continue my goat-milk runs to the friend at the Görden (see page 3-6). This time I could use the railway bridges over the Havel and the Silo Kanal (see Figure 6 of Chapter 1), both bridges survived the war just that there were only the rails to walk on, or one had to jump from tie to tie.

Within the family, I could observe the beginning of the same rivalry between Joachim and Manfred as I had experienced between Frank and myself in earlier years. Joachim was responsible for Manfred and did not like it when the little one had his own ideas. As was typical for the two of us older boys when we were small, Joachim and Manfred were always to be found hand in hand (Chapter 1, Figures 7 and 9). An interesting incident occurred 1949. Manfred was crying.

Everyone ran into the bedroom to see what happened. Joachim knew, if there was an 'investigation,' he would be at a disadvantage. Mutti would side with Manfred. He tried to save the situation by saying: "I am sorry, 'Kleiner' (little one), did I pinch you inadvertently?" As if an accidental pinch was likely! Everyone was sufficiently amused that the only concern was to mollify Manfred.

Overall, however, I found the two were better friends than the two of us had been. It helped that Mutti now had a full-time help, Frau Zaja. She was good to all of us and when Mutti was later not fully up to par, she managed to cook acceptably. Naturally, Mutti's meals were best. Frau Zaja had developed a different style of cooking for her family.

Vati's business was set back many years in automation, as he had predicted (see page 2-19), but was flourishing otherwise. Everything was hand-hauled and carried up the ladders on ones back again. What was not foreseen was that all the hand-carts and many of the simple tools were items of great interest to refugees and forced laborers trying to get home. Much was gone by the time we got back from Lütte, along with trucks, cars, and bicycles. All had to start from the bottom up. Outwardly, the business looked as before, as can be seen in Figure 11 and also Chapter 2, Figure 7.

Figure 11 *Vati's business after the war, photo ≈1952.*

The progress was not only hindered by the shortage of material, but also by the continued threat of expropriation as a capitalistic enterprise. Fortunately, Vati had kept in touch with his friends of the SPD who suffered during the Nazi time, and now were in positions to know the course of action before it was taken. Usually a few months before it became law, Vati knew the limit of number of employees which were permitted and could shrink accordingly to remain owner of a 'small business'.

He also knew how to get rid of the bothersome shop stewards of the communist labor union. Initially, there were three of them. One day they appeared in his office and said: "Boss we were told to check your books. Could we look at them?" Vati said: "Gladly, I always like to get help and advice." He called the secretary to bring the 'books' and let the "Betriebsrat" (shop stewards) alone in his office for about an hour. When he came back, he asked: "Now, what do you have to say? All in order? What can we do better?" The answer was something like: "Yeah, we don't really know what to say. To tell the truth, we don't understand what this all means." Now Vati's friendliness was gone. He said: "If your organization did not manage to train you to be of help running the business and understand that the books are in perfect order and everyone gets paid the wages set by law, get out of here and never ask for the books again. You know I do my best, and you often enough get extra food items if I can get them from the customers. You wasted three valuable work hours. Back to work." No complaint was lodged, but Vati knew, he had to get rid of them as fast as possible.

Shortly before the business had to shrink again, Vati put all three together on one job, and he, Frank, and Opa, checked on them not only once a day, as customary, but observed them when they arrived, had breaks, and left in the evening. In addition, their performance was monitored and carefully compared to the norms. After two weeks they had failed miserably. There were too many political discussions they had to have. Next day, they got their papers. Their answer: "You cannot fire us, we are the shop stewards!" But Vati had sufficient reason. New, more compliant, shop steward were elected, and the smaller size of the business was reached. Naturally, this continuous shrinking could not go on forever. It ultimately led to all of us leaving Brandenburg.

At home, life improved in very small steps. Not only the insufficient food rations caused problems, also the distribution system failed. For some months, there was no salt, then no matches, no candles, no string, etc. In the corner of our building was a general store of the "Konsum (-genossenschaft)" (consumer co-operative). This "Konsum" (Co-op) had its roots in the 19th century and was quickly reinstated in the Soviet Zone. It grew into one of the biggest organizations for consumer goods in the GDR. Only later was it rivaled by the HO (see page 3-11). If the "Konsum" did not have it, nobody had. Food rations in the West Zones were not much better during these first years. In many instances, they were worse.

In addition to Vati's efforts, in the second half of 1946 CARE packages arrived from time to time (Cooperative for American Remittances to Europe). The initial packages were US Army surplus, purchased in the US and directed to specified recipients. Our's were originated by Tante Lieschen in Hastings, NE. Letters were exchanged before describing what happened since 1939. Each of the CARE packages made us feel like Christmas. The meats and fat enriched our meals for weeks. Mutti enjoyed the coffee which she did not have for years. For us boys, the Hershey block chocolate was exiting. We had almost forgotten that things like this existed. As time went on, the packages were packed by Tante Lieschen herself, addressing our specific needs. I had a chance to write to Jim, who was only little younger than I. We exchange notes about our daily lives. What a difference! (See Tante Lieschen and Jim visiting in 1937, Chapter 1, Figure 12.) In school, we received "Schulspeisung," calory-rich foods like cheese, distributed by the YMCA in Brandenburg.

In 1946 we still had to spend time in the ruins of Brandenburg to recover bricks and to clear the rubble. These afternoons involved all students and teachers. Figures 12–13 and 18, below show examples of this extracurricular activity ("Trümmeraktion"). Most of the boys in these pictures are my class mates. Professor Geißler is shown in Figure 12, I can be seen in the center of Figure 13 with the beret, looking at Günter Ahrendts, missing in Figure 25).

In my spare time, *I soon set up a tiny laboratory in the basement*. In a corner, next to the potato bins, the barrel with herrings, and the coal pile, I had a three-foot wide wooden working surface, an old chair, an electric bulb for light, and a set of shelves for the reagent bottles. The main

Figure 12 *"Trümmeraktion" with Professor Geißler and some of my class mates in 1946.*

Figure 13 *"Trümmeraktion" of the 10th grade in high school, working near the Pauli Kirche.*

point was that Vati had good friends among the druggists. Any left-over laboratory equipment from pre-war times found my laboratory. In this corner were Erlenmeyer and round-bottom flasks, watch glasses, spatulas, an alcohol burner and a small balance, and my proudest possession, a Kipp hydrogen-generator. There also were a lab manual and a book to record the experimental details.

Fortunately, at that time. there were no safety restrictions on selling chemicals to the public, not even to minors. If the druggist thought I was reliable, I could get any chemical he had. Things like potassium cyanide, white phosphorous, and even radioactive uranyl acetate were in my collection. One day, I could lay my hands on a sealed can of about ten ½-inch balls of metallic potassium, protected in kerosene. A most interesting possession was a German army field-laboratory kit of about 15 bottles of reagents to test for all sorts of agents. All were stored in beautifully labeled, ground-glass bottles. This set was sufficiently expensive, so that I had to get Vati's separate permission. His question was: "Do you really need it?" My answer: "Not really, but it looks so nice, and I probably can learn a lot from it!" I got it. Another high-prized possession was a subscription to the "Chemische Zentralblatt," the German survey journal of the chemical literature. Every week, I would receive a several-hundred-page copy of abstracts of international articles. It was most interesting to read what chemists were doing in their laboratories. I am sure I was the only one in the city who received this journal. It did not bother me that the abstracts were often rather short. Most were too difficult anyway for me to fully understand.

It did not take long and *people in the house started complaining about the esoteric smells* and the occasional small explosions emanating from the basement. Vati found a good solution. In our house in the Werderstraße the basement apartment was cleaned and used for storage of roofing felt, nails, and other goods. There was no roof, but the high pile of rubble kept things reasonable dry (see Chapter 2, Figure 9). In one corner, there was a small pantry with two side closets, ideal for a 'big' laboratory. I could have it. The wall and ceiling were painted, and I got some help in building a lab desk and shelves. There were functioning electric light and water, and I even had a gas connection nearby, so that I could install my Bunsen burner. This burner was a left-over from before the war, probably because it was too expensive. It was not only functional, but had amenities like a small-finger regulator and shut-off, so that you could adjust the flame when working with both hands. I could even start bending glass tubing and seal ampules.

What experiments to do? Not only those I designed myself. I had the book by Hermann Römpp "Chemische Experimente die gelingen" (Chemistry Experiments that Work, 1939). One day, I worked with synthetic dyes. I spilled some. No problem, I washed myself carefully. Still, when I got home, Mutti was concerned: "You look so yellow?" It is hard to run a 'clean laboratory.' The effective dye had spread over my hands and face, and nothing could wash it off. I had joined the yellow race. Only after 3–4 weeks did the skin slowly peel and I again became a paleface (term American Indians used to describe 'whites,' according to Karl May, see page 1-25).

Most of the experiments were quite serious and scientific, but what do you expect a teenager to do when with other teenagers? Here are some of the experiments, *not* recommended. Today, these would draw much harsher penalties than a youthful prank which they were considered to be at that time. Potassium, when put on water reacts vigorously, develops hydrogen (and KOH) and, specially when the kerosene is not fully removed, it catches fire and may explode. What is the safe experiment when showing off? Drop a not cleaned ball of it from a bridge at night. From a safe distance it gives a beautiful firework. Simpler, but also not to recommend, is to fill an ampule with "Buttersäure" (butyric acid), which when broken, produces a noxious, awful smell of vomit or rancid butter. There are all sorts of places, where the ampuls could be dropped. It clears the room as effectively as introducing a skunk. Small amounts of explosive, as are used in percussion caps, can easily be mixed in the lab (carefully) and then loosely sprinkled on paved sidewalks. When someone steps hard enough on it with a metal heel or tip, it produces a mild jolt, explosive noise, and at night, a flash of light. For us this was big excitement when watching it from a distance.

The laboratory experience led to my getting a position *to show experiments in the "Volkshochschule,"* an organization best described as adult education. One could go there for a reasonable fee and be instructed in a wide variety of subjects. Courses were held evenings in the class rooms of a school. The teachers were from high schools or universities nearby.

The teacher from the girls high school, Dr. Kerstan, gave the chemistry course and asked me to assist. I had collected many of these experiments for a 90-page thesis, written for my graduation in 1949 with the title: "Versuche aus dem Demonstrationsunterricht"—meaning: Experiments, to be used in teaching). There were to be a total of 20 experiments. Each was pre-mounted and could easily be brought in a single trip from the preparation room to the demonstration bench. The photography session of Figure 14 was to document all.

Figure 14 *Experimenting in front of some students in the chemistry room of our high school (in 1948).*

After the first 19 were photographed, a 20th experiment was to show the explosion of magnesium powder when heated with potassium permanganate and sulfuric acid. It had worked several times before. It involved simply adding the ingredients into a reagent tube, put the Bunsen burner underneath, and duck quickly behind the hood, partially seen on the right of Figure 14. But this time, it did not work. Nothing happened! Finally, I wanted to give up and pull the Bunsen burner off the table since I did not know how to turn it off from the safety behind the hood. Just when I reached for the gas hose, the explosion occurred. Loud enough to be heard all throughout the building, and forceful enough to blow out all the plastic windows and to shower me with the glass shards of the test tube.

Help was fast on hand and I got a big, preliminary bandage around my head, and two of my school friends, seen in the picture, helped me to find the way to the nearest physician who was alerted of my coming. It took him about an hour to search all the bleeding scratches for bits of glass. It was impossible to get all of them out, but he hoped since there was enough bleeding, that there was no infection. Some of the left-over shards were slowly pushed out of the face many months later, and a hard object over my left eye still persists, it may still contain some. With much smaller band aids, I could then go home. Next day, all plastic window panes had been nailed back into place by the janitor, and I was on the job again. Experiment 20 was scrapped. Nobody else should try the same, they may not be as lucky as I was.

Another major undertaking was started by about ten students of my class. We had noted that there was a long, half-collapsed shed in the back of the school with four rowing boats still in place. Because of the many waterways in Brandenburg, middle and high schools close to the water had boat houses and, before the war, had sported crews for competitive rowing. Figure 15 shows a recent photo of the present "Regattastrecke Brandenburg." We were complete novices and our physical education teacher also knew nothing about rowing. In the new spirit of freedom, we found that one of our parents (who was already retired) had experience in boat building and repair. With his help we got permission from Herrn Büstrin to try <u>*salvaging the rowing boat wrecks for our use*</u>. Indeed, after a few months of work, two beautifully restored boats emerged from the four and enough oars were repaired for us to be able to start rowing. Since the storage shack was not repairable, we got permission to transfer the boats to the facility shown in Figure 16. A boat house near the water, close to the Dom. It belonged to the former "Ritterakademie," a high school for nobility, at that time in use as a second public high school.

Figure 15 *Regatta training on the Beetzsee in Brandenburg. Recent photograph, taken in 2002.*

The next item was to learn how to properly handle the boats. We found books in the library and learned enough, so that after sufficient training we hoped to enter regattas, a frequent sight in Brandenburg before the war and now again, as shown in Figure 15. One rule we did not abide by, and did not enter in our newly written manual, was: 'No mixed boy/girl crews.' We thought this would not be fair, and indeed, we had many good trips with mixed crews (see Figure 20, below). Girls made good coxswains and, if paired properly, could do well in balancing the stroke.

Figure 16 *Boat-house near the Dom (recent photo, of 2004, the faded club name indicates the later political leaning).*

My later specialty was with Peter Klomfas to row as coxless pair [two in a boat without coxswain and rudder, and only one oar for each ("Riemen")]. We had to coordinate the steering (sitting backwards) by proper timing and strength of the strokes. We were properly matched and could in this way easily outrun the other boats. Figure 17 is a contemporary picture where five of us went down the Havel to the Plauer See, but found the waves too high for boating and had to make an emergency stop after the boat was in danger of sinking. Two of us jumped into the water to maintain the boat afloat and direct it to shore without getting broad-sided by waves. After cleaning out the water (Figure 17), we snuck back into the Havel and undertook our usual exercise, to outrun the excursion steamers.

This effort of us ten independent students came just at the right time. Before 1945 we would have had to be members of the "Hitlerjugend" to use the boats (and at least I, was not). In March 1946 the FDJ ("<u>F</u>reie <u>D</u>eutsche <u>J</u>ugend," the Free German Youth), the

Figure 17 *Reconstructed rowing boat for four with coxswain at the shore of the Plauer See. The fifth crew member took the picture, I am on the right, at the bow, helping to lift the boat. Picture taken 1948.*

communist youth organization, was founded under the leadership of Erich Honecker.[1] In a very few years, the FDJ took over all organized sports, and again, I did not become a member and would have been excluded from rowing [see the faded name at the boat house of the "Betriebsportgemeinschaft, Einheit (BSG)" in Figure 16].

Initially, everyone in our high school class made use of the newly found freedom. We elected a "Klassensprecher" (speaker of the class) a position I held 1948/49 (see page 3-15). We also started a school newspaper, the "Observator." It functioned in form of a bulletin board with Eberhard Witte as editor. All of us could, and many did, contribute uncensored information. *The democratic rule in school and country, however, saw a fast demise*.

In 1946, all of the new political parties (see pages 3-8–9) had started newspapers. Of importance for us were the "Neues Deutschland" the paper of the SED (it grew out of the KPD paper, "Deutsche Volkszeitung" and the SPD paper, "Das Volk"). It is still active today as a left-leaning paper with no clear party association and readership, largely in the area of the former GDR. The "Märkische Volksstimme" was a local SED paper printed in nearby Potsdam. The CDU newspaper was the "Neue Zeit." In addition, the "Berliner Zeitung" was founded by the Soviet Military Administration. "Der Tagesspiegel" appeared under the license of the American Military Command. Initially all these and other papers from all over Germany were freely available in Brandenburg. "Der Tagesspiegel" represented our ideas best. By April 16, 1948, just two months before the blockade of Berlin (see page 3-28), it was forbidden to buy western newspapers and magazines in all of the Soviet Occupation Zone, and you better let no one know that you were listening to the RIAS, the Radio in the American Sektor of Berlin, which had started broadcasting in 1946. Barely three years of freedom of the press was all we had.

In October 1946, six months after the forced uniting of SPD and KPD to the SED (see page 3-9), democratic Landtagswahlen were scheduled (elections for the parliament of the "Länder," provinces or states and counties). We discussed these in detail in school, particularly with our physics and chemistry teacher, Dr. Karl Geißler, shown in Figure 18. He was active in the LDP and had hopes that the SED would be soundly defeated despite their overwhelming election advertisements and coercion. It was not to be. Although not agreeing with the unification of KPD and SPD, many of the old SPD members were afraid to switch party allegiance. For the state of Brandenburg, the votes were SED 38.5%, LDP 18%, CDU 26.7% VdgB (the union candidates) 4.3%. In Berlin, the SPD could refuse to join with the KPD and the Allied Military Administration

---

[1] (1912–1994). A fellow roofer, who left his apprenticeship in 1929. In 1930 he went to study at the International Lenin School in Moscow, USSR. After his return to Germany as a politician of the KPD, he was arrested in 1935 and kept until 1945 in the "Brandenburger Zuchthaus" (a state penitentiary). From 1971–1989 he led the GDR as General Secretary of the SED (see also page 2-12 and the Footnote of page 9-24).

permitted in all sectors voting for the SPD *and* the SED (for the City Magistrate). The result was much closer to what probably was the will of the electorate also in Land Brandenburg. The votes were SPD 48.7%, CDU 22.2%, SED 19.8%, LDP 9.3%. Although in Brandenburg the SED was the largest party, it did not have an absolute majority. In the next "Landtagswahlen" in 1950, one year after founding the GDR, all parties were combined as the "Nationale Front" and no separate votes were permitted ever again until 1990 as part of the "Wende" (see page 9-25, Footnote 2). Quickly the democracy had ended.

Figure 18 *Dr. Karl Geißler at the "Trümmereinsatz" in 1946.*

Efforts to indoctrinate the students were also made outside of the confines of the school. One tool was occasional lectures at the "Haus der Kultur." Here, politically schooled instructors lectured about themes of the day. In this way, a central instruction was possible with fewer (and better?) teachers. At the time of the elections, after returning from one of these meetings, we were locked in our class room. One after the other we were called out for interrogation by plain-clothes police. The questions were like: "Did you see anything written on the blackboard in the instruction room? What was it? Who wrote it? Now, write the following sentence on the board! ..." They did not find out who the culprit was. If they had, this student would have been severely punished, expelled from the school, and most likely incarcerated in the again flourishing political prisons, inherited from the Nazi time and newly built (see Footnote of page 3-39). We were all back in school on the next day. Then the secret was leaked. Someone had written the well-known verse: "Nur die allergrößten Kälber wählen ihre Schlächter selber!" (Only the dumbest calves grown, elect a butcher of their own), a slogan well known and circulating in these non-democratic elections.

This made it clear that we had to be on guard of informers. By 1945 the reign of the Gestapo had ended ("Geheime Staatspolizei," secret state police). Already in 1946 a secret police existed again. It grew into the Stasi (official name after 1950, derived from "Ministerium für Staatssicherheit, MfS," ministry for state security). Later the Stasi infiltrated every aspect of life. The short period of greater freedom had ended. We were back to the mode of passive resistance. Whoever was caught exceeding the permitted limits, could count on severe repercussions.

From time to time, we also were led to the viewing of movies made in the USSR. Some of these films were quite good, but most of them had a political overtone we resented. Being quite sensitive to these, we did not necessarily keep quiet in the dark of the theater to the irritation of our chaperones. In 1946 the East German film monopoly (DEFA, Deutsche Film Aktiengesellschaft) brought out its first film. It was included among our 'forced-to-see movies.' It was quite successful. Its title was "Die Mörder sind unter uns" (Murderers Among Us), directed by Wolfgang Staudte and featuring Hildegard Knef. Later DEFA movies also were often, but not always, filled with politics.

Cultural events were re-emerging quickly. Theater, opera, and operetta productions were developed, starting only a few months after the war. Since the city theater was destroyed, a hall at the "Adlerterrassen," was used (about 500 seats), along with the more modern "Stadthalle," suitable for musical performances and balls (about 750 seats). Figure 19 is a recent photograph of the theater

Figure 19 *Adler Terrassen, entrance to the theater. Picture taken on a visit of Brandenburg in 2006.*

hall of the "Adlerterrassen," a large outdoor restaurant at the foot of the Marienberg (see Chapter 1, Figure 6, Point 4). From the second season on (1946/47) until my leaving Brandenburg in 1949, Vati, Mutti, and I had season tickets for opera and operetta. For me, this was a new experience. Looking back to this time, I must admit that all the operas I heard at that time are still the ones I enjoy most when sung in German, which was the standard language for all operas in Brandenburg. The first opera I attended, was "Figaro's Hochzeit," quite delightful. The quality of the performances was better than one would expect in a provincial town. The mechanism to draw all better talent to Berlin was not quite developed in these early years. For the 1947/48 season, I even managed to get a season ticket for my first girlfriend. We kept on going to the performances even for some time after we broke up (as I also still visited with her father, who was a teacher in the special education school, see page 2-5).

My technical skills were occasionally called upon at home. Watches were at that time one of the most scarce items, just as it was difficult to find spoons. Both of these things had a great attraction for the Russian soldiers. The watch, to tell the time (after they learned that the watches had to be wound and set), and the spoons, to eat the soups common in their diet. Of Mutti's fine silverware, for example, most items were left, but no soup spoons. The knives were not thought of a quality worth the taking, but the spoons fit nicely into the boots.

Frank and I had received the same type of watch from Oma Siedlung when we demonstrated to her that we could tell the time (see page 1-12). Both watches had given up their life by 1945 and survived in Mutti's junk drawer. So, one day, I fetched a magnifying glass, tweezers, and a small jeweler's screwdriver and took both watches apart. There was a table full of pairs of identical gears, springs, etc. This was not the best way, because I had a devilish time to reconstruct the watches. The idea was, since I saw nothing obviously wrong with either watch, such as a broken spring, to compare every pair of parts carefully and pick the better one to put together a superior watch. After a long time, the job was done. The superior watch began ticking. Great! I did not think it was necessary to put the second watch together and just dumped all into a bag.

Soon I noted a problem. The watch did not keep time. Within a day it lost 4 hours! This made it useless. The only adjustment I could do was changing it by ±5 minutes per day. After a while, I gave up and went to Vati: "Vati, I tried to fix my watch. It works now, but it does not keep time. You know a watch maker who could fix it?" He did. As so often, one of his customers owned a jewelry store. The watch maker was glad to be of help. "You will have it back in no time, going precisely." I was doubtful, but, what did I know about fixing watches? Vati brought the watch back a few weeks later, running perfectly. But he also brought a story. After trying all the obvious, the watch maker should have admitted defeat. But a good craftsman does not admit defeat to a fellow craftsman from a different profession. What did he do? He counted the teeth of every gear, calculated the timing, and then concluded: One (specific) gear had four teeth less than it should have! How is that possible? He never came up with a solution, but being a really good watchmaker, he cut a new gear with the proper number of teeth and I got a watch that worked.

To complete the story, Frank had a newer watch that had also survived the war, i.e., he needed no new one. On the way to a date in 1948, he was asked by a good-looking soldier: "What time?" "Six o'clock." That was when his watch was confiscated. Now he was looking for his old watch. He found the bag, was mad at me for taking it apart, but did not feel up to putting it together. And I did not tell him that if he would get it together, one of the gears would have four teeth too many, and that our jeweler friend might still have a spare gear with the right number of teeth.

Another gadget I rescued, was Vati's electric shaver from the 1930s. It did not work, and I think he never thought it was working too well. I asked for it, and got it as a present. With little effort I found the loose wire and fixed it. It served me well until 1953 when I bought my first "Braun" shaver in West Germany, the best shaver manufacturer in the world (I think). From this time on, I owned a progression of ever more sophisticated "Braun" shavers. The last few models being self-cleaning were available at the local Wal-Mart. I never needed a blade or knife for shaving, except when working on patients in the hospital at Ingleside, near Hastings, NE, where I worked when attending my first college in the US (see page 6-7).

The year *1947 was the year when I went to dancing lessons*. These were customarily given by a private school, catering to students of the higher grades. It was the first official coeducational instruction for most of us. Much needed to be taught about the general behavior, from how to move chairs, ask for a dance, escort a girl back to her table, and hold her while dancing. Since at that time power outages were still quite common, the rule was that every student had to bring a candle. Also, on cold days, a piece of wood was necessary since the ballroom was not heated. Instead, it had a small temporary wood-burning stove around which we could dance, and the piano player sat not to far from it. Besides teaching the important dance steps, the school also organized balls. In this fashion we got acquainted with the girls of the parallel high school.

The total amount of spare time for such diversions was limited. Commonly each Saturday was reserved for the cinema. One other evening was reserved for the dancing classes. When the weather was nice on weekends and during vacations, we went out to the lakes for boating and swimming. All these activities were now shared with the girls. Figure 20 shows an excursion with two rowing boats. Sitting on the left is Dieter Heckert, on the right, Fritz Bockmühl. This suggests that the unrecognizable person in the back might is I. The one who holds the oar is Harald Müller. The girl on the left was my first girl friend, Eva-Marie Schmücker. The remaining four were from parallel classes and I forgot their names.

Figure 20 *Weekend of rowing. At the shore of the Plauer See on the island Buhnenwerder in 1948.*

Figure 21 illustrates that I also was interested in competitive canoeing, but I did not join the club for reasons, I do not remember. Most likely was that either I was not a member of the FDJ, or I did not have time for an extra activity. But, I enjoyed the first efforts to explore the small waterways around Brandenburg shown in the picture.

Figure 21 *My first effort to join the paddle club in Brandenburg, 1948.*

By the year 1948 we were well established in school and had some time to do additional things. My job as assistant in the "Volkshochschule" (see page 3-20) allowed me to look into this organization, directed by the "Pädagogische Kabinett" of the provincial government in Potsdam. Not only science was taught. Perhaps 5–10 other high-level courses were offered each trimester. I attended (no fee) courses on biology (insects) and physics (electricity), and registered (for a small fee) in Darwinism. It was interesting to hear lectures about these subjects, free of the pressure of being graded.

Also, it was common practice that when in 11th grade, *a high school class should or could put on a theater play*, to entertain the other school classes and the public. We had seen a few of these productions before and had a rather severe criticism for these efforts. Being not coeducational, the male and female roles had to be played by students of the same gender. We thought, this was not acceptable. In addition, we had two other problems. We were a math and science school, so we felt not qualified, except for our three or four 'experts' in German. Our whole class was broken down into groups of 3–6 'experts,' each group for one specific field of instruction. The ones who did not belong to any one of these groups, usually did not make the grade, so that at the end, only 19 passed the "Abitur" (final examination). Three to six male actors, however, where not enough to put on a play. The second, and biggest problem was "Herr Studienrat Hartmann" (see page 3-15). None of us liked him to be our mentor for a play. What to do? Science students think they are problem solvers. Was there nobody else who could help us produce and direct the play?

The lead expert in our German group was Fritz Weickert. He knew of Dr. Wolrad Rube, the former director of the Brandenburg theater, who was politically not acceptable anymore, and thus, pensioned off (see page 3-9). Fritz and I payed him a visit. We told him of our plight and asked if he could help us to produce a play, although we did not have too much money. He rose to the challenge, and he was a great teacher. Too bad that he was not our German teacher, I might have done better in German. We had suggested the comedy by Heinrich von Kleist "Der Zerbrochene Krug," a classic play from 1806.[1] The topic was the material for several movies, most famous was the 1937 Ucicky film with Emil Jannings. Several (unknown) operas were based on it as well.

The discussion led to the following results: 1. The play is too long for you. Let me shorten it without destroying its verse and sense. 2. You are not actors, your diction is atrocious, please enroll into the speech class in the "Volkshochschule." 3. He agreed with a plan of ours that every

---

[1] The broken pitcher ("Krug") arose from an effort of old judge Adam to seduce Eve. Eve's mother, Marthe, had seen young Ruprecht earlier in the room of her daughter, while Ruprecht had seen a stranger leave Eve's room through the window, breaking the pitcher. Marthe wants to save the honor of her daughter and asks the judge for help. The day of the inquest turns out to be rather awkward for Adam. His superior, Walter, is inspecting his office. Eve was forced to keep quiet by Adam's promise to keep Ruprecht from military service if .... All ends happily, except for Adam.

male actor had to find his own female partner (girlfriend?). 4. We had to strictly follow his direction and absences from rehearsals would not be tolerated. 5. He would control the play by being in the prompter box. We agreed to all and put together the cast shown in Figure 22.

With zeal we went to work, and soon, one could hear us recite the silly rhymes to sharpen our diction. Just think of the well known equivalents in English: "The rain in Spain ... ." Most difficult was pronouncing the consonant 'p' without moving the flame of a candle held in front of our lips and the proper sounding of the vowels (a, e, i, o, u), "Umlaute (ä, ö, ü)," and diphthongs (ai, au, äu, ei, eu).

The cast had in the main female role Frau Marthe, a young "Neulehrerin" (new teacher, quickly trained to fill the shortage of teachers). She and Fritz Weickert were the two most qualified. The play was shown in the Adlerterrassen (see Figure 19). It was performed three times and was a success as judged by the reactions of the audience. My playing the judge was not planned, but Dr. Rube thought, I would fit into this role better, so I had to switch with Fritz Weickert. In this way, I practically knew the whole text of the play by heart and felt quite secure in my role.

Wredowschule – Jungen
(12 stufig)
Brandenburg (Havel)
Arbeitsgemeinschaft Laienspiel

Der zerbrochene Krug

Ein Lustspiel in einem Aufzug
von
Heinrich von Kleist

Personen:

Walter, Gerichtsrat	Peter Klomfas
Adam, Dorfrichter	Bernhard Wunderlich
Licht, Schreiber	Dieter Heckert
Frau Marthe Rull	Ilse Döbert
Eve, ihre Tochter	Eva-Marie Schmücker
Veit Tümpel	Fritz Bockmühl
Ruprecht, sein Sohn	Fritz Weickert
Frau Brigitte	Christa Archut
Ein Bedienter	Günter Ahrendts
1. Magd	Waltraud Freudenberg
2. Magd	Marianne Schünke
Büttel	Eberhard Witte

Regie: Dr. Wolrad Rube

Figure 22 *The cast of our play.*

At the premiere, we thought we made a big mistake. One of the 'witnesses' had twice to say similar sentences, switched the two, and left out the first. The whole cast was quickly redirected by the prompter to leave out the intermediate part. We were heartbroken, but Dr. Rube said at the end: "Well done. Nobody will have noted the missing two minutes." Indeed, I asked many who had heard the premiere, they had neither noticed it nor thought that there was a missing clue.

After the generally agreed upon success, Herr Hartmann had to come to terms with us. His comment about the play was that we did a good job and he would have been glad to help us, but Dr. Rube surely had the greater experience. He forgave us, and we were glad not to have a German teacher mad at us at the Abitur (may be, he was a good teacher after all).

This effort *to be an actor for a few months was of great value to me* in my later career. Nobody who has to lecture in front of a larger audience should miss such rudimentary instruction in speech and acting we had to go through. Unfortunately, I had to sit through many lectures by fellow scientists who made serious mistakes, easily avoided by amateur 'actors.' The most common flaws are mumbling to the blackboard or screen instead of clearly projecting to the audience, not keeping the lecture properly paced and on time, and not having the text sufficiently rehearsed.

In 1967, when trying at RPI to lecture to 500 students with experimentation enhanced by in-room, TV (see pages 8-13–14), I was the only of the lecturers to regularly earn applause at the end of my lectures in Freshman Chemistry. No student left early from the back of the room. This was possible only by extensively practicing the experiments, lectures, and blackboard writing, followed by delivering all as if it were done for the first time.

In addition to the knowledge gained in science, the teaching in technical drawing, speech, and acting were of greatest value (not to forget German, English, and Latin). Unfortunately, Latin was not enforced, so that I must rely on my old texts to understand even the simplest phrases.

The split of Germany into a democracy and a dictatorship continued and was practically completed by 1949. In the middle of our preparation for the "Zerbrochene Krug" in June 1948, a currency reform was announced for the western zones. The old "Reichsmark" (RM) was exchanged 1:1 to the new "Deutsche Mark" (DM). Savings were reassessed 10 RM to 1.0 DM. Following within a few days an analogous reform was carried out in the eastern zone to include East Berlin (on order of the Soviet Military Administration) by identifying the 'new money' by paste-on exchange coupons. After printing of enough new money, the old was withdrawn from circulation. To distinguish the initially equivalent currencies, they were simply called "Westmark", DM-West, and "Ostmark," DM-Ost.

One day after the "DM-West" was introduced in West Berlin, its blockade began. There was no access to or from the city from West Germany for people, food, gas, coal, and other goods. The Soviets hoped to starve and freeze the West Berliners into submission to join East Berlin. Two days later, the US airlift began, and the Soviets left the Allied Military Administration in Berlin. A few months later, the German city government split and West Berlin decided on new elections without the SED. The blockade and airlift continued until May of 1949 with West Berlin remaining independent of East Berlin.

The immediate problem for our school play was its financing. *Was there enough new money available* in the audience to sell tickets? We produced a 'Literary Evening' on June 26, 1948 as a fund raiser for the initial cost of the theater production. Students and parents of all grades were invited to the school. Both, the schools for girls and boys participated with songs, choir performances, recitation of poems, and readings. The take was less than expected, but sufficient to continue the preparation of our play. By scheduling three performances of the "Zerbrochene Krug," we ended without loss and could pay our debts in the new "DM-Ost."

With the economies of the FRG and GDR severed, the woes of the two countries changed. West Berlin became the shopping paradise for East Germans. The exchange rate between the currencies began to soar, at times as high as 10 DM-Ost to 1 DM-West. When the *Marshall Plan of a European Recovery Program* (ERP), (named after the US Secretary of State) was signed by President Truman in 1948, the Soviet Union refused to participate, as did East Germany (as decided by the SED). From 1948 to 1952 almost 13 billion US Dollars were expended on 17 European Countries. The most support went to the United Kingdom ($ 3.3 billion), France ($2.3 billion), West Germany ($1.4 billion), Italy ($1.2 billion), The Netherlands ($1.1 billion) and Belgium ($0.7 billion). As a result of the Marshall Plan, and, even more so of the relaxations of restriction to industrial development and forced exports, fast growth occurred in the FRG. In the years 1948 to 1952 the industrial production increased by 35% and the agricultural production surpassed pre-war levels. The poverty and starvation of the earlier years disappeared. All Western Europe embarked upon an unprecedented growth. The standard of living increased dramatically.

While the part of the Soviet Union affected by WW II had been similarly destroyed, the eastern USSR had seen a rapid industrialization during the war. Instead of the Marshall plan, large reparations were collected from Finland, Hungary, Romania, and East Germany. The GDR was, in addition, forced to ship large amounts of supplies to the USSR. Estimates are, that the USSR collected as much help from its vassal states as would have come from the Marshall Plan.

The end of the blockade was followed by founding of the "Bundesrepublik Deutschland, BRD" (Federal Republic of Germany, FRG) out of the three western zones (May 23, 1949). The "Bundestag" of the FRG was elected shortly before. In the east, the "Volksrat" (People's Council)

Figure 23 *Wilhem Pieck proclaiming the GDR 1949.*

proclaimed the "Deutsche Demokratische Republic, DDR" (German Democratic Republic, GDR). Figure 23 shows the event on October 7. The "Volkskammer" saw a no-choice election in 1950 and lasted until 1990 (see page 3-23).

This was the time in Germany when we thought a *'United Europe'* would be the best solution for the future. With it, the problems caused by the redrawing of the borders in Europe (as shown in Chapter 2, Figure 16) could have been defused. By creating a common economy with free migration of the populations to areas where they were needed, old adversities might slowly be forgotten. It was not to be at that time. Now, some 60 years later, progress has been made, but by far not enough to reach the ideal we thought possible in these days of my high school time.

The final decision of what to do after the Abitur was now due. For Vati, he wanted to know if I really knew enough about chemistry. He got help from his business friend, Herrn Eilert (see page 3-10). Herr Eilert had studied a few semesters of chemistry (and failed?). On a visit, he was to ask me unobtrusively some test questions. No need to say that these proved only that *he* knew little about chemistry. He quickly became a strong supporter of mine. He suggested to Vati: "You must send him to the university!" But I had not quite made up my mind! There still were the possibilities of engineering, mathematics, and physics. Mathematics and physics, I excluded since in Brandenburg the only mathematicians and physicists I knew were high school teachers. High school I did *not* want to teach, and I could not even dream of being able to teach at a university.

How about the engineering? Did I not like all types of machinery? (See the pages 3-11–12.) In the Potsdam accord of 1945 (see page 3-9) and the later published limits imposed on the industrial potential after WW II, I could read in the newspapers: 'All German armed forces, as well as munitions factories and civilian industries that could support them, were to be abolished. Further, all civilian industries are to be severely restricted.' The first plan signed by the Allies in 1946, stated that: 'German heavy industry was to be lowered to 50% of its 1938 levels by the destruction of 1,500 plants. The cap on steel production was set at about 5,800,000 t per year, equivalent to 25% of the prewar production level. Steel plants thus made redundant were to be dismantled.' Germany was to be reduced to the standard of life it had known at the height of the depression in 1932. Automobile production was set to 10% of prewar levels, and so it went on through levels for aluminum, magnesium, and other metals. On February 2, 1946, Brigadier General William Henry Draper Jr., chief of the American Economics Division reported from Berlin: "Some progress has been made in converting Germany to an agricultural and light industry economy." He explained that Germany's future industry and economy would need large imports of raw materials and food to maintain a minimum standard of living. This could be compensated by exports of coal, coke, electrical equipment, leather goods, beer, wines, spirits, toys, musical instruments, textiles, and apparel to take the place of the heavy industrial products. This would have created a situation worse than the dictates of Versailles after WW I (see page 1-2). It is likely that the consequences could have easily been similar. For me, it left no room for an aspiring engineer for heavy machinery. Germany would become a third-world country (see page 5-10 and also the comment on page 8-61).

Naturally, to become a politician or government employee ("Beamter") was also excluded by me because of the bad experiences made in Europe in the first half of the 20th century. There was no role-model left after so many politicians had been deflated. One of the jokes in East Germany went as follows: "Why are here so few crimes committed when compared to the western countries?" The answer: "Here are no criminals left—they are all in the government!" (See also page 9-29).

But why then did I want to become a chemist? First, the experience I had in the last years in school pointed in this direction. Second, many of the businesses in Brandenburg and vicinity had connections to chemistry. With great interest, I visited the printing firms making lithographs and etchings in addition to their standard printing. All these processes contained fascinating chemistry. Another local small business using easily understood chemistry was making silver mirrors. Silver nitrate solutions were reduced to a continuous, bright metallic film precipitated on glass and protected from the environment with a lacquer in the back. There even was a chemical-technical laboratory to serve the local industry. Then, one should not forget photography, which interested me and was based on chemical processes. Vati's Kodak camera (6×9 cm) with a Schneider-Kreuznach lens from the 1930s survived the war, and I had constructed a mask, so that I could take two pictures on one negative (films were scarce). In time, I also understood its chemistry and acquired the equipment to develop my own pictures. Finally, there was also the reestablished toy manufacturing in Brandenburg.[1,2] VEB "Patent," was a sponsor of our high school class ["Volkseigener Bertieb," (People-owned firm)]. On visits, the use of plastics in the manufacture of toys was discussed and formed one of my experiments for the high-school thesis (see page 3-20). We also made an excursion with the "Brandenburgische Städtebahn," a narrow gauge railway to the cities in the north-west. About 15 mi from Brandenburg, in Premnitz, there was a rayon factory.[3] Here I could see for the first time a large plant for synthetic fibers. They had not only exquisite

---

[1] Formerly "Ernst Paul Lehmann," the world-famous "Patentwerk," founded in 1881 in Brandenburg to make lithographed tin-plate action toys. After being disowned, the owners fled to Nürnberg in the FRG to reopen their business in 1951. See the beautiful book: H. Schwarz, ed., *Bewegte Zeiten* (Ernst Paul Lehmann Patentwerk, History of a Toy Factory). Tümmels, Nuremberg, 2003 (Publication of the Nuremberg Toy Museum, Volume V, 320 pages, ISBN 3-921590-90-6).

[2] Founder: Oskar Wiederholz,1905 "Lineol," famous for its hot-pressed toy armies, barnyard scenes, and zoo menageries out of sawdust, linseed oil, chalk, and animal glue. Hand cleaned and painted. Lineol was disowned 1946 and moved in 1949 to Dresden where it continued under the name "VEB Lineol-Plastik Dresden" for many years. See example toys in Figure 18 of Chapter 1.

[3] In Premnitz there was originally a nitrocellulose manufacturing for explosives. In 1919, after WW I, it was closed, but research led to a rayon process. From 1930 on, "Kunstseide" (viscose rayon) was produced. In the 1940s, nylon 6 was introduced ("Perlon," for parachutes). After WW II, much was taken to the USSR as reparation. By 1949, it was made into a VEB.

manufacturing facilities, but also a fully equipped chemical laboratory. There, I saw the first glass electrode for pH measurement (acidity determination). The technicians were surprised to find a high school student who knew what a pH was and who even wanted to know how their glass electrode works (they could not explain it fully). This was by far the best excursion we had in support of my choosing a profession. Other, more fun-excursions by the high school, went to Berlin to see the art in Schloss Charlottenburg and an opera performance (by choice, Offenbach's Orpheus in Hades).

Third, early in 1949 a former student, Horst Bonhoff, then a chemistry student in Berlin, was invited to our highschool to tell us about his experience in Berlin. He did a nice job and invited me to join one of the weekly chemistry seminars at the newly named Humboldt University, HU. I went, and loved the lecture, listened to by about 100 chemistry students and professors. Afterwards, there was a long discussion with critical questions. I even remember the subject. The lecturer proposed to feed alcohol to cows for the caloric intake and pointed out the ease to ship alcohol via pipelines. He had found that when using topinambur (helianthus tuberosus, or Jerusalem artichoke) for the (incomplete?) fermentation, it kept the cows from getting excessively drunk. This was somewhat nonsensical, but a good base for discussion. On my walk back to the train station, I saw on a ruined building with a sign directing to a chemical equipment store. Indeed, in what looked like a half-destroyed factory, there was a counter with a big catalog with many pictures of accessories for the chemical laboratory. I leafed through it for quite a while, when the salesman finally said: "You do not really have to look through the catalog. We don't have most of what is in the catalog." We got to talk, and when he realized my interests were, he said: "I don't need these catalogs anymore, you can gladly have one. By the time you enroll in the university, we may be able to help you." Happily, I lugged the big book home.

Fourth, from these experiences, I concluded not only that chemists were less dependent on the strongly curtailed heavy industry. Chemists also had interesting jobs at all levels of formal education. Some of these, such as the assistants in the laboratories, I felt I could do already. With a few semesters of additional study, I could go into a bigger laboratory and, perhaps do more independent work. With a PhD, finally, I could do research or supervise or design chemical plants. At that time in Germany, every chemistry study included chemical engineering courses and PhD chemists were the leaders of chemical industry. So, my decision was made, and Vati, having been convinced by my teachers and Herrn Eilert, was fully in support of me.

The high school final exam, the "Abitur" was to be the last hurdle to start a study of chemistry at the Humboldt University. In June 1949 the grades, labeled preliminary, were received. They were quite good, und could be used to fill out applications to enter the University in fall. To make up for my not being a member of the communist youth organization, the FDJ, and to overcome the "numerus clausus" (fixed maximum of students admissible), my chemistry teacher, Dr. Geißler (see Figure 18) wrote a special assessment of my high school thesis. A transcript is shown in Figure 24. He did an outstanding job, being involved in the LDP (see pages 3-9 and 3-22), he knew that perfect grades were not enough. He emphasized my humble beginnings, (downplaying my father being counted as bourgeois, a member of the middle class, whose children were not wanted in higher education) and stressed the favorable result the communist school reform had produced (see page 3-13). And at the end, he pointed out that I was active in society by supporting the work of the VEB. All this was submitted to the Volksbildungsamt (office for people's education) in Brandenburg. Conferences about my application were on the 25th and 27th of June, and I was told, after September 15, the answer would come directly from Berlin.

Beurteilung
der Jahresarbeit des Bernhard Wunderlich
"Lehrreiche Versuche aus dem Demonstrationsunterricht."

Wunderlich lieferte mit dieser Jahresarbeit einen lebendigen Beweis für die Notwendigkeit und Richtigkeit unserer demokratischen Schulreform. Als Sohn eines Handwerkers hatte er ursprünglich nur die Mittelschule besucht und wurde erst im Herbst 1946 bei Auflösung dieser Schulart an die Oberschule übernommen. Bis dahin hatte er sich noch nie mit Chemie beschäftigen können. Mir fielen von den ersten Wochen an neben seiner echten theoretischen Begabung sein ungewöhnlich großes experimentelles Geschick verbunden mit einer ausgezeichneten Beobachtungsgabe und Findigkeit auf. In der folgenden Zeit konnte ich mich darauf beschränken W. von verfrühten hochschulmäßigen Studien abzuhalten, ihn auf die Fruchtbarkeit einer Vertiefung in unseren Schulstoff zu überzeugen und ihm sebstständiges Arbeiten zu ermöglichen. (Einsatz im Rahmen der Volkshochschule!) Die Jahresarbeit zeigt, daß sich W. weit über das normale Maß des in der Schule üblichen hinaus vertieft und eine überdurchschnittliche Begabung bewiesen hat. Im Versuch 18, der Kunststoffsynthese für unser Patenwerk, den VEB "Patent" wird der schulische Rahmen gesprengt.

Die Arbeit sollte nicht durch eine Note bewertet werden, sondern nur durch eine Tat: Die Ermöglichung einer sofortigen Aufnahme des Fachstudiums.

Brandenburg/H. den 11.4.49

Figure 24 *Recommendation for admission to the Humboldt University based on the high school thesis (see Figure 9), including proof of extra-curricular activity and demonstrating a degree of social consciousness in addition to the above-average talent in chemistry.*

Next came the "Abitur" itself (state exam). Several written exams were required. One in German (a composition), one in mathematics (a lengthy set of problems and proofs of suppositions), and one in the major field, chosen by the student. For me, naturally, this was chemistry. In order to exclude any cheating, the exams were given in a big room reserved for drafting. There were enough large tables in this room, so that everyone had his own. This placed one far enough from the neighbors that it was impossible to read their papers, neither left, right, in front, nor back. One could be excused to use the restroom, but only one at a time.

A fool-proof examination system? Not quite. Some students had hiding places on the way to the restroom to look up handbooks, etc. One also could carry hidden mathematic equations. Two friends of mine desperately needed help in mathematics. We arranged for long, thin, black strings between our seats. After I had solved the mathematics problems, I wrote two more solutions. I was too proud of my unique solution to give it away, and it was safer this way too. The extra solutions were attached to the strings and hauled-in by the recipients. Both passed the written mathematics exam. For Fritz Bockmühl, it helped him to abandon his career to become a gardener and take over his mother's well established flower shop. He, instead, studied medicine and later was a very successful physician. After the written exams, we had to wait for the day of the oral exam.

Wednesday and Thursday, July 20 and 21, were the days of the exam. Dr. Dickhoff, the representative of the "Landesregierung" (provincial government in Potsdam) was introduced to us in our class room, and then we were told what the examination topics would be. Everyone was examined in his subject of choice (chemistry for me) and one to three other subjects. Our director, Herr Büstrin, came to me and said; "We could not find a field in which we need to examine you because of an uncertain grade, so I suggested my subject, history, in addition to chemistry. You know, that even in history one can have graphs to visualize things, so I think you will do well." I knew immediately he wanted to examine me about the French revolution which he had presented only a few months before by drawing a half-circle on the board and listing the steps from the beginning to the height and end. I had about half an hour to try to recall these steps. One at a time, we were called out of the class room to be questioned in front of several of our teachers and Dr. Dickhoff who mainly listened, but could join into the questioning at any time. All went well, and my grades were as shown in Figure 9. In chemistry, I was asked to discuss the various experiments in the thesis I had submitted, and in history, the beginning, height, and end of the French revolution.

Of the 20 students in our class, 19 passed, as shown in Figure 25. Only at that time did we know for certain that we had graduated from high school, we had reached the "Reife" (matriculation standard, see Figure 9). Looking back, this having to be able to be conversant about all topics taught over the last three years in an oral examination made us try to understand the subject matter. Today such broad knowledge is often missing. Students frequently learn (cram) an extraordinary amount of material to prepare for a specific exam, but only a short time later, they (of necessity) have forgotten it. It is too much to keep in mind (and the student knows, it will not be needed for later exams). When discussing study habits today, I think about this experience and often tell my students, anything you have forgotten later, you may as well not have learned.

Only when you are at the peak of your profession, will you need quick recall. This is helped by understanding the overall information. Only the basic facts are required in this case. Looking up details and recovery of forgotten facts, or generation of new material, can wait for later stages of a developing project.

What does one do after the Abitur? Celebrate! A get-together with parents, friends and teachers was held in the assembly room of the Rathaus in the old town, shown in Figure 26. In its garden, the group picture, Figure 25 was taken. This Rathaus is now the center of town, after the Rathaus of the new town was destroyed (see Chapter 2, Figures 1, 18, and 21). Our "Abi Ball" was held somewhat later in the "Stadthalle" (see page 3-23) for all four high schools together. At least in our family, school graduations

Figure 25 *Our Abitur class in 1949. From left to right:* ***Top row:*** *Myself, Hartmut Grätz, Peter Klomfas, Dieter Heckert, Norbert Pauligk,* ***Middle row:*** *Wolfgang Ruperti, Harald Müller, Erich Schultze, Hans Rumpf, Ernst-August Taege, Peter Menzel, Werner Thieme.* ***Bottom row:*** *Gerhard Schadebrodt, Heinrich Selzer, Rolf Strobelt, Prof. Geißler, our home-room teacher, Horst Rückert, Fritz Bockmühl, and Fritz Weickert. (Günter Ahrendts, missing, see page 3-18; and Peter Clausen, failed).*

Figure 26 *Rathaus with Roland in the old town. Photo taken in 2004.*

were not as great an affair as they are today (see Figures 30 and 31 of Chapter 11). No caps and gown, no gifts or cards of congratulation, and no assembly of relatives. The graduation was the self-understood completion of the chosen school. There were no graduations from Kindergarten, primary or secondary school that warranted special attention. Similarly, on my receiving the PhD from Northwestern University in 1957, only Tante Lieschen, having lived long enough in the US, sent a card and gift. Everyone else was proud of my achievement, but it was nothing like a special birthday or anniversary. Our "Abi Ball" was arranged by the students and united all school friends. The dance lasted into the early morning, but was far from a formal 'High School Prom,' as customary in the US.

Among the rowing partners (see pages 3-20–22) we decided to have a last, relaxing vacation before the university opened. We undertook a round trip of 150 mi. Five of us, shown in Figures 27–29, rowed a boat for four with coxswain, starting from the boat house in Brandenburg (Figure 16), past Potsdam and Berlin to the Wolziger See, near the hometown of Fritz Weickert. His father, conveniently, owned a butcher shop and could help us with our provisions for the trip home. First, we had to go through the lock at the end of the Silo Kanal (Chapter 1, Figure 6, NE corner) and be raised to the "Oberhavel." This was done together with several of the large barges which were again on their way to Berlin. The blockade was lifted three months earlier. Occasionally, we could hitch a ride as shown in Figure 27. Especially in bad weather, this was very much appreciated. The countryside looked as beautiful

Figure 27 *Bumming a ride between Brandenburg and Potsdam on our rowing vacation in 1949.*

as ever and had occasionally something to offer along the line of vegetables, potatoes, and fruit. The pleasure boating was not yet redeveloped, and only few of the outdoor cafes on the shores of the Havel were open for business, but it was easy to get permission to camp. Ration-cards were still needed to shop for food, but excesses on the fields were free for the taking. The next interesting stop was near Babelsberg where the East German film industrie (DEFA) was located . We had entered by then the "Teltow Kanal," a waterway bypassing Berlin. There we found the completed movie set, ready for "Die Blauen Schwerter" (the blue swords, the "Meißener Porzellan" trademark, a movie about Johann Friedrich Böttger, the inventor of the first European white porcelain in 1709). It contained the fake "Jungfern Bastei," a medieval castle. We 'knights' could not resist to take our portrait as shown in Figure 28.

Figure 28 *The crew in front of a movie set in Babelsberg by Berlin. Left to right: myself, Horst Rückert, Heinrich Selzer, Werner Thieme, and Fritz Bockmühl (Was there an apple tree handy? 1949).*

When we got more into Berlin's suburbs, many bridges were still destroyed and bypassed by temporary structures. This meant we had to pick our way carefully through the half-immersed steel girders. Fortunately none closed our path completely.

After three days we reached the "Wolziger See," our goal. Fritz Weickert, was expecting us and joined our camp at the shore of the lake, as shown in Figure 29. After 8 days we were back in Brandenburg and it was slowly getting time to watch the mail for a letter from Berlin, letting me know that I was accepted.

This period of 1945–1949 was the most important one for my future. During this time I changed my goal from becoming a farmer, held for almost 10 years, to possibly studying the fields of mechanical engineering, mathematics, or physics, and settled finally, after detailed analysis of the opportunities described on pages 3-29–31, on becoming a chemist. If one reads the recollections carefully, one finds that already on my first 'consulting' for the "VEB Patent" I was introduced into the world of polymers (plastics) which became so important to my later career and started already then with research in my laboratory, supported by an effort of studying the literature. This subject I did not leave in the subsequent 60 years. The details about the scientific progress do not fit into this narrative. It is displayed in my professional publications. In Appendix A, the data base of the ≈600 scientific publications is given, along with a list of my coworkers. In Appendix B, a review of the major advances that arose out of these publications is written up, and Appendix C lists all lectures given outside the home universities. The material developed to display my insight into "Thermal Analysis of Polymeric Materials" is reproduced in the publications #500 and 560 of Appendix A.

Figure 29 *We reached our goal, camp on the shore of the Wolziger See. In the center is our host for a few days, Friz Weickert.*

It was also clear to me that later in 1949, I would have *to leave my hometown Brandenburg*, the two over 1,000-year-old cities which were combined in 1715, as depicted with its seal in Figure 30 (see also page 2-1). The proudly displayed nine towers did not protect the city from the enemy on the ground or from the sky. Even the stern knight, "Roland," had to shift from the Neustadt (Chapter 2, Figure 1) to the Altstadt (Figure 26, above). In the seal this move from right to the left is not made.

It was not clear that I had to change not only my place of residence, but later, in 1953, had to officially also change my hometown. This was after we became political refugees in the FRG. As such, our passport could not list a hometown in the GDR, because on future entering of the territory of the GDR, I would be an immediate suspect of crimes against their government. Not the least one of which would have been "Republikflucht" (fleeing the country), a crime punishable with a prison term of at least five years. *My second hometown, thus, was Werbach an der Tauber*, a small, 750-year-old village of, by now, about 3,500 population, shown in Figure 31. Our family was relocated there after I had already left to study at the Goethe University of Frankfurt (see Chapter 5).

Figure 30 *Seal of the joint cities of Brandenburg. Old city, left, new city, right.*

Figure 31 *Our second 'official' hometown, located in the FRG, a small village in Baden-Württemberg.*

To complete my list of hometowns, I have to mention *Hastings, NE, as a third hometown*. Not an official hometown, but one by choice, selected in 1954. Figure 32 shows the street through the center of the city. This was the city Tante Lieschen went to when she left Germany in the 1920s. We tried before to go there directly from the GDR. The US Consulate even granted us visa in 1951, but only if we all had FRG passports (the GDR did not have an immigration quota). All efforts to get such a passport failed. Our emigration had to wait until after we had a new hometown in Werbach. Ultimately, our whole family ended up there as shown in Chapter 7 with

Figure 32 *Second Street, Hastings NE, photo 1955.*

Figure 21 and Chapter 11 with Figure 1. Vati and Mutti lived in Hastings for the rest of their lives (until 1985 and 1999, respectively). Frank found his start in Vati's business there, then transferred to Lincoln, NE, and finally moved to South Carolina. Joachim and Manfred grew up in Hastings where they arrived at ages 10, and 8. Joachim taught at Hastings College and still lives there with his wife Laura, while Manfred spent most of his professional life with his family in Omaha, NE. He was successful in the insurance business. My short stay in Hastings is recalled in Chapter 6.

What did we learn about the Nazi regime is the last question to be tackled in the present Chapter. It is clear from Chapters 1 and 2 that I grew up with this dictatorship. Without personal freedom, I had to learn to hide my opinions from others. The majority of people I had contact with, were resigned to passive resistance. The government set up by the "Nationalsozialistische Deutsche Arbeiterpartei" (NSDAP, usually shortened to Nazi Party) became rapidly and increasingly more oppressive. Of the people who had embraced the known goals of the Nazi party, the worst were the opportunists, who found no qualms in later changing direction. There are examples of Nazi sympathizers in East Germany who later strongly supported the SED in a similar manner. Anyone, who after 1933 openly opposed the government, could count on being incarcerated (see also the pages 1-14 and 1-31). In addition, there were an increasing hate-propaganda and persecution of Jews, and, finally, all races not Aryan.[1] Everyone was aware of this. The ultimate fate of people once imprisoned, however, was known to only few. At the time, even the free world with their outstanding intelligence network and a free press seemed not to have accepted the magnitude of the wide-ranging atrocities. In fact, refugees from Nazi terror could count on little help and protection from the rest of the world.

By now one slowly finds scholarly, well-documented treatises about the various conflicts. All had to conclude that *the 20th century was the bloodiest, cruelest, and most destructive known*.[2]

---

[1] The Oxford Dictionary definition: "**>n. 3** (in Nazi ideology) a person of Caucasian race not of Jewish descent. **>adj.** relating to or denoting the Aryan people or their language. ORIGIN from Sanskrit *Arya* 'noble' + -an (suffix forming nouns and adjectives meaning 'from')."

[2] See: "Historical Atlas of the 20th Century" (http://users.erols.com/mwhite28/20centry.htm). Specific for this paragraph, the "30 Worst Atrocities of the 20th Century" are summarized in (http://users.erols.com/mwhite28/atrox.htm), and the detailed military and civilian casualties are listed and sources compared in (http://users.erols.com/mwhite28/warstat1.htm). To quote from the summary: The blood letting "divides neatly into two parts—Eastern and Western. The Eastern Hemoclysm began with the overthrow of the Manchu dynasty in China in 1911 leading to 38 years of Civil War and a Japanese invasion. In 1949, the bloodbath of the interregnum gave way to a greater bloodbath as the Communists consolidated power under Mao (who died in 1976). When seen as a continuum, this phase of Chinese history was a 65-year nightmare which took some 75 million lives. The first sparks of the Western Hemoclysm were the Balkan Wars (1912–13) which quickly ignited the First World War. This brought down four of the most powerful monarchies in Europe, leading to a power vacuum which was eventually filled by the Nazis in Germany and the Communists in Russia, who came into conflict during the Second World War. The death of Stalin in 1953 finally extinguished the Western Hemoclysm after the loss of some 80 million lives." The accounting of the casualties in the East and West amounts to 155 million! On a percentage basis of lost population, the European countries suffering most were the USSR and Poland with ≈20% each.

The military weapons of mass destruction, WMDs, ranged from the ever more effective weapons for the aggression and defense in the front lines of WW I, to the technological wonders of WW II. They ranged from experiments with poison gases to aircraft, submarines, rockets, etc. New for the 20th century was the development of the terror attacks on the civilian population which before were largely left untouched, except for hunger, disease, and incidental losses as a consequence of wars between the armed forces. In the 20th century, systematic methods were developed to terrorize the enemy by direct military action against civilians, as described on pages 2-6–8.

The means of civilian mass destruction developed in the 20th century were similarly continually increasing in effectiveness. It started with the Second Anglo-Boer War 1899–1902 in South Africa when the term concentration camp was first used by the British for the detainment camps of 154,000 white Boers and black Africans (Zulus).[1] The horrific concentration camps were finally epitomized by the Nazi death camps. Other camps of mass destruction were the Gulag ("ГУДАГ," labor camps and colonies, administered by the Russian secret police, NKWD) and the prisoner of war and internment camps under Russian, Japanese, and German control. To the civilian mass destruction count also Mao's untold victims to hunger during the Cultural Revolution, and the ethnic cleansing in many parts of the world. The last use of poison gas, by Saddam Hussain in 1987–88, was against 40 Kurdish villages, which killed and maimed ten thousands of civilians.

Unfortunately, the efforts taken to guarantee a better 21st century seem not sufficiently effective. The fear of terror attacks on civilian population centers has increased again. Well-known examples are the terror attacks on September 11, 2001of four, highjacked civilian airplanes on the Trade Center in New York City and the Pentagon with 2,996 civilian victims, the continuing suicide bombings in the Middle East and South Asia which have likely claimed more than 100,000 victims. Adding the threat of proliferation of WMDs of nuclear, biological and chemical nature (NBC) and the genocide in a number of nations by more primitive means, lets one believe that the 21st century may see a continuation of mass casualties and has a great potential of even bigger disasters.

In the spirit of this book, *let me recall how I became aware and incensed about these happenings*. The experience about the bombings has been described already in Chapter 2 (pages 2-2–3, 7–8, 17–19, 22–23, 26–27, and 30). They naturally were of immediate concern to me. The first details about any executions of unarmed prisoners of war, surprisingly enough, came from the Nazi Press in 1943. I clearly remember, as a 12-year old, the shock I felt when viewing the details of the mass graves of over 4,000 Polish officers massacred in the woods of Katyn. The report was at that time verified by a commission of 12 European forensic experts with their staffs. They determined that the deaths occurred in early 1940 when the area was under Soviet control. After the war, these experts (except for a Bulgarian and a Czechoslovakian) reaffirmed their 1943 finding of Soviet guilt. Until 1990, the USSR continued to deny their responsibility. Only then, was it finally acknowledged that their secret police had committed the massacres and the subsequent cover-ups. Next were the large-scale executions of the people involved in the assassination attempt on Hitler in 1944. These were well publicized in the German press, and I even knew that some of these, and many earlier political prisoners, were held, in addition to the convicted criminals, in the "Zuchthaus" in Brandenburg (state penitentiary). This facility was at that time one of the most modern prisons in Europe, built from 1927 to 1935.

---

[1] "A report after the war concluded that 27,927 Boers (of whom 24,074 [50% of the total Boer child population] were children under 16) had died of starvation, disease, and exposure in the concentration camps." Quote from: http://en.wikipedia.org/wiki/Second_Boer_War.

The atrocities in the concentration camps, focused toward Jews and East Europeans, were later communicated to me by newspapers and movies. *Only as the Nürnberg trials proceeded*, became the full magnitude clear. It was such, that I had difficulties accepting it. How could anyone get involved in such crimes? Even in Brandenburg they were committed while I was growing up.[1] *On our visit to Poland in 1972 we saw the concentration camp* in Auschwitz-Birkenau (30 mi from Kraków, see page 8-60). It brought the horrors into focus by pictures, actual records, and relics It also documented that in the end, the perpetrators made a major effort to erase the evidence. This means to me, they knew the crimes they were involved in. The long-time camp commander, Rudolf Höß, testified at the Nürnberg Trials that 1.1 million people had died there during his stay, a number now displayed on the memorial plaque in the State Museum, a UNESCO World Heritage Site since 1979 (what a heritage!). Many were killed in gas chambers, others died by systematic starvation, forced labor, the lack of disease control, individual executions, and cruel medical experiments. After testifying in Nürnberg, Höß, was retried in Poland and hanged at the site of his crimes.

Let me finish this dark paragraph with a quote by G. B. Shaw from Act IV of his 'Caesar and Cleopatra' (Caesar answers Cleopatra:) "*... And so, to the end of history, murder shall breed murder, always in the name of right and honor and peace, until the gods are tired of blood and create a race that can understand. ...*"—Let us hope, that the new race does not have to await the intervention of the gods. The known gods have in this connection been remarkable inactive throughout the 20th century. The greatest chance missed for self-help of mankind, I think, was to limit the Nürnberg trial to what it did so well, to uncover the Nazi horrors and punish wherever laws could be found. But should it not also have, for example, included charges against the leaders responsible for the bombings of exclusively civilian targets? For example, Field Marshal Hugo Sperle, commander of the London Blitz, was accused only (and acquitted) of using Russian POWs for the building of military facilities. If there were no international laws at that time to charge specific persons of the use of the WMDs listed on page 38 on civilians, should the deeds not at least have been publicized and a larger effort been made to amend the Geneva Convention more comprehensively than was actually done? And finally, should not a case have been made to persecuted all future occurrences immediately by a permanent court? This is the only proven way how cities fight crimes today. Should not the world do likewise? The UN should have the mandate to discover, fight, and punish such deeds and not leave it to the politics of separate nations.

---

[1] From the book "Das Jahr 1945 in der Stadt Brandenburg" (referenced in the Footnote of page 2-26), the following statistics is quoted in translation:

- "From the more than 12,000 forced laborers (in Brandenburg) from all over Europe, 1,111 died (many of these in air raids because of insufficient shelters).
- Probably as many as 1,807 political prisoners or victims of the fascist justice were executed in the "Zuchthaus" (state penitentiary).
- Probably 9,772 inmates from mental institutions were killed with nerve gas within the Euthanasie Program in the "Alte Zuchthaus am Nicolaiplatz" (Old Penitentiary at the Nicolaiplatz)."

After 1945, the facilities in Brandenburg were used by the Soviet Military Command to house some members of the Vlasov Army [the ≈50,000 Russians that served in the German army under General Vlasov. Vlasov himself was executed 1946 and all members of his army were sent to the Gulag and were never heard of again, even those captured by American forces]. German civilians, perceived enemies of the USSR were interrogated and housed in the "Alte Zuchthaus" and other city buildings. The GDR housed political prisoners in Brandenburg's "Zuchthaus" until 1975.

CHAPTER-04

Humboldt University in Berlin, 1949–1953

My application to the Humboldt University (HU) was submitted on June 24, 1949 to the "Volksbildungsamt" (office for people's education) in Brandenburg. This was the premier university of the newly founded GDR (see pages 3-16 and 29). The university was founded in 1810 by the liberal, Prussian educational reformer and linguist Wilhelm von Humboldt.[1] Over the years it has become the model of many western universities. Figure 1 shows its main building "Unter den Linden #6," as it appeared in 1999. The university was given its present name on February 8, 1949. In 1948, when the political freedom was curtailed in the university, a rival university, the Free University, the FU, was founded in West Berlin. Today, both universities continue in the city with approximately 40,000 students each. A third university in Berlin is the Technical University, TU. It was founded in 1879 and was also located in West Berlin. Today it has about 30,000 students, and its list of alumni and professors includes eight Nobel Prize winners.

My application was supposed to be acted upon by September 15. By the beginning of September, seemingly all other students had already received their admission or rejection. I had heard nothing and slowly became concerned. With Vati's support, I traveled in the morning of Friday, September 9 to Berlin to check on my application. The train went from Brandenburg to Potsdam through the GDR. Potsdam was the station to change to the "S-Bahn" ("Schnellbahn", elevated, electric train). The S-Bahn crosses Berlin in several directions and also contains a ring around the city. From Potsdam, the S-Bahn went through West Berlin, at that time fully accessible for East German travelers, to the Friedrichstraße. From there I could walk to "Unter den Linden." Overall, this trip from Brandenburg, which I later had to make hundreds of times, took about ... (continued on page 4-3)

Figure 1 *The main building "Unter den Linden 6" after the reconstruction, showing the statue of Alexander von Humboldt. The statue of Wilhelm von Humboldt is on the left of the entrance (not shown).*

[1] Berlin University, renamed 1828 to "Friedrich-Wilhelms-Universität" (after Friedrich Wilhelm III, King of Prussia, who reigned from 1797–1840 and donated the first building, see Figure 1). This footnote continues on page 4-2 with historic pictures and a list of Nobel Laureates.

Continuation of the Footnote 1 from page 4-1:
Historic Pictures. **Top:** The main building. **Center:** The "Chemisches Institut, Hessische Straße 1–2." Left in 1945, right one of the restored floors. **Bottom:** The director of the chemistry institute.

Above, a historic, print from 1883 of the Humboldt University. Right, burnt-out shell after WW II, in 1945. See Figure 1 for a recent view.

Professor Erich Thilo, HU 1949

List of the Nobel Prize winners in Chemistry, Physical Chemistry, and Physics:
(year of NP, and years at HU)

Chemistry: Henricus van't Hoff (1901, 1896–1911); Emil Fischer (1902, 1892–1919); Adolf von Bayer (1905, 1866–72); Richard Willstätter (1915, 1912–16); Otto Diels (1950, 1904–1916).

Physical Chemistry: Friz Haber (1918, 1911–33); Walther Nernst (1920, 1905–32); James Franck (1925, 1911–20); Peter Debye (1936, 1935–39); Otto Hahn (1944, 1907–34).

Physics: Wilhelm Wien (1911, 1892–96); Max von Laue (1914, 1906–1943); Max Planck (1918, 1889–1926); Albert Einstein (1921, 1915–29); Gustav Hertz (1925, 1917–25); Erwin Schrödinger (1933, 1910–33); Werner Heisenberg (1943–45); Erwin Walter Bothe (1954, 1925–30); Max Born (1954, 1915–19).

1½ h plus waiting time in Potsdam. First, I went to the central administration of the university "Unter den Linden #6" (see Figure 1). The university was still not fully restored. Except for one wing, only one or two stories were usable, the rest was still burned out and in the process of reconstruction. I found the Information Office and could have guessed the answer. Without looking at any files, I was told: "The admission process is not completed yet, it is useless to enquire."

Next, I inspected the university, such as the lecture rooms, which for the mathematics lectures had flexible blackboards which could, with a crank, be continuously pulled up to the ceiling, then it turned over and moved down in the back to appear at the bottom for erasure when the loop was filled with writing. In my later lectures, given by Professor Grell, it took him about 40 min to have the first equations appear again. Large bulletin boards where fixed on the walls of the corridors. There the professors of the various departments announced their lectures. Next to the HU was the huge building of the "Staatsbibliothek," also still pockmarked with many bullet holes.

By then, the lunch hour was over and I went to the Ministry of Education. Dr. Geißler, my chemistry teacher, had given me a letter of introduction to Frau Möller-Krumbholz in the "Referat für Oberschulen" in the DVV, the"Deutsche Volksbildungsverwaltung" (German Administration for People's Education). He thought she would be of help, and he was right. Unfortunately, she was on a business trip, and I was asked by her secretary to come back on Monday the 13th. In this office I was treated much more politely, and I felt this was the place from where I could expect help.

Getting home at night, I gave a full report to Vati. His comment was: "Write down a summary of all the important information that you were given. I am sure you will have to make a few more trips and you must be able to remember all what was said and promised before." I still have the original notes which were then typed by Frau Ladwig, Vati's secretary. Below is a translation of these with a few additional comments. Alone, I probably would not have been successful. When the critical discussion was coming up, Vati came with me to Berlin and he also asked all his friends to possibly find connections to the TU or FU, which would have meant to leave the GDR. There were 18 more trips to be made before I was admitted.

A two-month effort was necessary to change the rejection to an admission to the HU. But first, I did not know that there was a rejection to come. Here are the steps, each connected with a new train ride out of Brandenburg via Potsdam through West Berlin to East Berlin:

Trip 02 (9/13) Berlin A. HU, IO (Information Office), As on 9/9: "The admission process is not completed yet, it is useless to enquire."

B. DVV, Frau Möller-Krumbholz: She was given the information from Dr. Geißler and a copy of the "Reifezeugnis" (see Chapter 3, Figures 24 and 9). She promised: "In case something does not work out with the admission, please immediately come to me, we will fix it ..." An attempt to contact Professor Thilo, the director of the "Chemische Institut" in the Hessische Straße, was not successful, he was on vacation. After a Möller-Krumbholz intervention by phone, I was invited back to the HU, IO.

C. HU, IO: "All acceptance letters will be sent out by 9/15. The rejections may take about two weeks longer. It is impossible for us to search now."

After waiting for one more agonizing week, it seemed certain that no acceptance was forthcoming. On the following Monday morning, I was back at the train station in Brandenburg (see Figure 6 of Chapter 2) to travel to Berlin.

Trip 03 (9/20) Berlin A. HU, IO: "Your requests cannot be answered, you must wait for the notification." But by now I knew that there was a special, separate "Zulassungsbüro" (Office of Admission), ZB.

B. HU, ZB: "We do not have your papers among the admissions."

C. DVV, Frau Möller-Krumbholz: "On a business trip, return uncertain."

Trip 04 (9/23) Berlin A. DVV, Frau Möller-Krumbholz: "On a business trip, return uncertain."

Trip 05 (9/27) Berlin A. DVV, Frau Möller-Krumbholz: "On a business trip, return uncertain." On second questioning: "Return uncertain, probably not before the middle of October." This did not sound too encouraging, but it also did not seem to be the truth, so I went to the next higher office, the "Schulabteilung."

B. DVV, Schulabteilung: "Frau Möller-Krumbholz is on vacation, return uncertain, her address is unknown." This meant I had to get the address from somewhere else. I found what today would be called the office of human resources: "Her address, in West Berlin, Schlachtensee, Lagard Straße 13."

C: Lagard Straße 13, Frau Möller-Krumbholz is not at home, expected late. I left a note with a description of the day's results and promise to return 9/28.

Trip 06 (9/28) Berlin A: Lagard Straße 13, Frau Möller-Krumbholz: "Why do you come to me? You have to see Professor Thilo. I cannot give you an introduction. But, try also to gain admission at other universities like Greifswals, Jena, etc."

B. Chemical Institute, Prof. Thilo: "Professor Thilo is for a few days at the University of Leipzig."

To leave no opportunity untried, I discussed with Vati that Greifswald would be the best alternative choice, so two days later I had a train ticket to Greifswald, 175 mi north of Brandenburg, and in case I would get admitted there, it would be close to Tante Grete in Stralsund.

Trip 07 (9/30–31) Greifswald A. Student Information, Bahnhofstraße: "You need to go to the "Zulassungsbüro," ZB (admissions office) in the Domstraße, Herr Jürgens."

B. ZB, Herr Jürgens, after my description of the situation: "Are you a member of the FDJ?" "No." "What is the occupation of you father?" "Roofer." "Does he have his own business?" "Yes." "How many employees does he have?" "7." "Then it is impossible that you will be admitted in Greifswald. You must first be active in the FDJ for one year." "Who else can I petition to be admitted?" "Tomorrow a representative of the "Landesverwaltung" visits (administration of the "Land," province)."

C. Next day, I sat from 9:00 a.m. to 1:30 p.m. in front of the ZB. Then I was admitted to the representative of the "Landesverwaltung." He said after inspection of my file: "In case you had applied first to Greifswald, an admission might have been possible. Now it is impossible. Work for one year as a volunteer in a laboratory and engage yourself in the FDJ."

Conclusion from this visit: Waste of time! The provincial universities are even more restrictive than the one in Berlin. Back to work on the HU admission!

Trip 08 (10/3) Berlin A. HU, ZB: "Information cannot be given before Oct. 10."

B. HU, Chemical Institute, Prof. Thilo: "I cannot help you." After looking through his note book: "I do not have your name among the applicants. I first need exact information. Go to Professor Steinitz, the Dean of Students."

C. HU, Office of the Dean of Students: "Professor Steinitz is not in Berlin."
D. HU, Chemical Institute, Prof. Thilo, tries to get a hold of Professor Noak. This was only successful in the afternoon. "Herr Professor Noak did not find your name in his files. You must go to the DVV to search for your files."
E. DVV: Closed. On finding someone who was willing to help. Call to Frau Piltz of the Student division: "Your files have been processed and forwarded to the university."
B. HU, ZB: closed. After searching for anyone: "Your files have been processed, result unknown."
G. HU, Chemical Institute, Prof. Thilo: "Please go to Herrn Dr. Hall of DVV." By that time it was time to catch the train back to Brandenburg.

In Brandenburg *I received the rejection notice*! Dated 9/19, mailed 9/29, but without a stamp, requiring a surcharge, so that it arrived only on 10/3. Reason for rejection: There are only a limited number of admissions, try again in 1950 or 1951 after working in a laboratory, *or find a different profession.* The address of a vocational councillor was given. No, I did not give up. At that time it was not so easy for the HU to openly refuse admission for a qualified student for political reasons. One year later, it would have been, by then, I would have had no chance!

Trip 09 (10/4) Berlin A. DVV, Dr. Hall: "I have no jurisdiction, I only work with medical students. Go to Dr. Karl."
B. DVV, Dr. Karl: "I have no jurisdiction, please go to Frau Piltz."
C. DVV, Frau Piltz: "I have no time, go to room 216."
D. DVV, Room 216: "Your admission is processed by Frau Piltz. Your file is in the office of Frau Orbanz."
E. DVV, Frau Orbanz was not present. Fräulein Ritter gave the following information: "Your file has been processed. To achieve a late "Immatriculation" (admission), Professor Thilo must work with Frau Piltz." This was the first glimmer of hope. At least there was a possibility of late admission!
F. HU, Chemical Institute, Prof. Thilo: "Please come back in the afternoon for an oral exam."
G. HU, Afternoon Chemical Institute, Prof. Thilo: An exciting one-hour discussion of what is of interest in chemistry. Professor Thilo was obviously impressed and interested in my joining his laboratory (see Figure 2). He was very friendly and I felt welcome. Later, when I was in the laboratory, he was the "Herr Professor" and the student etiquette required me not to approach him until having progressed to work on my thesis (which I never did at the HU). When four years later, I was in difficulties again after moving to West Berlin (see the Preface), he again showed an extraordinary degree of courage by providing my exam papers in violation of GDR laws (see page 4-34).
H. DVV, I delivered the exam result, a copy is given as Figure 2, below.

This was a big step forward, but, by the end of the week, I was back on the (rail)road again:

Trip 10 (10/7) Berlin A. DVV, Frau Piltz: "Go to Fräulein Ritter."
B. DVV, Fräulein Ritter: "Your file is being processed. You got a high recommendation from Professor Thilo. You will be informed."

Trip 11 (10/11) Berlin. It brought the same result as trip 10.

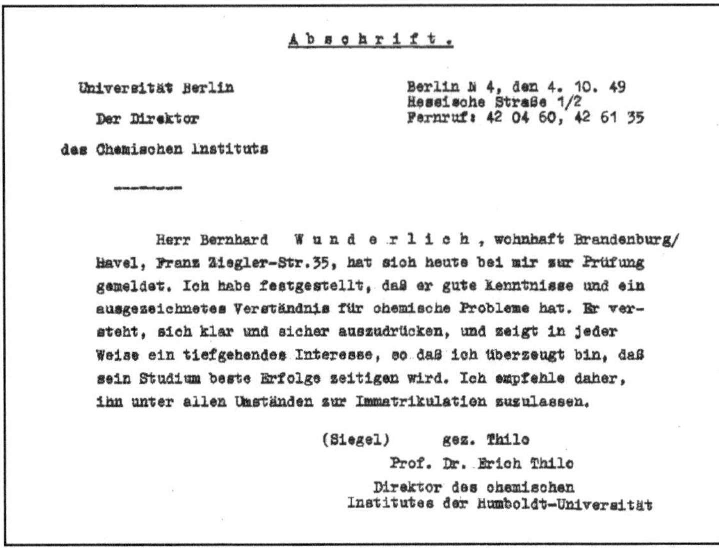

Abschrift.

Universität Berlin

Der Direktor

des Chemischen Instituts

Berlin N 4, den 4. 10. 49
Hessische Straße 1/2
Fernruf: 42 04 60, 42 61 35

Herr Bernhard Wunderlich, wohnhaft Brandenburg/Havel, Franz Ziegler-Str.35, hat sich heute bei mir zur Prüfung gemeldet. Ich habe festgestellt, daß er gute Kenntnisse und ein ausgezeichnetes Verständnis für chemische Probleme hat. Er versteht, sich klar und sicher auszudrücken, und zeigt in jeder Weise ein tiefgehendes Interesse, so daß ich überzeugt bin, daß sein Studium beste Erfolge zeitigen wird. Ich empfehle daher, ihn unter allen Umständen zur Immatrikulation zuzulassen.

(Siegel) gez. Thilo
Prof. Dr. Erich Thilo
Direktor des chemischen
Institutes der Humboldt-Universität

Figure 2 *Copy of the result of the examination by Professor Thilo, suggesting admission to study at the university under all circumstances!*

By now, it was clear to me that not all was won yet. There was probably a continuing and serious discussion at the HU between the new political forces who did not want to admit me, and the old traditionalists who still believed there was some political freedom left and I should be admitted. Since I did not know the outcome, additional avenues had to be explored. One was to check the TU in West Berlin, a not so easy step, since this meant an open break with the GDR and could cause repercussions to the family left in Brandenburg. Still, Vati had friends in Brandenburg with connections to the TU and the FU. They suggested that in case of a politically-based denial, an admission might be possible. The connection was kept open, but not yet pursued since it meant to become a political refugee.

Another possibility was to go for a year to the just opened university in Potsdam ("Brandenburg Landeshochschule"). So I went after the negative result in Berlin on that day to the "Ministerium für Volksbildung der Landesregierung" in Potsdam with the following result: "We are closed because of the admission of the new students. The semester starts today." But, I knew at least one person in the Building, Dr. Dickhoff, my examiner at the Abitur. I found his office. His answer to my story: "I remember your exam in Brandenburg. All I can do, is to send you to the Hochschulabteilung." He called and I was admitted, but the result was a final: "It is impossible to accept you now, we are completely overbooked." End of trip. When I got home, Vati decided we should wait a week and see if we get a result by then. Next, we *both* went to Berlin.

Trip 12 (10/18) Berlin, together with Vati. It led to the following discussions:

A. DVV, Frau Orbanz: "The file is still being processed. Wait for notification." After telephone inquiry to the "Studentenreferat" which was closed to the public: "The assessment from Professor Thilo is not needed any further, it can be picked up on the 21st... ."

B. HU, Chemical Institute, Prof. Thilo to Vati: "I have given your son a proper assessment of his abilities (see Figure 2). The rest is up to you. Ask at the information division of the Dean of Students."

C. HU, Dean of Students office, Herr Forkert (assistant to the dean) to Vati: "Your son was placed on the middle group of students. If he wants to be admitted next year to the HU, he must bring better support material. I cannot understand how the DVV could keep your son hoping for admission. All places are occupied."

Trip 13 (10/21) Berlin A. DVV, Frau Orbanz, returned the assessment of Dr. Thilo (Figure 2) and said: "The file is still being processed." On reading the comment of Herrn Forkert from 10/18 to her: "I have no idea what Herr Forkert could mean with 'better support material.' You will have to ask him yourself."

B. HU, Dean of Students office: "Herr Forkert is in a conference, please wait." On my questioning the secretary what Herr Forkert might have meant with better support material: "Crucial is not only the academic achievement, but also the organization you belong to (FDJ, SED?). Also, children of laborers and peasants are preferred for university study."

C: HU, Admissions Office: "We have no influence and are not informed about your case. Only Herr Forkert can give information."

D. HU, Dean of Students office, Herr Forkert: "It is inconceivable how Professor Thilo could have written this assessment (Figure 2), why the DVV has not given you an immediate rejection and still holds out hope for you to be admitted!" After another check of the grades (good), the social circumstances (not so good), as well as the political activities (nonexistent), he continued somewhat gentler, but guarded: "I see no reason why next year, an admission could not be possible. A binding promise, however, I can not give." I was released with the comment that Professor Steinitz, the Dean of Students, could probably be reached in the next week. Was that another glint of hope since now all cards beyond academic competence were on the table? This was still Berlin, where just a few hundred yards away the democratic West Berlin was located. I could go any time to West Berlin and by presenting my case, cause problems for the reputation of the HU.

Trip 14 (10/28) Berlin A. HU, Dean of Students office: "Professor Steinitz is not present." On my return in the afternoon, I was told to return on Tuesday (11/1), then he would have his office hour.

Trip 15 (11/1) Berlin A. DVV, Frau Orbanz: "Your files are now processed for a third time. Wait until you are notified." I left a newspaper article about my connection with the "VEB Patent"(pages 3-30 and 35), to strengthen my application.

B. DVV, Frau Möller-Krumbholz, yes, she was back in her office. After reading the assessment of Professor Thilo: "Under no circumstances must you give up now. Go to the Minister for Education, Herrn Wandel, and see if he can intervene for you."

C. DVV, Office of Minister Wandel, Frau Herzog: "The minister is in Warsaw and will return probably on Thursday (11/3)." After my explanation of the request I would have of the minister, she checked with Frau Orbanz and said to me: "Wait, as Frau Orbanz said, but you might check with the Department Head for University and Science, Dr. Hall."

D. DVV, Dr. Hall was not in the house.

E. HU, Dean of Students office, Professor Steinitz, after examination of my file: "For final admission, the judgement of the dean of the faculty would have been responsible. Before, all applications were divided into two categories, 1. Children of laborers and peasants (with property under 12.5 acre), working intelligence [*not thinking*—my comment], and independent craftsmen (under three employees). 2. Children of others [like I]. From the first group a fixed percentage was accepted. This does not mean that no one of the second group was accepted." After a telephone call to Herrn Forkert: "I can give you a minuscule chance. It seems likely that one study place in chemistry may be vacated and you may be considered for it. It is, however, inconceivable for me how the DVV could give you any hope of admission. No later than Monday (11/7) you should be notified."

Going home, I knew, I had some hope to have softened the sides who did not want to admit me, the university administration and the DVV. Professor Thilo's intervention, representing the academic side, might become successful (and my tenacious probing may have helped). Certainly I could not leave things to take their course now. Two days later I was on the train again:

Trip 16 (11/3) Berlin A. DVV, Office of Minister Wandel, Frau Herzog: "The minister is very overworked, I will see if I can clarify this matter." After an insistent inquiry in my presence to Frau Orbanz about what happened in my case: "Professor Thilo also has already been making an inquiry. Nothing seems to be decided yet. Professor Steinitz has not yet gotten in touch with Frau Orbanz. If there is a vacancy, there are no objections from the side of the DVV!"

B. DVV (school division), Frau Munin: "I cannot understand how one can be so bureaucratic!" Again, a call to Frau Orbanz who by now was quite upset. Frau Munin: "Still being processed, go to Dr. Hall or Herrn Krahn."

C. DVV Office of Dr. Hall: "He cannot talk to me."

D. DVV Office of Herrn Krahn: After checking my files he went personally to Frau Orbanz for a ½ h conference with the following result: "Frau Orbanz called Dean Steinitz, in my presence. The position at the university did not become vacant, but despite this, they will both try to admit you!"

Trip 17 (11/7) Berlin A. DVV, Frau Orbanz: "Come back tomorrow, nothing is decided yet. All is very difficult. In case you are not admitted, you will get a guaranteed admission for the next year."

B. HU, Dean of Students office: "The dean cannot be reached."

Trip 18 (11/8) Berlin A. HU, Dean of Students office, Professor Steinitz: "I think your application has been decided." After checking with Herrn Forkert: "*You are admitted!!*"

Trip 19 (11/9) Berlin HU, *Immatrikulation as stud. rer. nat. in Chemistry*! ("studuosus rerum naturalis," student of natural sciences).

This hurdle was overcome. Thanks go to my supporters: Professor Geißler (see Chapter 3, Figures 12 and 25), Dr. Geißler (see Chapter 3, Figure 18), Professor Thilo (see page 4-2) and the upper officials of the DVV, among them Frau Möller-Krumbholz who was still a good friend for the next years, the people in the office of Minister Wandel, and even the Dean of Students, Professor Steinitz, but most of all, to Vati, who directed all, and came to Berlin when it was most critical.

Now, *I quickly had to become a "Berliner*." The emblem of the city was well known to me and is shown in Figure 3. The semester had already begun. First, a room to live in had to be found. The university had practically no student dormitories. Fritz Weickert from my high-school class could help. In the house next to the one he lived in, a lady needed to lease a room since housing was still rationed. She became my landlady: "Fräulein Schmidt,"Am Friedrichshain 10 (see page 9-51). Figure 4 is a picture of her from this time period. Her parents had died, and she had to share her two-room apartment with me. Vati came with me to sign the lease after I had settled all the formalities. He needed to see who would look after me. He was somewhat taken aback when he noted that she was only a few years older than I, but seemingly trusted the arrangement. The rent was about 40 DM-Ost.

Figure 3 *City flag of Berlin, photographed in 2004.*

Food was still rationed, but as university students we got ration card B, the second best. Eating could be done in the "Mensa" (student cafeteria) for 0.60 DM-Ost using the ration card, and sometimes without it, but at a higher price. The quality was, with a few exceptions, mediocre. I quickly found out that taking the train home every Friday and coming back on Sunday night gave me two days of good meals at home, and Mutti had usually enough prepared for me to have things to eat for Monday and Tuesday. This left only Wednesday and Thursday to my own 'cooking.'

Figure 4 *My landlady, Fräulein Schmidt, Am Friedrichshain 10.*

The public transportation from the Friedrichshain to the university was convenient. Bus 9 left from the "Königstor," about 900' from the house, and reached the university (Figure 1) in about 20 minutes (via the historical "Alexanderplatz," the "Rote Rathaus," the "Museumsinsel," the "Invaliden Dom," the "Opera," and the Zeughaus,"). I could continue up the "Friedrichstraße" to get close to the "Hessische Straße" from where with ½ mile walk I would reach the "Chemische Institut." In all, it took about 45 min. Figure 5 is a picture of the backyard of the institute, showing the large windows of one of the student laboratories and the bridge leading to the private quarters of Professor Thilo.

Next, I put together **the list of courses I wanted to try in the first semester**. I received a "Studienbuch" and started listing the lectures of interest. Unfortunately, the chemical institute did not yet have enough space to let us start with the laboratory practice. The institute is shown in Figure 5 as it appeared in 2004 during a rehabilitation of the building. In 1949, the top floor (off the picture) was still not rebuilt from the destruction during of the war (see the page 4-2).

At that time, the study of chemistry in Germany was centered around laboratory instruction. Any delay in starting of the laboratory meant that the eight semesters for the PhD had to be extended. Naturally, I wanted to use the initial extra time to learn as much as possible. There was no strict study plan. Only the laboratory and a small number of auxiliary courses in physics, mathematics, and mineralogy were required. Figure 6 is a list of the courses I selected.

To be registered in a lecture, one appeared at the first meeting and had the professor sign you in ("Antestat"). At the end, one was signed out ("Abtestat"). This was the record you had of your study. No exam was given, unless there was separate, graded practice connected to the lecture. For this, a separate certificate was issued, the "Übungsschein." On the right of Figure 6, tuition was assessed: 182 DM-Ost, independent of the number of hours. The 31 h listed for each week was considered a very heavy load.

Figure 5 *The "Chemische Institute" Hessische Straße 1–2. Photographed during refitting in 2004.*

<u>How did one learn anything without exams during the semester</u>? The assessment of progress was measured only by comprehensive exams and the completion of the "Übungsscheine"! There were up to five oral exams. They were held separately for each candidate by the professors in the fields of study. The first exam was the "Zwischenprüfung", to be taken after four semesters (or later). Then there were the two exams connected with the "Diplom," and the last two, were the qualifying and final exam for the "PhD." The final Diplom and PhD exams covered mainly the progress one had made in the chosen area of research.

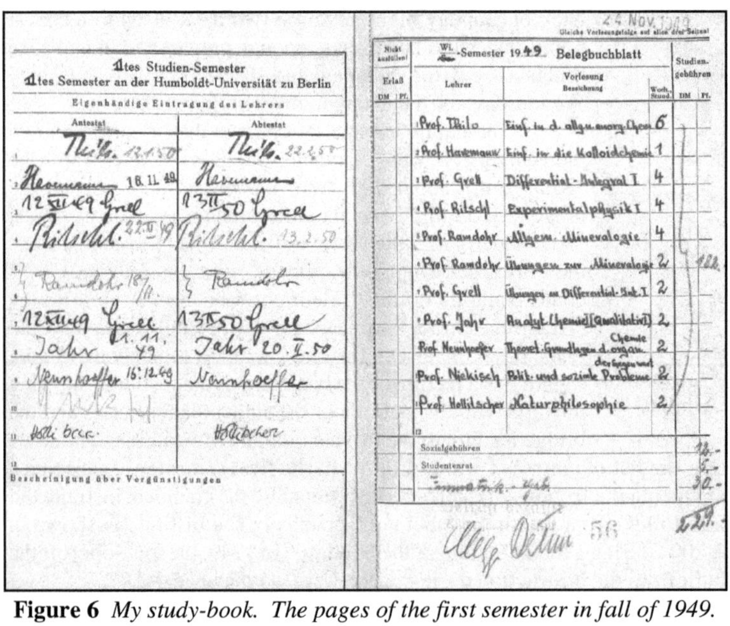

1tes Studien-Semester
1tes Semester an der Humboldt-Universität zu Berlin
Eigenhändige Eintragung des Lehrers

Antestat	Abtestat
Thilo 12.1.50	Thilo 22.2.50
Hevemann 18.11.49	Hevemann
12.XII.49 Grell	13.II.50 Grell
Ritschl 22.XI.49	Ritschl 13.2.50
Ramdohr 18/11.	Ramdohr
12.XII.49 Grell	13.II.50 Grell
Jahr 1.11.49	Jahr 20.II.50
Neunhoeffer 16.12.49	Neunhoeffer
Hollitscher	Hollitscher

Bescheinigung über Vergünstigungen

24 Nov. 1949
Gleiche Vorlesungsfolge auf allen drei Seiten!

Wi.-Semester 1949 Belegbuchblatt

Lehrer	Vorlesung Bezeichnung	Woch. Stund.	Studiengebühren DM
Prof. Thilo	Einf. in d. allg. u. anorg. Chem.	6	
Prof. Havemann	Einf. in die Kolloidchemie	1	
Prof. Grell	Differential- Integral I	4	
Prof. Ritschl	Experimentalphysik I	4	
Prof. Ramdohr	Allgem. Mineralogie	4	
Prof. Ramdohr	Übungen zur Mineralogie	2	122.
Prof. Grell	Übungen zu Differential- Int. I	2	
Prof. Jahr	Analyt. Chemie [Qualitativ]	2	
Prof. Neunhoeffer	Theoret. Grundlagen d. organ. Chemie	2	
Prof. Niekisch	Polit. und soziale Probleme der Gegenwart	2	
Prof. Hollitscher	Naturphilosophie	2	
Sozialgebühren			12.-
Studentenrat			5.-
Immatrik.-Geb.			30.-
			229.-

Figure 6 *My study-book. The pages of the first semester in fall of 1949.*

The "Diplom" exam was introduced only in the 1920s. At that time the study period to the PhD was slowly getting longer than the eight semesters originally scheduled, and industrial firms had difficulties to staff their laboratories. The "Diplom" is, perhaps, similar to the masters degree in the US, but always has a research project, A "Diplomchemiker" was from then on an accepted, academically-trained chemist. In our case, not getting into the laboratory in the first two semesters, meant at least 10 semesters to the PhD. This, however, did not account for the very difficult laboratory work which usually also needed extra time. When I left the HU, I had taken only the "Zwischenprüfung," and was working toward the "Vordiplom," the qualification for the Diplom.

In the week after my admission ("Immatrikulation"), I was sitting in classes, practically from morning to evening, as can be seen from Figure 6. *Most important was the introduction into inorganic chemistry* by Professor Thilo. He was a great lecturer and demonstrator. The big lecture room with about 400 seats was banked highly, so that everyone could see the experiments. To get a seat, one had to be there early. There were not only the chemists who took the course. It was customary at that time that any student could attend the lectures. There were some, who might have had an interest in chemistry, and many, who had heard that the lectures were outstanding.

Professor Thilo's lectures excelled in being up to date and exciting. The experiments were well prepared and executed by two senior assistants in white coats. The demonstrations were informative and often spectacular. On the first day of lecture he had two blackboards filled with titles of books. One set he recommended, the other, he said, we should avoid reading. My book by K. A. Hofmann (see page 3-11) was listed, and was highly recommended in a newer edition.

Of the other lectures, the '*Theoretical Foundation of Organic Chemistry*' by Professor Neunhoeffer had caught my eye (third from the bottom in Figure 6). Obviously, since I had never heard lectures on organic chemistry, this was what I needed—I thought. When I entered the room, it was almost empty. After a while Professor Neunhoeffer appeared, looked at his class and said: "I need at least three in the class room to be permitted to lecture." We were only two students. So he counted, one, two, and, pointing to himself, three: "That is enough!" The course could be given. In fact, there was a sequel in the next semester which I also took. This was the first set of lectures for me which showed the connection of chemistry to quantum mechanics, a subject I had a short introduction to in high school and I owned a small book about it [Walter Weizel, "Electronen, Atome, Moleküle," Volk und Wissen Verlag, Berlin, 1948/49 (since 1991 Cornelsen Verlag)]. We were never more than five students during the two semesters. When I went for my "Antestat," Professor Neunhoeffer looked at the brand new "Studienbuch" and said 11th semester (see Figure 6), I should have seen you before. My answer was, no, I am new at the HU. Indeed, the other student was much older and was grinning about my answer since it could be mistaken for 'I am an upper semester transfer student.' Unnecessary to say, I learned a lot and, without an exam, I did well.

The lectures by Professor Ramdohr, a quite well known mineralogist, were given in a lecture and seminar room of the famous "Naturkundemuseum" (Invalidenstraße 42). Figure 7 shows on top the beautifully restored facade of the more than 125-year-old building. Inside, one could see the biggest dinosaur skeleton found by then, and many other fossils. In particular, I liked that of the first bird (see right bottom). Also, there is an extensive mineral collection which was used in connection with the lecture. The pictures of Figure 7 were taken when I was going through the museum with Stefan, my grandson, and all our family. He was as excited about the minerals as I was when on my way to the lecture room in 1949. Perhaps for me, this was the first step to my later interest in the solid state. A second step occurred in 1953, with a lecture on scientific photography. In this lecture,

Dr. K. Meyer, a well-known visiting lecturer from industry introduced me to the study of the effect of the light on the defect structure in silver nitrate crystals. Two more steps were necessary to point me to my final field of interest, best described as the 'solid state of macromolecules.' These consisted of finding a book on modern solid state physics in the library of the refugee camp in Weinsberg (see page 5-2), and finally, the interest in the solid state was tied to the field of polymers with my thesis work at Northwestern (see the pages 6-19–21).

The lectures on mineralogy in the "Naturkundemuseum" had one other advantage. Next door was the student cafeteria built for the "Landwirtschaftliche Hochschule," (agricultural university). It was commonly called the "Ziegen Mensa" (goat cafeteria). Eating was somewhat better there than in the other student eateries. One of the meals, I even liked. They prepared very tasty and fresh fried herring ("grüne Heringe"). Perhaps, I would have ended up in this agricultural university had I continued my first ambition to become a farmer. The possibility to study agriculture, however, had not occurred to me in 1946 (see page 3-9).

Figure 7 *The "Naturkundemuseum" in Berlin. Top: The facade of the building. Bottom: Stefan with an interesting iron meteorite, and right, the first bird as a fossil in slate. Pictures taken on a 2004 return visit.*

The other courses taken in the first term were chosen to reach my goal <u>*to know and understand all there is in chemistry*</u>. To understand this ambition, I must admit that after having passed the "Abitur," just about four month before, I had a secure feeling to know much about science and engineering, and all that I did not know, I was convinced, with effort, I could learn. This feeling, however, faded quickly. As my studies advanced, I noted that the huge amount of knowledge available was beyond anybodies capabilities. Learning details about an increasing number of topics had to be abandoned in order not to lose the chosen focus, and increasing numbers of subjects were also too difficult. Note furthermore, that course 10 in Figure 6 did not have an "Abtestat," it shows the beginning of my passive resistance to taking politically motivated courses. I only went to this lecture once!

The next step in my study was the swearing-in ceremony of the new students on November 25, 1949 ("Verpflichtung"). This was a festive ceremony in the Great Hall in the "Unter den Linden" building (Figure 1). It was my second Friday as a student in Berlin and ended with the trip 'home,' to Brandenburg via Potsdam. To prove our status we were issued an <u>id</u>entification card (ID) which in front, bore the following inscription in Russian, English, French, and German: "Bearer of

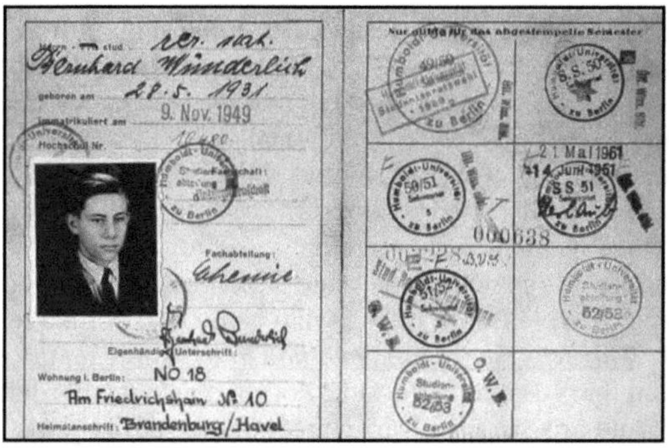

Figure 8 *Photocopy of the student ID, issued to give safe passage to and from the HU. (The original was returned to the HU after leaving the GDR in 1953).*

this is student at the University of Berlin. It is requested to let him pass unmolested and to support him during his studies." A copy of the back side is shown as Figure 8. It helped considerably in the multiple border crossing, of which not all were fully legal (for the last one, see the Preface).

The first semester brought two events of interest in addition to the studies. The first was *a Christmas Party, organized by Fritz Weickert* for all graduates of the "Goethe Schule" (boys and girls) who ended up studying in Berlin or Potsdam. Since I had the largest room and my room even had an old upright piano, it was the chosen location for us. My landlady (see Figure 4) was invited also, and had no objections. A former teacher (German and Philosophy), Frau Scherzer, well known to students of both schools, was also invited. We added two tables and 10 chairs and the party could begin with about 14 participants. It began with a formal part of reading of poems and literature, music, and some Christmas food (for that, ration card coupons were collected from everyone). As the evening progressed, an enormous amount of information was exchanged about life at the universities. Since we were sure that we were all of the same persuasion (i.e., no political activists were among us), the discussions went on and on.

About midnight, my landlady left (too much discussion for her). All others who were expected at home got to the last S-Bahn or train. This left about six of us who continued the party. Early the next morning, we went for a small breakfast to the room of Hartmut Grätz (see Chapter 3, Figure 25) who lived in the same house as I. This was followed by a visit to West Berlin ("Bahnhof Zoo") to eat something more substantial, sightseeing, and shopping. Ultimately, at about noon, we brought Frau Scherzer to the train for Brandenburg.

What to do with a half day? We went to Fritz Weickert's room in the house next to mine and continued our party. In the late afternoon something was thrown into his mail box. To our surprise, it was a chocolate bar dropped by one of our friends from West Berlin for one of the girls, who was still with us. He did not know her address and thought Fritz would deliver the gift. From his note, it seemed that he liked her. Unfortunately, he had left already by the time we saw the connection, otherwise he could have delivered the gift himself and joined the residual party. Naturally, we all shared in the chocolate. By midnight, again, we finally were all tired enough to bring the remaining guests to the train station and sleep into the Sunday. Characteristic of this party was that there was no alcohol served throughout, and naturally, smoking was not allowed. Nobody missed it. The main result of this party was to cement this group of Brandenburgers so that more frequently we could be found on Friday evenings in the same train compartment from Potsdam to Brandenburg.

The other event was the "*Studentenratswahl*" *(election of the student council)* on February 9, 1950 (see the stamp on the student ID, Figure 8). This was the first (and only) seemingly democratic election I could partake in at the HU (see also pages 3-22–23). We were to elect 20 of 40 candidates, most of them were members of the FDJ, but not all. A total of 7,300 students could vote. The debate of how we could vent our dissatisfaction was fierce and had strong support from the West Berlin student organizations. Much of the information was broadcast to us by the RIAS (the Radio in the American Sektor of Berlin, see page 3-22). It turned out that the lead candidate could only combine 20% of the votes, but still, he was declared 'elected.' Only 2,700 votes were cast for the candidates (37%). In protest, all others either left invalid ballots (38%) or stayed away from the election (23%). The next elections were cancelled, and no other was held. The FDJ fully took over the council and in the future, determined its membership internally.

Privately, my life had settled into a routine. Monday to Friday study, then going home for the weekend. Friday afternoon, going shopping to West Berlin to buy what could be afforded in food and other items not available in the GDR. *This was contraband and had to be smuggled into Brandenburg*. Most difficult was the way from the S-Bahn platform in Potsdam past the border guards to the train station. Similar problems could arise in the reverse direction on getting goods into West Berlin. I slowly became an expert in this new 'occupation.'

A special problem was to move my laboratory (see the pages 3-18–19) to Berlin. Two large wooden crates were built, all was packed carefully in excelsior. Vati gave me one of his workers along, who helped to move the crates. At the luggage car of the train, the conductor, after checking our ticket, immediately said, without permission ("Warenbegleitschein"), you will not be able to get this in Potsdam into the S-Bahn for Berlin. But he accepted the crates. In Potsdam, we moved the first crate to a spot on the platform where we knew the luggage compartment of the S-Bahn would stop. When we turned to get the second crate, I saw a border guard approaching. It was helpful to have had some acting lessons. My face lit up and I greeted the Vopo ("Volkspolizist," people's policeman) with my most friendly voice: "Could you please help us and watch the crate? We must get a second one." He gladly obliged, and when we arrived with the second one and had thanked him profusely, he left. He did not think it was possible that we had anything illegal in those crates. On the "Alexanderplatz," we got a luggage transporter to bring the boxes to my room. I could now do limited experiments even in my room. The boxes could serve as a lab table and storage shelves. The room was big enough, and there was a balcony for smelly things.

On the weekly trips home, I always carried my oversize briefcase for the books and clothes, and, particularly at Christmas time, for the forbidden goods. The border control was approached without any apparent apprehension. The usual question on passing was: 'Any "Westware" (western goods)?' Answered with noncommittal, 'no, I am a student going home with dirty laundry and to get supplies for next week.' Usually nothing was checked. If, he asked to look into the briefcase which bulged with dirty clothes, the conversation continued casually. As he approached in his search the hidden oranges, butter, chocolate, or, may be even small bottle of champagne (common items which were not available in the GDR), I would say 'there are a few ... for my baby brother, mother, etc.' This 'honesty' was sufficient to be forgiven. The border guards had a reasonable degree of leeway. Things got more stringent as the ratio of DM-West to DM-Ost became larger.

On these trips back and forth to Brandenburg *I also got to know Heidel better* (see Chapter 11, Figures 13–16, 18, and 19), my future wife. She was a student of law and also had only been able to get a late admission to the HU. We seemed to have a good number of common interests, and when not, we were willing to learn. When I asked her to come along skating one weekend, it turned out, she did not know how, and had no skates. Fritz Bockmühl (see Chapter 3, Figure 25), riding in the same compartment, helped out. His mother had an extra pair of skates, so it was settled—except for the weather. When Heidel called to check if the skating was still on, it had turned too warm overnight, so our date was postponed. On such short notice, I did not have any idea what else we could do. A couple of weeks later, traveling the route Berlin/Brandenburg, we found a good movie to go to on Saturday evening (February 11). One must remember that the trains at that time were not fully repaired from the war. They were not heated and always crowded, so one hade to move closely together. In addition, there was no lighting and at the Berlin latitude it gets dark rather early during the winter. From this movie visit on, we spent as much time together as possible. Our paths had crossed earlier on several occasions, such as at a dance-school ball, and during common school functions, and naturally, at the past Christmas party, but initially there seemed to have been no common liking. Figures 9 and 10 are two pictures of this early time of getting to know each other better in 1950.

Figure 9 *The two of us outside Brandenburg.*

In Figure 9 we managed a picture with a self-timer on Vati's old camera. The flooding is typical for the spring and fall around Brandenburg. Figure 10 shows Heidel on a visit to my apartment in Berlin in August 1950 on the occasion of a trip with my brother Joachim to the "Berliner Zoo." It shows also the state of the once beautiful park, the "Friedrichshain." By then, it was called "Mont Klamott," a huge, mountain of rubble, transported from destroyed houses by the narrow-gauge construction steam trains, mentioned on page 3-10. Overall, not many trees had survived WW II and the later shortage of heating material. The regrowing took about 50 years (compare to Figure 49B of Chapter 9 and pages 9-51–52).

Figure 10 *Heidel Am Friedrichshain, on my balcony.*

At this time we had our schedules well coordinated, so that we could spend time together in Brandenburg as well as in Berlin, and the trips back and forth were made together. In Brandenburg, we could go to the movies or take walks to the outskirts of the city, which together we discovered anew. When the weather was nice, Mutti might prepare a package with cake which we could enjoy in an outlying restaurant. With these provisions we did not have to use ration coupons for mediocre commercial cake. Even today, there is rarely a problem in restaurants in Germany to just sit down and order only a soft drink, tea, coffee (beer or wine).

When it got warmer, Vati had bought a narrow-bodied sailboat (unfortunately without sail, which could not be obtained easily at that time). So, the mast and keel of the boat were removed, and it was rigged as a rowing boat for one. The other had to handle the rudder. Frequently we took my brothers Joachim and Manfred along, who were 6 and 4 at that time. It was not always clear who was the chaperon for whom. Occasionally, we would meet Frank with his girlfriend canoeing on the water. This led to a friendly competition among the four brothers who had the better boat. The memorable final comment of Joachim at the end of one of those bragging contests was 'but we have water in our boat and you don't!' This statement could not be topped.

In Berlin there was a large range of theaters, music performances, cinemas, and not to forget the large "Waldbühne," an amphitheater used as an open-air cinema and variety theater. It was built for the 1936 Olympics for an audience of 20,000. Its programming was designed to show movies which were not available to the East Berliners, and to present politically-motivated variety shows. On such days we arranged our return trip to Berlin with a stop at the "Waldbühne." Here, as in most other West Berlin theaters, the tickets were priced 1:1. "Ostmark" and "Westmark" were equally accepted. The only restriction for us was that we had to dress during the cold months such that nothing needed to be left at the check room. There, the fee was in DM-West. In DM-Ost, this might far exceed the ticket price. Wednesdays in Berlin, Heidel came often to my apartment to do the cooking. I quickly found out that Heidel liked cooking and learned fast to become an expert. In her tiny room in the apartment of Frau Oesterreicher in the Schröderstraße 13, the cooking privileges were limited to warming things.

Figure 11 *On May 9, 1950, in front of the "Invaliden Dom" in Berlin. In the center Gabi, to the right Fritz Weickert with one of his girl friends.*

Parties like for Christmas 1949, mentioned above, were not repeated, but on May 9, Heidel's friend Gabi (God), living in Erfurt, came on business for a visit to Berlin and could spend some time with us and our friends. A picture from a sightseeing excursion around the Humboldt University is shown in Figure 11.

Later in the month, my 19th birthday was celebrated in Berlin with Fritz Weickert and girlfriend, Hartmut Grätz, and the two of us. My birthday fell on Whit Sunday ("Pfingsten"), a holiday for us, followed, as usual, by a work-free Monday. This was also the weekend of the first "Deutschlandtreffen der FDJ," an effort at a meeting of representatives of all the youth of Germany, to show off the political 'advances' in the GDR.

During our summer vacation in 1951 an even bigger meeting was held, the 3rd "Weltfestspiele," organized by the World Federation of Democratic Youth (WFDY). The WFDY was founded in 1945. These games are still held today, now in a four-year cycle, and usually in communist-leaning countries (Caracas, Venezuela, 2005; Minsk, Belorussia, 2009). In August 1951 in East Berlin, they had delegations from ≈100 different nations. Everything possible was undertaken to make it a success. Heidel and I had spent the vacation in Brandenburg and Berlin and this was a chance to meet with some of Heidel's cousins who attended the meeting. The photograph of Figure 12 was taken on this occasion. Naturally, we did not only show our guests East Berlin, but also West Berlin, something of great interest (and some danger) to our visitors from the GDR.

The second semester went until the end of July 1950. Figure 13 shows the excerpt from the study-book for this summer term 1950. The theory of organic chemistry with Professor

Figure 12 *Group photo in Berlin-Gesundbrunnen with cousins of Heidel. From the left Henning with future wife Irmgard, and Herbert (see page 11-3). Next to us in the center, Heidel's friend Gabi and Herbert's friend, whose name I have forgotten.*

Neunhoeffer continued for a second term. The introductory lecture in organic chemistry ("Einführung") was much simpler, but forced the rote learning of reactions and structures needed for the start in a new subject. The mathematics lectures were coupled with exercises ("Übungen"), straight-forward because of the preparation I had from high school. Finally, a major effort was made in the second term to get the physics background solidified, and doing the required laboratory ("Praktikum"). The total credit hours had increased somewhat, to 35.

But the sensation of the term should not be forgotten. It was the lecture series entitled: 'About Physical Chemistry' (#1 in Figure 13). The eight best-known physical chemists, largely from the FRG, had made it their goal to continue the tradition of the Humboldt University to be one of the premier universities in physical chemistry (see page 4-2). Professor Eucken started the series. He was the best-known thermodynamicist in Europe. My special interest in this topic can be traced to his lectures. Thermodynamics became the basis of much of my future research (see Appendix B). All subsequent lectures were of equally high standards.

At the end, this special lecture series was written up as a book and made available to us. Also, textbooks and advanced treatises printed in the FRG could sometimes be bought in the GDR. When a case was made by our professors for us needing a book from the FRG, it sometimes was imported and became available at a low DM-Ost price. A number of these books are still on my book shelf. To my surprise, I found one of those titles in 2008 in an antique mall in Bavaria. I inspected it, and it was inscribed by my fellow student, Werner Schmidt! (See the pages 4-34–35.)

The reason for <u>*making an effort not to lose the books one first learned from*</u> needs some further explanation. Having once understood a subject, it

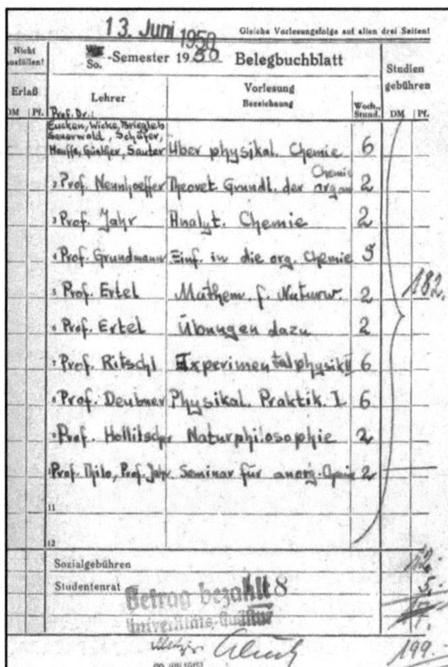

13. Juni 1950 — Gleiche Vorlesungsfolge auf allen drei Seiten!

So.-Semester 1950 Belegbuchblatt

Nicht ausfüllen! Erlaß DM Pf.	Lehrer	Vorlesung Bezeichnung	Woch. Stund.	Studiengebühren DM Pf.
	1 Prof. Dr.: Eucken, Wicke, Brieglеb, Sauerwald, Schäfer, Hauffe, Günther, Sauter	Über physikal. Chemie	6	
	2 Prof. Neunhoeffer	Theoret. Grundl. der organ. Chemie	2	
	3 Prof. Jahr	Analyt. Chemie	2	
	4 Prof. Grundmann	Einf. in die org. Chemie	5	
	5 Prof. Ertel	Mathem. f. Naturw.	2	182.–
	6 Prof. Ertel	Übungen dazu	2	
	7 Prof. Ritschl	Experimentalphysik II	6	
	8 Prof. Deubner	Physikal. Praktik. I	6	
	9 Prof. Hollitscher	Naturphilosophie	2	
	10 Prof. Thilo, Prof. Jahr	Seminar für anorg. Chemie	2	
	11			
	12			
	Sozialgebühren			12.–
	Studentenrat			5.–
				199.–

Betrag bezahlt 8
Universitäts-Quästur
29. JUNI 1950

Figure 13 *"Studienbuch," second semester.*

is easy to refresh one's memory using the familiar resources. Using a new book, one again has to make all connections between the facts. Perhaps, one interesting example is my set of two small, bound, hand written booklets which I started in high school. It now contains hundreds of mathematical formulas and geometrical constructions. This collection contains everything I ever needed for trigonometry, analytical and spherical geometry, differential calculus, series, statistics, error analysis, and algebra. It has been tested countless times, so that I can trust it to give me quickly and precisely the needed information.

It is regretful, that today textbooks are often only loaned to students. Every time the student returns the books (or sells a book as a used book), he or she loses a part of the prior learning. This poor practice is supported by frequent testing of only well-advertised portions of recently taught material without future comprehensive exams. I found it most annoying when in the 1970s the most common questions from students became: 'What will be on the next exam?' The body of knowledge is dynamic and must be constantly refreshed and enlarged. Testing should best be done infrequently. The exams should then always be comprehensive and expand as the learning progresses. The study by the student for such exams is best be done with all original resources on hand.

In the middle of July 1950, we were surprised that one day, the famous equestrian statue of *Frederick the Great ("der alte Fritz") had disappeared* from the middle of "Unter den Linden," not far from the university. Frederick the Great is the most admired Prussian king. At that time, the decision was made by the SED to downplay the Prussian heritage in the GDR. Friedrich Ebert, City Mayor of East Berlin, had ordered the removal of the statue to Frederick's palace "Sanssouci" in Potsdam, near Berlin. The surreptitious removal, however, became a public spectacle. The transport broke down on the Alexanderplatz and hindered the traffic for several days before it could be sent on its way. A picture of the statue after its reappearance from "Sanssouci" was taken in 2004. It is shown as Figure 14. By 1981, the "alte Fritz" was rehabilitated in the GDR. Prussian history was appreciated again. As one would have expected, in Brandenburg the large statue of Frederick II was also removed at that same time. The city administration, however, took the order to remove Prussian heritage more seriously and destroyed the equestrian statue, so that it could not be resurrected later.

Figure 14 *Statue of Frederick II after its return to Unter den Linden.*

The summer vacations in August and September 1950 were spent largely in Brandenburg. Heidel, who had done volunteer work in the summer 1949 in the court house, shown in the background of Figure 1 of the Preface, to 'strengthen' her application to the HU, now spent one month working at the "Ernst Thälmann Werft" where fishing trawlers were built for reparation (export?) to Russia. Another internship like this, at the city council, was done by Heidel later, all to prove her strong ties to the working class and to display the proper socialistic attitude. In this respect, I was lazy in the summer of 1950. I only worked on the better understanding of all the new material learned during the first two semesters. This gave us enough time together on weekends and allowed occasional return trips to our apartments in Berlin.

An important date in our family was *October 10, 1950, the silver wedding anniversary of Vati and Mutti*. The third semester had just started, when going back to Berlin on the weekend before the anniversary, Mutti asked me if I could come home on Tuesday with Heidel for the celebration and to get to know Heidel. Frank with his girl friend Dagmar were also invited for the first time. In addition, Lenchen and Wilhelm Heise, a couple Vati and Mutti knew from the 1920s, was invited, as well as Tante Grete and her girlfriend Inge, who was known to me from the visit in Stralsund (1940, see page 2-3). We all could get together on short notice. The year 1950 was still not the time for big parties, but Tante Lieschen had sent a CARE package with a large, canned ham and other food stuffs from the US, and Vati had found a good wine for dinner. For this time, it was a gourmet meal as it was rare to be had. The party was spent with much story telling about the 1920s, and was a pleasant introduction for the two girl friends, who later became Vati and Mutti's daughters-in-law. At around midnight, Frank and I brought our girlfriends home. Both lived on the "Görden," a good distance away, as seen in Figure 6 of Chapter 1 (point 13, Haydnstraße). For Heidel and me, it meant early next morning to get back to Berlin for a two-day week.

Now, I should explain why this first invitation of Heidel and Dagmar to the anniversary party was such a special occasion. It probably took a big effort on Mutti's part to convince Vati to agree to the invitations. Frank, after he returned home towards the end of 1945 (see pages 2-24–25 and 3-8) had worked with Vati and completed in June 1947 his "Gesellenprüfung" (bachelors exam). Since he had graduated from middle school, he could now enter the "Landeshochbauschule Brandenburg" for the Sommer semester 1948 to study engineering and architecture. As long as both of us lived at home, *Vati maintained strict rules* for our independent social life. He always wanted to know where we went, and there was, depending on our age, a sliding curfew.

There was a problem with Frank staying out beyond the curfew. He did not come home very much later than midnight, which for him was the curfew, but being 15–30 min beyond the limit was frequent. It was enough to upset Vati. I really could not understand his problem. Although I did not go out as often as Frank, and my curfew was probably about 11:00 p.m., I made it a point to be home 15 min earlier. Of the few times I went to parties where I liked to stay longer, I remember two. One was a birthday party with friends. By 10:00 p.m. I was on the phone with Vati: "Vati the party is rather nice, could I stay a little longer?" His answer: "Who is there?" "Fritz Bockmühl, Fritz Weickert, Peter Klomfas and our girlfriends." Vati knew all of us, so he asked: "How long do you want to stay?" "Until 3:00 a.m.—perhaps??" "All right, but do not stay any longer!" He was very happy, when I was home by 2:00 a.m., the time I wanted to be home in the first place.

The other occasion was our party after the successful theater performance, described on pages 3-26–27. It was at the house of Eberhard Witte and we (our mothers) chipped in with some food, and Mutti even got two bottles of "Alkolat," a fortified wine of 35 proof, the latest invention in East Germany in 1948. For this special occasion, I was allowed to stay as long as it lasted (6:00 a.m.), but Mutti and Vati knew all the 'actors,' and it was just a few houses around the corner.

Initially we were encouraged to bring our girlfriends home, and Vati was happy to get to know them. After a while, however, he found out that, in his opinion, Frank changed his girlfriends too often, and he did not like it when he was talking to friends or customers about a girl friend which by then was a 'has-been.' From that time on, he set up a simple rule: 'I do not want to get to know your girlfriends anymore.' This automatically applied also to me. While Frank still tried to arrange it that he 'accidentally' had movie theater seats close to Vati and Mutti's, I heeded his wishes. Thus, we were surprised when being invited to the anniversary party *with* girlfriends.

A further rule which caused problems was: 'No engagement or marrying before you earn your own money.' After being in the US, these rules applied also to Joachim and Manfred, and they had another one: 'You have to pay for your own car.' I will come back to some of these rules from time to time. In principle, now, I find them more acceptable than when they were made, particularly since exceptions were possible. Coming home only on weekends and having had Frank showing me how not to deal with these rules helped much in my getting along with Vati splendidly (most of the time). One must remember, that Vati had ran away from home at this age, and he met Mutti when she was ≈15 years old and they got married four years later.

Following Vati's rules, I also never had to ask for money. All basic needs were provided for, but no more. As seen in Figures 6 and 13, the tuition was low. Besides these expenses, *my monthly budget was 150 DM-Ost*. Adequate to live on, considering the added help during the weekends. Any additional expenses for books were paid without questioning.

From the fifth semester on, I had free tuition and managed to get a monthly scholarship ("Stipendium") of 170 DM-Ost. When I brought Vati the good news and said I did not need his monthly help anymore, he surprised me. He said: "I do not want to have the communists pay my son. I will still give you your 150 DM-Ost each month. What you do with the money from the university is your business." I thought it over and decided: 'I will not support communism, and so I took their money.' It was used for special purchases. Suddenly, I was 'rich!'

How did my scholarship come about? If you were the son or daughter of a laborer or peasant, you always got a monthly base stipend of about 150–200 DM-Ost. From the winter term 1951/52, however, one could also get an additional "Leistungsstipendium" (stipend for good performance). There were no exams for chemists in the first semesters. The professors made an effort to keep the old customs, i.e., my "Abitur" counted for my performance (Chapter 3, Figure 9). That someone like I could get admitted to the university was not considered when the rules were established. It was thought all students follow the new study plans, be members of the FDJ, and had taken courses in 1. Fundamentals of "Marxismus und Leninismus," 2. Russian, 3. German (mainly Marxist literature), and 4. attend regularly the meetings of the "Gesellschaft für Sport and Technik" (a political organization to acquire military prowess). I did not partake in any of these four. As long as I did not get a stipend, I got away with it. In fact, after the ordeal of getting admitted, *I was at a university which gave me good-old-fashion academic freedom*. All the special FDJ meetings Heidel and my friends told me about, I was not invited to attend, and I followed the knowledge gained from the "Jungvolk" (see page 2-9): Don't let anybody know I am here.

The 3rd to the 8th semester were characterized by inorganic laboratory work. It meant to spend full time at a permanently assigned laboratory space. The laboratory in the Hessische Straße 1–2 was open from 8:00 a.m. to 7:00 p.m. For its location and degree of destruction and rebuilding, see page 4-2 and Figure 5. Any spare minute was spent in the laboratory to 'cook' one's preparations and make qualitative, and quantitative analyses. The remaining lectures, time for lunch, any shopping, etc., meant leaving the lab desk. Often it was possible to have the neighbor watch extractions or dissolutions and do multiple tasks. The arrangement with Heidel was that she would see me on Wednesday afternoons in the laboratory and get my key to cook the evening meal for both of us, and be ready when I came home 'after work' to the Friedrichshain. Late in the evening, often after the last bus had left, I walked Heidel back to her room in the Schröderstraße 13, an approximately 2½ mi round trip. By the time I left in 1953, I had finished all inorganic laboratory requirements and passed the "Zwischenprüfung" described on pages 4-10–11 with the top grade.

The outstanding first exam after the qualitative analysis laboratory immediately promoted me from one of the 'nobodies' to one of the 'somebodies.' I soon should advance to "Famulus" (assistant) to the laboratory instructor who was a PhD chemist or a candidate working on his PhD thesis. But this promotion was only considered. The FDJ group leader hoped the position would go to one of his members. Earlier one day, one of my friends (Dieter Toursel) joined me after lunch and said, he would appreciate it, if he could accompany me for the next few weeks when going to and from lunch. *He had been asked by the FDJ group leader to convince me to join*. He also said, if at all possible, you must *not* do it. You will never again have such an undisturbed freedom if you join the FDJ. We became even better friends and discussed interesting chemistry during those walks. Finally, he reported that I was a hopeless case, I could never be convinced!!

In1965, when flying from London to Aberdeen, Scotland, I met Dieter again on the plane. Both of us were going to the 'First International Congress on Thermal Analysis' (ICTA, see also page 8-23). He was sent to gather information for the company he worked for in the GDR, and I was professor of chemistry at RPI in Troy, NY, on my way to give a lecture (see Appendic C, page C-6). Again, he was in difficulties. He told me on the plane that he was worried. He had lost his party-observer who had to accompany him. They were shopping in London the day before, but the observer must have missed the plane. This would give repercussions at home. Also, in case he showed up, I should pretend not to know Dieter anymore. It was forbidden to make new contacts with western scientists. I got a quick update of life in the GDR and gave a brief summary of my experiences. It was clear, I would never have survived the GDR. (The FDJ was right, I did not fit).

To *summarize the remaining courses at the HU*, in the 3rd semester (winter 1950/51) I still had to finish my physics laboratory (6 hours per week, see item 8, Figure 13). The only other lecture course with exercises was on the 'Structure of Matter' by Professor Rompe of the Physics Department. The fourth semester (summer 1951) was largely taken up with an illness of mine, to be commented on next, and the 5th and 7th were only spent in the laboratory. The 6th semester (Sommer 1952) saw two more 4-hour lecture courses by Professor Thilo, one an introduction, but different from the first semester, the other, an advanced course. Both were highly interesting. A third course was part two of the introductory lectures on physical chemistry.

Due to the illness during the 4th semester, I had missed the first semester of this course. They were given by the well-known Professor Havemann whom I heard before in his lecture on colloid chemistry (listed in Figure 6). He was a politically very active scientist. Since 1950 he held a seat in the "Volkskammer" (people's chamber of the legislature of the GDR), and finally he was a recipient of the National Prize of the GDR (in 1959).[1] This heavy political activity left little time

[1] (1910–82) He studied chemistry in Berlin and Munich (1929–1933) and worked then in the Kaiser Wilhelm Institute in Berlin. In 1932 he joined the communist party, and worked for the resistance until his incarceration in 1943. He received a death sentence, but was permitted to work on his research in the penitentiary of Brandenburg, where he was freed in 1945. After the war, he was the head of the Kaiser Wilhelm Institute in West Berlin, but was dismissed for protesting against the atomic bomb in 1947. After this incident, he became professor at the HU. In 1963 he was dismissed from the HU also. He had published an article "Dialectic without Dogmatism—Natural Science against Communist Ideology." He continued his work as a socialist critic and was put under house arrest in 1976. He was rehabilitated by the communist party in 1989 (after his death).

for the preparation of introductory lectures. His earlier, special-topic lecture was acceptable, it showed his great knowledge of the subject of which something stayed with me and is closely related to my later work on microphases and nanophases (see Appendix B). For his introductory course in the summer term 1952 he was simply not prepared and made numerous mistakes so that it was "unter aller Kanone."[1] Perhaps he was to be forgiven. Personally, I found the outstanding and the worst lectures most interesting. The outstanding ones are obvious, but I also learned much from the worst, since I paid close attention and read a lot in my textbooks, so that I was well prepared about the material to be lectured on and could detect the problems in the lecture.

In the summer term 1953 I took two courses of more applied topics. The first dealt with 'Scientific Photography' by Dr. Kurt Meyer, director of research at the VEB Agfa in Wolfen (formerly the Agfa-Wolffen, maker of photographic film, and producer of the first true color film in 1936), later also Professor of Physical Chemistry at the HU. The second was 'Chemical Technology' by Professor H. H. Franck (also director of the Institute for Applied Silicate Research, Berlin). Something special was the course 'Molecular Physics' by Professor A. Deubner with a required colloquium by the students, designed for advanced study in physics. Since I could not quite handle the theoretical topics because of lack of mathematics, I chose in my colloquium to talk about 'Monochromatic Molecular Beams,' based on a survey of the international (mainly English) literature. This was my first lecture given in front of a critical audience of scientists. It turned out, that I made the top rank in the class. That had never been accomplished by a chemist before.

<u>In February 1951 I had to interrupt my study</u>, as mentioned above. I came down with a severe sore throat. Our physician knew about my earlier bout with diphtheria in 1942 (see the pages 2-15–16) and did not want to take any chances, so I ended up again in the isolation station of the hospital in Brandenburg. Ultimately, the diagnosis was *Angina Plaut-Vincent*, a relatively rare form of tonsillitis. It is a bacterial infection and the hospital had no specific medication against it on hand, but they let Vati know, that the at that time novel penicillin was a sure chance for a cure. Vati checked all his contacts, and within a few days, penicillin was requisitioned. The cure was not instantaneous. All together, I spent 3–4 weeks in the hospital and then needed a number of weeks to recover. During this time in the hospital, I had again visitors in front of the window (see the pages 2-15–16). Not only Mutti, Vati, Joachim, and Manfred made regular visits, but specially welcome were Heidel's visits. She even tried also to make it sometimes between classes. Of the little gifts one brings to the hospital, one of Heidel's was by far the most memorable. She brought a single, generous slice of fresh, perfectly ripe pineapple. It was smuggled through the border control in Potsdam from West Berlin (despite its heavenly smell). For me, it was the first taste of a fresh pineapple, and today, it is still the standard against which I judge fresh pineapple. Only since 1988, in Knoxville, TN, can we frequently obtain pineapple of such perfection.

By the time I was reasonably well again, the summer semester had started and I had been given a leave of absence. Since at that time all my work was in the laboratory, this could be continued for the second half of the 4th semester as soon as I was able to stand at the desk.

Other things happened at home earlier in 1951. While I was in the hospital, <u>Frank and Dagmar got engaged</u>, before his study at the "Bauschule" was completed. This, and that the engagement was arranged while I was in the hospital, was against Vati's rules (see page 4-19–20).

---

[1] A humorous German transmogrification of 'below all canon,' in Latin: "sub omni canoni." The canon being the grade scale of performance in Latin, not a 'cannon' (German: "Kanone").

Begrudgingly he went with Mutti to the small engagement party in Mrs. Hanf's house. When, however, it became clear that the marriage was planned already for April 16, before Frank would have had his first pay check, and nothing could be done to reach at least minimal compliance, the connection of Frank and Dagmar to Vati broke completely.

The marriage took place as predestined. Heidel and I were not invited, although we would have found a way to go. From our family, Tante Grete came from Stralsund (see page 2-3) and tried to broker a peace. But even she did not seem to notice me. Tante Grete always had a favorite nephew who would be doted upon until some offense made her choose the next. This seemed to go according to age. Initially Henry was it, followed by Bertie (Tante Else's oldest children). Then, by 1951, it was Frank. Finally, Heidel and I had a period of favoritism (see Figure 32 of Chapter 5). This lasted until our first visit in Germany from the US. Then I had to attend a meeting for the largest part of a visit she had managed to get permission for from the GDR, she thought I should have had more time with her. Then, her interest turned to Klaus (see page 11-2, top center).

Frank then completed the "Bauschule"at the end of the winter semester of 1950/51 with a BS degree in architectural engineering. His first job was as a design engineer, at "VEB Projectierung, Brandenburg," followed by the "Entwurfsbüro für Hochbau, Potsdam," and at the "Entwurfsbüro für Industriebauten, (Ost) Berlin." By then he had moved with Dagmar, Eberhard, and his mother-in-law to Klein Machnow, Teltow, a suburb between Potsdam and Berlin, located in the GDR. There, I could visit them in 1953 to help in their preparation to flee to West Berlin.

On February 9, 1952, their son Eberhard was born and this brought a slow reconciliation with *Mutti, who had been rather ill in 1952/53*. The problems within the family, the political pressure, and the closing-in of our impending leaving of the GDR were probably to blame for what was loosely identified as chronic circulatory disturbance ("Kreislaufstörung"). Finally, on May 26, 1953 Frank left the GDR for West Berlin after a few visits with Eberhard in Brandenburg. For Heidel and myself, Mutti's illness meant that much of the weekends from Berlin were spent by Heidel in our apartment, taking over Mutti's duties. Frau Zaja, her daily help (see page 3-17), and sometimes also Frau Ladwig (see page 2-10), Vati's continuing secretary, covered the weekdays. Heidel got to know our family in this way. In particular, she had to take over the cooking under Mutti's direction from her bed. Heidel got an 'A+' from the whole family for her performance. She followed the instructions so well, that everyone agreed that it tasted as good as the boss' meals (or better?). Even when Manfred went for his first school day in the Jahnschule (see Figure 13, Capter 1), Heidel was his stand-in mother and had to deliver his "Schultüte."

During this time, Vati managed to reconstitute a prewar automobile, as shown in Figure 15. Since no automobiles were built or imported for the general population, it had to be obtained illegally. Vati had an opportunity to buy a wreck of a pre-war automobile from the police. Its motor could not be repaired, so Vati found an exchange motor in West Berlin and had the number of the old motor (illegally) transferred to the

Figure 15 *Vati's "DKW Meisterklasse," in 1952.*

Figure 16 *Wendoche, Fläming.*

new one. The body was also badly damaged, but the DKW body was largely made of plastic-covered plywood and could be fixed by a good cabinetmaker. The figure shows the result. Weekend trips still had to be clandestine. They had to be linked to business trips. Naturally, Vati could also not load the family in front of the house since our personal "Stasi Spitzel" (spy) would report such activities. We went on afternoon walks, and somewhere, around a few corners, we were picked up by Vati, who had 'accidentally' to go in the same direction. Book had to be kept for all travel and was accordingly doctored to always be in order. Figure 15 shows such a clandestine picnic in 1952 with all six of us. Yes, we all fitted in, 4 (women and children) in the back, Vati and I in front. Besides exploring the woods around Brandenburg, we also made some further trips. Well remembered are day-trips to Dessau at the Elbe river, and a three-day vacation trip at Easter to the city of Treseburg/Bode in the Harz Mountains to (April 13–15, 1952).

Before the winter term in 1952/53, Vati arranged for Joachim, Manfred, Heidel, and myself a month-long *vacation at the one-family village of Wendoche*, shown in Figure16. It was a farm owned by the Häberle's in the Fläming hills near Lütte, the village where we spent some time in 1945 as refugees (see pages 3-2–4). The Häberles took-in guests during the summer months. It was not quite clear, who was babysitting whom, but all four of us had a great time. I had bought an old plate camera and so, I could make somewhat better pictures. Naturally, Heidel and the countryside were the major motif, as seen in Figure 17.

On the weekends, Vati and Mutti were coming to 'visit.' Figure 18 illustrates the most successful picking of mushrooms we ever have accomplished. After a nice rain in the evening, we found the proper location so that the close to 50 pounds of the most delicious "Steinpilze (boletus

Figure 18 *Mushroom hunting in Wendoche, 1952.*

Figure 17 *Vacation with Heidel.*

edulis)" were collected in only one hour. But the time was not just vacation. Heidel had spent the month of August in Neuruppin for an internship, required for her study of law, and now had to use the remaining month to prepare for the next semester. Similarly, I utilized the early morning hours to review my books for the first chemistry exam ("Erste Zwischenprüfung" after completing the "Praktikantenschein I" in qualitative analysis). On October 22, 1952, I passed the exam. Besides studying for the exam, I had to study about reactions needing to be understood for the quantitative analytical lab for the "Praktikantenschein II" (completed by June 15, 1953).

On reading about the progress of my studies, one may wonder *why did it take so long until I finished the inorganic laboratory*? More than three semesters of almost full-time work was used for the qualitative laboratory, and the rest for quantitative analysis. The *qualitative laboratory* consisted of a set of 15 inorganic preparations, five demonstration experiments, and seven spectral analyses. This was followed by ten qualitative group analyses, each with up to ten different cations and anions to be tested for. The analyses were to be performed in the presence of all prior possible unknown ions. Finally, two full unknowns with about 75 possible ions were to be completed. This does not sound so much, but in case of any mistake, the preparation or analysis was not counted, and if the correction was not acceptable, two instead of one replacement had to be executed under the same conditions. Anyone who did not quickly become an expert chemist built up a backlog of work sufficiently large to suggest a change in field of study. My time was considered little better than average. The following *quantitative laboratory* consisted of five titrimetric measurements, four gravimetric analyses, one mineral to be determined qualitatively and quantitatively, one potentiometric analysis, and one complete alloy analysis. This was followed by one final analysis by titrimetry and one by gravimetry. This was to be done with reagents which one needed to prepare oneself. Commercial reagents of guaranteed composition were not allowed. Again, analyses had to be correct within rather narrow error limits. As soon as one completed these two major chemistry laboratories and all minor work in related fields, one could start preparing for the "Vordiplom," described on page 4-11. By the middle of June 1953, I was ready to do so.

The new study plan developed for 1951/52, was to have a six-week industrial practice to be added, along with the study of the new, non-scientific topics outlined on page 4-20. The HU professors in the natural sciences did not yet adhere to this study plan. The laboratories and examinations did not follow the new plan. Also, there was still a close connection to the West German university professors (see page 4-17). We students also were free to attend lectures and meetings at the TU and FU. Most instructive for me was a meeting on physics ("Physiker Tagung") held at the TU in 1953. Here, I could see demonstrations by Dr Ernst Ruska (then of Siemens AG, later professor at the TU) of the enormous magnifications possible by the imaging available in electron microscopy. Even more spectacular were the field-electron and field-ion microscopes, demonstrated by Professor Erwin Müller of the TU (later at the Pennsylvania State University in the US). Here I could 'see' for the first time images of tungsten atoms in molecularly sharp crystal tips and see the motion of larger phthalocyanine molecules adsorbed on such tips.

The enforcement of the new study plan was taken over by the office of the dean of students "Abteilung für Studienkontrolle" created 1950/51 (later called "Studienabteilung"). This was directed by "Herrn Forkert," a political functionary with little science training whom I knew from my effort to gain admittance to the HU (see the pages 4-7–8). In return for the new stipends, discipline was exhorted for the politically inspired lectures (but not the science courses!). "Stipendienentzug" (revocation of stipend) was used to enforce the compliance of the students. My first bout with Herrn Forkert in this connection occurred before the start of the 6th semester in April, 1952. Heidel, when picking up her stipend, looked also under my name and saw a note: 'Needs approval from the "Studienkontrolle." '

I was forewarned, but also had heard about the fate of three physics students. They, when confronted with their lack of study of Russian, had stated, they did not think that they should study the language of the so ill-behaved soldiers of 1945. All three were arrested when leaving the office, and ultimately dismissed from the university. When I entered Herrn Forkert's office there was no trace of defiance in my manners. I just asked: "I am sorry to have to come to you, but it looks that I will not get my stipend for the next months?" His answer was in a quite official tone: "I have a note here which says you did not attend the lectures in Russian. This is not tolerable." With my saddest face I replied: "I am sorry, I tried to learn Russian, even in high school, but it is not possible, the pronunciation is so difficult. I just cannot get my mouth to make those sounds. If I would spend the time I would need to even do a poor job in Russian, I could not succeed in chemistry anymore. It certainly would not be in the interest of our society to have a student poor in Russian *and* chemistry." To my big surprise he commiserated with me and said: "I know exactly what you mean. I have the same problem you have, and I do not do so well myself. Try to do better in the other subjects!" He meant the other politically oriented lectures, but I took it to mean chemistry and related fields. He gave me a note for the "Questur" (cashier). I thanked him, and then picked up my money. To my surprise, when I later had a chance to look at my "Übungsscheine" (certificates of exercises, laboratory experiments, and other activities), there was one for Russian. I had neither registered for, nor attended the lectures. The grade was "nicht zu beurteilen" (cannot be assessed), given on May 5, 1952. Since this was *not* a failing grade, thus I could get my stipend. (Did he conspire to keep me at the HU? If so, despite all, he must have been a friend in the office of the Dean of Students). These "Übungsscheine" were not given out, so that one had no documentation of the study until one was officially released ("exmatrikuliert") from the HU.

At the beginning of the 7th Semester, in November 1952, the same exercise occurred with Herrn Forkert. This time I was delinquent in the 'Marxism-Leninism' course. My excuse: "This is such a poor lecture, I cannot waste any time on it!" This was a bit risky, but before he could react, I explained: "I went to some of the lectures (I lied, I did not go, I only heard about them). There I found the active FDJ members in front and almost asleep. They knew the material and were bored. Students like I, who would have liked to learn, sat in back and had no chance to follow the lecture. The lectures were aimed between us two groups of students, good for neither." Herr Forkert's answer: "Yes, I have heard the same from other students. I will have to look into this matter. See that you can learn the material on your own. Sooner or later you *must* take an exam in this subject." He was right, but for the moment, again, I got a reprieve and stipend for another semester.

A third reprimand came before the end of semester 8. This time, I got an exam date for Marxism-Leninism at the end of the summer term, the middle of June 1953. Earlier in 1953 all students living in West Berlin were dismissed from the HU. It certainly looked like *the possibility of an apolitical study at the HU was coming to an end*. An effort of mine to possibly switch over to the TU in West Berlin without leaving East Berlin was made for the winter semester 1952/53, but was not successful. At this time the number of refugees processed daily in West Berlin, arriving from East Berlin and the GDR was 1–2,000. The only two possibilities left for me to stay at the HU were to study for the forthcoming exam and possibly join the FDJ and give up my independence, or to join the constantly growing flood of refugees.

Heidel's study of law also became increasingly unattractive. It required to study the new laws of the GDR which, just like the Nazi laws, made the lawyer a tool of the oppression. Finally, Vati's business was increasingly in danger to be disowned, or it had to be decreased to three employees, too small to be profitable. The choices of Vati were to gift the business to the state and work out a deal to become the manager of what would then be a "VEB" (state-owned business) or to leave as a refugee. All these unpleasant alternatives were discussed each weekend. Naturally, no one could act fully independently either, anytime someone left for West Berlin, the remaining friends and family members were scrutinized. Any hint of having known about the defection would lead to severe penalties for the ones left behind.

In fall 1952, I had saved enough money from my stipend to buy *the new, 35-mm, single-lens mirror-reflex camera (SLR)* named "Exa" in the HO. It was quite versatile with exchangeable lenses. This model was built mainly for export by the world-famous IHAGEE[1] firm. Their premier camera was the "Exakta," the SLR competition in the 1930s to the even more famous 35 mm "Leica" which had a separate view finder. My first attempt at 35-mm photography is reproduced in Figure 19. Now I could really take good photos and develop them myself. As a Christmas present, I got a special leather case that could hold two cameras, a tripod, extra films, etc. It is still in working order today.

Figure 19 *First ice, observed in Brandenburg, 1952, Exa, 35 mm camera.*

Slowly, with the increasing standard of living, the Christmas celebration took on some of its old glory. Particularly, there were now two considerably younger brothers who were appreciative of new toys. For the 1952 Christmas, I fixed the residuals of Frank's and my old Märklin H0 trains for Joachim and Manfred (see page 1-24). The rescued trains and rails from the rubble of the Werderstraße were modernized with some new parts, bought in West Berlin and new transformers since by now

[1] "Industrie- und Handelsgesellschaft," IHG, (in English pronounced and written 'eehagay'). A company in Dresden, founded in 1912 by Johan Steenbergen (1886–1967), a Dutch citizen. The company was destroyed in 1945 during the bombing of Dresden (see page 2-17–18). Johan Steenbergen, with his Jewish wife who was a US citizen, left Germany in the last moment in 1942 for the USA and returned after the war to West Germany as a consul of The Netherlands. After the war, the camera production was restarted with few left-over machines. It could not be disowned because of the foreign owner, but was not supported either. It continued production until 1969. Efforts by Steenbergen to create a new company in West Germany failed. (Note, that an "Exakta" was used by Jimmy Stewart as the photographer in Hitchcock's 'Rear Window' in 1954, and another one is still in my possession since 1954, bought from my first stipend as a refugee in the FRG).

we had 220 V AC current. In addition, I built a train indicator and control box, so that one could see the location of the train on the track by lighted section indicators. Also, Vati had built with help of a friend cabinetmaker a toy grocery store, as he had for us some 18 years earlier. By then, naturally, Heidel was invited for all the holidays. When I asked Mutti some weeks before Christmas what may be a suitable gift for Heidel, I got a surprise answer: "How about an engagement ring?" Mutti would never have suggested this on her own. Vati must have given his agreement before. Had he loosened his rules? Yes, he had. I think times were volatile and he wanted to make sure that the two of us had some stronger base to weather the upheavals obvious for the near future. When I brought this amazing news to Heidel, we both agreed, it was high time to get our parents together to work out details for the future. On Christmas day, *it was decided our engagement should be celebrated at Easter* which in 1953 was on April 5.

Figure 20 *Hans and Inge Gottschalk.*

The year 1952 was closed out by us with a New Year's celebration with Inge and Hans Gottschalk in Brandenburg. They had gotten married earlier in the year and now had their own apartment in Brandenburg. Hans was close to finishing with his MD. Figure 20 shows a picture taken at the party. They were to remain our friends in the GDR and never broke contact with us, even when it became a crime to develop new connections to the western world. They had to register our address with the police in order to receive mail. All through these times until the reunification in 1990 (see the pages 9-25–26) we also met occasionally in the eastern European countries by their surreptitiously arranging vacations to match the dates of our international scientific meetings in Budapest, Prague, Bratislava, and even the GDR (see Chapters 8 and 9).

We always had to be aware that all mail to and from Hans and Inge was opened by the "Stasi" (state security service, see page 3-23). To find when they might have confiscated one of the letters, we numbered the correspondence. Finally we also found a way how to smuggle US Dollars to them on postcards without ever being caught. Once the western currency was in their hands, they could go and buy goods with it in the special stores (see page 3-11), even deposit it in the bank (without exchange into the low value DM-Ost). The GDR was so depleted of foreign exchange, that the possession and spending of western currency was no crime, just the act of acquiring it was forbidden. But, how do you hide money on a postcard? We wrote in a letter to Hans, whom we had made into a collector of vacation postcards: "When you get the next post card, cut off ½' on one edge it increases the value." He immediately understood. We had bought two identical (preferably thin) postcards, put the money between the two cards and bonded them together with double-stick tape. Then, I reshaped the two cards with a paper cutter, so that a perfect match was reached. It worked for over 30 years.

After reunification, Hans and Inge could even visit us in the US. A special victory since the same police officer who forbade Hans a visit into the FRG four years earlier, had now to give him the passport to the USA and politely wish him a good trip (see page 9-33). I think that such change from dictatorship to democracy with a peaceful revolution, when only the major criminal elements in government are subsequently punished, is also a development of the 20th century. Before, such revolutions were always bloody. Peaceful changes to a less oppressive government were also observed during the breakup of the USSR and the other communist dictatorships. Even the end of the Cultural Revolution in China was relatively peaceful. Perhaps, there is hope for the future?

The efforts to prepare for the flight out of the GDR accelerated in 1952/53. Initially there was an effort in 1949 to join Tante Lieschen in the USA. She had agreed to sponsor our emigration, but this initial attempt was unsuccessful (see page 3-36). We knew, once in the FRG, we could try to reopen our immigration request. Photos such as Figure 11 in Chapter 3 were taken to document the property that needed to be left behind. In the meantime, Vati had rented a large closet in an apartment in the Passauer Straße in West Berlin, not far from the "Bahnhof Zoo." There we stored easily transported goods. From then on, I always went from Brandenburg to Berlin with a full suitcase. Vati's and my clothing were no problem, but Mutti's, Joachim's, and Manfred's would have looked suspicious on inspection. Since by that time I was a well-known traveler on this route and had experience in smuggling, I never had to open my suitcase. At "Bahnhof Zoo" in West Berlin, I would get out, lug the suitcase the mile to the Passauer Straße. At the same time, I would check for mail that could not be sent to Brandenburg. To make things simpler, I left the suitcases, and Vati bought new ones for next week's trip. This was not always easy since suitcases were not readily available—but Vati had good friends. Any letter or money which needed to be transported was kept in a tight wrap at the bottom of butter or lard, pressed when warm into a container, so it looked like the food I took back and forth, as did many students. *In the end, I had moved over 20 suitcases of goods* which ultimately ended up in the US, as shown in the Figure 21 of Chapter 11.

Figure 21 *Border-marker as seen when leaving the American Sector, before reaching the Soviet Sector.*

On going home for the weekends, some *time was spent in the American Sector of Berlin* to shop for goods not available in Brandenburg. A sign that alerts one of the borders within Berlin is reproduced in Figure 21. Little danger of inspection existed there. In Potsdam, located in the GDR, the way to and from the "S-Bahn" passing though West Berlin to the train from and to Brandenburg, in contrast, was always heavily controlled. Three delicate occasions were the following: When I had to get some rum to Brandenburg to speed-up Mutti's circulation (see page 4-23), I took a 2-liter acid bottle with a western industrial label, washed it, and filled in the store with rum (after tasting). Then, I sealed it professionally. A chemistry student must get his chemicals! Indeed, I never had to open the bottle.

On another occasion, close to Christmas, Heidel and I were on the way home. This time, the check started already in the "S-Bahn," one station before reaching Potsdam. Several border police entered into every coach. They systematically had all luggage opened. Heidel, sitting next to me said (sadly?): "Now I am going to see my Christmas present!" My answer: "If you do, you do not get it because it will be confiscated." Everyone had to open every piece of luggage, briefcase, and handbag. My big briefcase was easily visible beneath my knees. The policemen looked several times at my briefcase and than into my face, but I did not indicate any emotion. They probably thought the other fellow had already looked it over. Perhaps, they thought, such a nice and docile student could not possibly smuggle. Heidel did not see her Christmas present, but later got it!

The third occasion arose when I went through the inspection in Potsdam together with Vati on our way back from West Berlin. While I looked like a student, he looked like a rich, well-dressed business man with rim-less glasses. Since he also had a full briefcase, there was no question, he must be smuggling, and the way he dresses, he must be an enemy of the people. He got stopped, his passport was confiscated and he was held for a detailed search. Naturally I (with the larger contraband) went through unchallenged. I deposited my briefcase in the train to Brandenburg with friends and went back. Vati was still standing there with a group of about 10 others, all to be searched. I joined the group and talked briefly to Vati telling him that he would probably miss the train and that I would let Mutti know he would come with the next one. Then I bent down picked up his briefcase which I had myself positioned next to, so that it could have been mine, and left. When Vati came home with the next train, he said they picked him over in the greatest detail. They could not understand why they had arrested him when he had *no* luggage. Having been to Berlin and bringing home nothing, that was unheard of.

Figure 22 *The guests to our engagement party in addition to Figure 18, Chapter 11. Upper row: Frau Ladwig, Dagmar and Frank, Tante Marie. Bottom Row: Irmchen and Henning, and Gabi and Rolf. Not shown are Onkel Rudi and Frau Zaja. Easter 1953.*

At Easter, our engagement was celebrated with relatives and friends of the family. There was Heidel's and my family, seen in Figure 18 of Chapter 11. Additional family members were Frank and Dagmar, Tante Marie and Onkel Rudi, as well as Irmchen and fiancé Henning,. Also, Heidel's friend Gabi and her boyfriend (now husband) Rolf Bögel and Frau Ladwig and Frau Zaja were at the party of 18. Most of the additional guests are shown in Figure 22. All got along well. Food, drinks and other items were slowly approaching pre-war times, although it was an effort of many weeks for both families to provide for the festivity. The engagement rings, for example, were made by a local jeweler, but the gold had to be found within the families.

Shortly thereafter, Heidel left for a practice study-semester with her student group in Neubrandenburg, a town about 100 mi north of Berlin. There she was to learn the daily routine in the court house, rotating through the different departments. Visiting back to Berlin or Brandenburg was only by permission. For me, this last term at the HU (summer 1953) meant to work doubly hard, to finish the quantitative analyses and attend a few lectures in applied Chemistry and theoretical physics (see pages 4-20, 22, and 25). About one month after the engagement party, Mama and Papa had their silver wedding anniversary on May 8, 1953, a Friday. For such an occasion, Heidel had no difficulty to get a few days of vacation to travel from Neubrandenburg.

A similar group of relatives and friends as for the engagement party was assembled, but for the immediate family, it was not quite such a joyous occasion. In Neubrandenburg, the learning of the practice of law had developed into requests to write indictments for crimes against the state. First as exercise, but soon it was planned to actually use the indictments written for actual cases. They naturally followed the, 'new law' of the GDR, and violated the 'old' accepted German laws. This would cause problems in the future if we wanted to leave the GDR.

The agreed upon solution was that *Heidel would not return to Neubrandenburg, but register as a refugee* in West Berlin to await how things would develop. Heidel could stay initially with distant relatives of her's in West Berlin, and I could bring clothing, etc. in the well tested manner. Naturally, nobody should have known her defection, not even Mama and Papa. To indicate that this was the case, Heidel sent a good-bye letter to them once she had settled in West Berlin (as proof that they did not know about it before). The letter did not arrive! The next letter was sent registered, indicating Heidel's concern that they had not written to her in West Berlin. This letter arrived. Soon thereafter, Heidel's parents got a letter from Neubrandenburg, informing them that their daughter had not returned from her vacation and that she had taken along the identification card allowing her to enter the court house, a criminal action. She was to return it as soon as possible, or else .... As soon as we heard of it, Heidel arranged for return of the papers. No further action was taken.

The actual route of defection of Heidel was from East to West Berlin by bicycle without luggage through the "Brandenburger Tor" on May 11. I visited her in the evening and brought her suitcase. Vati heard shortly thereafter from Frank that he would also leave for West Berlin with Dagmar, Eberhard, and his mother-in-law (see page 4-23). In his job in Berlin he was having political difficulties. Ultimately, he also was recognized as a refugee in the FRG.

The next problem concerned Opa. He was ill for some time. We found out later that he had colon cancer. He died on June 7, 1953 in Brandenburg and was buried in the "Krematorium" (see Figure 2 of Chapter 2). Tante Lorchen remained for a number of years in Brandenburg, but then moved to relatives and friends of hers in Berlin (East?), transferring also Opa's ashes.

The next step was to move Mutti, Joachim, and Manfred to West Berlin to Tante Minna and Onkel Paul. This was early in June, and I was accompanying them. It was surprising that nobody stopped us. Is it not suspicious when on a warm day a lady in a fur coat and two little boys with their school bags and an older son with two heavy suitcases enter the train for Berlin? The excuse of a summer vacation might not have worked. This left Vati in Brandenburg to secure the business.

For me, *the exam in 'Marxism-Leninism,'* which was forced upon me (see page 4-26), was to be held on June 12. By now, I knew that I had to fail this exam to have a stronger case to gain the status of a political refugee with rights to get financial support for continued study in the FRG. The goal was just to fail, which would ultimately cause dismissal from the school, and not to aggravate the examiner so that he would have me arrested. The exam also had a minor goal, because in order to get paid for the next two summer (vacation) months, I had to bring proof that I had taken the exam (though not necessarily passed?). This was to support my initial stay in West Berlin. I arrived in the office next to the exam room and was greeted by the FDJ group leader who was to take care of me. He was very much concerned that I should do well (to pass also meant for him to have done well). He showed me where to randomly draw two exam questions. One of which was to be my initial exam topic. I drew: 1. "What book did Stalin write in 1938 and what is its importance?" 2. "What is the importance of the Red Army?" Both questions would have been easy for me to get something acceptable to say about (perhaps even get a good grade!). Stalin's 1938 book is the History of the Communist Party of the Soviet Union.[1] And about the Red Army, everyone in the GDR could repeat some of the political propaganda in our newspapers (trying hard to undo the 'order of the day of the Russian Marshal Zhukov when occupying Berlin,' quoted on page 2-28). The group leader looked at the topics and felt relieved. He said: "You have ½ h to prepare yourself before you are called." My answer: "I do not need any. When the examiner is ready, I am ready."

After a little while, I was called. A professor and two assistants were handling the exam. I put the papers with my titles on their desk. The examiner was very friendly and said: "I think I know what subject you are going to talk about." My reply: "Yes, I don't know what Stalin wrote in 1938.—But I also have difficulties saying much about the Red Army." He remained friendly and joked: "The Red Army is so big. You must be able to say something about it!" I, naturally, could joke too: "Yes, but you do not want me to say something about every soldier?" His somewhat less friendly reply: "Let me help you a little and put you on the right track. Perhaps you can tell me why Stalin could say in 1941 when the German troops invaded the Soviet Union: 'The Red Army will ultimately be the victor?' " I put on my most stupid face and replied: "Did Stalin really say that? I think I remember Hitler was saying the same thing about the German "Wehrmacht" in 1941. What is the difference? The Red Army is there to defend the Soviet Union, the American Army, the USA, etc." His reply was brief, but not angry and while writing a big 'unsatisfactory' on the paper in front of him: "What grade did you think you would get?" My answer: "I did not expect a '1' but also not to fail." He followed with a long sermon of how workers and peasants were trying their best to give me an opportunity and support for my study and how I must repay them. Before leaving, I handed him the note about my stipend, to be paid after the exam. He said nothing, but signed it.

I quickly made it to the "Questur," got my money and took the next "S-Bahn" to West Berlin to report to Heidel who was waiting for the result of the exam with Peter Menzel, an old school friend of mine (see Chapter 3, Figure 25). Everyone was happy that things went as planned. Since it was a Friday, I then took the train to Brandenburg, not to get back to East Berlin until 15 years later during my sabbatic leave at the University of Mainz (see the pages 8-39–49).

On this sabbatic in 1968, we visited friends and relatives of Heidel's in West Berlin and chanced, as American citizens, a brief afternoon in East Berlin via Checkpoint Charley. Figure 23, however, was generated in 2004, another 36 years later. It is rebuilt at the same location as a reminder of the divided city. Although all is authentic, including the widely used GDR automobile, a "Trabant," parked in front and a guard, properly dressed. Not true to 1968 are the tourists, the marks on the pavement indicating from where to take the best pictures, the clean facades of the houses, and the sign to the "Mauermuseum." In 1968 the "Mauer" was still bitter reality and the East-Berlin side with the "Mauer" was heavily fortified so one could not crash through the gate.

---

[1] Since we do not throw away books, Heidel's 1952 copy of the book (in German) is still on our bookshelves. Just that today, it is usually thought that the book was not actually written by Stalin, but by a commission of the Central Committee of the party. In accord with this, our book has no author listed! It is stated only: "Edited' by a commission of the Central Committee."

Figure 23 *Checkpoint Carley as it appeared in 2004.*

Back in Brandenburg I made final plans with Vati. He had sold his car, so we had to walk to the office. On Monday afternoon, when crossing the Wilhelmsdorfer Straße, we accidentally met Tante Else. We did not have much contact with her and Onkel Richard since 1941, so Vati took this opportunity to report the recent changes and our intention to leave soon. It was always important for relatives to be prepared for the interrogations that were certain to follow the defection of anyone within the family. Later, we found that, in time, Tante Else and Onkel Richard also left Brandenburg and ultimately settled in Victoria, Canada where we still have contact with two of my cousins.

With my stipend for two months from the HU, I wanted to invest in a new Zeiss microscope which was to arrive on Wednesday, June 17. This was an object that could be bought in the HO department store and was highly sought after in West Berlin. I hoped in this way to have a favorable exchange of DM-Ost to DM-West. Then, later in the week, we were to join Mutti in West Berlin. All suitcases were moved, and the trip should have been without problems.

On Tuesday June 16, we heard over the West Berlin Radio RIAS (see page 3-22) that there was some unrest among the workers in East Berlin about work quotas. Knowing how closely the Soviet Command usually kept order in the streets, we did not think much of it. The next morning, when I wanted to pick up my microscope, the city of Brandenburg was taken over by the revolt. It was not about the newly set work-norms anymore, it was about the lack of freedom. In the Preface, the happenings of the 17th of June which brought us prematurely to West Berlin are detailed.

By the 17th of June we were refugees again. The typical bureaucracy had to be dealt with. First to register, along with, on that day, thousands of others who had to leave East Berlin or the GDR. This was followed by soliciting of detailed information about the circumstances of our leaving. In order to be recognized as a political refugee with the rights of a citizen of the FRG, the facts were examined by a court. If the overwhelming reason was economic hardship without immediate danger for one's life, the refugee status was denied, and, when above 24 years old, one was asked, if possible, to return to the GDR. If this was connected with danger for ones life, one could remain in the refugee camp, but would not get additional financial support, i.e., one had to find one's own job and could not get government-supported furniture, housing, etc.

Fortunately, we could stay with Tante Minna in Berlin-Rudow, near the border to the GDR (see Figure 3 in the Preface) and did not have to live in one of the crowded refugee camps. There was now plenty of the time to rest and plan. Figure 24 illustrates results of my spare-time photography. Heidel had also moved to Tante Minna, so that our plans could be coordinated.

For German bureaucracy, having a fiancée meant nothing. *Heidel had arrived earlier and was flown out earlier*, and since she had another last name, she was sent to another "Land" (province). Her court proceedings were postponed to be taken up again in the refugee camp Westertimke. near Bremen. On July 15 Heidel had to report to the refugee camp near the airport

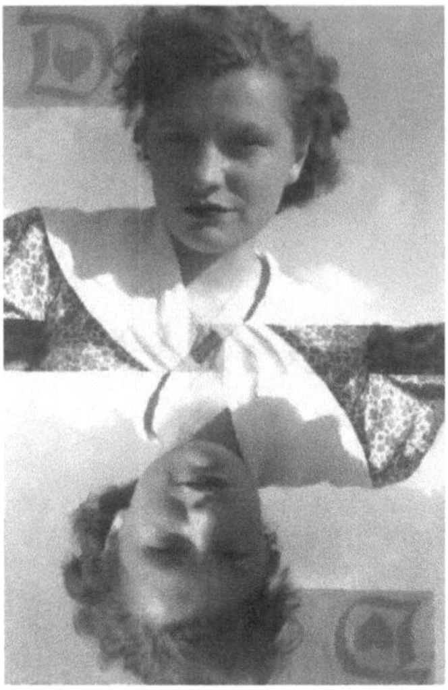

Figure 24 *My Queen of Hearts*

Tempelhof to await her plane to Bremen. In Westertimke, the recognition as refugee was granted on August 18, coupled with a transfer to the "Land Hessen." Through the camp in Giessen, the referral to the refugee camp Hanau was made on August 21. During all this time, connection to me was kept by mail.

Back *in West Berlin, the case of recognition as refugees of Vati and myself was combined*. Figure 25 illustrates one of the many discussion and planning sessions held with Onkel Paul. One of the pressing items was to put the luggage in order which we had stored in the Passauer Straße (see page 4-29), so that it could be shipped as soon as there was a permanent address. For me, I had no proof of my study at the HU, except for the study book (see Figures 6 and 13). Since after the first two terms no "Antestat" and "Abtestat" (see page 4-10 and Figure 6) were required with the new, stricter control of the students, this book had little value except to prove that I had paid the first semesters and then had a scholarship.

I needed to get my "Übungsscheine" (certificates of completion of exercises and laboratory). These were kept under lock and key at the HU (see page 4-26). Again, I tried the early recipe. Go to the top and see what happens. The borders to East Berlin had been opened again in early July. Thereafter, I wrote a letter to Professor Thilo, stating that due to the events of June 17 I could not return to the HU, but would like to continue my study somewhere in the FRG later this year. Could he help me in getting my "Übungsscheine?" He could. A few days later when I was in the Passauer Straße, sorting luggage, my friend and fellow chemistry student Werner Schmidt, rang the door bell. When I opened, he said: "Nice to see you, but you know, you did not see me! Here are your 'Übungsscheine.' It took some difficult maneuvering, but we got them all. Good bye and I wish I were where you are." Werner Schmidt was one of the best students and held the lowest assistant position in the laboratory ("Famulus"). I thanked him, but he was in a hurry to get back. Now I had all my 12 documents and could go to any other German university to continue my study. The difficulties involved in getting all, I could deduce from the fact that Werner Schmidt had to personally sign the last "Übungsschein" in chemistry in addition to Professor Thilo. It was dated June 15. The regular assistant could probably not be trusted with such illegal help.

Figure 25 *Vati, Mutti and the two "little ones" with Onkel Paul, discussing our future in the FRG.*

In 1968 I met Werner Schmidt again. We were on an excursion to West Berlin as US citizens and as such, half-day visitations to East Berlin were permitted (see page 4-32). The "Mauer" (wall) had been built in 1961 and provided sufficient 'safety' for East Berlin. We drove at that time our brand-new Mercedes with visitors licence plates. The plates were not valid in the GDR and we had to purchase East German plates every time we crossed the border. This done, one might ask: "Who in East Germany can drive a brand-new Mercedes with East German licence plates?" Answer: "Only a high party-official."

This conclusion soon became obvious. We parked in front of the "Chemische Institut" (see Figure 5) and I marched full of confidence to the doorman (better, the 'guard'). Entrance to the laboratory was permitted only with proper ID. I asked for Professor Schmidt. Yes, he was in, but he was giving a seminar at the moment. I was prepared to wait, but the guard said to me (without asking for any identification!) I bring you to the seminar room. I am sure, for you, he wants to be interrupted. We went, Dr. Schmidt was interrupted—and almost fainted when he saw me. When we were alone, he quickly said: "Let us get out of here before anybody else sees us. You should never have gotten permission to enter the institute." We left via the back entrance.

Sitting in my car, we had an hour-long discussion of life in the East and West. He was teaching and also was also in charge of the chemistry library. His problem was that occasionally journals of the American Chemical Society (ACS), bought with scarce foreign currency, went missing and he had no means to replace them. Fortunately, I could help. When I went back to the US, the ACS gladly mailed him the missing issues without cost to him, and minimal expenses for me. Some gifts we had brought along from West Berlin were left. From then on, we stayed in occasional contact. After reunification, he invited me on June 6, 1991 for a lecture on "Molecular Dynamics Simulation and Thermal Analysis of Macromolecular Crystals." It was held as an "Institutscolloquium" at the famous "Institut für physikalische Chemie" in the Bunsenstraße. Professor Schmidt gave the introduction with proper details of how I was forced to leave in 1953.

Early in August, our acceptance proceedings as refugees were completed. Figure 3 in Chapter 5 shows a reproduction of my recognition as a refugee which was given to me after arrival in the refugee camp. Next, our flight to Frankfurt (Main) was arranged. The final destination was the refugee camp for the Land Baden-Württemberg, located close to the city of Weinsberg (documented in Chapter 5 with Figures 1 and 2). Close to the day of flight , we had to report to the refugee camp near the airport Tempelhof to await a suitable plane. This camp was in a large, old factory building with many bunk-beds with straw-mattresses in its large halls. We occupied a small corner. After a stay of about 24 h in this depressing environment, our plane was ready on August 10, and I finally had my first airplane ride (for the first unsuccessful attempt in 1938 see page 1-20). Unfortunately, the plane was not pressurized. On landing in Frankfurt, my sinuses hurt so badly that I never wanted to fly again. After a stay at the 'Rescue Mission' at the train station, we finally were bussed to Weinsberg. There, I learned how to function in a refugee camp. Figure 26 is a reminder of the wait for the 'flight' from Berlin.

Figure 26 *Photo of an eagle living in the "Berliner Zoo" taken while we waited for the 'flight' to "Frankfurt am Main," in West Germany, the FRG.*

CHAPTER-05

Goethe University, Frankfurt, Federal Republic of Germany, 1953–1954

Getting settled in "Weinsberg an der Weibertreu" was rather simple. From the train station we were bussed to the refugee camp. This was a former camp for workers building the nearby Autobahn, and later working in the nearby industrial firms. Figure 1 is an overview of the barracks from a photo, taken when we went exploring the neighborhood. Joachim is surveying our new home. In Figure 2, a post card is reproduced with a picture of the village (sent to Heidel introducing our location and also trying to keep contact). On August 11, 1953, we registered in the refugee camp Weinsberg.

Figure 1 *Refugee camp in Weinsberg, Württemberg.*

Figure 2 *Close-up of Weinsberg with the castle "Weibertreu." It is named after a sad happening on December 21, 1140 when the castle had to capitulate. It was decreed that all men should be executed. The women were permitted to take along only the valuable possessions they could carry on their backs. They all managed to carry their husbands, thus the name.*

Since both, Vati and I were both accepted as refugees, we could also register with the state employment office. On the back of the official acceptance form as a refugee, the 'legal' reasoning was given as is reproduced in Figure 3. It presents a very brief summary of my discussion in the Preface and on pages 4-31–33. The state representative in the camp assured me that all would be done, to get me to another refugee camp so that I could attend the University of Heidelberg by the winter semester.

With this encouraging information, we settled into the camp routine. Each of the buildings seen in Figure 1 was divided into two large rooms with utilities in the middle. About six families could be accommodated. We 'owned' a corner of three bunk beds with blankets and pillows provided. There also was a small wardrobe, but most of one's possessions remained in the suitcases under the bed. Chairs were there, but not enough for everyone. For the meals, there was a cafeteria. I don't remember any poor meals. The one item on the menu that was new was "Fleischsalat," a cold salad made of diced sausage, like bologna, mixed with pickles and some mayonnaise. This I liked, although the people who had stayed longer in the camp were complaining. It was served too often and it was only a way to get rid of the sausage that could not be sold in town. I did not fully agree. Being quite sensitive to spoiled meat, I could attest that it was of quite good quality. Each day we received pocket money of 25 "Pfennige", enough for one postcard (as Figure 2) plus stamps.

What to do next? Joachim quickly found a friend, and off they went playing. Vati discovered that Frank, Dagmar, Eberhard and mother-in-law were in the same camp, so that Mutti could, for once, enjoy her grandchild. In the camp library, I found an up-to-date text on solid state physics. In the next few weeks, I learned a lot about dislocations and defects in crystals. Otherwise, I could check once a day for mail. Had my transfer papers to a camp and the University in Heidelberg come? What was Heidel doing in the meantime?

As long as the weather was nice, long walks were also of interest, but what else to do? Should one just wait? It did not take me long to find out that something had to be undertaken to speed things up. First, however, I went to town to look for a job. There were no laboratory or similar jobs. When I inquired about my transfer to Heidelberg for about the 10th time, I was told, perhaps not the winter semester, but later. To remain for half a year in this camp was not to my liking. By the end of August, Heidel had made it to a refugee camp in Hanau, near Frankfurt (see page 4-34). Frankfurt had a university! It naturally would be best if both of us went to the same university, so I filled out an application, and Heidel delivered it together with hers.

B e g r ü n d u n g:

Der Antragsteller ist Student für Chemie und war an der Humboldt Universität immatrikuliert. Er will keiner Massenorganisation angehört haben.

Zur Begründung seiner Absetzung aus dem sowjetische besetzten Teil von Berlin führt der Antragsteller an, daß er auf Grund seiner politischen Indifferenz zu einer "Sonderkonsultation" in Gesellschaftswissenschaften bestellt wurde und diese Prüfung nicht bestanden habe. Nachdem sein Bruder die DDR verlassen hat, wurde er noch schärfer beobachtet.

Die Ereignisse des 17.6.53 zwangen seinen Vater, die sowjetische Besatzungszone zu verlassen. Bei Feststellung aus dieser Tatsache befürchtete der Antragsteller Zwangsmaßnahmen gegen sich. Ihm blieb keine andere Wahl, als die Humboldt Universität und den sowjetisch besetzten Teil von Berlin zu verlassen.

Der Sachverhalt wurde glaubhaft vorgetragen. Der Vater des Antragstellers fand unter der Reg.Nr. 21o 242 die Notaufnahme. Unter Würdigung der Gesamtumstände mußte dem Antragsteller zuerkannt werden, daß er sich in einer Zwangslage im Sinne des Notaufnahmegesetzes befand, da zweifellos außer der Exmatrikulation weitere Maßnahmen gegen den Antragsteller eingeleitet worden wären. Diese unter den gegenwärtigen Umständen abzuwarten war dem Antragsteller nicht zuzumuten.

Die Verhandlung erfolgte in Anwesenheit des Antragstellers.

gez. Unterschriften

Derzeitige Berliner Anschrift:
Berlin-Rudow, Str. 2oo Nr. 25

Figure 3 *Legal reason for my recognition as a political refugee.*

Figure 4 *University of Frankfurt with Heidel. Winter of 1953. The chemical institute was in the next street.*

Pretty soon I received *an invitation to come to the "Johann Wolfgang von Goethe Universität"* in Frankfurt[1] for an interview. The main entrance to the University is shown in Figure 4. The 200 mi round trip was not cheap, so I chose the train, known as the "schwäb'sche Eisebahne," a slow series of local trains. I left in the morning, but arrived only in the late afternoon in Frankfurt after an enjoyable, long trip through the countryside, changing trains two or three times. Heidel met me at the Hauptbahnhof (see Figure 11, below) and we explored the "Senckenberg Anlage," a park which led to the University. Heidel had gotten a job at the University, transferring student information to Hollerith punched cards.[2] Since it was getting late and we were 'homeless,' we spent the night in the waiting room of the train station with a small meal and a soft drink. Eating and waiting was permitted there, but not sleeping.

Next morning, Heidel went to work, and I went to my interview. There was no problem to be admitted, and as an accepted refugee, it should also be possible to get a scholarship and stipend. All seemed to be easy. Heidel and I met again in the cafeteria where Heidel commonly ate, being employed by the university. The food was a good bit better than that at the HU. In the future, it would become our main meeting place at noon for the week-days when both of us were students at the university. They made a particularly good pea soup (the cheapest dish). Also, the tableware caught our attention. It was new, of stainless steel and modern in design. After Heidel finished her work, we went to her refugee camp in Hanau, which was for women only. The guard at the gate made it clear that I certainly could not stay there overnight, despite the fact that Heidel's two room mates said they would not mind my sleeping there for one night before my train next day.

What was I to do? I could not afford a hotel. The guard had experience in cases like that. He said, do you have at least one DM-West? Yes, that I had. Then, he said, you can go to the police station around the corner, tell them that you have no place to stay for the night, and they will put you up. They did. What I did not know was that *I was incarcerated for the night*. The bed was clean,

---

[1] Privately founded in 1914 to establish a link "between dynamic research, the legacy of the Enlightenment, and the conscious commitment of the citizens of Frankfurt to the fundamentals of international trade and industry." Its motto, a quotation from Goethe: "Wenn du eine Antwort verlangst, must du vernünftig fragen." ('If you require an answer, you must ask sensibly.')

[2] First practical punched cards for data processing. Credited to Herman Hollerith, who used Jacquard's punched cards, long applied to represent patterns for weaving, for data gathered in the American census of 1890, and to read and collate these automatically. He founded the 'Tabulating Machine Company' in 1896, one of three companies that merged to the 'Computing Tabulating Recording Corporation,' CTR, renamed in 1926 to IBM, 'International Business Machines.'

a bit hard, but I was sufficiently tired to sleep well. But, the door of heavy steel bars was locked, and when I wanted to get out in the morning, I could not. Processing of prisoners was at 10 a.m., not before. So, I waited, signed the release papers, and paid my one DM-West. Then I was a free man again. After a nice and sunny Saturday with Heidel, I returned to Weinsberg, full of hope that I could continue my study starting with the winter term of 1953.

In the camp, the waiting time was getting excessive for me. They were looking for men willing to work on road construction. Why not? I needed some money to get started in Frankfurt anyway. A few day's later, I was on the road gang. The pay was good, but the job was terrible. The earth was clay. The job consisted mainly of shoveling the clay from one place to another. Yes, I was strong—I thought. But the others in the gang seemed to be able to do more. I was dead-tired in the evening. This was even harder than moving sacks of grain (see page 3-3). Vati commented when I complained: "Carrying responsibility is by far easier than carrying dirt." Fortunately, three days later, I got the admission to the University. Heidel had been moved by then to a refugee camp in Frankfurt-Griesheim, Stroofstaße 33–38, near to the "Farbwerke Höchst."[1]

Just a few days later **I was on my way to the refugee camp in Frankfurt-Griesheim**. During the war, this was a camp for forced laborers. Heidel found out that I could stay there, but not quite legally. She was told, there were free beds and nobody would check. Figure 5 shows a typical scene in the camp and Figure 6 illustrated the drab interior. By leaving my place in the camp in Baden-Württemberg without official transfer, I did not have the right of a place in a refugee camp in Hessen. I had to look for my own place.

But first, I needed a job. The beginning of the winter term was still several weeks in the future and I knew that many students were employed in the laboratories of the Farbwerke Höchst for the summer. So, I walked the ≈2 miles to the impressive '**h**' of the company buildings shown in Figure 7. The enormous factory complex is next to the small town of Höchst. At that time, there were ≈30,000 employees in what was colloquially called the "Rotfabrik" (red factory). The name came from the Alizarin Red which was one of the first products by the, then, "Teerfarbenfabrik Meister, Lucius & Co." Just as I had found out in my own laboratory (see page 3-19), working with dyes, easily changes the color of your skin, so the workers developed a reddish tint of their skin. Now, the methods are sufficiently improved, so that everyone's color has long returned to normal.

Figure 5 *Typical hut in the refugee camp in Frankfurt-Griesheim. The children did not mind the drab environment. Picture taken in September 1953.*

[1] One of the three largest chemical companies in Germany (Farbwerke Höchst AG, BASF SE, and Bayer AG), all formed in the second half of the 19th century. Today, the Farbwerke Höchst are fully absorbed into the French concern Sanofi-Aventis, headquartered in Paris. In 2007, these three mammoth chemical and pharmaceutical concerns had about 100,000 employees each.

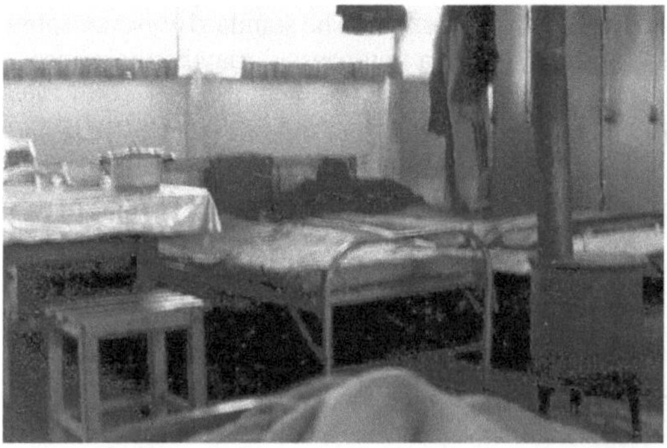

Figure 6 *View from my bed in the refugee camp in Frankfurt-Griesheim. Cooking had to be done on the stove shown on the right. Photo of September 1953.*

As one may have guessed, when I inquired about a job in the employment office, the answer was: "The semester has been over for a few weeks, all students applied for jobs long before that. Now there are no additional student positions until early in 1954." This was disappointing. I had thought there would be no problem.

When I was standing back at the gate, the guard must have seen my misery. We got to talk and he explained to me the inner workings of the company. Suddenly, a well-dressed man went through the gate with a friendly "guten Morgen" for the guard. The guard turned to me and said, this is one of the directors. Run after him. If anyone can help you, it is he. I did. He took the time to listen to my story and first gave me the same information. The student jobs are a learning opportunity and can only be entered at the beginning of the vacation. There are lectures as part of this program and these are half over by now. But, he said, this is not what you really need. You need a well-paying job as a "Hilfsarbeiter" (an unskilled, temporary worker). He went with me to the nearby employment office, and soon I was processed for the job, including the medical exams.

Figure 7 *Signs of progress at the Farbwerke Höchst in 1953. The employment office was to the right of the "***h***" formed by the bridge and adjacent building.*

By noon, I had all the papers and was hired as of the next Monday to work in the pilot plant ("Versuchsanstalt"). Since I really wanted to see the interior of the chemical factory and get the best possible pay, I had signed up for shift work, alternating between day and night shifts. Each lasted from six to six, 12 hours, but three times a week only, followed by an 8-hour shift and a three-day long 'weekend,' which could fall anytime within the week to make the total working time 44 h. I liked this arrangement, since either the days were used fully for work and sleep, or they were entirely free. The first period of work was from September 28 to November 14, cutting the first two weeks of classes in order to gain more income. In 1954, I spent two more such periods at the "Rotfabrik:" March 8 to May 7 and July 28 to October 14. Later, the bus shown in Figure 7 brought me back and forth between Frankfurt and Höchst.

On Monday, September 28, I walked to work for the first time. At work, I got the appropriate safety-clothing and shoes and a locker to keep my things until quitting time. Next, a time card was issued to record time-in and time-out, and a brief lecture was given by the director of the "Versuchsanstalt." He was very helpful, explaining the detailed working of the section to a 'fellow chemist' and also gave me a good introduction to the full-time workers. The standard crew accepted me, fully realizing that at the moment I was one of them, but in a few years, I may be their boss.

After they found out that I would not tell on them, they let me know all the secrets, of how they made life acceptable during the night shift. They certainly did not come to work fully rested after eight hours of sleep and ready for action, as I did. They had to fill-in a few more hours of sleep to make it through the shift of 12 h. Working during the night permitted me to do not only menial work, but also tackle the most difficult jobs. This was practical training for a chemist at its best. It was great fun to balance a three-storey distillation column with several exit points. I always had a notepad with me and weighed not only the ingoing and outgoing chemicals for the entry into the log book, but also the available intermediate weights, for my notepad. There, I found out how much of what was lost during the operation and where it went, in other words, evaporated into the air, or was dumped into the river Main, flowing nearby. (Today such losses are not permitted any more.)

Other things I learned were how to solder lead pipes, roll tall compressed gas cylinders with one hand at the top and twirling it with the other in a straight path (just like the garbage cans, see page 1-20) and finally, judge the big 1,000 L (250 gallons) steel barrels. If these huge barrels are filled with liquid, you can move them by pushing slowly, never lifting any unnecessary weight, and watching that on down slopes, they do not run away. If, however, the content was frozen, the solid plug needs lifting to roll. This is quite difficult to accomplish. Dangerous, however, is the violent forward motion once the center of gravity has passed its maximum elevation. This makes the barrel run away and break through anything in its path, particularly when moving downhill.

Evacuation drills were required periodically, and actually needed occasionally. It is an interesting experiment to try to run away from a poisonous gas leaking from a line break. You may think it is just a hundred feet down the stairs to reach the door, but trying to run without breathing makes you quickly lose consciousness. (I tried it.) It is better to have a gas mask handy.

After a while, I found out that I was one of the few who got to the punch clock later then the check-out time. At quitting time, there was a long line of workers, trying to coax the clock to jump ahead, nobody wanted to stay longer inside the fence than the end of the howl of the siren. Overall, I had fun, could learn a lot, got well paid, and (most importantly) knew it was not forever. I learned from comparing it with road building (see page 5-4), that a measure of liking a job is to note how time flies and that you never need to ask: "Is it quitting time yet?"

Getting back to the 'home,' shown in Figures 5 and 6, reminded me that as fast as possible I needed to find a place to live. The standard rooms for rent were too expensive, but the university could help. They offered me a stay for the rest of the summer in the student dormitory for foreign visitors in the Mertonstraße. In German universities, dormitories were not common for the general student body. Everyone had to find his own, private accommodations. So, both Heidel and I started to look for a permanent place to live and not to be homeless when the semester started. In short order, Heidel found a building under construction in the Mendelssohnstraße 81, shown in Figure 8.

<u>*This house was much too fancy for us*</u>. At the bottom floor were business accommodations and a store. The next four floors were good-sized apartments, but, as was customary at that time in the FRG, up front, you had to pay a substantial "Baukostenzuschuss" (contribution to the building cost, which resulted in a lower rent). This was impossible for us. But the information board about the building going up indicated in small print that the top floor consisted of 18 rooms for rent without "Baukostenzuschuss" (see the dormers in Figure 8, nine on each side). Was it possible for us to rent a new, unfurnished room? This should be cheaper than a small furnished room or

Figure 8 *Our room, top floor, second from the left in back. Mendelssohnstr. 81, near the "Palmengarten."*

apartment. As soon as possible, we appeared in the apartment of the owner, Herrn David Taic. He was a Jew who had survived the holocaust and was now quite active to get his old houses repaired, and he also built new ones. We were the first to inquire about a room. He seemed to like us as tenants, but we were not married! We told him that we planned to get married as soon as we had a place to live and had collected all the necessary papers from the GDR and the FRG. He was agreeable and suggested a monthly rent of 50 DM-West, something we could afford. We immediately signed the papers and got to work to make the room livable. It had an area of about 10×15' with a sloping ceiling. Two bath rooms were located on each end of the long corridor. We added two amenities. A sink within the room, and a fluorescent light replacing the incandescent bulbs. I had calculated that the lower use of electricity would easily make up the purchase price within a year.

We were now both working full time and not so poor anymore. After the Hollerith-job at the university was over, Heidel found full-time employment in the office of "le rouge baiser," the well-known lipstick company.[1] Heidel was well liked in the office and always could return in semester breaks. Her boss, Herr Freyer, even liked me to join the firm. He suggested, I should change my field of study to cosmetic chemistry. In the meantime, he used my services for making advertisement photos of the "le rouge baiser" display windows in the city. One example is illustrated in Figure 9. Naturally, for such photos I needed an electronic flash as they had just come on the market.

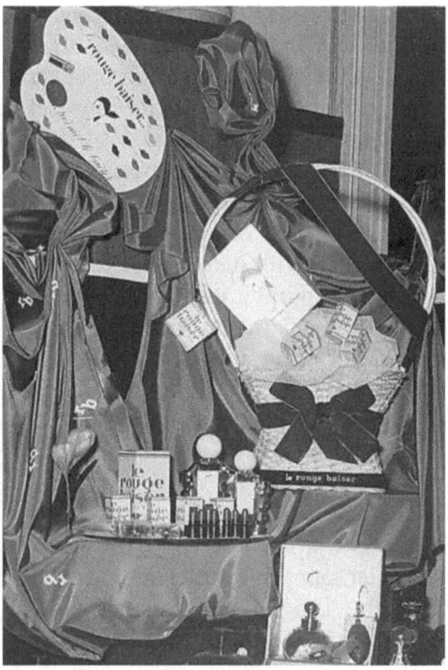

Figure 9 *Display-window. My first experiment with an electronic flash, 1954.*

[1] On their 2007 web-site one can find the following statement: "Rouge Baiser is arguably one of the best known names in European make-up: a lipstick specialist since 1927. This sophisticated brand tops the fashion stakes in the 1950s and becomes the hallmark of the stylish, emancipated and alluring femme fatale. With packaging like jewelry, and products inside featuring ground-breaking formulas tested and created in the firm's in-house laboratory."

Figure 10 *Frankfurt "Hauptwache" as it appeared in 1954.*

Soon I also had earned enough money to buy an upgrade for my small Exa to the "Exakta Varex" described on page 4-27. Finally, color photography was new and needed to be tried out. Figure 10 brings an example showing the effort in 1953/54 to make Frankfurt rise from the ashes again. All of these expensive 'toys' of mine were largely paid for with the money coming from the side-line job as a photographer.

Figure 10 is a picture of the city center near the "Hauptwache." The part of Frankfurt between this city center, the "Hauptbahnhof," reproduced in Figure 11, our house, Figure 8, and the University of Figure 4, was the area we knew best. On going to work, this was extended by the road the bus took to and from Höchst (30 min each way, see Figure 7).

Early in November we moved to the Mendelssohnstraße 81 (see Figure 8). Heidel's friend Gisela (Ostrowski), whom she had met in the refugee camp and who had introduced her to "le rouge baiser," was in a similar situation. She was engaged to Günter (Fricke), a dental technician. On Heidel's suggestion, they also rented a room, just across the hall from us.

How did we find the proper furniture? Checking in the camp, we found that the city gave help to buy basic furniture for refugees moving out of the camp. Yes, Heidel would qualify, but I would not. I was registered in a different province. So, we ended up with limited new furniture: One wardrobe, one chair, one table, and one single bed, all for a very reasonable price. The second chair we got from Gisela, who insisted on (and could afford) two soft chairs. With the bed, we made do with by putting our suitcases in back of it, thus a few inches were gained. During the day, the bed served as a sofa.

Figure 11 *Frankfurt Hauptbahnhof, photographed in 1953.*

Additional purchases were the following, an old chest of drawers with a hotplate on top (it passed as a kitchen, see Figure 27, below), a used FM radio (on a covered box, it was our entertainment center, see Figure 15, below), a shelf in front of the window (it served as refrigerator during the cold season, see Figure 26, below), and a coffee table (see Figure 24, below). These completed the furnishings. When looking out of the window, one had the view of Figure 12.

Figure 12 *View from our window facing to the back of the Mendelssohnstraße late in 1953. Not all war damage was repaired as well as in our new house.*

On November 11 we celebrated Gisela's birthday, and our move into our respective rooms, as well as Heidel's and my upcoming continuation of the study of law and chemistry, respectively. The study was to begin on November 15. Since this was an important milestone, we all went to Kempinski's, at that time the premier café in Frankfurt, near the "Hauptwache." The next time we saw Herrn Taic, he said: "I see you made quick friends with your new neighbors. I noticed you at Kempinski's." Was this Frankfurt so small?

Now, that the five months of life as refugees were over, one might ask the question: *Why did we work so hard to get out of the camp?* Would it not have been easier to stay within the safety of the camp and let bureaucracy take its time to place me at the University of Heidelberg, and Heidel into the University of Frankfurt? In time, we probably would both have been able to finish our degrees.

We both had learned quickly that the refugee camps were a charity, and it was our goal to be independent and to stand on our own feet as quickly as possible. We were glad to accept charity when absolutely necessary, but it was our goal to be able to take control of our life again.

By no means did all refugees in the camp have the same attitude. The living conditions were primitive, but the camp offered security. What if you lose your job? You can't simply go back then. The camp offered a warm place to sleep, three sufficient meals, and usually even some minimal pocket money. It supports one important human trait most perfectly, the *laziness*.[1] Very little needs to be done in the camp to keep it livable. Also, one could find a suitable job without leaving the safety of the camp. This would support the other human trait, the *greediness*.[2] There were many who developed a certain luxury in the camp through outside work by buying an expensive radio, a motorcycle, some furniture, new clothes, mattresses, etc. But, they delayed leaving the safety of the camp. They had developed the 'mentality of a refugee.' They mistook charity which gave them the 'comfort' of the camp as a gift, due everyone in this position to promote their *welfare*.[3]

Looking to refugee camps worldwide, one finds that they fail badly if they only warehouse the people without promoting (and sometimes even forcing) the hard work needed to get on one's

---

[1] (Oxford University) Dictionary definition: "The unwillingness to work or use energy."

[2] (Oxford University) Dictionary definition: "The intense and selfish desire for wealth, power, or food."

[3] (Oxford University) Dictionary definition: "The health, happiness, and fortunes of a person or group," but also "the action or procedure designed to promote the basic physical and material well-being of people in need, chiefly the financial support given for this purpose."

feet again. Naturally, this takes efforts on both sides of the fence of the camp. The established population supporting the camp must be willing to accept the refugees, and the refugees, in turn, must be willing to adjust to the new environment and exert themselves to fit in, if necessary with jobs not up to expectation. As students, we found great support from fellow students and the university administration. The general bureaucracy, in contrast, was afraid we would take jobs from the local population. The economic upturn in the FRG was considerable since the partitioning in 1949 (see page 3-28), but it had not reached a level such that the 1–2,000 refugees crossing the borders every day could easily be absorbed. In the next year, 1954, we felt more welcome in the US.

With the political stresses of the GDR removed from our lives, *our interest in world politics had decreased considerably*. What remained was to understand the political developments causing the relaxation of the harsh conditions put on Germany at the end of WW II (pages 3-29–30).

The Cold War accelerated continually from its beginning in 1945 (page 2-23). In 1949, both the FRG and the GDR were created (pages 3-28, 29) and the US joined the North Atlantic Treaty Organization (NATO). The FRG became part of it in 1955. The Warsaw Treaty of Friendship, Cooperation and Mutual Assistance (Warsaw Pact) was also created in 1955 and included the GDR. The Cold War had divided the WW II enemies of Germany into two groups, with Germany part of their conflict and located right at the center. Each side relaxed the economic restrictions in their share of Germany and made it into an important part in their Cold War effort. Economically, both sides tried to rebuild. The Marshall Plan (1947, see page 3-28) helped western Europe. The Council for Mutual Economic Assistance (COMECON or CMEA), formed in 1949 in Moscow, governed eastern Europe. Militarily equilibrium was approached in 1949 when the USSR attained the atomic bomb. The Cold War and ended only with the thawing in the 1990s (see also page 9-23).

In 1952, the US, Great Britain, and France had officially ended the state of war with Germany (and separately also with Japan, where it meant the end of occupation). With the FRG added into the NATO, the division of Germany had gained a completely different format from what was proposed on its inception at the Allied conference in Teheran in 1943 (see page 2-18). The split was now between a communist dictatorship in the GDR and a capitalist democracy in the FRG. Obviously the Germans had little choice in the division. From the history of the first Reich (see page 1-1, Footnote 3) one notes that Germany consisted for a thousand years of many Kingdoms and Dukedoms. Within these, for example, the Prussians did not like the Bavarians, and vice versa. This means that the division was not the main problem, the missing freedom in one part, was.

The 'Third World' changes[1] were followed by us even less than the Cold War developments, although in the 21st century they seem to govern all politics. Much of the decolonization of the Third World falls into this time period. Especially troublesome turned out to be the 1949 declaration of the Peoples Republic of China (PRC) by Mao Tse-tung, followed by the annexation of Tibet by

---

[1] Today 'Third World' is synonymous with developing countries, regardless of their political status. The French demographer Alfred Sauvy coined the expression in an article in *L'Observateur* of 1952 by analogy with the 'third estate,' the commoners of France before and during the French Revolution—as opposed to priests and nobles comprising the first and second estates, respectively (a similar division of estates was used in Great Britain: The heads of the Church, Lords, and Commons). Like the third estate, Sauvy wrote, "the third world is nothing, and it wants to be something." The term Third World referred during the Cold War to countries unaligned with either the NATO bloc (1st World) or the Warsaw Pact (2nd World).

the PRC in 1950 and the recreation of the Republic of China by Chiang Kai-shek on the island of his refuge in Taiwan, also in 1950. The Korean war (1950–53) ended in a two-Korea solution. At the 1954 Conference on Indochina, the originally French-dominated Vietnam was also divided (after the defeat of the French) into a northern, communist-dominated Democratic Republic, and a southern, US-supported Republic, and Cambodia and Laos became independent kingdoms. This led to the well-known, disastrous result of the Vietnam War, mentioned again on page 8-52.

Other problems arose in the Middle East when Great Britain, the US, and the newly created UN (see page 2-23) failed to resolve the Arab-Israeli conflicts, causing the declaration of independence by Israel in 1948. The British and Soviet invasion of the Iran in 1941 to prevent an allying of Iran with Germany was followed in 1953 by a joining of the US in the overthrow of the elected Prime Minister Mossadegh and reestablishing of the Shah (Mohammed Reza Pahlavi). The Shah, then, became increasingly autocratic in his effort to modernize Iranian industry, which finally led to the Islamic Revolution (1978/79) and further unrest in the region.

The review of these problems makes it clear that my decision not to enter politics based on personal experiences on page 3-30, would have also been reached from a broader study of the politics in the First, Second, and Third World Countries in the 20th century. (See also page 9-29).

Back to November 16, 1953, the date of the festive *admission to the "Johann Wolfgang Goethe-Universität" for Heidel and me*. What a difference from the ceremony at the HU four years earlier! (See page 4-12). We also attended it together, but in 1953 barely knew each other.

The problems of the first semester in Frankfurt were to get the new university to recognize the prior study. For Heidel, this meant of her eight semesters at the HU, only three were recognized. The law in the GDR had changed so drastically that the semesters 4–8 were declared not acceptable. Heidel's law study which was almost finished in Berlin, was reset in Frankfurt to the 4th semester. After coming to the US, it was even worse, continuing her law study would have meant starting again with the first semester (but conceding a bachelor's degree).

For me, things went better. Chemistry is an international profession. I had a second interview, this time with Professor Royen, the director of the "Chemische Institut." He thought my "Übungsscheine," discussed on pages 4-10, 26 and 34, were sufficient to bypass all the laboratory work and apply for the "Vordiplom" (see page 4-10–11 and 25). His laboratory assistant, however, noted that the number of quantitative analyses done in Berlin (see page 4-25) were only half that required in Frankfurt. My argument that our error limits were half that required in Frankfurt and it was not permitted at the HU to use commercially guaranteed reagents, did not impress him. Professor Royen finally made a compromise: "Do one analysis of each type. If it is correct to *our* standards, you need not do the additional ones." I had to do one single gravimetry, one gravimetric separation, six different titrations, and one full analysis. By the time I started my second work period in the Farbwerke Höchst (page 5-5), I was finished. I had bought all new reagents at, for us, a large expense. For the first time, someone had done *all* analyses without a single mistake! It certainly helped to have had a tough lab in Berlin. Immediately, I was accepted as an 'expert' and all my reagents were in high demand. By selling them, much of their initial cost was recovered. Another laboratory I completed was a 14-experiment quantitative physical chemistry laboratory with written extensive laboratory reports which still sit on my bookshelf.

Then I could go to the "Vordiplom." It was completed on June 1, 1954. The exam started in physics with Professor A. Magnus. Since I never had heard his lectures, it was expected to be difficult. At the outset, he asked me questions about optical microscopy. This lasted for about half an hour and I estimated that I did all right, but not outstanding. I stumbled over the discussion of the limit of resolution of the microscope. When he wanted to conclude the exam, I knew something had to be done to achieve a better grade. So I asked: "Can we not go on with the discussion of microscopy and talk about electron microscopy and the new field-ion microscope?" He was interested and had not seen a demonstration as I had at the TU in Berlin the year before (see page 4-25). We spent another half hour discussing the pro and con of the new methods. I left, having made a new friend. A year later, he wrote an important letter of recommendation to Northwestern University. My grade in this trend-setting exam was a 'very good.'

The next exam was in physical chemistry. Professor Hartmann knew me better from his lecture ande laboratory. Grade, 'very good.' This was followed with inorganic chemistry by Professor Royen. He was quite familiar with me from the admission interviews. Grade 'very good.' The last exam was organic chemistry. Since its main study was to be in the future, it was the easiest, particularly since I had started the study of organic chemistry in Berlin with a lecture in "Theoretical Foundation of Organic Chemistry" (see page 4-11). Professor Wieland, well known among organic chemists, ended the exam by handing me an inscribed copy of the 34th edition of his book, initiated in 1894 by Professor Gattermann: "Die Praxis des organischen Chemikers," a sign, that I had passed *all* four exams with the grade of 'very good.' I still have the book on my shelf and it helped in the following year with the organic laboratory at Hastings College. There, I could finish all preparations in record time by being well prepared (see page 6-14). The "Vordiplom" is shown as Figure 13.

<u>*This left only course-work and the beginning of the thesis*</u> research for the "Diplom" (masters degree). The semesters winter 1953/54 and sommer 1954 saw me attend the following: Advanced Inorganic Chemistry III & IV of Professor Royen (4 h, each), Advanced Physical Chemistry Parts I & II by Professor Hartmann, Inorganic and Organic Engineering by Professor Winnacker, and Theory of Solids, given by Professor Hund. All were given by professors of world renown. This heavy course-load in Frankfurt, taken in addition to the laboratory work, helped me to be freed of all of the initial courses when later entering the graduate school at Northwestern University (see page 6-21).

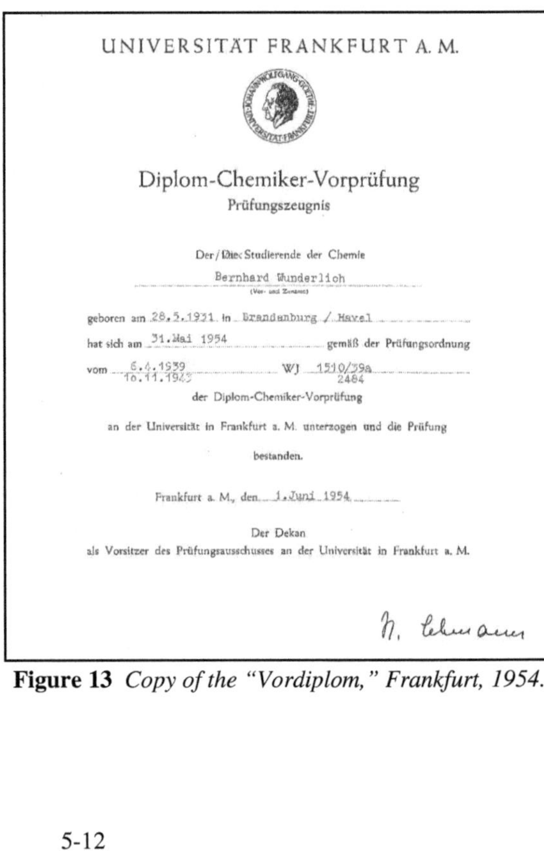
UNIVERSITÄT FRANKFURT A. M.

Diplom-Chemiker-Vorprüfung

Prüfungszeugnis

Der/Die Studierende der Chemie

Bernhard Wunderlich

(Vor- und Zuname)

geboren am 28.5.1931 in Brandenburg / Havel

hat sich am 31.Mai 1954 gemäß der Prüfungsordnung

vom 6.4.1939 / 10.11.1943 WJ 1510/39a / 2484

der Diplom-Chemiker-Vorprüfung

an der Universität in Frankfurt a. M. unterzogen und die Prüfung

bestanden.

Frankfurt a. M., den 1.Juni 1954

Der Dekan

als Vorsitzer des Prüfungsausschusses an der Universität in Frankfurt a. M.

Figure 13 *Copy of the "Vordiplom," Frankfurt, 1954.*

But I get ahead of my story. Before doing much of the laboratory work in quantitative analysis, I had to complete several important personal items. First was the "<u>*Exmatrikulation*</u>" <u>*from the Humboldt University*</u>. We had left in a hurry on June 17, 1953 and got recognized as political refugees (see the Preface and pages 4-35, 5-1, and Figure 3), but we certainly would like to stay in the (apolitical) graces of the Humboldt University, a university which certainly would outlast the GDR episode, although nobody could know that it would last longer than the Second and Third German Reich. So, on October 17, 1953, Heidel and I wrote letters to the "Studienabteilung" (see page 4-25), asking what we could do to be properly "exmatrikuliert" (removed from the student register). They promptly answered to Frankfurt that we needed to submit the "Studienbuch" (see Figures 6 and 13 of Chapter 4), the student ID (see Chapter 4, Figure 8), and a, to be filled-in, request form. We sent all and received the officially certified "Studienbuch" back with the remarks "gelöscht" (cancelled), January 13, 1954, with the comment: 'Nothing detrimental is known about his/her conduct.' It looked like everyone concerned was happy about the way things went. On April 30, I even got a letter from the "Chemische Institut" with a certificate of my having passed the "1. Zwischenprüfung," which I needed for the "Vordiplom." All was in order, and in 1968, I even was invited by the "Akademie der Wissenschaften" of the GDR to lecture at their "Faserforschungsinstitut" (see page 8-46–48).

Next, we had to get the papers together to get married, as we promised Herrn Taic (see page 5-7), and at the same time, also <u>*work on the emigration to the United States*</u>. The American Consulate was in Frankfurt near the "Palmengarten," just a couple blocks from our house, so that we were the liaison to do all the initial legwork. The immigration papers went under Vati's name, as they did in our first attempt,(see pages 3-36 and 4-29), and it was no problem to add Heidel.

But for us, the decision to go with Vati to the US was difficult. In Frankfurt we had managed to have a new beginning. For me, the study was practically guaranteed to the PhD degree. It would take another 3–4 years. For Heidel, it would have taken 3 semesters to the "Referendar" exam and, perhaps one more year to the "Assessor" exam. To give all this up for the uncertainty of becoming an immigrant? Vati's argument, however, was convincing. The Soviet Union is too close for comfort. Look at Joachim and Manfred. They are better off being as far away from a communist dictatorship as possible. The USA would most likely be the last place that would become communist, and there was a good chance that the US will win the Cold War. In addition, he said: "Try it for a year. If you can't get started by then, you can always go back." Could we? We checked. Yes, we could resume the fellowship after a year in the US. So we agreed, and by now, when asked about a return, we say: "The trial-year is not over yet!" (See the pages 6-14–16.)

Frank had in the meantime also left the refugee camp in Weinsberg to accept a job as construction manager in Reutlingen (some 60 mi south of Weinsberg). He joined us two years later in the US (see page 6-30), while his mother-in-law stayed in Heilbronn, close to Weinsberg.

Before getting all the papers in order, <u>*I had to travel once more to my new hometown in Werbach*</u> an der Tauber—about 50 mi NE from Weinsberg (see Chapter 3, Figure 31). Vati and family had been moved from the central camp in Weinsberg to a smaller one in Werbach, and later to their final apartment in Großrinderfeld. The mayor's office in Werbach had the jurisdiction over my getting a passport for travel to the US, as well as the papers to get married. While in Werbach, I could also try out the new color film. An example can be seen in Figure 14.

Many questions were discussed when I was in Werbach. Vati was permitted to look for a new roofing business in the framework of a loan supported by the "Lastenausgleich" (equalization of burden). A possibility arose in Freiburg. Vati went there to interview the widow of a roofer who had a business comparable to Vati's. Vati liked the property, but to get to know the typical Freiburgers, which would have been vital for a successful business, he thought, was difficult to

Figure 14 *Mutti, Vati, and Manfred in Werbach.*

accomplish. The local dialect was hard to learn, and so was the habit to do business in a wine cellar, rather than in a beer hall. Seemingly small things, but when it goes to rely upon someone to fix the roof, you want to be sure you can trust the craftsman.

He finally had decided that it would probably be better to go to the US and start new in the small town of Hastings NE (see Chapter 3, Figure 32). Indeed, the business habits there were also different, but it made little difference that, initially, he did not even know the language. The quality of his work was quickly established by word of mouth: "There is this new immigrant from Germany who still does the work as they always did in 'the old country.' He will do a good and honest job." After a short time he had more work than he could handle.

When I got home to Frankfurt from my second hometown, Werbach, we could settle our future plans. **Getting Married** had now the number one priority. We could not expect any relatives to show up. Naturally, our new friends Gisela and Günter (see page 5-8) had to be our witnesses, but since they also were getting married, we did the same for them, and together we made the smallest wedding party possible for two couples.

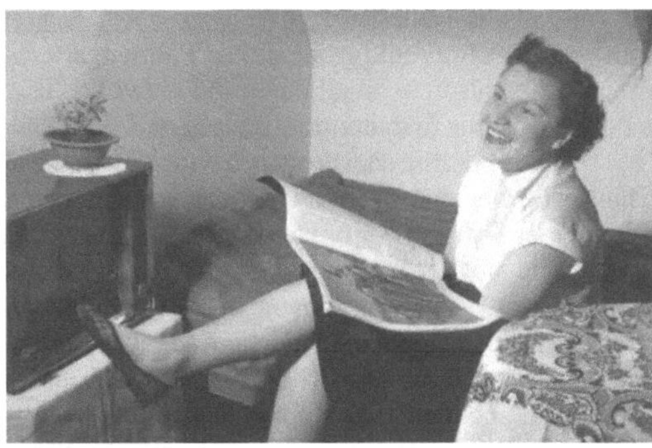

Figure 15 *Heidel, awaiting my return from Werbach an der Tauber in 1953 (in our entertainment corner).*

HEUTE
HATTEN
HOCHZEIT

BERNHARD WUNDERLICH

ADELHEID WUNDERLICH

GEB. FELIX

28. 12. 1953

FRANKFURT/MAIN
MENDELSSOHNSTRASSE 81

Figure 16 *Our wedding announcement.*

Because the waiting period for the civil wedding in Frankfurt's famous "Römer" was 14 days, and we were not aware that weekends and holidays were not counted, the wedding day, thus, was pushed to December 28. Our announcement is reprinted as Figure 16. We managed to cut the cost of it by convincing the printer to use the identical layout for our and the

Fricke's announcements, just the names needed to be changed. Before, we wanted to have a honeymoon in Switzerland over the Christmas holidays. Besides the change in date, however, a problem arose.

The stipend for our study that was to be paid monthly, starting on October 1, 1953. It was not paid yet! As time went on, our savings from the vacation jobs shrunk. The two-week honeymoon was changed to one week, and ultimately to just three days in the nearby town of "Königstein im Taunus," visible from our window. Finally, we even were not sure if our finances would last to January 1. Extreme savings were necessary.

On the wedding day, captured in Figures 17 and 18, Günter and I walked to the "Hauptbahnhof" to hire two taxis (we would not squeeze into one). Then we walked home, to await the taxis to get us to the "Römer" (Figure 18) and after the ceremony, to the bus station to go to Königstein (Figures 19–21). There we had booked two rooms and a fancy evening dinner for the four of us. We were somewhat disappointed that the receptionist did not want to check our brand-new marriage licences. Perhaps, it was obvious. One more surprise, when we paid our taxis, my bill was a good bit lower than Günter's! I, for reasons clear from page 3-12, had hired a diesel Mercedes, Günter, one with a gasoline engine which is less economical. Otherwise, the two taxis were

Figure 17 *Ceremony with lilac in the Römer.*

Figure 18 *The "Römer" in Frankfurt. The ceremony was held in the second floor center. The exit was over a ceremonial bridge from the restored old part to the new wing on the right which had to be fully rebuilt.*

Figure 19 *Exploring the castle with wife Heidel on the next day, on December 29.*

Figure 20 *The village of Königstein seen from the castle ruin Falkenstein, originating from them 14th century.*

of the same type and color (a shiny black). On the next two days the four of us explored the castle and the city. The days went by much too quickly.

The third day was already New Year's Eve. This time, we celebrated in Gisela and Günter's room as seen in Figure 22. It shows that there was still enough to eat. Much had changed since the New Year's celebration in 1952/3 with Hans and Inge (Chapter 4, Figure 20).

In January the stipend was finally paid and we received a 'big' check, the stipend for four months! All at once, our financial worries had disappeared and now our careful savings paid off. This was Christmas in the New Year. We needed clothes, I 'needed' an electronic flash which would have improved Figure 17, a week before, and we had seen an even more modern version of the stainless steel 'silverware' we admired in the student cafeteria (see page 5-3). It was a purchase important enough to take pictures of and is shown as Figure 23. Note the ergonomic design (using the word introduced in the 50s). The knife has a longer handle to transmit more force to the smaller blade. The fork has the tines shortened, so that one can scoop gravy with the fork. Naturally,

Figure 21 *At the castle in Königstein on Dec. 29, 53.*

Figure 22 *The two newly-wed couples Dec. 31, 1953.*

there had to be a butter knife, a fish fork and knife, and look at the two soup spoons. One is more rounded for the more liquid broth, the other elliptical for a thicker soup or stew. Naturally, we bought a service for 12. As a chemist I was quite convinced that silver is not suited for 'silverware,' it tarnishes too quickly, and who likes to eat Ag_2S? I had to convince Heidel of that, especially since, if anything, the stainless steel 'silverware' was more expensive than silver. As we had calculated from the heavy use in the cafeteria (counting the washing cycles a day, etc.), it should last for 25 years or longer. Indeed, it is still in daily use today, after more than 50 years, and looks almost as good as on the first day. Only for a fancy dinner party do we use our impressive gold-plated tableware. It was a gift from Papa after he spent most of his time with us in Troy, NY. If I was not favoring silver, he thought, gold would be acceptable for the tableware for a chemist. He was right, just that it is mechanically too soft to be in daily use or to go into the dishwasher.

Figure 23 *Our steel 'silverware' brand name Pott.*

Now we were ready for guests and a normal life in Frankfurt. Yes, things could get a bit crowded in a one-room apartment. But our bed-couch could easily (?) seat six, as you can see in Figure 24, taken when Fräulein Katz and her fiancé were visiting together with Gisela and Günter. Remember, we had still two chairs to spare (see page 5-8). This came-in handy when later in the year Vati, Mutti, Joachim, and Manfred came visiting to appear for the final visa conversation at the American Consulate, around the corner from our house. They did not only visit, but also stayed overnight. The floor area was just enough to place everyone, with Mutti having been given the bed. We had a great time, specially appreciated by Joachim and Manfred.

Figure 24 *Visitors at Mendelssohnstraße 81.*

How was cooking and baking done in this efficiency apartment? Quite simple, see Figure 25. The "Torte" was produced by Heidel without need of the non-existing oven. The recipe calls for commercial "Bahlsen Kekse" and the filling and glaze takes only the available hotplate (page 5-8). Note that the percolator is self-powered, increasing the efficiency of the kitchen, and in addition, the shiny chrome exterior helped in my commercial enterprise, photography. The freshly developed pictures were pressed with a squeegee to the surface of the percolator, and then the water was brought to a boil. In no time the pictures were dry and they also were perfectly glossy. Figure 26

Figure 25 *Table set for coffee and self-baked cake.*

Figure 26 *Gisela and Günter as guests for supper.*

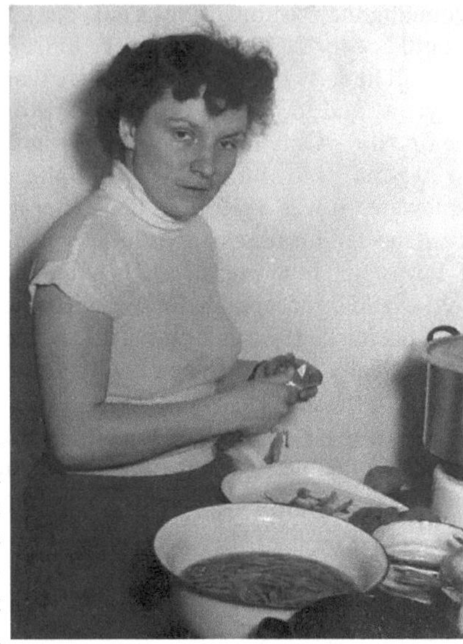

Figure 27 *Work in the kitchen corner.*

illustrated supper guests inspecting the special cold cuts and cheese from the "Käs' Petri." In the back, you can see our only window illustrating the outside shelf. It had to serve as our 'refrigerator.' (See also page 5-8). The dessert and coffee are waiting in Figure 25. The lawyer-housewife in action, not quite ready for guests, and not quite happy to being photographed, is shown in Figure 27. You can see the 'kitchen corner,' quite compact, but well organized.

The visitor in Figure 28 was special. It was Heidel's oldest friend Gabi (see Chapter 4, Figures 11, 12, and 22). She had been able to get permission to travel from the GDR and arrived at the railway station in Frankfurt. It was her first trip "zum Westen." Relative to the life in the GDR, much had changed. For us, the most important change was the freedom we now had. There also was no shortage of any goods anymore. It was only important to have sufficient money to buy things, not important for the casual visitor.

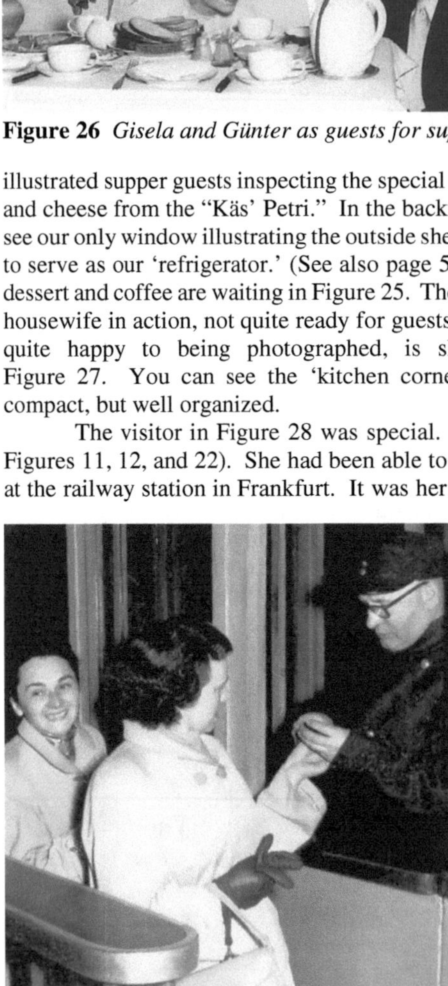

Figure 28 *Gabi arriving in Frankfurt, 1954.*

Standing at the curb and seeing the cars go by, many of new construction, we wished we could afford one. At least a two-seat BMW "Isetta"or a "Kabinenroller." Our justification was that this would get us back faster to and from the university or work. On our first sabbatic, about 15 years later, many students could afford a small or used car. Did they use it to study or work more efficiently? No, we found that being motorized, meant now that they could stretch the weekend into Tuesday morning and enjoy life more.

By now we were quite certain that we would leave for the US later in the year. Tickets were already reserved on the MS "Gripsholm" of the

"Norddeutscher Lloyd" on the third of June, 1954, only a few days after my passing the "Vordiplom." The cabin number was 217, the passage, $180 per person.[1] The official postcard of the ship shortly before its sale to the "North German Lloyd" in the Swedish harbor of Gothenburg is shown in Figure 29 (Figure 17 of Chapter 7 shows its final appearance). Tante Lieschen had arranged that the passage for the whole family was a loan to us, so that this expensive trip could be paid back after, hopefully, earning more in the US. The proposed departure date was October 26, 1954, sailing from Bremerhaven and the arrival was estimated ten days later. The "Gripsholm" of 18,600 "BRT" (GRT) was not one of those fast 'Blue-Riband' steamers of the North German Lloyd of the 1930s, needing less than five days for the crossing. (See the model in Figure 20, Chapter 1.)

Figure 29 *MS "Gripsholm" after its refit in 1950.*

<u>*The last few months in Frankfurt passed rather rapidly*</u>. The exam fell in the second period of work at the Farbwerke Höchst. This was followed by the second semester of study, outlined already above (see page 5-12). The big worry about money was over. The decision to leave for at least one year was made. This also meant, I did not as yet have to search for a professor to work on a suitable thesis. I could enjoy the semester. We also had some more time to discover the town. Figure 29 shows a picture of the "Senckenberganlage" which connected to the university and the railway station. The last work-experience was from July 28 to October 14. Its extra income served to update and expand our wardrobe, so that we did not have to start buying clothes when arriving in the US.

Figure 30 *One of our last strolls in Frankfurt in the fall of 1954.*

Then everything needed to be packed. For this purpose, we had two wooden crates built, assuming that these could come in handy later to store the winter potatoes, as we were used to do. They served well to get our things safely to the US, but there, potatoes were bought as needed, never stored at home for long periods anymore. The two boxes can clearly be seen in Figure 21 of Chapter 11. An itemized, complete

[1] The MS Gripsholm was a Swedish American Liner, built 1925 in Newcastle, England. She was the first transatlantic liner with diesel engines and operated on the Gothenburg–New York route. Its speed was ≈17 kn (≈20 mph ≈30 kmh). (Compare to the SS Bremen, Figure 20 of Chapter 1). She was refitted in 1950 with a new bow and funnels. The MS Gripsholm was sold to the North German Lloyd in 1954 and renamed Berlin. When we used her, she had not yet changed her name.

packing list is also still in our archives in Knoxville. Smallpox vaccinations needed to be renewed and various certificates were to be obtained, showing that we owed no taxes. Finally, Mama and Papa got permission from the GDR to visit us shortly before we left. Figure 31 and in Chapter 11, Figure 20, are reminders of this bitter-sweet occasion. Mama's last words made it even harder: "I know I will not see you again, I will never get into a plane or on a ship for a visit."

All that was left now was to coordinate the luggage. Vati und Mutti, with Joachim, and Manfred left directly from Großrinderfeld, and we, from Frankfurt, hoping that we would see all the baggage together in New York at the pier. It worked, as shown in Figure 21 of Chapter 11. Also, the gifts for the Zarobans were to be organized. For Onkel Richard and Tante Lieschen, I helped to purchase of a "Contax" 35 mm camera and a pair of "Zeiss" binoculars. Also, a cuckoo clock was on the list, and a big box of "Harzer Käse" (cheese). For Marilyn, Vati had his rather valuable violin (which he had not played for 30 years) overhauled to perfection. All arrived safely, except the cheese. This cheese ripens rapidly and then becomes rather smelly (tasting exquisitely), but after one week at sea, we could not stand the smell in the cabin anymore. We had to feed the fish with the cheese. We know by now how much Tante Lieschen must have waited for the cheese. It is very hard to get in the US and if you have once been introduced to it in Germany, with time, you miss it more and more. We do not forget to eat it on every one of our trips to Germany or Austria.

The final steps in Germany are documented with Figures 32 and 33. We said good bye to our friends in Frankfurt. Our apartment was rented to a colleague of Heidel's from "le rouge baiser" (Fräulein

Figure 31 *A last visit from Mama and Papa in fall 1953. The picture taken in the Senckenberganlage.*

Figure 32 *Meeting with Tante Grete from Stralsund who managed to see us off in the Immigrant Hotel of Bremerhafen. The evening before our embarkation.*

Figure 33 *Leaving from the pier of the Norddeutscher Lloyd in Bremenhafen. A last look to Germany from the deck of the Gripsholm, October 26.*

Pointner), who also bought all of our furniture. Then came the train ride to Bremen, and on to the harbor, "Bremerhafen." There we stayed for a day in a low-priced hotel for emigrants, arranged for us by the shipping line. To our surprise, Tante Grete awaited us. She had used all her connections in Stralsund to get a week's permission to travel in the FRG to see us off and to visit Frank in Reutlingen. On the next morning all was ready. We stood at the railing and waved good bye.

The trip to the US started to get serious when entering the rather rough British Channel. The steward was helpful and suggested that the waves would not be so choppy on the Atlantic. He was right, but the waves were bigger and made one feel even worse. Another early bit of misinformation was that children do not get seasick. Quite the contrary. Joachim and Manfred were sick most of the days, with a notable exception, the day of the children's party. We grown-ups had a constant fight of going to eat or go on deck to feel better in the fresh, cool wind. Everyone missed a number of meals. I tried my best to capture a few pictures, as reproduced in Figures 34–36. They speak for themselves and also give a reason why it took us many years before we got again onto an ocean liner. The 11 days were not very pleasant. It did not help that the crew of the ship did their best to be of good spirits. Someone should have told us that when crossing the Atlantic, it is better to choose the calmer months of the

Figure 34 *The family united on the breakfast table on one of the quieter days. Joachim is still not recovered.*

Figure 35 *A game of shuffle board. It was possible only on the few calmer days.*

Figure 36A,B *The Atlantic Ocean, as it was most of the days on our crossing in November 1954.*

year. Now we could imagine how much worse the early emigrants must have felt on their much longer trips by sailing ships. But, ultimately we made it. On the evening of November 5 we arrived outside New York Harbor. The sea was quiet all at once. The ship anchored and waited for docking and debarkation on the next morning. We were impressed by the lights of the city and the moving headlights of the hundreds of cars driving on the highway along the shore of Long Island. The evening meal was, for once, eaten and enjoyed without the feeling of seasickness. Both ladies of our party had made reservations at the beauty parlor on the 5th, to enter the new country in their best appearance. The joy of the beautiful hairdo was tempered, however, when it was discovered that on entering American territorial waters, the prices were readjusted to New York level, almost breaking the banks of both families.

The next morning we were on deck to watch our entry into the harbor and to see the sunrise through the skyscrapers, as shown in Figure 37. It was a most impressive view. Our thoughts went back to the great luck of the whole family through the war years to have survived the bombings and the final fighting (Chapter 2), the great effort to survive the time after the war, described in Chapter 3, which led to the rise of a new, but now foreign-dominated, dictatorship, as detailed in

Figure 37A,B *Arrival in New York City Harbor in the Morning of November 6, after 11 days at sea.*

Chapter 4, and finally the lucky escape in 1953 as summarized in the Preface. The last year had shown some recovery as described in this Chapter, but now life was to change drastically once more. It must have been a particularly anxious moment for Vati. Instead of refugees for the last year, we now were immigrants. How would life be in the US? Vati was over 50 years old! Was he too old to learn the new language? Could he provide a comfortable start and a good education for Joachim and Manfred? Would he ever be able to start a new business of his own again, or would he have to work for the rest of his life as hired hand? For Mutti the question was how would she get along with her older sister. Initially there was great anticipation to meet Tante Lieschen, but what about the future. As children, the two did not get along very well, and now we all were deep into debt to Tante Lieschen and Onkel Richard. Just as Frank was trying to dominate me, Tante Lieschen and Mutti had a similar relationship. How would the living together with their families work out?

For Heidel and me, it was much simpler. We knew the language (we thought) and we could go back if we did not succeed. Our place at the university in Frankfurt to finish the study would still be there for another year, as was arranged at the end of the semester (see page 5-13). For us, it was more of an exiting adventure to explore the 'New World' we had heard so much about.

CHAPTER-06

Study in the US, 1954–1958

Early in the morning of November 6 we arrived in New York City. The view from the ship was breathtaking (see Chapter 5, Figure 37), although—I had thought the skyscrapers would be taller. Only when standing at the foot of one, looking up, did I get an impression of their gigantic size.

The immigration formalities were done on board of the ship. Ellis Island, as a processing center for immigrants, was abandoned the year before, i.e., as soon as the papers were checked and the ship securely docked on the pier, the debarkation started. On the pier, Tante Lieschen and Onkel Richard were waiting, and later, they were joined for a while by our cousin Dick with wife Lola. He was with the Air Force and stationed in Watertown, NY. They all had seen off my other cousin Jim the day before. He went overseas for a one year of duty with the Army in Germany.

It took some time until we collected all luggage. The stowers had sorted all pieces according to large labels of the first letter of the last name, attached in Bremerhafen. After about an hour, we missed only one suitcase. Finally, Vati got the idea, they mistook a W for an M. Indeed, that had happened. I went to the letters M, and there was our missing suitcase, standing almost alone.

The customs inspector immediately got the idea that we were immigrants. He asked what is in these big boxes? They showed too prominently (see Figure 21 of Chapter 11). I answered, I have a list. He did not want to see it, nor have them opened, but he pointed to the late suitcase and said: "Open this one, please." Vati opened it. It mainly contained Joachim's toys. After one look, we could close it and he signed off on all. Next, Onkel Richard got the agent from the express office who marked all for shipping to Hastings, NE, our new hometown in the US.

Then, *with two taxis we went to the Pennsylvania Station*[1] (Penn Station). Several views of it are given in the Figure 1. The gate and colonnades with the famous eagles are seen in the upper left. The huge waiting room is in the upper right figure. Not far from the clock, at 1:45 p.m. of November 6, you can see our party, a close-up is shown at the lower left. One can make out hat and head of Mutti, Vati with his leather coat, and Tante Lieschen and Marilyn with Heidel trying to

---

[1] Penn Station was the monumental entrance to New York City. It was opened in 1910 when sufficiently strong electric engines could move the trains directly into the city via tunnels under the Hudson River. The station was made of pink granite, decorated with colonnades of Corinthian columns, arranged in Doric order. It was the largest indoor space in New York City, with an enormous waiting room inspired by the Roman Baths of Caracalla (built in Rome A.D. 212–216 during the reign of Emperor Caracalla). Its arriving and departing trains were stationed at two separate concourses. In 1963, the structure was demolished (creating an international outrage).

Figure 1 *Penn Station in New York City. Upper left, the impressive pedestrian entrance to the station. Upper right, the huge entrance hall, which was modeled after the 'Baths of Caracalla.' Bottom left, our party, awaiting the departure of our train to Chicago, the 'Broadway Limited'. Bottom right, a coach of the California Zephyr, such as carried us to, Hastings, NE, later photo.*

Figure 2 *Tante Lieschen and Mutti in the train.*

straighten out differences between Oxford and Midwestern English. Joachim and Manfred are hidden behind Vati and Marilyn, respectively, one can make out their black caps. Onkel Richard was getting tickets, and I was taking the picture. Marilyn was the first person we tried to have conversations with who did not speak German. It turned out, that we understood less than we thought. Already on the ship, we found out that British movies we enjoyed, but American ones were often difficult to follow.

By about 3:00 p.m. we had settled in the comfortable seats. Mutti and Tante Lieschen, seen in Figure 2, had naturally a lot to talk about, to catch up with their respective lives. For us, we were waiting for a decent meal, which was not to come. Tante Lieschen had brought a large bag of sandwiches from the hotel they had stayed in. We found out quickly that the diet in the US was much different from what we were accustomed to. The bread was not the heavy sourdough "Brot" we were used to, but a soft, spongy, tasteless support for all sorts of things, like salad and raw vegetables, which do not belong on a sandwich. So we did not eat very much and slowly built up real hunger.

The countryside flew by the windows and certainly looked different. First the sprawling city, then a wide expanse of fields and woods, looking little tended. And the telephone poles! They demonstrated that they did not need to be straight and of identical length. The distances between the cities were much greater than we were used to, and the houses in the countryside were not collected in villages. Finally, there seemed no building code for houses. They were not uniform in number of stories, type of roofs, etc. And always, there was a large volume of automobile traffic.

The streamlined train traversed the 900 mi between New York and Chicago in 16 hours and we arrived early the next morning at Union Station in Chicago. The next train, to Hastings, NE,[1] was the California Zephyr, of which a car is shown in the lower right of Figure 1, photographed at a later time in a train museum. A perfect replica of the Zephyr graces my Märklin electric Z-scale train set (see page 1-24). This train held the speed record in 1939 and went from Chicago via Omaha and Denver, to end in Emeryville, across the bay from San Francisco, a distance of almost 2,500 mi. In Europe one could have crossed a dozen nations over such a distance. *Hastings, NE, was reached 650 mi into the trip of the Zephyr* at about 3:00 a.m. The train left Chicago only in the late afternoon, at 3:30 p.m., so that we had a whole day's sightseeing.

Impressive were the skyscrapers, traffic, and Lake Michigan, which looked like an ocean. We went to the Field Museum of Natural History[2] after Vati managed (without knowing any English) to get us each a hot dog at the station shop. This was more to our taste than the sandwich, but, again, the bun also was no "Brötchen." It was too soft and had no crust. The hot dog was acceptable, but a bit small for our appetite. We spent quite some time among the fossils and dioramas. By the time we met the California Zephyr, we were very tired world and history travelers.

---

[1] Hastings, NE, is a city with a population of about 20,000, still in the central time zone, located south of the Platte River. The city was founded in 1872 at the junction of two railroads and named for Col. D. T. Hastings of the St. Joseph (MO) and Grand Island (NE) Railroad. It is connected to the rest of the US via Interstate 80 (17 mi north of town, close to Grand Island). The California Zephyr still runs the same schedule, once a day in each direction. The US-highways 6 and 34 east-west, and 281, north-south run through town. It has a small airport. The main gateway cities are Lincoln, NE (100 mi east) and Omaha, NE (150 mi east-north-east). Hastings College, a nationally renowned four-year college is located at the edge of town, founded in 1882. It is a private, coeducational, residential, Presbyterian-related, liberal-arts college. During WW II, the largest Naval Ammunition depot was located near Hastings on 50,000 acres on farm land.

[2] The Field Museum, founded in 1893, is named after its major benefactor, Marshall Field. It is a first class museum illustrating "art, archeology, science, and history." Close by are the Shedd Aquarium and the Adler Planetarium which we inspected briefly. They form the Lakefront Museum Campus. All three are among the finest in the world and attract the most visitors coming to Chicago.

There was little to see from the train in the darkness. The unending Great Plains had to wait until we traveled through them by car. Ultimately, we arrived in Hastings, dead tired and ready for bed. Our cousin Don with wife Betty and Tante Lieschen's friends (the Sealys) met us at the station on Monday, November 8, our first day in our new 'hometown' (see page 3-36). Our discovery of Hastings had to await the very late morning and the following days.

The first breakfast in the US was a feast. To our surprise, the bread we scorned on the train tasted pretty good when toasted. All other breakfast items were to our liking too, and for some time breakfast became our favorite meal. Hastings, NE, we noted has a strictly rectangular arrangement. On First Street, there was the train depot which formed the backbone of the city (see Figure 15, Chapter 10). Second Street is the main business street (see Figure 32, Chapter 3). Next, to the north, was a residential area. Tante Lieschen lived on 216 W 12th Street in a beautiful two-level house. On the 'other side of the tracks' (less desirable for living, according to Tante Lieschen, our main source of information), was 'South Street.' Further streets were labeled A, B, C, etc. The north-south streets were named. The Burlington Street (US 281) goes through the center of the city. The highways from the west, US 34 and 6, both go to Lincoln, NE, come into the city on J-Street, with US 34 following Burlington Street before turning east in Grand Island which is the next city, about 20 mi north of Hastings. There was little need for a GPS (Global Positioning System).

Driving through the city, we had the impression of *a city whole and untouched by the terrible past* we had experienced. People were able to benefit from their life's work. Most lived in one-family houses within a park-like environment without fences. Most impressive to me was a neighbor across the street who owned a pretty house within a well-tended lawn and garden. He held a lowly position as a brakeman with the railroad, but he could afford this 'villa' (see page 1-8).

Tante Lieschen and Onkel Richard's house was somewhat larger, for the bigger family. To accommodate us, they had remodeled the house. Mutti and Vati jad a newly added bedroom, Heidel and I had a room in the basement, and Joachim and Manfred occupied a corner of the basement.

Of our cousins (see page 11-2, center bottom), I had met Jim when he was on a visit in Germany in 1937 (see Chapter 1, Figure 12) and we just missed him in New York. We briefly met cousin Richard (Dick) with his wife Betty, as remarked on page 6-1. Cousin Marilyn was about 14 at that time and still lived at home. She took on the job as English tutor to Manfred and Joachim, by then 8 and 10. Both of them had found it very funny when we had tried to teach them some English on board ship, but now they found out that it was serious business. Cousin Don worked in town at the Montgomery Ward store and his wife, Betty, was a nurse, teaching at the hospital.

Onkel Richard had immigrated to the US in 1922 as a tool and die maker. He started to work in Omaha, and since 1941 was Factory Superintendent of the Manufacturing Division of the well-known Dutton and Lainson Company[1] in Hastings, NE. In time, he also had brought his mother and Tante Lieschen to Hastings. A picture of all three is shown in Figure 6, Chapter 11. In Figure 3,

---

[1] The company was founded in 1886, when William M. Dutton and two friends started a small harness and horse collar business which grew into one of the largest manufacturing plants in the nation for horse collars. In 1920, Harry Lainson started a wholesale house and expanded the manufacturing division, resulting in diversification and growth that has led to the company we see today when Dutton-Lainson is one of the country's leading manufacturers of quality products for the marine, agricultural, industrial, and automotive markets throughout the world.

he comes home from work in his 1952 Hudson, for us an impressive automobile.

<u>*In short order our luggage was delivered and our move was completed*</u>. By then, Manfred was frustrated by not being able to converse freely with everyone. Vati had brought his radio (see Figure 14 of Chapter 5). Before leaving, he had it converted to 115 V and fitted with an US plug. When it was unpacked, Manfred's face lit up and he proclaimed: "Now I can finally listen to the children's hour on the radio." After he found out that even the radio could only 'talk English,' he knew, it was better to get to work and learn the language.

Figure 3 *Onkel Richard returning home from work in his Hudson Hornet, 1952, a highly regarded car.*

Learning the language was accomplished quickly. Next day, both Joachim and Manfred were enrolled in school. The teacher spoke no German, but was quite aware of what had to be done and made sure both were placed into the proper grade for their age. In no time they had adapted to the new environment, and by the next school year, they were doing well.

Heidel was quickly integrated into the household chores, as shown in Figure 4. What was I to do? The solution was that **Hastings has a College**, as indicated in Footnote 1 of page 6-3. Its Logo is illustrated in Figure 5. Tante Lieschen was quick at hand to introduce both of us to the College Administration, and in short order, we had a scholarship to 'audit' the fall term, which was already in full session.

It took me a while to find out what it meant to 'audit.' I registered to the surprise of everyone in the following subjects: 'Qualitative Analysis,' 'Physical Chemistry,' 'Differential Calculus,' 'Differential Equations,' and 'Physical Optics,' a total of 16 credit hours. On purpose, I stayed away from non-science topics. Improving the language was my goal, I knew that getting into the humanities would be hard because they were designed for US students beyond high-school level, which I was not, and not to teach me the language.

Figure 4 *Tante Lieschen and Heidel at work.*

The level of instruction in the sciences and mathematics was much lower than I was used to. The language spoken was a different problem. Especially in chemistry, I found that in the first and last five minutes of each lecture some general conversation was held between the professor and his students which I absolutely could not follow and could not take part in. Only much later did I find out, this was a discussion of homework problems. As a result, I did none. I never knew of homework beyond high school. When we got into the subject matter and the teaching was supported by equations on the blackboard and drawings, I had no difficulties anymore. In fact, the students found out faster than the professor that I was a good source for information and solutions to their study problems.

Figure 5 *Hastings College Logo. "Pro Rege."*

By the time the final exams rolled around, I certainly made sure that I took all of them although, by auditing, I did not have to take any. Especially interesting was the chemistry exam. It was a standardized multi-choice exam by the American Chemical Society, requiring the marking of an answer sheet, and grading by a master punched-card key. I was the last student turning in my answer sheet. It was difficult for me to read all 100 questions and the five possible answers for each. My reading in English was just not fast enough by then. But I finished, and Dr. Fleeharty wanted to find out immediately how the new student did. When he put the key over my answer sheet, all was white, i.e., no answer was correct! He and, even more so, I, was surprised. Then he saw that the card was not matched properly. A ¼" shift was necessary. Now we were even more surprised. All the answers were correct! "How is that possible?" He asked. "I told you only the first 50 questions were to be answered. We did not treat the first half of the material in the exam, this is due only in the next semester!" Naturally, by then, I would not take the second half of the course.

HASTINGS, (NEBRASKA) DAILY TRIBUNE, SATURDAY, NOVEMBER 20, 1954

A New Home in Hastings for the Wunderlichs of Germany And They'll Be Welcomed at Openhouse Here Sunday

Recently arrived from Germany are the Wunderlichs, shown relaxing at the home of Mrs. Richard Wunderlich's sister, Mrs. Richard Zaroban. From left, on the divan are Mr. and Mrs. Bernhard Wunderlich, Mrs. Richard Wunderlich and Mr. Wunderlich. Seated, at the left on the floor, is Manfred, eight, and his brother Joachim, ten. They and Bernhard are sons of Mr. and Mrs. Richard Wunderlich.

Figure 6

__Next was our official welcome in Hastings with an 'Open House,'__ properly announced by Tante Lieschen in the local newspaper with a picture which is reproduced as Figure 6. A short article summarizing our trip from Germany was appended. We were very well received by the many neighbors and friends, so many, that we hardly remembered any.

In contrast, practically all of the guests at the open house remembered us. From this day on, it was difficult for us to walk the 12 blocks from home to the college. Pedestrians were hard to find, except in downtown. Every longer distance was traveled in those beautiful cars. As soon as we started our walk, the dogs would greet us when passing the houses, and it would not take long until a car would stop and we were offered a ride.

The next step was for Tante Lieschen to find a roofing firm in Hasting where Vati could start to work. Although it was more difficult for Vati and Mutti to learn the language, they did remarkable well, and Vati was quickly accepted as a valuable worker. He certainly had the most difficult time, since in Germany he did not have to do any of the physical work anymore. Now, he had long, hard working-hours each day and the climate was also much more extreme than in Brandenburg. It was back to start working at 4 a.m. in the morning to avoid the hottest hours in the summer, and in winter, work was often not possible because of the cold and snow. But he made the best of it, and he remarked that it might help him to stay fit and healthy.

This left one last job for Tante Lieschen, who could always be counted on when one was in trouble. We had no money. How could we find a job in addition to the college study? She arranged with Mrs. Welch in the financial aid office of the college for an interview at the *Ingleside*[1] *State Psychiatric Hospital*. Figure 7 is an early picture of the facility. By the time Heidel and I went to the interview, the whole ground was covered with large 50-year old shade trees. We were well received, given a number lengthy, for us rather difficult, questionnaires to fill out. On the next Monday, we started our jobs. For that, we had to have proper, white clothes. They were bought in town, using almost all of our wealth of $35.

The pay was $0.85 per hour, despite a countrywide $1.00 minimum wage. To satisfy the college requirement, we could work shifts from 3:00 to 11:00 p.m. The work was six days a week with one day off, substituting for the Sunday. Our free days were Tuesdays. A typical monthly combined net paycheck for the two of us was $282–291, depending on wether there was a holiday, with a higher rate of pay, in the month or not.

Figure 7 *Ingleside Hospital near Hastings, shortly after its original construction in 1890.*

---

[1] In 1887, the Nebraska legislature had appropriated the sum of $75,000, to be used for a 'State Asylum for the Incurably Insane' to be located at Hastings. The city ultimately donated 630 acres, one mile west of the city limit. The first structure was the three-storey brick building with a central tower, shown in Figure 7. In 1891 north and south wings were added and in 1902 a North Annex was erected. In 1904, an amusement hall was built where dances and entertainments were held for patients. Later, a cottage for gardening and farming was erected. A medical surgical building was added in 1926, and in 1938, a psychiatric hospital. The first patients were received in 1889. Ultimately, as many as 1,200 inmates and resident employees made up the population of Ingleside, which had its own post office. It is known today as 'Hastings Regional Center.' The 1920s saw first attempts at treatment with electro-shock. In the 1930s and 1940s, fever therapy, hydro-therapy, and insulin-shock therapy was begun. In the early 1950s, the biggest breakthrough in treatment came with the discovery of psychiatric drugs, which included tranquilizers. From that time on, it was possible to unlock wards and start a program of rehabilitation of the inmates.

On our first day of work, I distinctly remember the funny feeling the two of us had when waiting in the white clothes in front of Tante Lieschen's house for the hospital bus at about 2:30 p.m. We checked in the administration building, and each of us was assigned someone who brought us to our respective buildings. All buildings were connected by underground tunnels. My building was one of the older ones, matching in style the one in Figure 7. I was brought to the third floor. All windows were barred. My set of keys was for the doors of this building only. My guide opened a door for me to enter and called: "Here is the new one!" And slammed the door shut.

There were about 40 men of all ages, dressed in blue, sitting along the walls of a large room. The patients, I was supposed to handle. It was a terrible sight, and the smell was not very pleasant. In a little while I saw two people in white, like myself. After introductions among us in white, all information about the job was given. I was told that there was nothing much to do until dinner time. This waiting for the next activity was the major part of the work. It allowed me a good amount of time for studying. Figure 8 shows our trio of attendants on duty. We were frequently together, always working the same hospital ward. We got along well, and I am still in occasional contact with George.

Figure 8 *Three attendants at Ingleside, L to R Sheppert (Shep), myself, and George Mohlman, 1954.*

Dinner meant to line up the patients, see that none were missing, and to march to the cafeteria. My own first dinner was in a separate cafeteria, but on the first day, I was at a loss about the choices. Much of the food did not look familiar, and if it did, I did not know the name. As a result, I got my tray filled randomly. When I tried it, all tasted awful. I especially remember a terrible cold noodle salad which I thought was hot. Who eats cold noodles? Almost the only thing acceptable was a glass of milk. All else went back. On my way out, I saw Heidel coming in, and I could warn her about the choices. She did slightly better. But we both lost considerable weight during those first months in the US.

Next morning, we told Tante Lieschen: "This is not a pleasant job. We will be back for dinner, we are not going back to work." We could see that she was upset about our decision. What would happen if we would not support ourselves? As our sponsor she was financially responsible for us, including the return ticket if things did not work out! There were no illegal aliens at that time. Soon we let her know that this was our day off, we would be back to work on Wednesday.

Both George and Shep made proper fun of my pronunciation and showed me how to improve. The big problem of a German is the proper shaping of tongue and mouth for the 'th' (which sounded to me like an 's') the 'r,' and the 'l.' Also, there was the large number of words which had a completely different spelling and pronunciation. I was very thankful for their lessons.

Since the ward I was attached to had mainly incorrigible patients, I hardly ever saw a nurse or a physician. Most medication and examinations (if any) were handled in the morning. Our main jobs were to round up everyone for dinner and occasionally bring a selected group to movies or dances. In the evening, there was a fixed bed-time call, followed by a check to see that no one had disappeared and the proper number of patients was present when the night shift took over. This bed check was not as easy as it sounded. Everyone slept in one big room, but constantly someone was coming or going. And starting to becoming bilingual, counting "eins, zwei," three, four, "fünf," etc. made it easy to miscount with the needed additions and subtractions. On nice days, an afternoon stroll was arranged in the park, which meant continuous watch that none would wander off. If all would decide the leave at once, there would have been little chance to go after any. To get things to work smoothly, a number of patients were graduated to helpers and they handled all recalcitrant patients. This left the small jobs for us, like to hand out of the bags of 'Bull Durham' tobacco which the patients could either roll into cigarettes or smoke in the provided corncob pipes.

The patients could easily be divided into two groups. One group was not far from normal. They were committed when being found drunk or were pleading insanity in the commission of a crime. In case no one who would sign them out and be responsible, even the drunkards would stay for a long time. The ones pleading insanity needed to pass a thorough psychological examination in addition to someone taking responsibility. The other group was, indeed, mentally ill. They often said nothing, were not understandable, or did not make sense in their conversation. Fortunately, there were enough drugs by then so that violent or overly noisy behavior was not common.

Let me mention three 'crazy' stories. One was about our oldest patient. He was in his 70s and was committed at age 20. He knew all about the early construction phases of the hospital (see Footnote on page 6-7) and where to find obscure shut-off valves, tunnel lightings, etc. He also had a position of trust, i.e., he could leave in the morning and work in his garden plot and sell the harvest to the employees, so that he could build up some savings to spend in the store and cafeteria. His English was not very good, he was born in Greece. One day, he went missing. About two weeks later, he was back on his own. With some difficulty we could extract that he had traveled by train to Denver, CO, to see a Greek Orthodox Priest for advice on how to be buried in Greece. I am not sure what was arranged, but he was happier after he came back, just that he now frequently mumbled 'so many cars and people' (in Denver). He had not seen any big city in his life.

Another story was of a 'funny patient.' He was always happy, but tried to do many things he could not. One day, he wanted to light his pipe while standing. He held the match two inches above the pipe, and jumped up several times to get the pipe lit. He was happy, when we helped with the second match to move the arm to the pipe. So, it was only inadvertently for our entertainment.

The third story concerns a surprise I had after I left for Chicago. When I returned a few months later at Christmas time, I visited the ward to see Shep and George. There were still the same patients. Most did not react at all to my presence since I did not wear a white uniform on this visit. This applied mainly to the patients which I thought were somewhat more responsive. When I talked to them, they said yes, I remember, you were a 'patient' here some time ago. But the one patient, I would have classified as most catatonic, who never said a word, his eyes lit up when he saw me, and I knew that he remembered me fondly (perhaps since I never yelled at him at bed time).

Heidel's work was at the hospital. It was somewhat more demanding. In the admissions ward, the physicians still hoped to cure the patients. This meant, she was involved in the terrible electric shock treatments and had much more to do with medication and injections. Despite both of us had still to improve our English, it was quickly found out that we did a good job in writing the daily log. What we lacked in speaking ability, we made up with written English. Many of the attendants were farmers and not too eager to keep books. Because of automation, farming was not a full day's work. For them, the extra income from Ingleside helped to make ends meet.

After the immediate needs were taken care of, life got settled in December. One day, Tante Lieschen asked me: *Had I already registered at the post office for the draft?* This was a sensitive topic since I had heard before that new immigrants from Germany were drafted within a few weeks of their arrival and shipped promptly back as occupation forces to Germany. I had been very careful to read all the fine print in our immigration papers. No form I signed said anything about registering for the draft. My feelings about the army are described in pages 1-27 and 2-20–21. This does not mean that I am a pacifist or conscientious objector. I just don't like the life in the military or similar organizations. So I replied to Tante Lieschen: "Who told you to inform me about the registration?" She replied: "No one, I just think everyone must do so." So, I informed her about my point of view that the army is a dire necessity and I admire those who choose to serve, but also about my phobia of being in such organizations. The question did not come up again, until I applied for citizenship.

We also had to learn some different customs. One day when I appeared in my short leather pants, I was informed by Tante Lieschen that this is not proper attire, not even in the back yard. Although the pants still exist, I have never worn them since. (Possibly they do not fit anymore?). Another rule of hers was that women of good reputation do not go into a liquor store. When friends who were teetotalers visited, not only were no drinks served, but all bottles were hidden so that they could not be revealed by accident. This was much stricter than our rules which held that everybody was permitted to abstain or to drink, it only was not considered acceptable to get drunk.

There also were rules about religion. It made little difference to which church one went, as long as there was one. There were more than 30 churches in Hastings. Vati found that he could not get a job working on any of the churches unless he joined one. The First St. Paul's Lutheran Church on Burlington Street fitted his needs best. On our later visits to Hastings Caryn and Brent were also baptized there. It was interesting to see that the finances of the churches were independent of federal tax collection. Largely voluntary contributions by the parishioners sufficed for their support.

Figure 9 *Christmas at Tante Lieschen's in 1954.*

Soon, Christmas was coming, the holiday most dear to Germans, and most important for us was the celebration on Christmas Eve (see page 1-22–25). The Christmas tree, to our surprise, was up days before Christmas, and increasing numbers of beautifully wrapped packages accumulated underneath the tree, as shown in Figure 9.

We had to work on Christmas eve, and as usual, the hospital bus got us home by about midnight. We went into the house full of expectations to see everyone and wish a Merry Christmas, but the house was dark. All were asleep already. This we could not understand. It was not even possible to get something to eat, since the kitchen was the sleeping quarter of the little Pomeranians of Tante Lieschen. They would have woken up the whole house with their barking if we would have gone past them. This certainly was the low point of our stay in Hastings. None of the festivities on the next morning made up for it. But, more important were our first Wedding Anniversary and the plans we had to make for the New Year, 1955.

The last two years certainly had brought major changes into our lives. After the successful fleeing from the dictatorship of the GDR, making a new beginning in the FRG, and getting married in 1953, 1954 was fully determined by the new life started with the generous help of Onkel Richard and Tante Lieschen. We will always be thankful for their help in every aspect of our coming to the US and supporting all the stages of getting started in a new world for not only Mutti, Vati, Joachim and Manfred, but also for the two of us. First, the US had a major role in freeing us from the tyranny of Hitler during WW II, and within these last two years we had gained freedom and now got a jump-start in a country which welcomed us with open arms and never made us feel as former enemies. In fact, we felt more welcome in the US than in the FRG (see page 5-10).

Our plans for 1955 needed to address advances in several directions: First, we needed our own place to live, second we needed a car. Beyond this, we had to save enough money to pay for the second term in Hastings College, and to reduce our debt to Tante Lieschen. Finally, we had to find a good-quality University we could afford and, hopefully, go to in the fall. All these points forced us to construct a super-tight budget. Every penny was accounted for.

Why was a place to live so important? We needed more privacy to do all the work that was to be done in connection with the study, and Tante Lieschen certainly had her house full with Mutti, Vati, and Joachim and Manfred. Also, since we now earned money, we contributed to the household, i.e., it would be no added expense to cook for ourselves. In fact, we could save by organizing our meals more effectively around our sometimes crazy schedule. Again, Tante Lieschen was of help to find a suitable accommodation with Mrs. Mizen in E 4th Street, close to the college (see Figure 22, Chapter 11). The apartment was half the lower floor consisting of a furnished living room, bedroom, kitchen, and bath, all for $50 a month. We got along well with Mrs. Mizen and also her other tenants, the Lewandovsky's who lived upstairs, and also worked at Ingleside.

Why was a car so important? Time was precious and not sufficient for all the things we wanted to do. Waiting for the bus getting us to and from the hospital wasted at least one hour each day. Then, after we moved, going shopping was difficult for Heidel. The next supermarket was almost a one-mile walk and trying to shop only once a week meant to carry a large load home. Buying closer-by would have meant higher prices. In a city were everyone did their business by car, it was difficult to do without. So, again, Tante Lieschen helped. 'Ovi' Watson, the father of her daughter-in-law was a car salesman for Chevrolet. In no time he had us matched with the car shown in Figure 10. When I mentioned that I didn't know how to drive as yet, he said, no problem. Get yourself a learner's permit and when you pick up the car tomorrow, I will show you how it is done. Indeed, that is how it went. I got a crash course in how to drive (though not literally), and there was our car on 4th Street. I needed a licensed driver sitting beside me to operate the car. Fortunately, Vati had an international driver's licence and I had found out that this was sufficient. So,

Figure 10 *Our first car, a used 1949 Chevrolet, bought in Hastings for the grand total of $500.00.*

Figure 11 *The Platte River between Hastings and Grand Island. Photographed on an outing in 1955.*

Figure 12 *Family-outing along the shore of the Platte River. Photo near Hastings, NE, spring 1955.*

Figure 13 *A flowering cactus in Nebraska, 1955.*

Vati quickly taught me the tricks of the trade, and it took only a week or two until I thought I could chance trying the driving test. Yes, I passed on the first try. Hastings with its relatively sedate population of defensive drivers was a good place to learn. Not much later, I could repay Vati's efforts. When he needed a Nebraska driver's licence, I was permitted to help him as a translator to understand the written exam.

Now, we could explore Hastings in more detail. Figure 11 shows the first trip north, to the shores of the Platte River. The main attraction in this area are the sandhill cranes, migrating in February to April from Texas, New Mexico, and Mexico to their nesting grounds in Minnesota, Canada, and Alaska. We had a chance to see them "en masse" only on a later visit. The whole family fitted into the car in these outings during 1955, as indicated in Figure 12. It now became also easy to get together with Onkel Richard and all cousins in his cabin on a small pond, visit George Mohlman in Blue Hill, a small village south of Hastings, etc. Figures 13–15 show some typical Nebraska views, a bunch of cacti blooming in the dry and sandy spots, one of the huge feed lots for the cattle, and the backbone of Nebraska agriculture, the man-high corn, which today also is made to 'feed' the automobiles.

By now I was convinced I should apply to graduate school. The knowledge at my command was that Nebraska had a fine University in Lincoln, NE, only about 100 mi from Hastings. We had been invited there for a recital of cousin Jim's fiancée to-be, Barbara, and her mother, Mrs. Jones, owner of a local drugstore. For us, this visit in Lincoln and the university was amazing for two reasons besides being impressed by the university and the outstanding piano recital. First, after the recital, Barbara was presented with a brand new 1955 Chevy by her mother and grandmother. It was the most beautiful car I had seen to that day. Second, we were all invited to a dinner at the casino of the air force, near the commercial airport. They served us our first steak. It was bigger than the plate it was served on, and came with a minuscule baked potato. In our experience, this huge amount of meat would have amply fed a family of four. It was impossible for me to finish it, although it tasted great, and we certainly were not overeating at that time. But, I got used to even such culinary chores quickly, and very much appreciated the learning experience.

Figure 14 *Cattle feed-lot near Hastings, NE.*

Fortunately, Jim had investigated universities for himself before he decided to join the army first. The catalogs of these universities were studied carefully by me, and I found that *only the graduate school of Northwestern University*[1] *seemed to have what looked like a world-class chemistry program*. This university I chose to apply to.

There was a problem with my going to graduate school, I had no bachelor's degree, although I had studied chemistry for five years in Berlin and Frankfurt. This needed special attention for the preparation of my application.

Figure 15 *Nebraska corn field. Easy to get lost in.*

My professors at Hastings College gave me outstanding recommendations. In the second semester in Hastings College, I had registered for lectures in organic chemistry (6 h) with a weekly laboratory (3 h), and vector analysis (3 h). All ended with top grades among both, the majors of Hastings College in mathematics and chemistry. The organic laboratory, scheduled in the afternoon,

---

[1] Northwestern University is located in the cities of Chicago (downtown) and Evanston, adjoining Chicago to the north. It is a private, nonsectarian, and coeducational university, founded in 1851 by Methodists from Chicago (including John Evans, after whom Evanston is named). Northwestern opened in 1855 with two faculty members and ten students. Today's student enrollment is about 8,000 undergraduates and 6,000 graduate students. Northwestern competes in the 'NCAA Division I' of which it was a founding member. The Technological Institute, which houses the Chemistry Department, and other science buildings are located on the north side of the campus, close to lake Michigan. The main library was the Deering Library, built 1933, honoring Charles Deering from Deering Harvester Company (for donating most of the funds needed for its construction). In the 50 years since Heidel and I left the university, much expansion has taken place.

The organic laboratory, scheduled in the afternoon, I could always finish in half the allotted time with the use of my "Gattermann-Wieland" (see page 5-12 for the importance of this book). This meant, I did not have to miss work (see page 6-7). To document the adequacy of my earlier chemistry training, I asked my former Professors Royen and Magnus to write letters of recommendation, which they promptly did (see page 5-12). To confirm my progress in English, I carefully translated all transcripts, exams, and laboratory certificates. By the end of February, all was submitted. As before, the hopes were set on an application to this one university only.

Besides the work at the Hospital, as before, my camera occasionally helped in earning some money (see page 5-7–8). I found a job photographing merchandise for a local catalog store. The suppliers would come to a hotel in Hastings with their samples. The company representative would pick the shoes, bags, etc., he thought they could sell and I would immediately take photos, develop them, and deliver the finished pictures on the next day to the local printing company, the Cornhusker Press. There, a small catalog was assembled, to be distributed to the farmers. Also, for an irrigation company, I took pictures of their plant for advertising and stationery. On a return to Hastings, some 20 years later, I saw the photo still in their ad in the 'Yellow Pages.' They must have liked it.

In March 1955 I was admitted to Northwestern University to study chemistry. The question of how to finance such expensive school was solved only a few days later. The Chairman, Professor Burwell, offered a teaching assistantship with a stipend of $1,300 for the academic year, in addition to full tuition. Just as in the year before, all loose ends were secured in good time to leave Hastings at the end of August. The main effort was now to earn enough money to settle all old obligations.

By the time we went to Chicago, we had settled Tante Lieschen's loan, paid off our car, covered the tuition at Hastings college, and had saved $500 for the start at Northwestern University. To follow Heidel's original profession, law, would have meant for her to start at the beginning again because the US law is based on the completely different Anglo-Saxon law (see also page 5-11). At Hastings College, Heidel had already taken mainly courses for secretarial skills, and after I was accepted at Northwestern, the university offered Heidel a job in the Deering Library (see Footnote on page 6-13), so that our finances were secure.

Before describing our taking off to the big city, I would like to answer the question *why we never even considered going back to Frankfurt*. I do this on hand of an old audio tape, I made Christmas 1956. On this tape Frank, Dagmar, and Eberhard, who had decided to join us in 1957, were given a summary of our Christmas festivities (see Figure 6 of Chapter 11 which was photographed on this occasion). All of us, Vati, Mutti, Joachim, and Manfred, Heidel and I, as well as Tante Lieschen, Onkel Richard, Jim with Barbara (married by then), Marilyn, and Grandmother Zaroban contributed to the tape. All came to the conclusion that it would be a good idea for Frank to join us in the US and all said that they themselves never regretted their move. Even Grandmother Zaroban agreed with this assessment, still speaking in good German ("Hier ist es viel besser ...").

On the tape it was again and again expressed by everyone how welcome Frank's family would be. Indeed, we discovered this friendliness toward us, particularly in the Midwest, not only within the family, but also from complete strangers, and later also toward visitors of ours from Germany. This made living in the US so much easier than in the FRG. Vati made a point that it was important to leave 'the volcano Europe.' He thought that the Cold War had a good chance to heat up. Unfortunately, he did not see the reunification of Germany some 34 years later and the (possibly temporary) democratization of Russia. He also emphasized, as did Mutti, how small a fraction of one's income was needed for the basic food, and that jobs were easy to find. Vati mentioned that to live, he had to spent over 30,000 DM-Ost (≈US $7,500) in the GDR annually, an enormous amount (excluding black marked activities, which did not appear in his bookkeeping). These were the expenses for his family of six. With an earning ratio of US $ to DM close to 1:1, the official exchange rate of Dollar to DM-West was at that time 1:4.2 (for the fluctuating illegal course of the DM-Ost see page 3-28). Heidel and I had spent a frugal $30 a month for the two of us for groceries. In Chicago, this still did not exceed $40. This meant, that much more money was available for the luxury items like new automobiles (we had to buy a new car just before Christmas, see pages 6-28–29), refrigerators, washing machines, televisions, and tape recorders (we had acquired one shortly before Caryn was born, to capture her early voice) and stereo equipment (which we bought in Chicago, see page 6-17). In short, even with a lower income, one could have a considerably higher standard of living.

Figure 16 *The automobiles of the three Wunderlich families, in the drive of Vati's house in Hastings, NE.*

Vati had similarly just moved from being a helper in the roofing business, to owning his own business and house. Two years later, in 1958, he had secured his business and had paid back most of the loans needed in the beginning, and I had started the first teaching job at Cornell University.

Figure 1 of Chapter 11 shows the family at Christmas 1957, including Frank and family. Figure 16 illustrates the car park of the family on this visit. Each family had an automobile, and Vati, in addition, owned a pickup for the business, displaying proudly his old logo (see Figure 26 of Chapter 1, and 11 of Chapter 3). But Vati also pointed out on the tape that craftsmen were poorer in their skills and, in his opinion, many things were made to a lesser standard of perfection to reduce the cost.

Interesting were also the comments of Joachim and Manfred. Both assured Eberhard that it would not be hard to learn English. Inadvertently, they demonstrated this with a diminished skill in German. When talking was quick, English was faster and without an accent, having learned from the teacher, not from us. They remarked that the school was more fun in the US. They enjoyed the much longer summer vacation and that there was little or no homework (but longer hours in school). Finally, they assured Eberhard that they knew the Kindergarten teacher and that she was looking forward to his arrival. In addition, they gave a full accounting of all the new fauna and flora and the changes in the school sports. Animals like gophers, raccoons, skunks, and blue jays were not known to them before. Living in Nebraska, they were already fully indoctrinated with the importance of football and needed to explain to Eberhard the difference to German "Fußball" (soccer).

All these observations were naturally also considered when Heidel and I pondered our future. The biggest hurdle, once we were in the US, was naturally to complete my studies. This had to be accomplished without having any money of our own, or family resources to support us. Remember Vati's rules on pages 4-19 and 20. This had to be weighed against a guarantee of support when going back to Frankfurt. As things went, money was scarce, but a solution was always in sight, so that we never needed to consider returning to Germany. Initially, we would envy the American students their support from home, they did not have to work as hard besides studying. But after a few years, we had caught up, and on the way had learned an important lesson for life. We never lived beyond our means, however small they were, and in the future, we always increased our expenses less than the growth of our income would have permitted. After the first new car was paid off in 1960, we never again bought anything on credit and did not borrow any money, except for the mortgages on our house in Troy, NY, and later in Knoxville, TN. The last mortgage was burned in 1991. For us, we had identified 1945 as the point zero in our adult life which prematurely started at that time as is described on page 3-1. It was a zero, indeed, in standard of living and personal possessions. From then on, life improved steadily and we never had to look back and reconsider our choices. As remarked before, the trial year in the US is still not over today (see page 5-13).

Early in July, Professor Malcolm Dole of the Chemistry Department of Northwestern University made *an offer to change my teaching assistantship into a 12-month research assistantship* for academic research on 'thermochemical properties and the effects of high energy irradiation on high polymers,' supported by the Office of Ordnance Research of the US Army (unclassified). This paid monthly $200 for a period of 12 months, with one month of vacation. In addition, it could be started any time. Tax and tuition had to be paid, but the research could be applied to my thesis work. He added two welcome restrictions: No smoking in the laboratory, and work was expected on Saturday mornings. I immediately accepted for the 1st of September.

Next, we gave notice to Mrs. Mizen, and on August 27, 1955 we quit our jobs at Ingleside, collected our seven days of vacation pay, prepared our car to travel with all our possessions, and started our move to Chicago on Sunday, the 28th of August. The trip almost went well. This was our first travel in daylight through the unending corn fields in the flat lands of Nebraska and the up-and-down hills of rural Iowa bordered by the Missouri on the west, and the Mississippi on the east.

Somewhere in Iowa, in the middle of nowhere, the car gave out. A cloud of steam arose in front, and when I investigated under the hood, I found a broken fan belt. The cooling of the engine had stopped. Before we could consider how to get to a service station, a car stopped. The driver said: "You have a broken fan belt and may have difficulties getting a new one for this car on a Sunday. May be, I can help you." He was a mechanic from the next village, about 10 mi distant. He drove me into the village while Heidel stayed with the car. He knew all possible sources for fan belts and got people to open their shops on Sunday. In short order, we found and bought the proper belt. Then he drove me back and installed it properly, free of charge. With many thanks to him, we continued on our trip. All had taken little more than an hour.

In the late afternoon of *August 29 we arrived in Evanston* and parked in front of the impressive Technological Institute. I found Professor Dole in his office, and was welcomed. He insisted on us staying with him for dinner and the night and phoned Mrs. Dole to announce the guests. After he had finished his work, we went to his home.

There was a small complication, however. He went to work by bicycle, i.e., the professor went in front by bicycle and the new student followed by car. We did not feel quite comfortable in this arrangement, but there seemed to be no other easy solution. Some 10 years later, Professor Dole's habit of going by bicycle to the university brought national attention for him. After being warned several times that a few weeks earlier the Sheridan Road to the University was forbidden to bicycles, an over-eager policeman arrested him and brought professor and bicycle with a paddy wagon to the station. Northwestern, however, has a good law school, and so it made the national news when, instead of being fined, he had to be granted an exemption from the city ordinance. He could from now on continue to legally use the bicycle going to and from Northwestern.

Next morning, after breakfast, we called one of Heidel's girlfriends, Dorle Mendheim (née Radikowski) from Brandenburg, who had married an American business man. I also knew her family. They lived with their youngest daughter Eva in the same house we lived in (Figure 5 of Chapter 3 in the first-floor apartment on the left). By 1955, Dorle had brought Eva and her older sister Lotti also to Chicago. Dorle immediately offered that we could stay with her and she would also find an apartment for us. In short order, we found a suitable place in the last block of North Sheridan Road (#7729), the main thoroughfare going along the lake shore from Chicago, through Evanston, on to Wilmette, Winetka, and all the way up to Racine, about 60 mi further north (which is already in Wisconsin). A picture of the house is reproduced as Figure 23 of Chapter 11.

Living in Chicago and attending Northwestern in Evanston was of advantage. The rent ($50) was less than in Evanston, presently a city with a population of about 75,000 and part of Chicago's affluent North Shore region. The drive along Sheridan Road to the university went close to and partially along the shore of Lake Michigan. It was pleasant, and took less than 15 min. In front of our house, I estimated that daily about 80,000 cars passed by, but we got used to the noise quickly. Evanston is the headquarters of the Woman's Temperance Union, and as such was a dry city. The first street in Chicago, Howard Street, as a result, was dotted with liquor stores. The whole area was home to a large number of recent European immigrants. Walking along the lakeshore in the evening, one could hear a wide range of foreign languages and little English. We moved into the apartment by the first of September. We had one room with an inner-door bed, an efficiency kitchen with a breakfast corner, bathroom, and a sun-porch, facing the street.

Next we needed furniture. This should be easy. With the measurements of the apartment we went downtown to a big furniture store in the 'Loop,' recommended by Dorle. For hours we went from floor to floor. Either it was too expensive, or it was not to our liking. All we found was an outstanding stereo system by Grundig, a company we knew from Germany. In the US, it had a much lower price. We bought it, and it was delivered the next day. It can be seen in Figure 17.

By late afternoon, we were tired of shopping and gave up. On our way home, beyond the end of the Lake Shore Drive, on N Broadway Street, we suddenly saw a much smaller store with modern furniture. We stopped, crossed the street and stood in front of a window with gorgeous pieces of furniture, just to our liking. We must have stood there for quite some time until a salesman came out and asked us if he could be of help. Our answer: "No, we need to furnish our small apartment and have just come from the big store where we found nothing suitable. Your furniture is much to our liking, but we cannot afford it." As a good salesman he answered: "How many rooms do you have? How much money do you want to spend?" When we told him the size of the apartment and the available $500, he promised to work with us and find the pieces to our liking without spending more than that. With doubts, we went in. He sat down with us, drew a picture of the

Figure 17 *Our apartment, N Sheridan Road in 1955. Note the bright windows on all sides of the sun porch.*

apartment and slowly filled it with the furniture, as shown in Figure 17. The bill was $500. All pieces met our enthusiastic approval. It started with a modern steel and plastic kitchen table with a Formica top and four chairs to fit into the breakfast corner. The chairs had soft backs and seats with beautifully patterned and black and red designs. Next was the cabinet shown on the left of Figure 17; the lamp was a good-bye gift from Heidel's coworkers from the hospital. The furniture designer was well known (Paul McCobb of Massachusetts). The two-tone appearance was our invention, made in the store. The cabinets came in black or blond. One was too dark, the other to light. So I moved the drawers of one to the other and we were amazed by the effect. It did not cost any more. The book bench was the suggestion of the salesman. Many of the books are the ones brought from the GDR. The fancy radio (one of the first with electrostatic tweeters) was mentioned above. On the other side of the room was a simple couch (black frame and orange cushions and bolsters). The two steel-frame chairs were by the Knoll design firm (founded 1938) and probably now cost more as antiques than when they were new. The cocktail table in the center was a Californian design with a space for bottles and glasses in its wide leg in back, and a storage shelf for magazines as seen on its left. This was the most expensive piece and the one we originally had caught sight-of in the display window. It needed drastic discounting, and finally, was sold to us as demonstrator (it had suffered slight sun damage from being in the window). But, we still needed floor covering. An oriental carpet (even an imitation) was out of the question. The solution was cheap. You see the long, black and white, paper-woven floor covering, cut to proper length for our apartment. It lasted for almost 30 years! That was it.

By the time we finished, it was after closing, and we had the feeling we made an advantageous purchase and met a helpful and skilled salesman. The radio became obsolete when we lived in Ithaca, and we parted with the kitchen table, floor covering, and sofa, when we moved to Knoxville in 1988. The Paul McCobb line cabinet and bench, and the Knoll chairs stayed with us. When we moved to Ithaca in 1958, we found a furniture dealer who ordered additional Paul McCobb pieces and Knoll chairs matching ours. All this furniture is still functioning today.

The delivery of the furniture took several weeks. *In the meantime, we lived out of suitcases* and used the wooden radiator cover in the sun porch as table. The apartment was freshly decorated for us new tenants and we could choose contemporary colors to match our furniture. We chose a different color for every wall. When first inspecting the apartment, I had noted a faint smell of gas and Heidel pointed out that the last tenants had left the stove in a sad (and dirty) condition. She got to work, but only a few days after cleaning it to perfection, we got a brand-new one. The rental agency took my nose seriously and replaced the stove for safety's sake. All we had to do now, was to wait for the furniture. Fortunately, the inner-door bed saved us from sleeping on the floor.

Before the furniture was delivered, we had the first visitors. George Mohlman and wife Rosalie came from Hastings for a weekend-visit to Chicago. They had to join us on the floor, but could listen to glorious music from WFMT, our favorite radio-station in Chicago. For sightseeing, George preferred me to drive. The traffic in was too much for him and the side streets too narrow. I seemed to have learned enough in the few months in Hastings not to be bothered by big-city traffic. To maneuver in the streets with two solid rows of parked cars, leaving barely enough space for the traffic, I had found a secret: To go safely through the narrows in oncoming traffic, one looks into the distance to judge whether one can pass or not. If there is enough space, then one just goes and never checks again. Closer to the problem spot, one does not have as good a perspective and may leave too much space on the side one looks at, and cause a scrape on the other.

On September 1, it was time to go to work. I was introduced by Professor Dole[1] to Bill (William) Howard who had not passed his qualifying exam and was now finishing his MS degree. He studied the influence of ionizing radiation on polymers by adiabatic calorimetry. The calorimeter consisted of a cylindrical adiabatic jacket with a diameter of about two feet. It was connected to vacuum and cooling pipes, fed-in from the crawl-space underneath the floor. Figure 18 shows a colored schematic taken from my thesis. The electric and electronic gadgets are indicated and their connections marked. All controls were collected on a big control panel and table, filling half the length of the room. Originally, the calorimeter was designed as a twin calorimeter, i.e., all controls were there twice. Bill was not on too good of terms with 'his' instrument, but was to teach me.

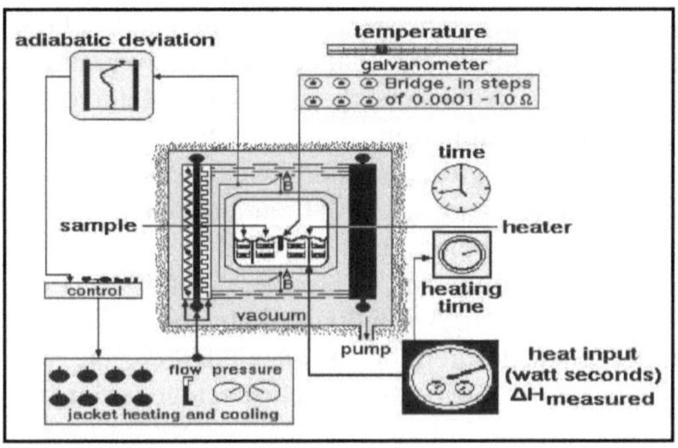

Figure 18 *The adiabatic calorimeter. Description published in the Rev. Sci. Instruments,* 26, *689 (1955).*

Things often went wrong, requiring him to start over again. The measurements were done with increasing temperatures, each step taking one to two hours. The longest time was spent on the measuring of small temperature drifts to establish trends for extrapolation into the actual heating sequences, which lasted only 10–20 min each. During heating, the temperatures inside the calorimeter and the big jacket were raised as closely in unison as possible. All the heat inputs, temperatures, and temperature deviations had to be recorded as a function of time. For every step, there were up to 200 numbers to be recorded. These were needed for calculation of the heat capacity of the sample as it changed with temperature. The sample, of ≈ 1.0 lb, was contained in thin silver trays, with a platinum thermometer in the center. It sounds and is complicated, and I will not describe the many experiments, calculation details, and interpretations.

References to the research carried out over the years, starting with this PhD thesis, are represented by the close to 600 publications listed in Appendix A. A more detailed review is given in Appendix B. Finally, perusal of the 3,000 screens of the Computer Course on 'Thermal Analysis of Materials,' Publ. #560 in Appendix A, is a challenge to understand all there is to learn in this, my field of expertise. It has led me to give more than 1,000 lectures worldwide, starting in 1957. The Appendix C lists the titles and locations for most of these lectures, and all of them after 1962.

---

[1] M. Dole, 1903–1990, see: "My Life in the Golden Age of America." Vantage Press, New York, NY, 1989. A few milestones: PhD, 1928, Harvard University (Electrochemistry), Postdoctoral work at Rockefeller Institute (with D. A. MacInnes, working on electrochemistry and the glass electrode) and a summer with Professor Debye in Leipzig, Germany, were he investigated the viscosity of electrolytes. From 1930–69 he was at Northwestern University, interrupted from 1944–45 by WW II (working on chemical warfare in Utah and isotope separation in Oak Ridge, TN). In 1969 he was appointed Welch Professor of Chemistry at Baylor University in Waco, TX, a position he held to 1982. His major work dealt with electrolyte solutions, isotopes, and polymers.

I slowly learned about Professor Dole's past accomplishments (see Footnote of the previous page) and experienced his admirable style of guiding students so that they could grow on their own into the expertise needed to succeed in the chosen field of study. One must know that the subject he had selected for me, that of 'polymer science,' was not taught at that time at any university, and thus, was not included in any of my university training. There was some early interest of mine in polymers (plastics, see page 3-30 and 35) and I combined it quickly with the study of solid state physics in which I also had earlier interest and training (see the pages 4-11–12, 5-2, and 5-12).

Later, I learned from my Chinese students that one did not only have a family genealogy as given in Chapter 11, but in parallel, there is a scientific one, and *one should honor one's scientific 'grandparents'* as much as the family ancestors. Only the latter ones are listed on pages 11-2–4. Naturally, my science genealogy includes mathematics and science teachers through high school. I must note that there was not a single teacher in these fields whom I did not like. I owe gratitude to all of them. Their help was described in Chapters 2–3. Next came the succession of university professors who in Berlin, Frankfurt, Hastings, and Evanston were not only often world-class scientists, but also always outstanding teachers and had been greatly supportive (see Chapters 4–6). With Professor Dole, I could establish links to international experts, as will be pointed out from time to time in the continuing text. One's research professor being, as it is called in German, the "Doktorvater" (father of the PhD student), I later was not only the "Doktorvater" for 40 students, but also the 'stepfather' to 57 postdoctoral students, as listed in Appendix A. Many of these are by now also teaching at universities, creating scientific 'grandchildren,' 'great-grandchildren,' etc. The succession of science generations is faster than in families. Such thoughts, however, go far into the future and were still beyond my expectations when I arrived in Evanston.

First, there was the goal *to tame the beast of an adiabatic calorimeter*. Second, there was to find out what courses I still needed and wanted to take, then, third, there was the sequence of exams. For the first half year, the calorimeter was tackled with respect. By watching Bill's technique I was trying to see why sometimes things did not work as expected. By the time Bill finished his measurements in April 1956, I was left to my own devices. During this time, Bill and I had gotten to know each other well and we also met sometimes outside the laboratory, as shown in Figure 19. After he left, I finished a lot of work over the summer. It represented about half of my thesis research.

The main experimental technique I developed for using the calorimeter was not ever to shut down a measurement when its somewhat unreliable tube electronics acted up. The controls for the other half of the twin calorimeter were still there. This meant, there always existed an idle, second half of the calorimeter controls. As soon as I could tinker with the monster, I added switches and wiring, so that in a surprise shutdown of the controls for one twin, I could flip to the other. By doing so, I lost at best one point that should have been measured. During the

Figure 19 *Visit at the Howards' apartment in Evanston, 1956. They gave us an introduction into the Southern style of living, talking, and hospitality.*

continuation of the experiment, I then had enough time to fix the first twin, so that in case of a rare second stop, I could flip back to the first. I never again had to interrupt a measurement.

A specific problem with my thesis samples was that within their broad melting range, I could not stop a series of measurements at increasing temperatures. Overnight, the samples changed noticeably. This was an important scientific observation in itself, and had to be studied in detail. Typically, 20 to 30 h runs were now scheduled to cover the melting range *without* stopping. Since there was no second student to work a second shift, I had to stay overnight and try to stay alert. But, the important problem was solved. There also were no breakdowns during those marathon runs. Later, I developed a technique to measure continuously through the melting range. I reconfigured the discontinuous adiabatic calorimeter into the first, scanning, quasi-adiabatic calorimeter.

Next, the courses that needed to be taken had to be established. Professor Dole had gone over my list of courses taken earlier. He suggested who could judge which courses of the prescribed lecture sequence for a physical chemist I needed to take. By seeing all professors involved, I was released from almost all typical first-year graduate courses, which meant, that my five years of study in Germany, finishing with the "Vordiplom," (see Figure 13 of Chapter 5) was practically counted as a 'masters degree,' and I had a two-year minimum for attaining a PhD. The limit could, naturally, only be reached if the research went flawlessly. Such had never been accomplished before.

In fact, during the later stages of my research, it became a race between my friend Werner Haag, to be mentioned below, and myself who could finish in less time (and just as much between our two research directors). The professor of Werner's was Professor Pines (in the field of organic chemistry and working in the area of catalysis). Werner had started with a completed "Diplom" from the University of Tübingen in Germany and would have finished similarly fast, but his thesis work, and especially the writing, needed more time.

There were three quarters in each academic year and one in the summer. After the first quarter, when I took also one course in organic chemistry and audited one in inorganic chemistry (on catalysis by Professor Burwell), I only needed to take advanced courses: Six quarters of Advanced Physical Chemistry, and one quarter each of Kinetics and Statistical Thermodynamics. Most were passed (after a written exam) with a grade of 'A.' All this was done in addition to every spare minute spent on working in the laboratory, a habit I had developed in Berlin and Frankfurt.

At Northwestern *all exams, except the one for the PhD, were written exams*. Outside the exams for the courses, there was a qualifying exam, passed in June of 1956, which gave me the best grades and a $200 reward for the best presentation. This also resulted in being awarded the Carbide and Carbon Fellowship for 1956/57 ($2,100 plus tuition and fees). In addition, there were two language exams. The German exam was obviously no difficulty, a discussion with the professor in charge certified that I was competent in German. The next was a French exam. This was more difficult since I knew no French. The exam consisted of a written translation, so I looked for a book on scientific French and learned translating without any idea about pronunciation. Spending a couple hours each day for two months was enough for a first try which, to my surprise, succeeded. By October 1956, all language exams were completed. This left the dreaded 'cumulatives.' Three-hour written essays on randomly chosen scientific topics, for me in physical chemistry. They were started in the fall of 1956. By July 1957, all needed 'cumes' (5) were passed. Seven exams were given each year, I failed 2, so I completed all in one year (barely). Finally, I was officially notified on May 2, 1957 that 36 credit hours were transferred toward my study from the Humboldt and Goethe University in Germany, and no further courses were required.

Comparing oral and written exams as I experienced them in the German and US periods of my university training, I came to the conclusion that oral exams are more productive, particularly if they are increasingly more comprehensive. Written exams can also be of value, but only if they are also used to evaluate the literary ability and penmanship of the student in the field of the exam. In mathematics and the sciences, they also are useful to check on the ability to reason and derive arguments, as was done in the cumulative exams.

In German high school tradition of the 20th century, I experienced the impossibility of supervision of written exams to avoid improprieties (see, for example page 3-32). This alone, made it necessary to switch to oral exams when entering the university in Berlin. When first coming to Hastings, NE, we were touring a high school and learned the difference to German schools. We were impressed by the much larger degree of honesty among the American students. To them (at that time!) it seemed natural not to engage in cheating and to report it, if noted elsewhere. Unfortunately, this trait has not endured. Today, major efforts are necessary to assure fair assessment through written exams.

The value of an oral exam is that it develops the ability to answer questions as they arise daily in one's professional life when discussing research planning, observations, or results. In my own experience of assessing oral exams, I found, it takes only 10–20 min to find the level of competence of a student. Since my typical oral exams were set for ½ h, the rest of the time was used to test for possibly increasing the initial grade through a series of higher-level questions. It rarely succeeded. Another advantage of oral exams is that exam-fright is easily detected. One can calm a student via a normal conversation before moving to questions revealing his knowledge. In written exams, a student who is 'bad in writing exams' is hard to distinguish from a poor student. Over the years, I found, for class of less than 20 students, oral exams are more efficient. For this number, it is a bigger effort to develop written exams and grade them (particularly, if they are poorly written).

After 1½ years, in April 1957, Professor Dole came into my lab and mentioned that I should think about writing my thesis. I could not believe my ears—but, I got to work immediately. This very day, on the way home, I rented a modern electric IBM typewriter with proportional letter spacing, and Heidel had the day thereafter part of Chapter 1 to type. She did an expert job to type in the evening what I wrote during the day. This was the main reason for my quick success. The figures to illustrate the measurements were also no problem. I had sufficient training from high school and my hobby to make technical drawings helped also (see the pages 1-18 and 3-12). In fact, at Northwestern, illustrating theses for fellow students led to some additional income. Professor Dole had a large drafting board in the laboratory and I had brought a full set of lettering templates and special drafting pens from the GDR. The earnings from these drawings for others paid for the rental of the fancy typewriter and the special paper I had chosen to achieve good quality and quick duplication of my thesis. Xerox copying did not exist then, but with the special paper, no carbon copies needed to be made. Direct and economical copying was possible.

The title of my thesis was: *Thermodynamics of the Copolymer System Poly(ethylene terephthalate-sebacate)*. Late in May 1957, I put the finished, 134-page thesis for a first reading and corrections on the desk of Professor Dole. I am sure he was surprised. We had not discussed any details of the thesis, except for the experimental results, as they emerged. He was fully satisfied and found that no corrections were necessary (he must have liked my German-English style of writing). Since the committee could not meet until late in the summer, the exam was postponed and I had time to drive to Hastings, NE, in the summer to take two weeks of vacation, to prepare for the exam, and take Frank and his family to their destination, as will be described below on pages 6-30–31.

The PhD (Philosophy, Doctor of) was officially conferred on August 31, 1957, precisely two years after my starting work for Professor Dole. A copy of the diploma is shown in Figure 20.[1] In fact all went so quickly that by that time, I had not seriously thought about what to do next. By the 31st of August, however, it was too late to find a job for the 1st of September. Again, Professor Dole had earlier thought of a rescue. He had arranged to ask the Department and Dean to appoint me as an *Instructor in Chemistry in the College of Liberal Arts of Northwestern University* for the academic year 1957/58, beginning September 1, 1957. My income jumped at that time by 100%, to $4,600. The appointment letter had already been transmitted to me by Dean Leland on May 7.

Figure 20

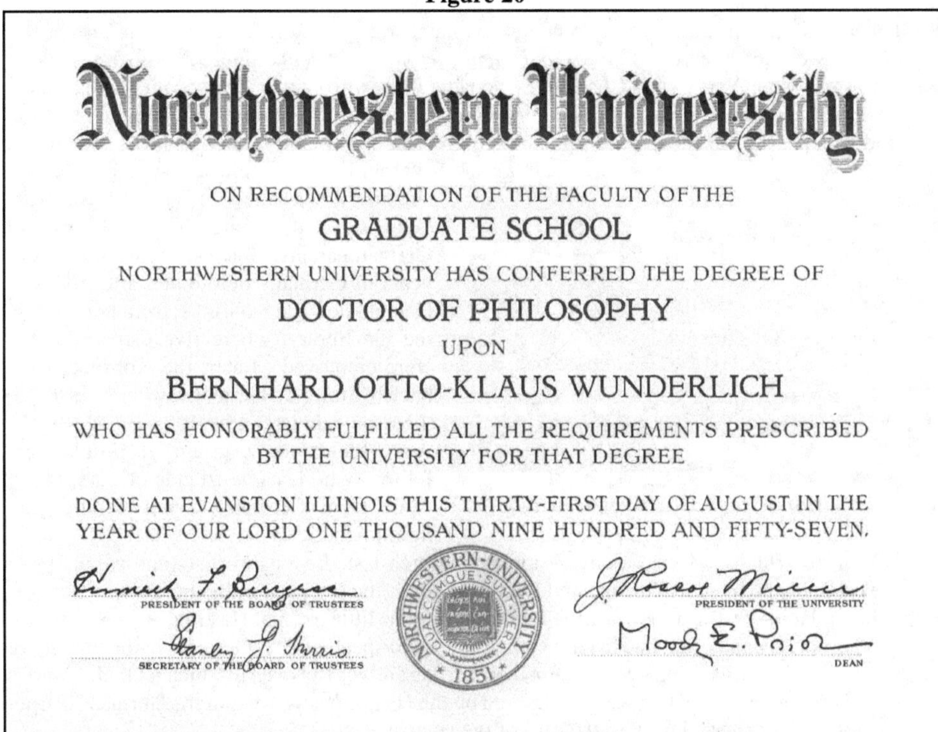

Northwestern University

ON RECOMMENDATION OF THE FACULTY OF THE

GRADUATE SCHOOL

NORTHWESTERN UNIVERSITY HAS CONFERRED THE DEGREE OF

DOCTOR OF PHILOSOPHY

UPON

BERNHARD OTTO-KLAUS WUNDERLICH

WHO HAS HONORABLY FULFILLED ALL THE REQUIREMENTS PRESCRIBED BY THE UNIVERSITY FOR THAT DEGREE

DONE AT EVANSTON ILLINOIS THIS THIRTY-FIRST DAY OF AUGUST IN THE YEAR OF OUR LORD ONE THOUSAND NINE HUNDRED AND FIFTY-SEVEN.

PRESIDENT OF THE BOARD OF TRUSTEES

SECRETARY OF THE BOARD OF TRUSTEES

NORTHWESTERN UNIVERSITY QUAECUMQUE SUNT VERA 1851

PRESIDENT OF THE UNIVERSITY

DEAN

[1] The phrase on Northwestern's seal is "Quaecumque sunt vera." Translated from the Latin: Whatsoever things are true (from Philippians 4:8). In the center of the seal on the pages of the open book is written in Greek "ho logos pleres charitos kai aletheias." This is translated to: The Word ... full of grace and truth. (This phrase comes from the Gospel of John 1:14.) Both the Latin and Greek phrases express the values of the University's founders, recalling their Methodist heritage.

During the study, our private life also saw changes. *Heidel started work in the Deering Library*, on September 1, 1955, the same day I started my laboratory experiments. She did secretarial work in the order department, just a few buildings south of the Technological Institute. This meant, we both left together after breakfast, and usually had our lunch together, which was carried from home. The problem we had originally with the unsuitable, soft American bread (see page 6-3) was solved already in Hastings by Mutti. She started to bake her own sourdough bread, and delivery was made to us through Joachim and Manfred. In Chicago, however, one could buy reasonable German-style bread, as well as cold cuts, etc. Still, after our next visit to Hastings, Heidel brought some of Mutti's sourdough as starter to Chicago, and has ever since baked our own, dark, rye bread, with only occasional refreshing of the bacterial colony from Germany. By now, our mini-bakery also supplies Caryn and Brent approximately every one to two weeks with fresh bread.

Soon we also had visits of *guests and friends in our apartment*. Figure 21 illustrates one of these occasions when Evelyn and Charlotte, who Heidel knew from the library, came with a friend to visit. Larger groups could also be accommodated, as shown in Figure 25 of Chapter 11 where our good friends, the Haags and Kohfinks, were greeting Frank and family in 1957. I had met Werner Haag as a fellow graduate student in chemistry (see page 6-21). He also was from Germany and Hanna, his wife, was of German parents. She had spent a number of years in Germany before and after the war. Heidel knew the Kohfinks, from her work in the the library, where five German ladies were employed. Later, the Kohfinks went back to Stuttgart, Germany, where we visited them on our later travels (see page 9-21). Also, there was a group of postdoctoral fellows who became friends of ours. They came from Germany, Great Britain, and South America.

Figure 21 *Visitors in our apartment, Chicago, 1955.*

With the Haags, we had a close friendship which lasted over the years and we also met Hanna's parents who lived in Chicago and introduced us to the typical immigrant families from the 1920s. I found it interesting that while the immigrants in the little town of Hastings were very much integrated in society and often had even lost their ability to speak perfect German, in the big city of Chicago the immigrants aggregated and kept their heritage alive. But even in Chicago, their German traditions were frozen, i.e., when we were invited by the Haags to a function of the German "Turner Verein," we felt transported back to the turn of the century.

In these immigrant organizations, the language, songs, and games came from the time when they arrived in the US. To us newcomers many, terms of their language were antiquated. In addition, many English words had crept into their German. Particularly, words were assimilated which were conveniently short or pertained to things not common during their life in Germany. One used words like fan, car, can, and the fridge, instead of "Ventilator, Automobile, Büchse, und der Kühlschrank," respectively. Especially humorous were some of the words fused of English and German roots, not easily understood by anyone else, like 'eintraden.' And finally, one wrongly used German words of similar sound, but different meaning, like "eine Kanne," which is not quite 'a can,' "schmal" which is 'narrow' in English, not small ("klein") and certainly 'dish' is not "Tisch" ('table'). Even when listening to the German radio station, this altered language could be noticed.

Our own ability in English seemed to have made a quantum jump in Chicago. We were understood better. Things did not happen, as they did in Hastings, where I pulled up with my car at the service station and stopped at the pump, but the attendant did not know what I wanted. It did not help when I said, in what I thought was perfect English: "I would like to buy some gasoline." He was waiting for me to say: "Fill'r up." And this, with the proper Midwestern pronunciation of the "ll'r" (see page 6-8). It was also difficult for me to accept 'her,' because in German, the gender a car is neuter. Being used to a specific gender for every noun made it awkward to use the 'wrong' one for a pronoun. It happened often enough in Hastings that when buying things, despite speaking clearly and saying the right words, we ultimately had to resort to the *international sign language*.

On our first travel back to Hastings over Christmas and New Year in 1955 we were disappointed. The fact of being better understood did not come from our improved English, but from the Chicagoans having no problem understanding our accent. I still have a practice audio tape of one of my first lectures, documenting that much needed to be improved.

We were often told, *don't lose your accent, it sounds so nice*, but for us, only accents other than of German were acceptable. The German accent indicated insufficient effort to learn the language properly. Over the years we improved, but not so much that we are not asked frequently: "Where are you from?" Most interesting was the conversation we had with a waitress when in 1988 we arrived in Knoxville. After figuring out that we were from Germany, she declared us to be 'foreign' Yankees" (see page 9-1). On the other hand, we always were happy not to have kept as heavy a German accent as was heard on television when listening to Lawrence Welk of 'champagne-music' fame (who was born in Strasburg, North Dakota) and Henry Kissinger, our later Secretary of State (1973–77, who had to leave Germany at age 15).

From these examples we learned that we could not teach each other better English (this would lead to arguments about the correct grammar or pronunciation). And under no circumstances must we teach our children to speak English. The Public schools did a better job. We had noticed this already, when we saw Joachim and Manfred's progress at Christmas in 1956 (see page 6-15). Our children, Caryn and Brent, thus, where initially instructed by us in German. Heidel also took it upon herself to also teach them to read and write in German with help of school books sent from Germany. The writing was taught only after their English skills were established. As a result, they do well in both languages. They do not have an accent in English or German, but are somewhat short in German vocabulary. This deficit decreases considerably with every visit to countries where German is spoken. Caryn spent one year as an exchange student at the ETH in Zürich, Switzerland, and Brent was sent for ½ year by his company to Göttingen, Germany, to work with a German client, the "Gothaer Versicherungsgruppe." Both had little problems integrating.

An incident in Caryn's early schooling gives some insight into the mechanism of learning languages as a child. Her English teacher thought it was of interest to Caryn to learn a German evening prayer. Proudly, Caryn recited it to us, but with an atrocious American accent. When we gently told her: "Caryn, you know how all these words sound in German. Why don't you say them properly?" She was close to tears. She thought we said that her teacher taught her something wrong. Still, even at an early age, both, Caryn and Brent, quickly found out who of our guests was more comfortable with English and who with German. From a short contact, both knew when going to bed to whom to say 'good night,' and to whom "gute Nacht."

In 1955, one major change occurred in the laboratory. Professor Dole decided to study the effect of irradiation on polymers with samples produced on-site. Earlier samples, analyzed by Bill Howard (see pages 6-19–20), were irradiated elsewhere and then brought to us for analysis only.

Figure 22 shows the arrival of what was called the 'pig,' a lead-filled steel contraption that had a stationary bottom with a well for an intense, stationary, cobalt-60 radiation-source (γ-rays). The upper part could be moved with a crank, either to put the tower over the source to allow lowering a sample for radiation, or to move the tower to one side to work on the sample from below. Figure 23 was taken from a safe distance during the exchange of the cobalt from the shipping container, left, into the 'pig.' All operation was done from behind the lead brick wall (left), observed through mirrors. Professor Dole, by lying on the ground, could move the cover of the 'pig' from the loading to the closed position.

On completion, the 'pig' was moved into the laboratory and placed on a concrete pier which was originally built to isolate equipment placed on it from vibrations. These were thought to arise from the heavy machinery in the building. It turned out, the pier was not vibration-free. It picked up the waves of Lake Michigan, but it served ideally to support the heavy load of the 'pig.' Figure 24 of this series shows David Milner, a new, English postdoctoral fellow at work.

In David we could observe the quick transformation from a reserved, conservative British scholar to a typically open, relaxed American. We still keep in touch today, but when I visited him a number of years later in England, he had changed back again to fit into British society. Still, I had a nice time inspecting gardens and the countryside around Oxford (see page 8-48).

Figure 22 *Arrival of the 'pig.' From the left, Dr. Milner, Professor Dole, and the machine-shop manager. In the foreground, the safety officer, 1955.*

Figure 23 *The safety officer and Professor Dole getting ready to charge the 'pig' in the back of the Technological Institute. Note mirror and lead wall.*

Figure 24 *Dr. David Milner, 'playing' with the pig.*

Figure 25 *Vati and Mutti's house in Hastings, NE.*

After the first three months in Evanston, we traveled back to Hastings, NE to celebrate the Christmas Holidays 1955 and the New Year 1956. Since we left, Vati had earned enough money that they had just moved into their own apartment.

The new year, 1956, brought him some major changes which ultimately turned out favorably, but looked troublesome in the beginning. Vati had changed jobs to work for a builder, Mr. Fast, who was of German origin. He wanted Vati to set up a roofing company to have an added capability within his own company. After a while, he arranged for Vati to move into a house which was moved to a location he owned at 921 W 12th Street, just seven short blocks from Tante Lieschen. Figure 25 shows a later picture of this house. All looked good, but Vati discovered that he was supposed to share his profits with Mr. Fast and that the conditions of the mortgage on the house were not particularly favorable. Vati did not speak English too well at that time, but he had more business experience than Mr. Fast. With help of J. D. Brock, a lawyer, and Tante Lieschen, this business deal was dissolved and Vati was working on his own more quickly than he had intended. With some extra effort, Mutti doing the bookkeeping, as in the 1930s (see pages 1-8 and 33), and Tante Lieschen being the translator in difficult situations, all succeeded. Work was coming in, and Vati had to hire help and could afford new tools and a company truck. By Christmas in 1957, on our annual visit in Hastings, all was running well, as shown in Figure 16.

Figure 26 *Heidel in front of the Deering Library*

For us, the year 1956 brought a most important event. *Our daughter Caryn was born on the 20th of September, 1956* in Evanston, IL. Figure 26 shows Heidel in her maternity dress coming from Deering Library to meet me for lunch. Figure 27 was taken after a baby shower, arranged by Heidel's friends of the library. Finally, Figure 28 is Caryn in person (see also Figure 24, Chapter 11). Heidel worked as long as possible, and we became friends with the Good Humor Ice

Figure 27 *Heidel and the gifts from the baby shower.*

Cream man whom I had to find on the beach nearby to get the chocolate-covered vanilla ice, Heidel craved and I liked as well.

The university had hospital insurance only, i.e., we had to pay the obstetrician. After investigating who would be best, we made an appointment with Dr. West. After he assured us that all was in order, he asked me: "What are you doing? How much money do you make?" I answered that I was a student of Professor Dole with a $150 stipend, to increase for the next year. His answer: "I am in the same sailing club as Dr. Dole. I will tell him to give you more money, but I will charge you my minimum of $125 for the delivery and all check-ups." He took perfect care of Heidel and Caryn, and maybe, it contributed to my getting the Carbide and Carbon fellowship in September for 1956/57.

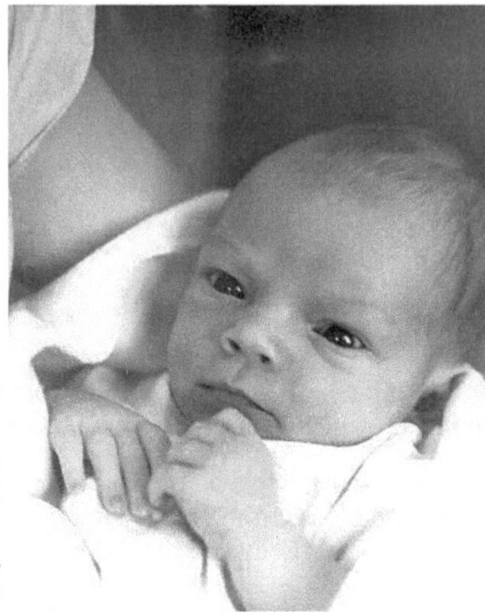

Figure 28 *One of Caryn's first pictures.*

The first few weeks all of us had too little sleep, but things settled down after the first major outing at the Chemistry Picnic in fall at one of the forest preserves west of Evanston. Figure 29 shows the expert advice being given to Heidel during the picnic by Hanna Haag. Through all this, research went on according to schedule.

Figure 29 *Hanna Haag and Heidel exchanging experiences with their babies at the Chemistry Picnic.*

Not long after the picnic, we suddenly woke up in the middle of the night, not because of Caryn, but because *there was an automobile accident in front of the house*. When I looked out, I saw police cars all over. In front of our house, a car had run into a few parked cars, somewhat behind ours. Then it had come to rest in the middle of the street. Nothing for *me* to get excited about.

Next morning, I discovered that when bouncing off the cars parked behind mine, the driver had also demolished my car's left side, so that the doors could not be opened. Fortunately, the police left the insurance information, police-report number, and address of the other party on my windshield so that I could discuss my claim quickly. I could still drive the car, but it was a difficult exercise to get in and out. The adjuster looked up the value of the car and determined that it was a total loss and he offered $350. I complained bitterly, I had just one month before bought new tires, and hoped that the car would last many more years. Nothing could be done, but, finally the adjuster said, you can keep the tires and the car, which you can use as a trade-in. That was it.

I needed a new car, and quickly. We looked at various used-car lots and at our finances. We decided it was not a good investment to buy a used car because we would not get much for our wreck. For new cars, however, we could afford the down payment, but not the monthly rates.

Chicago, being a big city, had many opportunities. At Z. Frank, one of the largest Chevrolet dealerships, we were offered a new 1956 Chevy for $1,900. After checking our cash flow situation, they also accepted a lower rate schedule until September 1957. On the next day, I could pick up my first new car. Heidel was somewhat hesitant, but what should go wrong? Something did. The bill of sales read the tank would have five gallons of gasoline. When trying to proudly drive Heidel and Caryn around the block, the car gave out. 'No gas,' was my diagnosis. The next gas station was not far to walk. But, I could not even complain. When carefully looking at the bill, it said '.5 gallons of gas,' meaning I could have gotten to the next filling station, but no further!

On December 14 of 1956 all my classes were over, and we could start our second visit to Hastings, to stay over Christmas and to be back on January 7 for the first day of classes. The new car was just a few weeks old and still needed to be broken in for another 200 more miles. We left at 5 a.m. and arrived at 8 p.m. after having driven the slightly more than 600 mi. Figure 30 shows our car during the return trip shortly before crossing the Missouri into Iowa. There was some snow on the road, but nothing a car with new tires could not handle with ease. Figure 31 shows the official family Christmas picture with the star attraction: Caryn. The picture was taken in Tante Lieschen's house in front of the fire place. Figure 6 of Chapter 11 shows the 'star' being admired by Tante Lieschen's family in Vati and Mutti's new house, shown also above in Figure 25. The year 1956 was a major milestone in our personal life in the United States, as described already on page 6-14–16. It was settled by then that Frank, Dagmar and Eberhard would join us in the US in 1957.

Figure 30 *Our new 1956 Chevy. With Heidel and Caryn on our way home from Hastings, in January.*

Figure 31 *Our family on Caryn's first Christmas in 1956. Picture taken in Tante Lieschen's living room.*

Having most of our family together in Hastings, *we decided to have Caryn christened in Vati and Mutti's First St. Paul Lutheran Church* with Frank (in absentia), Marilyn, and Mutti as god parents. The choice of a Lutheran christening was not only one of convenience and the church not raising objections of Heidel's being Catholic, it was precipitated the day Caryn was born.

When checking into the Evanston Hospital, Heidel disclosed our different religious affiliations. Although I had an ill feeling about it, it became worse, than I had feared. Shortly after Caryn was born, a nun visited Heidel. Instead of congratulating her on the birth of Caryn, she chided her about not having had a church wedding. Finally, she said to her face that she was an unwed mother and Caryn an illegitimate child. Naturally, Heidel was quite distressed. With somewhat more insight and diplomacy the nun might have been able to save three souls for the Catholic Church.

The question of our religions had come up as soon as we were engaged. The more traditional members of Heidel's family were starting to push towards a Catholic wedding. When I was asked, I must have displayed some of Vati's traits since I voiced my opinion that I had no strong leaning to either a Catholic or Protestant wedding; but if one of the families should insist, I would not oblige. As things developed, nobody was present to insist when we got married, and we also could not afford a church wedding (see the pages 5-14–15). Both of our children were christened in the Lutheran Church in Hastings and it was left to the god parents to guide their spiritual development.

The next milestone in our family history was the arrival of Frank, Dagmar, and Eberhard in Chicago, IL, on their way to Hastings, as discussed on page 6-14 and 15. They arrived in June 1957, after the spring term was over and I had delivered my thesis to Dr. Dole. This meant, I could take some vacation and we could drive together to Hastings. Earlier in the year, we had found that something needed to be done about the hot nights during the Chicago summer. The solution to open all windows led to an increased noise level which kept us from sleeping. Now, Caryn did not like those hot nights either, and with the impending arrival of Frank with family, we decided to invest in a window air conditioner. It certainly repaid its cost by cutting down on sleepless nights.

Frank, Dagmar and Eberhard arrived at Midway Airport, flying Lufthansa[1] from Frankfurt to Chicago with a Lockheed Super Constellation (see Figure 12 of Chapter 7). Figure 32 shows one of their first pictures in the new world with the Chicago's Loop in the background.

To be able to visualize the polymer molecules analyzed during my thesis work, Frank brought a big bag of atom models along, to be paid for by Professor Dole. When the stewardess asked Frank what was in the bag, he answered truthfully: "Atom models." Her answer was you will never get that past the customs inspectors. Anything 'atomic' may not be imported. So they renamed the things in the bag 'Eberhard's toys.' No problems occurred at the customs counter. A typical application of the atom models is shown in Figure 33. It represents the conformation of a nylon 6 molecule, $[(CH_2)_5–CO–NH–]_x$. The models have lasted over 50 years and illustrated many complicated polymer structures.

Figure 32 *With Dagmar and Eberhard in Chicago.*

[1] The "Deutsche Lufthansa AG" is not a direct (legal) successor of the German airline flying between 1926 and 1945 of the same name. It was founded after WW II in 1955 in the FRG and started the first transatlantic flights to New York in 1955 and service to Chicago and Montréal in 1956. "German Lufthansa" (East) was also founded in the GDR in 1955. The two airlines had, except for name, logo, and history, nothing in common. The East German Lufthansa, however, failed to get an international trademark and was pressured to be renamed to "Interflug" in 1958.

Figure 33 *A model of a segment of a nylon molecule.*

Figure 34 *Stopping at a motel on the way in Iowa.*

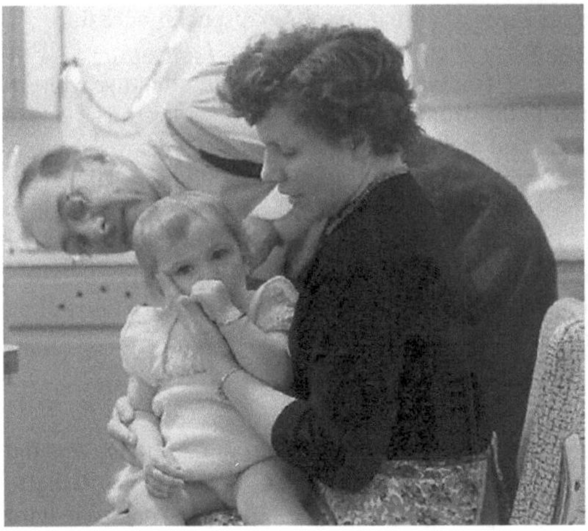

Figure 35 *Caryn's first hair cut, performed by Vati.*

After a few days of sightseeing in Chicago, we traveled to Hastings Nebraska. For the convenience of the passengers, we stopped on the way at a motel, also something new for the immigrants from Germany. The motel is shown in Figure 34. On the next day, we arrived at the 'family mansion,' seen in Figure 25. For the first time in five years the whole family was together again.

After delivering Frank, Dagmar and Eberhard and getting settled, Vati had to perform an important job, to give Caryn her first hair cut, documented in Figure 35. This started a tradition which was repeated some three years later when Brent needed his first hair cut (Figure 20 in Chapter 7).

Frank soon started to work with Vati, and by the time we visited Hastings again at Christmas time in 1957, he had, just like we had two years before, adapted to the new environment. Figure 1 of Chapter 11 gives a picture of the 10 members of the family reunited and well adjusted to their new life in the USA. Figure 16, above, shows that by then everyone was well motorized and had a good start. The last of such family pictures in 'normal times' was taken 20 years before (see Figure 9 of Chapter 1). At that time, the early Wunderlich family had only four members. In the meantime, the turbulent 20 years, described in Chapters 2–6 had passed.

In February 1958, Frank decided that his education as structural engineer was a safer basis to continue progress in the US and he accepted a job with the Lincoln Steel corporation in Lincoln, NE. His employment there involved engineering design of office and bank buildings, shopping centers, hotels, and a power plant. In 1967–70, he was the construction manager of the building of the new St. Elizabeth Health Center. On the completion of the hospital, he remained there as Director of Plant Services until 1979.

Next, he went from Lincoln, NE to Florence, SC, as Director of Planning and Engineering of the McLeod Regional Medical Center where he stayed until 1983. His position was that of the Department Head for Maintenance, Biomedical Engineering, and Buildings and Grounds, with responsibility for planning, budgeting and supervision of all construction and renovation. This was followed by managing of offices for tax preparation. His sudden, early death occurred in 1992.

My final step to graduate came after the visit in the summer of 1957 in Hastings. It consisted of the PhD exam and went without a problem. All questions of the committee dealt with the thesis subject. As we had calculated a year earlier, by the time I received the first paycheck, our bank account had reached the absolute minimum, $3.50. This should never happen again.

Teaching as an instructor at Northwestern was my new job in the 1957/58 academic year. It dealt with undergraduate physical chemistry lectures (3 hours weekly) and with supporting of the big freshman chemistry effort led by Professor Summerbell. My own lectures in freshman chemistry came in 1958 during the summer session. This was a vital experience for my later teaching. Key was to learn how to handle the large groups of students. It was important never to lose control of the group during lecturing and organizing activities. This was brought home during the well-organized instructions given to the new teaching faculty at Northwestern. This introduction and my excursion into acting, described on pages 3-26–27, were valuable experiences. Unfortunately, one frequently observes that university teaching is often carried out by professors that do not have such experience. Planning, execution, and analysis of the scientific research are often the only requirements for a teaching position. This alone, however, does not make a good teacher.

The atmosphere among the faculty at Northwestern was just as friendly and congenial as it was to the students and staff. The whole department met once a week for a bag-luncheon in the lounge where one of the members had to eat fast and then present his research results and problems or some other subject of interest to all. Subsequently, the presentrd material was discussed. My research was similarly analyzed and interesting suggestions were made. I never was a member of another department of chemistry where I knew as much about the research of everyone else. This is a rather valuable asset when advising new graduate students. In one of these meetings, I also was served my first pizza, brought along by Professor Frost as a treat for all on a special occasion.

At the same time, *I was invited to consult for the Toni Hair-waving Laboratory*.[1] This yielded a monthly retainer of $50. The laboratory of Toni was located at the Merchandise Mart in Chicago's Loop, an easy ride with the 'Elevated Train' from our apartment. Besides the structure of hair and the chemistry of sulfide-bond breaking and making during hair waving, the discussions concerned questions of the properties of the rollers included in the home-perm packages. They were made of polymers. Searching the literature on high-pressure extrusion of polymers and structure-changes occurring under conditions of making the rollers revealed that nothing was known about it. This was one reason for me to plan, as my first future research to study the effect of high pressure on polymers. Ultimately this was quite successful and led to the first growth of fully crystalline polyethylene with an extended-chain conformation (see Appendix B, page B-4).

---

[1] Home permanents had been sold since the early 1930s, but the Toni Company was the makers of the first successful, popular product (the company was acquired in 1948 by the Gillette Company). In their famous ad, they used a set of pretty twins with identically styled hair, one with a salon perm (value $15), and one with the home perm valued at $2.00 (and a refill for only $1.00).

Professor Dole made sure that *my research during this extra year at Northwestern* was largely independent of his own, although we frequently discussed the details. The research was largely theoretical and computational. It dealt with the description of what Professor Dole and I had called *cold crystallization* when it was discovered during my thesis work (see Appendix B, page B-2). The five research publications generated at Northwestern included my thesis work (Appendix A, publ. #3), the new independent work on cold crystallization (publ. #4), extra experiments I did in support of the overall heat capacity project (publ. #2), and the two papers Professor Dole generated on the interpretation of my data (publ. #1 and #5).

The new work on cold crystallization was based on calculations with the new electronic computer at Northwestern's Mathematics Department, an IBM 650.[1] To calculate the amount of crystallization on quenching and its link to the subsequent melting range which is governed by crystal size, I wrote a program of 700 calculation instructions. This program took a long time to run for one copolymer composition. Up to 45 min were needed for the iterations over the temperature range. When printing the first data output, I noted that there were pages and pages of either 0.9999 or 0.0000. It looked that there was something wrong, but in the middle of the pack of useless numbers, there were two pages where the data changed smoothly from no melting (crystal fraction 0.9999) to complete melting (crystal fraction 0.0000). I had guessed the range of change improperly and calculated over a too large a range of parameters. Subsequent runs were closer to 3 min.

As computers improved, simulations of molecular behavior could be extended. Ultimately, molecular motion in crystals could be followed. By cooperation with Dr. Noid of ORNL, (Oak Ridge National Laboratory) molecular dynamics calculations were made of up to 30,000 atoms for time periods of up to 100 ps (one picosecond = 10^{-12} s). These simulations established the defect structure and properties of semicrystalline polymers (Appendix A, publ. #350). Despite the enormously faster supercomputers of the 1990s, we used about 8,000 h of time on the Cray Y-MP computer[2] for the many runs made from 1989–1996 described in Appendix A, pages A-14–21, and B, pages B-18–19. The developments leading to online measurement of data and immediate data treatment are summarized on pages 8-64–65 and Section D in Appendix B.

In 1957, I joined the American Chemical Society, ACS. As my initial activity in the society, *I attended my first National Meeting of the ACS* (April 13–18, 1958, San Francisco, CA). There, I reported on my thesis work. Travel support came by coupling the meeting with an interview at the Shell Development Co. and a visit with Dr. Lau, a former postdoctoral student at Northwestern, who was then with RayChem Company. The presentation at the ACS went poorly. My talk was scheduled in a morning session, just before the break for lunch. I was prepared to take precisely the allotted time, but the earlier lectures were running late. The lecture before mine was given by Professor Mandelkern, a well-known polymer scientist. He went on and on with his lecture, so that

---

[1] The IBM 650 was one of the first main-frame computers. Shipping of these started in 1954. Over 2,000 had been built by the time they became outmoded in 1962. The hardware consisted of a console and a power unit, and the input and output of the data were by an electro-optic card reader and mechanical punch. The memory was an electromagnetic drum of 2,000 words.

[2] The CRAY Y-MP series of supercomputers was introduced in 1988. They featured 8 processors. Over 200 CRAY Y-MP systems have been built. It was the computer breaking the gigaflops barrier [(10^9) floating point operation per second]. Presently teraflops are possible (10^{12}).

my presentation could start only when it was supposed to end. Naturally, Professor Mandelkern did not stay for my lecture, and many of the audience also thought with his lecture, all interesting material was presented. Going for lunch was more important. There were only few people left to listen to my report. In addition, I had sufficient stage fright that at the end, I did not remember whether I had presented everything properly. I could only hope that I was prepared well enough that all went automatically. It must have, because the two people in the audience who assessed my qualifications for a job were sufficiently satisfied to arrange for offers.

In the future, the stage fright became less, but it never completely disappeared until the first few sentences were presented. A good preparation is to know these first sentences by heart. As the lecture progresses, one then must listen to one's own words and make sure that what was said was sufficiently explained. If not, quick correction may be possible. I heard in an interview of some famous actor that he always had stage fright before a play. If this were not so, he suggested, he would not have enough respect for the audience to do well. I fully agree with this statement.

This 1958 ACS Spring Meeting also concluded my job search. The finding of a suitable job with sufficient chances for advancement in the future was a major effort and started already in 1956. Company recruiters visited the university each fall to establish a first contacts with students likely to finish during the next year. Since at that time I was not sure if an industrial job was to my liking, Professor Dole and several other Northwestern faculty members had contacted several universities.

Overall, I had *employment discussions with ten companies and nine universities*. The job interviews meant travel by train or plane. This was the third time in my life I had to face travel by plane. The first attempt and actual flight in 1939 and 1953, respectively, were not very pleasant, as described on pages 1-20 and 4-35. By 1958, the planes were pressurized and this travel was actually enjoyable. In addition, the service was very good. Even the food served was often outstanding. I remember my first shrimp cocktail being served on a cross-country flight as being so good, that I have not been able to duplicate it anywhere. In fact, one may characterize the period from 1958 to 1980 as the best for flying. Since then, service declined, planes became uncomfortably large and crowded, waiting times got unpleasantly long, direct flights are rare, and changing a flight is difficult, costly, and not permitted to another airline. Where are the times, when, in case I finished early at a meeting or on consulting, I could just go to the airport with my ticket to find out what airline flew next, and have my tickets accepted? The same was possible when arriving at a stopover. At this time, when average seat occupancy was 50%, any airline was happy to accommodate you. The airline would even pay for a car rental when getting stuck close to home because of weather.

The first contacts with companies were made during the fall Term 1956/57 by Union Carbide (due to my holding the Carbide and Carbon Fellowship), Shell Development, and DuPont. The next year added General Electric and Eastman Kodak to the list of companies. Because of my instructor position at Northwestern for 1957/58, I postponed visits to early 1958, except for a drive on September 10, 1957 to the local Visking Company in Chicago, a subsidiary of Union Carbide.[1] The Visking research into other than cellophane polymers was spearheaded by Dr. Paul Lindenmeyer

[1] The Chicago-based Visking Company was founded in 1925 for the production of nonedible viscose cellulose and plastic casings and nettings used in packaging processed meat and poultry products. In 1957, it was sold to Union Carbide, and in 1986, acquired by Envirodyne, forming the again separate, global 'Viskase' Corporation, with production in the US, Mexico, Brazil, and France.

who became a good friend of mine. This research was exciting. A big surprise for me was also how Visking made the cellulose casing for skinless hot dogs! They were based on easily recognizable, unappetizing, striped casings. After processing the hot dog, the skin was peeled by machine to make it skinless. The distinctive pattern made it easy to spot any hot dog missed by the skinning machine. On September 16, 1957, I got my first job offer, details being held by Union Carbide until after the visits to one or two of the other Union Carbide companies.

The first contact with a university was also early in 1957. On October 1, 1957, I gave an earlier scheduled lecture at Loyola University in Chicago. This was too early for serious discussions about possible employment. Next, in January 1958, Professor Dole wrote to a number of universities he had contacts with and promptly, I received several invitations to apply. From Yale University in New Haven, CT, I applied by Jan. 27, but got a negative answer by April 25, 1958 (after I had already accepted the offer from Cornell University). Next, was a telegram on February 14, 1958 by Professor E. Bright Wilson, Jr., the Chairman of the Chemistry Department of Harvard University, Cambridge, MA. He made an offer of an instructor position based on the strength of the recommendation by Professor Dole, an alumnus of Harvard. This was my first secure job offer from a university. I could have stopped all further interviews at that moment, but ... . The University of California, in Riverside, CA, showed interest. Nothing resulted from my application before I accepted the Cornell offer. I had sent an application on March 5, 1958, invited on the suggestion by Professor Burwell. Telephone conversations and a visit by Prof. Pitts from the UC Riverside at Northwestern where positive. To Indiana University, in Bloomington, IN, I was invited to apply on an inquiry by Professors Burwell and Frost of Northwestern. Before a promised visit was finalized for April, I had accepted the Cornell offer. Finally, I gave a seminar on May 6, at Illinois Institute of Technology in Chicago, but this was too late to discuss possible employment.

At the company research laboratories and universities visited, I reported on my thesis research and the current work in progress. The abstract of the presented lectures may be of interest. It read as follows:

Cold Crystallization of High Polymers

Synthetic high polymers can crystallize in two very distinct ways: a) by cooling the liquid slowly below the melting point, or b) by heating the supercooled amorphous glassy polymer to a temperature where crystallization is first possible. The two crystallization temperatures of those two samples can be up to 200°C apart.

Situation a) is an approximation to the 'Equilibrium Crystallization' of Flory. The b) case of 'Cold Crystallization' could be treated by a statistical model, which agrees perfectly with heat measurements done with an adiabatic calorimeter.

Recognition of these two different mechanisms leads to the resolution of contradictory statements which have been made in the past.

The first visit in 1958 was to the University of North Carolina in Chapel Hill, NC, invited to apply to on recommendation of Professor Frost. The lecture was given on January 20, 1958. On this trip I had a first view of the beautiful southern part of the US from the plane and I cherished the mild January climate. Also, the spirit and level of the university were to my liking, but by March 17 I was told by letter that no offer could be made. Next came a 15 h overnight train ride in a sleeper from Chicago to Schenectady, NY, to visit and lecture at the General Electric Company Research Laboratory on February 7, 1958. After the interview, during the summary discussion with

Dr. Marvin of the Personnel and Administration Office of the Chemistry Research Department, I was told that he had a perfect record, every offer he made was accepted. His scheme was as follows: He told me that I could get an offer during the next month or two at a specified salary level. As soon as I would know that I would accept the position, I should call him and he would send the written offer in the next day's mail and this would keep his record intact. I never needed to make this call. Among the company laboratories I visited, GE made only number two. They had a good group of researchers in the polymer field, but who could, in 1958, compete with the number one company, E. I. du Pont de Nemours and Company, the DuPont, for which I later consulted for 25 years.

The Union Carbide operation at Oak Ridge asked for my application after their 1956 recruiting visit at Northwestern, but already on December 14 they sent a letter stating that no offer can be made from the Oak Ridge Laboratories because I had no citizenship. But, they forwarded the application to their New York City headquarter. On February 20, 1957, I was invited to their Bakelite Company in Bloomfield, NJ and Carbide and Carbon Chemicals Company in South Charleston, WV. All these were postponed until 1958 due to my staying at Northwestern. The visit to Bloomfield was on February 28, 1958 by plane from Chicago. Offers were made on March 20 1958, including also the earlier Visking visit. They made number three on the list of preference.

The next trip went by train from Chicago to Syracuse and by bus to Ithaca, NY, to visit Cornell University on March 11, 1958 (return by air). This visit went especially well. It was arranged via Professor Peter Debye,[1] with whom Professor Dole worked in 1929 for one summer (see Footnote of page 6-19). Despite being retired, Professor Debye attended my lecture and we had, for me, the longest and most productive discussion of any lecture I have ever given. In later years he spent several evenings in our house and I worked with him to transcribe one of his later lectures at Cornell for publication. An offer was made March 24. (Instructor of Physical Chemistry, 9-months salary $5,250– and a $500 summer research grant.) My request for a high pressure autoclave to 100,000 psi pressure and a temperature of 300°C was accepted (≈$5,000).

The long anticipated trip to the DuPont in Wilmington, DE, first discussed in 1956, took place on March 24–26, 1958. Visited were the Textile Fibers, Polychemicals, and Photo Products Departments in Wilmington, DE, and also the second location of the Photo Products in Parlin, NJ, near New Brunswick. By March 31, 1958, an offer was made to work for any department of my choice with a salary of $9,120. The visit made a big impression. This was the location where polymer science was generated and applied. Many of the people I met during these days remained

---

[1] Petrus Josephus Wilhelmus Debije (1884–1966). He later changed the spelling to Peter Joseph William Debye, born in Maastricht, The Netherlands. He held professorial appointments in Zürich, (Switzerland), Utrecht (Netherlands), Göttingen, Leipzig, and Berlin (Germany) where he was Director of the Kaiser Wilhelm Institut (now Max-Planck-Institut). From 1937–39 he was the President of the Deutsche Physikalische Gesellschaft. After leaving Germany in 1940, he was Professor of Chemistry at Cornell University and from 1942 until his retirement in 1950 he was chairman of the department and remained active as emeritus professor and consultant thereafter (a US citizen since 1946). Nobel Prize in Chemistry 1936. His many contributions to physics and chemistry include the development of the powder X-ray method (with P. Scherrer), the theory of electrolytes (with E. Hückel), the dipole contribution to permittivity, light scattering of macromolecules, and the theory of heat capacity. (See also the pages 7-3, 24–25, and 9-80.)

professional friends. This was the number one company. On the way home, I made a stop at Eastman Kodak Company, Rochester, NY. They also had contacted me on their recruiting visit to Northwestern in November 1957. The visit was on March 27, 1958. Their decision was to be sent by April 20, too late to be considered. The visit to the Shell Development Company in Emeryville, CA, in conjunction with the ACS meeting also went well. They also had an impressive research program in polymers, but by that time, my job choice was practically made.

Back home in Evanston, I had to make a decision between the two university offers and the best industrial offer. I certainly was attracted by the independence of a university professor in his research and teaching. With the help of Professor Dole, the following reasoning was made: If I cannot decide between an industrial and an academic job, it is better to choose academia. A later change from a university to a company is easier than the other way around. After having worked in industry for many years, it is hard to prepare basic lectures in chemistry or physics when not having kept up with the basics. In addition, it is also hard to live later with a lower salary. The salaries offered by Cornell and Harvard were 40% less than by the DuPont. The argument Cornell vs. Harvard was resolved by knowing that there was a smaller chance of gaining tenure at Harvard. In the past, practically all instructors at Harvard were mainly used to teach undergraduate courses and were not promoted. I accepted the position as *Instructor of Physical Chemistry at Cornell University*, starting in September 1958.

During this hectic year of search for a job, there was also continued travel also for our private enjoyment. Students always have periods of vacation during which we could undertake shorter or longer trips by car. One of the longer travels went to Toronto, Canada. Our friends from Frankfurt, Günter and Gisela Fricke (see pages 5-8 and 5-14–16 and Figures 22, 24, and 26 of Chapter 5) had also emigrated from Germany and settled in Canada, where Gisela's father had a business. Figure 36 was taken during a walk in downtown Toronto.

Figure 36 *Caryn and Heidel visiting Gisela on a first trip to Toronto, 1958, traveling by car from Chicago.*

At home, between trips, Caryn was enjoying the family life as much as we did. Figure 37 is an example of her learning how to jump on the sofa.

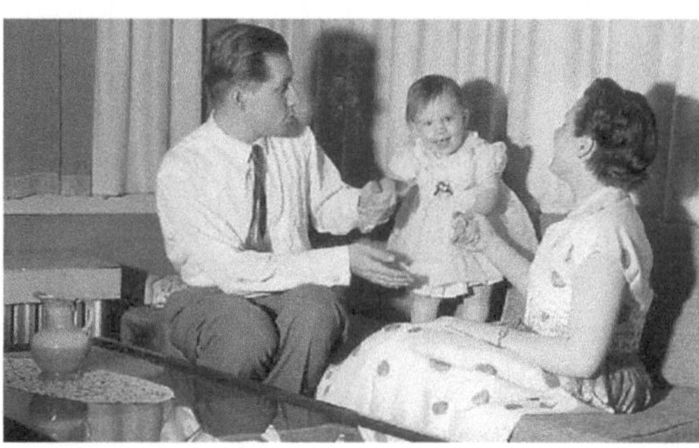

Figure 37 *Family time at home in Chicago, 1957/58.*

A day-trip went to the Mississippi (Figure 38) with our neighbor, Fräulein Rank ("Tante Bonbon"). She often helped us by babysitting with Caryn and usually brought some sweets, hence her nickname. Figure 39 shows the beautiful countryside around Fox Lake, north of Evanston. Figure 40, finally shows Caryn at the shore of Lake Michigan in the area of the Indiana Dunes Park.

Figure 38 *A trip to the Mississippi with Fräulein Rank, our German neighbor in Chicago, in 1958.*

Figure 39 *Heidel and Caryn at the Fox Lake, north of Evanston. A weekend day-trip in 1957.*

Figure 40 *Caryn on the shore of Lake Michigan at the Indiana Dunes. One of the last outings, shortly before the move to Ithaca in August of 1958.*

Figure 41 *A first look at Cayuga Lake in Ithaca, NY, with Heidel and Caryn after our move in fall 1958.*

Now it was time to prepare the move to Ithaca, NY. First, I traveled there by myself to look for a suitable place to live with help of the Cornell Housing Office. On this second visit, I discovered the beauty of the area around Ithaca and Cayuga Lake, as is as displayed in Figure 41. After discussing the possible choices of apartments over the phone with Heidel, we chose 18 Hawthorne Circle, a two-bedroom apartment making one half of a duplex, one hill removed from the university. It is located just a few hundred feet from the position of Figure 41. Another choice was an apartment in a house built on-top of a cliff, overlooking the gorge running through the campus, shown in Figure 6 of Chapter 7. The view was beautiful, it was almost in walking distance to the university, but we deemed the balcony not safe for Caryn's enterprising ways. Finally, there was an apartment, also in walking distance, but the landlord was used to students, and his condition of no parties and no alcohol on the premises were too restrictive for us.

After this trip, all that was left was to decide how to move. From Hastings to Chicago all our belongings fitted into our car. From Chicago to Ithaca, we rented a double-axle U-Haul which turned out to be a problem for our car, but we made it.

CHAPTER-07

Cornell University, 1958–1963

Over the Labor Day weekend of 1958 we arrived in Ithaca, NY, at 18 Hawthorne Circle, as shown in Figure 1. The trip was not without difficulties. We learned that a trailer cannot be easily moved by a passenger car, and that the great looking Uniroyal tires were not designed for such a job (at least at that time). Just as now, it was said then that the tires by Uniroyal combine the responsive handling of a performance tire with the comfort of a passenger tire. Thus, we had hoped for an "even wear and durability, exceptional handling, and great looks," all benefits still advertised today.[1] Just outside Chicago, we lost the right rear tire and stopped on the road side to replaced it with the spare. The side wall could not carry the load of the trailer and had broken in a way that defied repair. Fortunately, we bought a new tire as a replacement just a down the road. Not long thereafter, the left rear tire gave way and we went through the same exercise once more. Shortly before we reached Ithaca, the original spare on the right rear gave out also. All three tires had broken side walls. This simple, but expensive experiment proved that the load was excessive for the 'Uniroyals,' but not for the newly bought 'Goodyears.' After our arrival in Ithaca, we picked up the key for our new living quarters, unloaded the furniture and other goods from the trailer, and returned the U-Haul to the local dealer. He said: "Yes, I know, Uniroyals are not the tires to have for such a haul, their side-walls are too weak."

<u>*The duplex apartment had two bedrooms*</u> upstairs, and the living room and kitchen downstairs. Our one-room furniture from Chicago did not fully suffice. Caryn's crib could quickly be moved into one of the bedrooms and until we acquired additional furniture for her, she had a large area to play in. Our bedroom, however, was completely empty. Fortunately, our neighbors in the duplex, the Shulmans, owned a furniture store, and could quickly deliver a new bed. The duplex and the delivery of the bed are documented in Figure 1. Although the living room was sparingly furnished, we now had time to slowly improve our domicile.

Figure 1 *Our duplex (left half) in Ithaca, NY, shortly after our arrival. Our newly 'tired' car is in the middle, our bed is delivered with Shulmans' pickup.*

[1] As printed in www.uniroyal.com/overview/tiger-paw-touring/

Ultimately, we found a dealer outside of Ithaca who carried Paul McCobb furniture and could also order Knoll chairs, so that in time, we furnished a comfortable, modern home in our much improved living quarters. Under the back stairs, we even arranged a barbeque after we discovered the Cornell Agriculture Department's chicken barbeque recipe (see Figure 36, below). Today, we still enjoy it each summer. Cornell University,[1] as a land-grant college, had an outstanding Ag Department which locally sold many of their products. It was located about a mile east of the Chemistry Department, and besides enabling convenient milk runs to supply Caryn, and later also Brent, their staple also included an outstanding maple-walnut ice cream. A result of their research?

The countryside in and around Ithaca was much to our liking and we explored the 'gorgeous' sights. The Taughannock Falls State Park is shown in Figure 2. Its creek, as all others, ultimately reaches the level of the Cayuga Lake, the central jewel of the area, shown on Figure 41 of Chapter 6. During the initial months, every weekend was spent to explore the countryside around the lake and visiting the numerous State Parks. The highlights for us and our visitors were walks through the campus gorge, to Beebe Lake, Ithaca Waterfall, and the city park at the foot of the lake, or a drive to Buttermilk State Park, the two (upper and lower) Treman State Parks, and gorge in Watkins Glen State Park near Seneca Lake, the next Finger Lake to the west.

Figure 2 *Taughannock State Park in the Fall 1958.*

Also, the country roads offered many pleasant Sunday-afternoon drives. A few miles out of town, on the road towards Dryden, we discovered a shell of an old southern mansion in an open field. It had four massive Greek columns in front, and an exquisite, curving grand stairway inside. Besides the facade and roof, a brick basement, and the stairway which all seemed quite solid, nothing was left. It looked so

---

[1] Cornell, was founded in 1865 by the New York Senator Ezra Cornell and Andrew Dickson White (the first president of the university). Admission was offered to students, irrespective of religion or race. It is one of the earliest Land Grant Colleges of the US, based on the benefits of the Morrill Act of 1862 (and 1890). Its declared mission was "to teach agriculture, military tactics, the mechanic arts, and home economics, not to the exclusion of classical studies." The university was opened in 1868 with 412 men. Two years later, the first women were admitted, making it the first coeducational school among what came to be known as the 'Ivy League.' These ideals, unconventional for the time, are captured in Ezra Cornell's 1865 motto on the open pages of the book in the seal shown on the right: "I would found an institution where any person can find instruction in any study." The university expanded in the 20th century. Its student population grew to about 20,000, and the faculty to more than 3,400.

harmonious in the countryside that I seriously considered purchasing it in a few years, to ultimately restore this beautiful shell and make it into our dream-house.

After the move from Chicago, **a new research and teaching effort had to be established at Cornell**, a laboratory had to be set up, graduate students to be attracted, and research funding to be found. Shortly after we arrived in Ithaca, *I reported to Dr. Franklin A. Long*, the Chairman of the Chemistry Department. A contemporary postcard of the Baker Laboratory which housed the Chemistry Department, built in 1923, is shown in Figure 3. First, I was assigned an office. Since space was at a premium, I was moved into the vacant secretary's office of the suite that was designed for the Baker Lecturers.[1]

By 1958 the endowment of the Baker Lectureship was insufficient to keep a private secretary for the lecturer. The departmental secretaries had taken over this duty. Naturally, I did not mind sharing the small entry hall into the two offices with the Baker Lecturer. In this way, I got to meet the lecturer, and if he had a question, I was glad to help out. It allowed me to befriend several of the visitors so that I could discuss many aspects of international research.

Figure 3 *The Baker Laboratory, home of the Chemistry Department of Cornell University.*

---

[1] The Baker Lecturers in the Chemistry Department (from 1926) included 23 Nobel Prize winners in the 20th century. It is the oldest and most prestigious series of endowed lectures sponsored by a chemistry department in the US. The Baker Lecturers I met as my office neighbors in the fall (F) and spring (S) terms between 1958 and 1963 and their lecture topics were:

1958 F:	M. Calvin, University of California at Berkeley. "Recent Advances in the Chemistry of Photosynthesis."
1959 S:	C. A. Coulson, Mathematical Institute, Oxford. "The Size and Shape of Molecules."
1959 F:	R. Criegee, Karlsruhe Institute of Technology. " Organic Oxidation Mechanisms."
1960 S:	R. S. Mulliken,# University of Chicago. "The Interaction of Electron Donors and Acceptors."
1960 F:	F. A. Kroger, Philips Research Laboratories, The Netherlands. "The Chemistry of Imperfect Solids."
1961 S:	P. Debye,* Cornell University. "Molecular Forces."
1961 F:	F. S. Dainton, University of Leeds. "Radiochemistry and Some Topics in Reaction Kinetics."
1961 F:	M. Eigen,# Max-Planck Institute of Physical Chemistry. "Physical Chemistry."
1962 S:	R. Huisgen, University of Munich. "Organic Chemistry of Cycloadditions."
1963 S:	E. A. Guggenheim, Reading University. "Application of Statistical Mechanisms to Some Problems in Physical Chemistry."

---

* The Nobel Prize was awarded to the Baker Lecturer before the lecture series.

The Nobel Prize was awarded to the Baker Lecturer after the lecture series.

Figure 4 *High-pressure generator. The barrel with the pressure vessel immersed in 600 K hot oil is behind the brick wall.*

To find a suitable laboratory was of greater difficulty. Voluntarily, professors never give up any research space. They always think they need more than they have. I learned this later when I was on the building committee at Rensselaer Polytechnic Institute (RPI) and tried to project the growth of the department.

Since at Cornell I wanted to work on high-pressure research (see page 6-32), a major part of my laboratory was relegated to the basement to protect everyone else from the effect of accidental explosions. Indeed, I used extra precautions in designing the equipment which had to be rated at more than 5,000 times atmospheric pressure (5×10^8 Pa or 75,000 psi) at temperatures of close to 600 K. The danger, however, was not as great as it appears, since only liquids were used in the apparatus shown in Figure 4 (for details see Publ. #10, Appendix A). Liquids do not produce as large amounts of energy on pressure release as do gases. The most powerful pistol shot is driven by a peak gas pressure of no more than 14,000 psi. Containing gases, our equipment would have required an elaborate, explosion-proof bunker, placed outside of the laboratory.

An additional, smaller room for my laboratory was the old, decrepit lecture-demonstrator laboratory on the second floor. When it was not fixed up after several weeks, I just got a gallon of paint and got to work to the dismay of the relevant union. After I had accumulated several students, Dr. Long gave up part of one of his own laboratories.

During the next alumni reunion, one of Dr. Long's old students came into this lab and discovered that an old reagent bottle of his was still standing where he had left it some 10 years before. This demonstrates (a) that the need for laboratory space was not as urgent, and (b) that in university laboratories, bottles which look like part of research in progress are never touched by anyone who does not know the history of the bottle.

With the purchase of the pressure equipment, the set-up funding by the university (see page 6-36) was exhausted. Other equipment had to be scavenged from the attic where discarded or outdated instrumentation was stored. I found a Leitz microscope, used earlier in Prof. Flory's laboratory at Cornell (see page 7-19), and a precision Northrup-White potentiometer to measure thermocouple voltages. The microscope was similar to the one, I missed out on in 1953 (see Preface and page 4-33). It served to establish our work on optical microscopy and is still in my possession. The potentiometer was the heart of my first, self-built differential thermal analyzer (DTA). It was used for the first DTA of semicrystalline polyethylenes (see Publ. #7, Appendix A).

The initial teaching duties were light. The main assignment was the handling of the undergraduate Physical Chemistry Laboratory, and later also the giving of lectures on Introductory Physical Chemistry, which I broadened by including the special behavior of macromolecules. About 30–50 students needed to be instructed. In the laboratory, technique needed to be taught, supervised, and the write-ups of the experiments graded. All was done with the help of two teaching assistants. The laboratory equipment was not new. One of my first initiatives, thus, was to get Geiger counters and special laboratory wares for experiments involving radioactive chemicals and design a series of new experiments, covering nuclear physical chemistry. I applied for outside funds and was successful. The 'new' P-Chem Lab was soon liked better by the students than before.

For two summers, I taught a comprehensive freshman chemistry course for high school teachers in chemistry. It was supported by the Shell Oil Company Foundation. This 'Shell Program' rewarded about 30 outstanding teachers from east of the Mississippi river. A similar program was run at Stanford University for teachers west of the river. I could build on my experience from Northwestern University and the outstanding instruction in high school and at the universities in Germany (see Chapters 4–6). The daily lectures and discussions were great fun. The high school teachers were very enthusiastic and always ready to participate in lively discussions. The main question of theirs, however, was: "How can I make my students understand the various new concepts?" I discovered, this was their ultimate goal. They were *not* looking for understanding, just how to teach the subject (see also page 9-86). Twice did I find a teacher who liked true scientific inquiry. Both asked at the end for help to get a position in a research laboratory!

After I had started to teach physical chemistry lectures, a need arose at Harpur College in Vestal, NY, for a substitute physical chemistry professor for 1960/61. I accepted this special job. Harpur College was near Binghamton, about 50 mi from Ithaca (in 1965 it was restructured into the State University of New York, SUNY, at Binghamton). To cut down on traveling time, I combined the three weekly lecture hours and spent a whole afternoon with the ≈10 students. Again, a very exciting experience because of the great inquisitiveness of the Harpur College students.

Finally, in Fall of 1962, I was asked to give a series of ten lectures on Mechanical Properties of Polymers in an industrial setting at the Electrochemicals Division of the DuPont at Niagara Falls, a distance of ≈170 mi from Ithaca. During the winter, this often meant to brave a lake-effect snowstorm going through Buffalo. Arriving in Niagara Falls, the weather could be sunny, but on the way back, there still would be the same blowing snow in Buffalo.

Since I had not studied mechanical properties in great detail, it presented a good opportunity to improve my polymer knowledge and learn together with ≈20 industrial researchers about this important subject. The visits, which always took a full day, were sufficiently successful to yield an exclusive consultant arrangement with the DuPont which lasted for over 25 years. At least once each month, I would travel to one of the various laboratory and plant locations (see Appendix C, page C-4). This narrowed the gap between academic and industrial salaries (see the pages 6-36–37).

Next, funding had to be attained and students had to be found.[1] From experience, I can say that the professor may set the direction to interesting research, but the level, precision, and the overall accomplishments depend on the quality, efficiency, and hard work of the students.

---

[1] My support for 50 years of research amounted to about five million dollars. Most from government agencies, companies, and foundations. Most went to the support of coworkers. During the summers before 1988, I could also supplement my salary. The rest went for equipment, expenses, and travel. *Supporters were:* the Research Corporation (RC); the Advanced Reseach Projects Agency (ARPA); the Office of Naval Research (ONR); the National Aeronautics and Space Agency (NASA): the American Petroleum Research Fund; the Humboldt Foundation of the FRG; and most prominently, the National Science Foundation (NSF, usually through their Polymers Program); and from 1988–2001, the Distinguished Scientist funding (from the Science Alliance of UTK and the Div. of Mat. Sci. and Eng., Office of BES, U.S. Dept. of Energy). Some industrial support came from: Owens Illinois; General Electric; Armstrong; Hercules; Goodyear Tire and Rubber; Bridgestone Tire; TA Instruments; PerkinElmer; Mettler Toledo International; ICI Paints; Toray Research; Exxon, Research and Engineering; Shell Development; and Cargill-Dow LLD (Natureworks).

Before a student could be enticed to work with me, I spent time to design the high pressure apparatus of Figure 4 and worked in the library to gain a new understanding of the glass transition (Appendix A, Publ. #6). In 1959, support could be gained from the RC, ONR, ARPA, and NSF (see Footnote, page 7-5). It was very fortunate that there were enterprising students at Cornell who would chance their future on a newcomer who might not stay to guide them to their PhD.

First, I hired Jim Flood, a technician (without academic training) to get the laboratory going (see Appendix A). He helped to get things started and could pass the experience he and I gained from setting up and testing the equipment to the later coworkers. Mr. Kashdan also helped in the laboratory. He was a chemistry teacher in high school who used his summer vacation to gain laboratory experience and earn some money. Then, there were seven undergraduate students who worked for various periods in my laboratory. They mostly started by taking the 'Advanced' Physical Chemistry Laboratory, I offered to the better students. For me, it was easy to identify among these students good coworkers for a possible assistantship in my laboratory.

In 1960, the first graduate student knocked at my door, Mr. David Bodily. He was followed by Peter Sullivan. Then, in 1961, I hired a postdoctoral fellow, Dr. E. A. James. Finally, there was an experiment with a Japanese Company, Teijin, Ltd. of Tokyo. One of their deserving BS-degree-level employees, Mr. Tamio Arakawa, was granted a one-year leave of absence to study polymer science in the US. In cooperation with Cornell University, the company arranged for a Fulbright travel grant and several smaller industrial supports for the initial stay with me.

Because Tamio did well, his leave was extended, but when I mentioned during the second year that he had done so well in his research and had taken sufficient courses that, with another year and the prescribed exams, he might earn a PhD instead of a MS degree, the company balked at this and did not communicate with him further. When he returned in 1964 with his PhD degree, Teijin ignored the degree, and sent him back to his old job (and pay) as punishment. Only after a few years, when a new research lab was opened in another location, was he promoted and his degree recognized. Tamio certainly was an independent thinker and later also went his own way, leaving the Teijin Co for a British subsidiary of ICI in Japan, something that was rarely done in Japan.

Tamio knew little English and had been given a week's training in Hawaii so that we could communicate. After a week of Tamio's arrival, I got an assessment of this stay. It was quite negative. By then, however, he had told me that he was supposed to undergo, what he thought was superfluous training in social skills. Instead, he went to the library and studied subjects of interest for his future work with me. This earned him the poor grades by the Fulbright supervisors. He was one of my best students and was recognized by the Chemistry Department of Cornell as the outstanding student for 1963/64. From the beginning, he worked hard on his vocabulary. When walking on campus, he was noticed speaking to himself in English, reciting new vocabulary, stopping and looking up details when necessary. This made him quite proficient within a short time. Further progress in English, however, slowed when his wife was permitted to join him from Japan.

These 13 coworkers made up my research group at Cornell, overlapping in study duration, as can be seen from Appendix A. Every one of these coworkers contributed to one or more of the 28 publications which had their origin during this period from 1958 to 1963 (see also Figure 37). It was a very productive period of innovative research, as can be seen from Appendices A and B. Many of the early ideas were later picked up again and expanded upon. The three graduate students needed not to worry about the completion of their research. Drs. Bodily and Arakawa finished their PhDs in 1964 under my direction and with my commuting to Ithaca for the evaluation of the experiments, theses, and for the final exams. Mr. Sullivan, completed a MS degree and then moved with me to RPI, to earn his PhD from RPI in 1965. One of the undergraduates in my Advanced Physical Chemistry Laboratory course at Cornell, Mr. Michael Jaffe, also went to RPI, continued his research project there, and became a graduate student of mine, to get his PhD in 1967.

My earlier involvement and interest in global politics had decreased even more than stated on page 5-10. It had only been five years since we left the political upheavals in Europe, and not surprisingly, we enjoyed the newly gained freedom and the independence of an academic career. Also, material worries were overcome. Our habit to listening to the news each evening continued, and in 1958 was replaced by watching the television newscasts, but the important milestones of history did not reach the same level of importance in our lives as they had before.

What had happened in the meantime in Germany? In 1955 the state of war was ended unilaterally by the Soviet Union, and the GDR was given its sovereignty. The US, Great Britain, and France had officially ended the state of war with Germany in 1952 (see also page 5-10). The GDR became a member of the Warsaw Pact and created the "Nationale Volksarmee, NVA," the National People's Army (1956). The FRG had become a member of NATO, drawing the cold war border through the center of Germany. An uprising in Hungary in 1956 which started as a student demonstration and quickly spread over the whole country was crushed by a large Soviet force supported by tanks. An estimated 2,500 Hungarians died, and 200,000 fled the country. By 1957, a new government was installed by the Soviets, proving that none of the eastern European nations of the Warsaw Pact enjoyed sovereignty, including the GDR.

The suppression of freedom in the GDR had sent by 1956 the 1,000,000th refugee to West Berlin, our family contributing nine of these (see pages 4-33–34 and 5-10). This number of refugees increased again drastically in the early 1960s, so that ultimately on August 12, 1961 the 100-mile Wall was erected around the three western sectors of Berlin. Despite this being a violation of the Potsdam Agreement (see page 3-9), only a mild protest was offered by the Western Allies and the Wall was accepted as an inevitable fact of international politics. Escaping from the GDR was now nearly impossible and one estimates that about 200 GDR citizens lost their lives trying to attempt to scale the Wall (despite this, the official reason for the Wall was to dissuade aggression from the West). Similarly, the border between the GDR and FRG was closed. Figure 5 shows a picture I took after the Wall came down some 28 years later and the two Germanies reunited.

Figure 5 *The Berlin Wall as seen in 2003 in the north of Berlin ("Gesundbrunnen") where a piece is kept as a memory of the 28 years of imprisonment of a whole country. See also Figure 27 in Chapter 8.*

We kept contact with friends in the GDR over the years, mainly by mail (see pages 4-28 and 34–35). In order to receive mail from western countries, the people of the GDR had to register their contacts with the police and justify their correspondence. Relatives and old friends from the time before the Wall could write, but the mail was carefully scrutinized. Any new contacts needed full justification. Questions which had to be answered were, for example: Did the meeting occur during permitted travel? Why is further contact necessary? If any of the answers were not acceptable, the correspondence was denied. Anyone in a politically advanced position could not have any "Westkontakt" at all, not even with relatives. For example, when Heidel and I were on an official visit in Teltow, near Berlin, supported by the Academy of Science of the GDR (see pages 8-47–48), Heidel tried to contact one of her best friends from high school by telephone, but she would (could) not accept a phone call from a US citizen, even when it originated from inside the GDR! This caused considerable anguish for Heidel.

Other Warsaw Pact nations did not fare much different. In Czechoslovakia, an attempt to 'humanize' the Communist system in 1968 (known as the 'Prague Spring') under the leadership of Alexander Dubček as First Party Secretary was brutally suppressed in August. The country was invaded by a Warsaw Pact army of 200,000 soldiers and 2,000 tanks. The army consisted of units from the Soviet Union, Bulgaria, Poland, Hungary and the GDR (the old adversaries of WW II!). When in 1987 a spokesman of the Soviet Foreign Ministry was asked during a visit of Gorbachev in New York City, what is the difference between the Prague Spring and "perestroika" (see page 9-23)? He replied: "Nineteen years" (G. I. Gerasimov, quoted by M. T. Kaufman in the New York Times, of April 12, 1987). A greater bloodshed was avoided in Czechoslovakia by minimizing armed resistance. Following the invasion, a wave of emigration followed, estimated at 300,000.

Many of the Third World Countries followed an almost predictable fate (see also on pages 5-10–11). Without experience and tradition of maintaining secular democracies or constitutional kingdoms, various dissatisfied fractions of the population often chose the course of unrest, revolution, terrorism, or outright civil war to further their goals. Viable, even if small, communist-leaning segments of the population could always find support by the dictatorships of the USSR or China which by the mid 20th century had reached world-power status. Some of the satellite nations as well as Cuba, also gave considerable support to newly emerging communist governments. These events led ultimately in many countries of Asia, Africa, and South America to oppressive communist governments, patterned after Communist China, the ultimately failed Soviet Union, or Cuba. Similarly, individual freedom in religious or racially divided countries is often restricted and, furthermore, leads to interference or support across boundaries from like-minded countries.

These scenarios can also be used to further expansionary goals. The UN, as well as independent actions of the United States and other countries had only little success in countering these trends and often worsened the situation. The loss of life due to such conflicts, including ethnic and religious cleansing, continues at an alarming level beyond the 20th century, which was on page 3-37 called the bloodiest, cruelest, and most destructive of all centuries. The usually unsuccessful attempts to help by external military intervention, leading often to dividing of the countries along assumed suitable borders were rarely satisfactory. The humanitarian help needed under the auspices of the UN, charitable organizations, and other, not involved countries illustrate the failure of mankind to resolve such problems before they become human disasters.

This was my last larger excursion into the world of politics except for observations on the fall of the Berlin Wall on pages 9-23–25. Already in 1949, I had excluded politics from my sphere of study because of the early experiences with dictatorships in Germany (see page 3-30). This conclusion could only be reinforced by the summaries just given (see also page 9-29).

The effects of the various world conflicts on teaching in the US, however, need some additional comments. At the end of WW II, the USA was the only country which had a much stronger economic position than before. It had become a superpower. Although the war-material production dropped at the end of the WW II, the quick development of the Cold War (see page 2-23) caused a new arms race that made use of many of the technological advances gained during the war years. It saw its peak in the race for the moon and in the development of advanced electronics, materials, and computers. This all was supported by advances in the underlying science.

During much of the 20th century the number of scientists grew exponentially. Such an increase meant that one could state that about 90% of all scientists that had ever lived are still alive. With many scientists from all over the globe having moved to the US, it had become the center of the world's most advanced research. The GI Bill[1] created an unprecedented increase in size of the US universities. The bill was enacted to help veterans of WW II to readjust and reintegrate into society after their return home. Overall, the GI Bill funneled more than 50% of the veterans into a higher education at an inflation-adjusted cost for the tax-payer far exceeding the cost of WW II. This increase in students also enticed many non veterans to choose a college education, raising the fraction of students with college education in the US to the highest in the world and providing the needed jump in number of scientists and engineers. My benefit of this development was the large number of open positions for researchers in industry and academia. Between 1950 and 1970 college graduates in science or engineering could commonly choose among many positions, as documented on pages 6-34–37. My research director, Professor Malcolm Dole, called it 'the Golden Age of America' (see page 6-19). The National Science Board[2] called the 1960s the 'Golden Age of Growth' since in the decade from FY 1957–67 the budget of the NSF increased tenfold, from $40 to 465 million. It's second tenfold-increase was not reached by the fiscal year 2000.

My own experience with the teaching of veterans was positive. They were more mature than the students able to study without interruption after high school. But this maturity could make up only partially for having lost some of the youthful capacity of rote learning, needed to overcome the drudgery for the beginner to acquire the basics of a scientific career. In balance, I think that asking high school student to 'serve their country' before completing their education is a detriment to the student's ultimate accomplishments. It is better to allow an uninterrupted study from kindergarten to PhD. Unfortunately, today this is not always common practice. Also, most students have not discovered their field of interest until this early window of large-volume learning has past. It is common to find college students who want to 'find themselves' and think that it is of value to interrupt the study for one or more years. On page 8-42 the cost of such an extra year is discussed. It was a great advantage that I was pushed to make up my mind early, as described on pages 1-21–22 and 3-9, 29–31. With an early developed interest comes also the realization that the goal of life is not to take delight in a short childhood, but to prepare for an enjoyable profession for the long rest of one's life. In addition, children are resilient to even appreciate their early years under harsh conditions such as deprivation, war, and famine. Why not being enticed to learn?

---

[1] Abbreviation of 'Government Issue,' originally applied to anything issued by the US Army, now also used as epithet for a soldier. The G.I. Bill has the title: Servicemen's Readjustment Act of 1944 and was modified later in 1952, 1966, and 1972. Since the changeover to an all-volunteer force, the 'Veterans Educational Assistance Program, VEAP' fulfills a similar function.

[2] Established in 1950 by an Act of Congress as an independent governing board of the National Science Foundation, NSF, the premier supporting agency of academic research.

Figure 6 *View into the Cornell Campus Gorge, 1958.*

Our personal life in Ithaca, NY developed parallel to gaining of experience in research and teaching. As mentioned, the country-side close to Ithaca was much to our liking (see Figure 2, page 7-2, and Figure 41 of Chapter 6). This beauty also attracted a good number of our friends from as far away as Germany (former high school classmates of mine) and closer to us, Toronto (Gisela and Günter Fricke) and Chicago (friends from Northwestern). Figure 6 shows a view of the Fall Creek from the Stewart Avenue Bridge with the apartment house overlooking the fall, mentioned on page 6-38. Figure 7 is a view of the water fall in the upper Treman State Park. The water ultimately flows into Cayuga Lake. As the weather got colder, the Ithaca Waterfall, shown in Figure 8, made a spectacular picture and skating was possible on Beebe Lake and on a flooded ice rink in the City Park. At Christmas time, we spent the holidays, as usual, with the family at to our 'third hometown,' Hastings, NE, declared on page 3-36.

A visit in winter by Gisela and Günter Fricke with friend is caught in Figure 9 on an outing to the ice-covered Taughannock State Park. In Figure 2, above, the entrance to the gorge is shown. In addition to Heidel and Caryn, Figure 2 shows the Carters in the background. Dr. Carter was a postdoctoral fellow at Northwestern and visited us in Ithaca for a few days. Many other pictures of Ithaca are kept in our slide and picture collection of this time period from 1958 to 1963.

Figure 7 *Water fall, at the upper Treman State Park in fall, as visited in fall 1958.*

Figure 8 *Ithaca Water Fall during, winter 1958.*

All _our travel was within the US or Canada_ and was done by automobile, with the exception of some consulting trips by plane, so that by 1959 we had accumulated about 40,000 mi with our, by then, three-

Figure 9 *Gisela and Günter Fricke with friend, Walfried Gehlhart. Taughannock State Park, overview of the fall in the late winter of 1958/59.*

year-old Chevrolet. Not the age or the mileage made it an 'old' car. Keeping track of expenses, it turned out that at this age, the car had a sizeable trade-in value and by then, it had not needed any repairs. To keep it well-functioning and reliable for the future, one had to count on bigger cost for maintenance without the guarantee of always possessing a car ready to go every morning and be safe on longer trips. One either had to keep one's car much longer and chance uncertain repairs, or buy a new one. We decided to do the second, particularly because we expected a family addition in 1959 and similar to today's custom that larger families need an SUV (sport utility vehicle), we thought traveling with two small children would be more comfortable with a station wagon, as shown in Figure 10.

Figure 10 *The new Rambler station wagon with family above Cayuga Lake (after Brent was born on 9/22/1959).*

It was an interesting exercise for us *to find the best suited and most economical station wagon* in the US. The American Motors, AMC,[1] Cross Country Station Wagon of 1959 fit us best. Figure 10 shows the car. It was a station wagon that was not just a passenger car with the trunk enlarged upward, but it had a unique design of its own. Where could we buy it? In Ithaca we had an AMC dealer, but he would not budge from the list price. We taught him a lesson by looking into in the New York City newspapers.

[1] American Motors' three roots go to 1878, when T. B. Jeffrey sold 'Rambler' bicycles in Chicago. In 1897 he built himself a motor car. In 1900 he bought a plant in Kenosha, WI, to build 'Rambler' automobiles on a large scale. The second root was C. Nash, starting Nash Motors in 1916 by buying Jeffrey's company. Nash Motors became the biggest truck manufacturer during WW I. In 1938 Nash introduced the first air-conditioned car, in 1940, the first uni-body car. The third root was J. L. Hudson, founding in 1909 the Hudson Motor Car Company, famous for their early six-cylinder cars. For a Hudson of the 1950s, see Figure 3 of Chapter 6. By 1950, Nash built the first modern compact cars, the 'Rambler.' In 1954 Nash and Hudson merged to 'American Motors' with the model name 'Rambler' for their main car line. In 1987, Chrysler Motors bought the Company.

We soon found a big-volume dealer who offered discounts for his cars. By telephone, we made all arrangements and found out that our local dealer would have to service the new-car guarantee. Earlier, we had planned a trip to Cherry Hill, NJ, across the Delaware River from Philadelphia, PA, to visit our friend Werner Haag and his family (see page 6-24) who had just begun his job in the research laboratory of the nearby Mobil Chemical Company (since 1999 renamed ExxonMobil). On our trip we stopped in New York City. Figure 11 shows Heidel and Caryn in the city. The car was bought, and on our way back, it was properly prepared so we could pick it up. Only, we had forgotten to prepare Caryn for the new car. In Figure 11 she looks so proud exploring the big city and she enjoyed the visit to the Haags with their four children, but when we hastened to take our luggage out of the car and the dealer drove away with the old Chevrolet, she started to cry and was unconsolable even when the dealer came back with the brand-new Rambler. It was not ours! She wanted the old one. It took quite some time until Caryn appreciated the advantages of the station wagon with the spacious back that allowed her to sleep in comfort. The new car was also ideal for us. We used it to stay overnight (not comfortable) at Letchworth State Park (south of Rochester). Better was to go to drive-in movies, then fashionable. It saved us hiring a babysitter. Caryn and (later) Brent could go to sleep in back (with some prompting, after seeing the introductory cartoons).

Figure 11 *Heidel and Caryn in New York City for a visit to a new-car dealer. Walking along 35th Street.*

One more problem needed to be resolved. *How could we handle the increase in family during the first few months*? Mutti could not leave her family in Hastings for long. She still had to take care of Joachim and Manfred (15 and 13 by then). Also, her involvement in Vati's business was still considerable. How about Mama? She was still in the GDR! The Wall was not built yet (see the pages 7-7–8) and we could write clandestine letters to relatives in West Berlin and give notice to Brandenburg, that it would be nice for her to visit 'Gisela.' Mama knew then, that she had go to West Berlin for such a special letter from the US. In this way, plans were made for her to visit Tante Marie for an extended period of time (page 11-3 and see Chapter 11, Figure 17). Tante Marie and Onkel Rudi still lived in Hoyerswerder, Saxony (GDR), a sufficient distance from Brandenburg not to have people notice any missing visitors and report back to Brandenburg. It is the common duty in Germany, to register with the police whenever leaving home for several days and arriving somewhere else. Registration in a hotel, for example, automatically serves this purpose.

In this way, Mama visited Tante Marie and registered at the "Einwohnermeldeamt" for an extended period of time, leaving Papa alone in Brandenburg. After staying in Hoyerswerder for a week, she left for West Berlin and flew via Frankfurt to New York. Before this could be done, visa applications were made at the American Consulate, inoculations gotten, and pictures needed to be taken for the passport. Since nobody was supposed to know about the real purpose of the trip, all this needed to be done in West Berlin. Even when after the vaccination her arm became excessively sore, Mama could not go to a physician in Brandenburg. The biggest problem, we thought, was that Mama had told us when we emigrated to the US that she would never get onto a ship or into plane

Figure 12 *Idlewild, New York City's airport in 1959 (now Kennedy Airport) showing the take-off of a jet.*

to visit us (see the quote on page 5-20). This fear, however, was by then overridden by the prospect to get to know her grandchildren. Since the travel arrangements took more time than was available, a trip by plane was the only choice. After Mama arrived in Ithaca, her comment about the trip was: I will never go back by plane! So we made arrangements on a ship for her to return in early 1960.

Figure 12 was taken at Idlewild Airport in New York City (Kennedy Airport since 1963) when I picked up Mama to bring her to Ithaca. The plane was a few hours late, so I had time to explore the airport and take some pictures. In front of the terminal stands a big Lockheed Super Constellation, common for transatlantic service at that time, and also used by Lufthansa, the German Airline (see page 6-30). The Super Constellation flew with a typical speed of $\approx$200 mph. The plane taking off in the back is one of the first Boeing 707 jets. It caught my attention with its characteristic whine, and, at that time, its trail of black smoke on take-off. With the jets, the overseas travel time was ultimately cut to less than half.

Our son Brent was born on the 22nd of September 1959. The escort Brent got on the way home from the hospital is illustrated in Figure 26 of Chapter 11. Caryn was particularly happy at that moment. Before, she was not permitted to visit in the hospital and had to see her mother from the lawn through the window. To lighten her concerns, Heidel had written a daily letter which I could read to her.

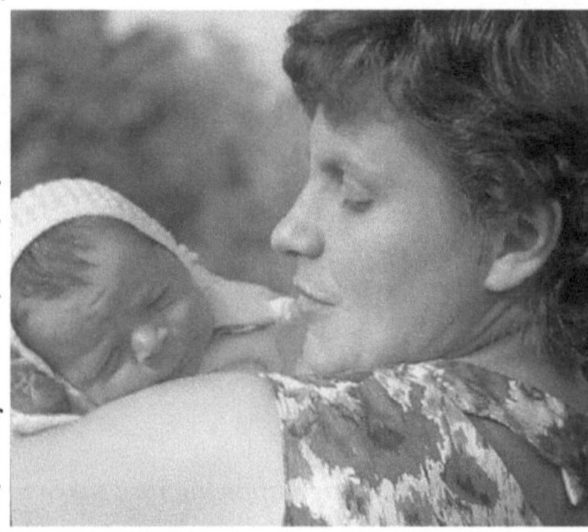

Figure 13 *Heidel and Brent in Sept. 1959.*

Figure 13 is one of the first pictures of Brent. Caryn was quite satisfied with her new brother and was carefully inspecting him on the way home from the hospital, as is shown in Figure 27 of Chapter 11. After a while, she was even happy to be his big sister and share her room with him.

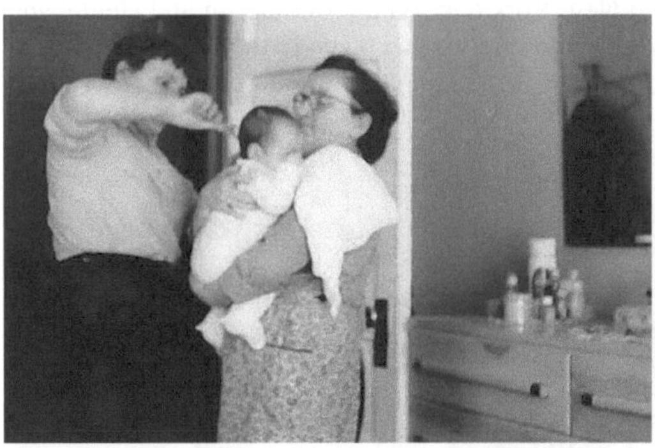

Figure 14 *The Mother and Grandmother taking care of baby Brent during our visit to Hastings, in 1959.*

Figure 14 shows one of the daily routines made much easier by the presence of

Mama. It also helped me, to go on with my long hours at Baker Laboratory. Weekends, naturally, were made use of to enjoy the countryside. Figure 15 is a picture taken when stopping at the parking lot of the lower Treman State Park with a nice overview of Lake Cayuga. The whole family could pose at the tailgate of the station wagon, Oma, Caryn, and Heidel with Brent, and myself as the photographer (see also Figure 10).

Figure 15 *One of the first weekend-trips with Brent.*

Brent developed quickly and the next picture, Figure 16 shows him quite content in his environment. At about the time this picture was taken, we got ready to travel the 1300 mi to Hastings, NE to celebrate the Christmas holidays and introduce Brent to the greater family. Both children were no problem on the long, two-day trip by car with the usual overnight stop in Chicago. There was enough space to sleep in the back of the Rambler, and they needed little entertainment to make the time pass quickly. On this occasion, Brent was also christened in the First St. Paul Lutheran Church in Hastings.

Figure 16 *Brent at age approaching three months.*

In February 1960, Mama left by boat, to first see her sister in Hoyerswerder again and officially end her half-year visit in Saxony. Nobody had suspected that she actually spent most of the time in the US.

Figure 17 illustrates her departure from New York City on the MS Berlin and Heidel and Caryn waving good bye. The MS Berlin was the same ship we had taken to the US. February, as November for our trip 5¼ years earlier, was not the best time for trans-Atlantic travel. (See also the pages 5-19–22 and Figures 29 and 35 of Chapter 5). After Mama's arrival home, she wrote to us that she did not enjoy the travel, and she would never go by ship again. As things developed, she and Papa visited us frequently in Troy after they had left the GDR as pensioners in 1965. For these, usually biannual trips, they would always take a plane as the lesser evil. The planes were, by then, exclusively the faster jets.

Figure 17 *The MS Berlin, taking Mama back home.*

By 1965, pensioners could officially leave the GDR with limited household goods, but had to leave behind their right of a pension from the GDR. Papa and Mama moved into the area of Köln, FRG, where they had some relatives. The FRG picked up their pension at the same level it would have been, had they worked always in an undivided Germany. This was a sizeable advantage for them because of the greater economical strength of the West German Mark (DM-West, see page 3-28). It, however, was not only a generosity toward residents of the GDR. It was based mainly on the "Alleinvertretungsprinzip," the exclusive mandate the FRG claimed over all Germany, resting with the non-recognition of the GDR and, thus, asserting legitimate authority over all Germans. This exclusive mandate was also the reasoning for our being granted refugee status and support for our study at the Goethe University in Frankfurt from 1953 to 1954, as described in Chapter 5.

Our next big step in 1960 was to apply for US citizenship. The five-year waiting period was over and we went to the Tompkins County Courthouse in Ithaca. After establishing our personal data, including the weights (we wish we had today), the question to me was: "What is your draft status?" My answer: "I don't have one, I am not a citizen yet." No, this was not so. Every man between 18 and 25 years of age had to register. Tante Lieschen was right with her reminding me on page 6-10.[1] My answer: "I am sorry, there was nothing about this in any of the paperwork covering our immigration," helped little. I was told: Nothing can be done about your application until you report to your draft board and get a classification.

On the next day I appeared at the Tompkins County Draft Board and had a friendly conversation. In 1960, there was only a Cold War which needed scientists, engineers, and in Ithaca, professors, not soldiers. After two weeks of waiting, I received my draft card. Classification: V-A (Registrant over the age of liability and never having been deferred). The rest of our applications to citizenship went without problems (Petitions # 2365 and 2366). We easily passed the rudimentary speaking, reading, and citizenship tests since we had refreshed our memory about basic US history and looked up our elected officials. A set of witnesses who could document our good behavior for all of the last five years was a bit more difficult to assemble because of the two moves, but someone could be found at each place of residence. On May 2, 1960, the "one hundred and eighty-fourth year of our independence," we became citizens in the usual court ceremony. Caryn and Brent, being born of German citizens, legally in the United States, had a choice when reaching 18 years of age to elect to become German instead of US citizens, but this was never seriously considered by them.

The year *1960 approached its close with an ice storm late in fall*. It left the whole area beautifully transformed. I took my Exakta to Cornell, to take Figure 18, a view from Baker Lab to the McGraw Tower.

As Christmas approached, Vati and Mutti decided they would come with

Figure 18 *The campus of Cornell after an ice storm.*

[1] Today, even undocumented aliens must register (see https://www.sss.gov/regist.htm).

Figure 19 *Christmas in Ithaca in 1960.*

Joachim and Manfred to Ithaca. Figure 19 shows the proper spirit with Caryn and Brent, and Figure 20 shows the obligatory first haircut of Brent's by his grandfather, just as it was shown with Figure 35 of Chapter 6 for Caryn in Hastings. The holidays were spent, as usual, at home and by visiting the many parks around Ithaca in their winter splendor (see also Figures 8 and 9, taken the year before). We managed even to get the first skating lessons for Caryn at the flooded rink in the Ithaca City Park. The especially snow-rich winter 1960/61 was followed by a particularly nice spring.

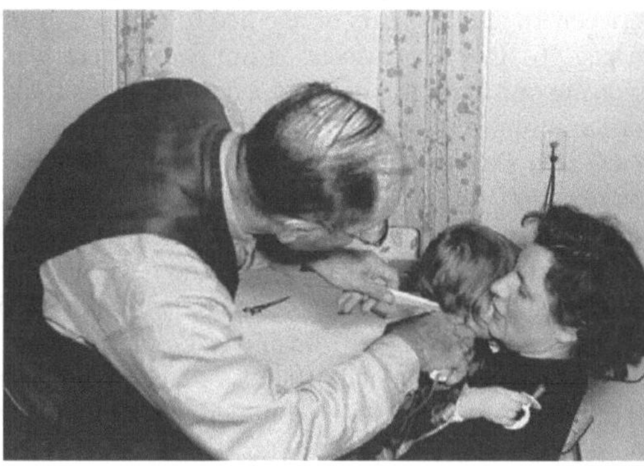

Figure 20 *The first 'professional' haircut of Brent's.*

When planning our travels for the next year, I found that I needed to be at the ACS Meeting in Chicago from September 4 to 8. We decided, may be we should, having traveled half the distance to Hastings, to complete this trip. The longer stopover in Chicago gave the opportunity to visit with our friends there (Fräulein Rank, see the page 6-37 Chapter 6, Figure 38) and the Mendheim family (see page 6-17). At Northwestern University, there was a special industrial meeting, organized by Dr. Pines. This allowed us to also meet with Werner Haag (see the pages 6-21 and 24), who participated in the Chicago Meetings.

<u>In Hastings the whole family got together again</u>, as shown in Figure 21. Frank and his family had by this time moved to Lincoln, NE, as mentioned on page 6-31, and on our way home, we could visit them there, to inspect their new house. Vati's business progress is visible by his bigger new truck, to be compared to Figure 16 of Chapter 6, and Joachim and Manfred were getting ready for their careers in the US. Joachim ultimately ended up teaching for many years at Hastings College (see also page 6-3, 5, and 6). He is Professor of Foreign Languages and Chair of the Department. Manfred, after starting as a

Figure 21 *Family get together in Vati and Mutti's house in the fall of 1961, Frank, Dagmar and Eberhard had joined us from Lincoln, NE.*

teacher and administrator in high school, changed to a successful career in the insurance field with Mutual of Omaha and is now largely independent with his own business. Naturally, we also visited all other relatives, as well as George Mohlman who by then had become a full-time farmer in Blue Hill, near Hastings (see Figure 8, Chapter 6 and pages 6-8 and 12).

One important duty of Heidel's was to enlist Mutti's help to construct a German "Schultüte" (see Chapter 1 Figures 13 and 15 and pages 1-14 and 15) for Caryn's upcoming first day of school in "Kindergarten." Figure 22 shows the clandestine production in Hastings, and Figure 23 proud

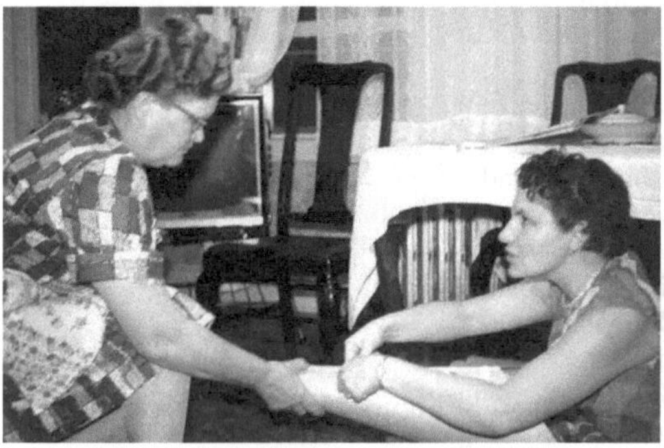

Figure 22 *Construction of Caryn's "Schultüte."*

Figure 23 *Caryn in front of her school in Ithaca on the first day of kindergarten 1961.*

Figure 24 *Our Christmas card for 1961.*

Caryn on the important day when the first formal instruction in English began for her. More about our effort to make sure that our children were properly instructed in both English and German, can be found on page 6-25.

Having set our visit in Hastings for the fall, we had to prepare for our Christmas of 1961 in Ithaca. Figure 24 illustrates the first step, the photograph used as the Christmas card, taken in downtown Ithaca while window shopping for toys.

In 1961, Brent was big enough to truly appreciate for the first time the spirit of Christmas, and even learn a small, traditional Christmas poem. Figure 25 depicts his receiving the (unwrapped) gifts in front of the Christmas tree he had seen in our house for the first time only minutes before, as is customary in Germany (see the pages 1-22–23).

Figure 25 *On Christmas Eve, 1961.*

As usual, when we were at home for holidays, we tried to invite students, friends, and colleagues who were far from their families. We appreciated this when we were students at Northwestern, and now tried to continue this custom.

In 1961, Professor Eigen, the fall Baker Lecturer from Göttingen (see page 7-3, Footnote 1) stayed through the Christmas holidays. On Christmas Day, he, his wife, and a Cornell graduate student from Germany were our guests. Figure 26 shows the dessert offering of Heidel and the entertainment Brent (and Caryn) could provide. We spent a very pleasant day with our company, as we did with many of our acquaintances in Ithaca on other occasions.

In fact, on the occasion of our Golden Wedding Anniversary in 2003 (see pages 10-9–11) we got an interesting letter about student invitations. Brent had 'borrowed' without our knowledge all our old addresses and written to everyone for comments, mementos, and congratulations. He and Anna assembled the returns into two large albums as a most special gift for us. The student at this 1961 Christmas dinner was from the Chemistry Department, Ms. Karen Köhler, married later to Professor Robert O. Pohl, at that time assistant professor in physics. She, wrote about her first invitation to our house: "... I was totally broke at that time (which you surely did not know!) and the teaching assistantship money didn't come till the end of the month. So the wonderful dinner you served was welcome in more than one way. The food was wonderful, and much appreciated. I also enjoyed the company and the warm atmosphere in your house which reminded me of home." After this initial invitation she was invited more often, including the 1961 dinner with the Eigens. In Figure 26. you may also want to compare the progress in Heidel's art of baking with the earlier Figure 25 in Chapter 5.

Figure 26 *Guest on Christmas Day.*

Overall, we found that the atmosphere among the colleagues in the Chemistry Department of Cornell was less congenial than at Northwestern, described on page 6-32. In Ithaca, all was more reserved, best described as more 'European.' We found that, perhaps, only one third of the faculty was interested in personal contact with the new members. These invited us to their usually quite beautiful homes and we could reciprocate with a get-together in our modest duplex. Some of these colleagues would even follow my later development and send letters of congratulation on special occasions. The others did not seem approachable and also had little contact with me on a professional basis.

My research topic had been from the beginning and is still that of the solid state physical chemistry of macromolecules as is discussed on pages 3-35, 4-11, 12, and 6-20. It was a sufficiently new topic so that hardly any chemistry department had a major effort in this field. The research of Professor Dole into the calorimetry and radiation chemistry of polymers at Northwestern was an exception. In fact, most of my later invitations to give lectures at other universities in the US did not come from departments of chemistry, but from departments of chemical or materials engineering, and in Europe, most of my invitations came from physics departments.

Why were the academic chemists not interested in macromolecules? The answer lies in the fact that in the 1920s and 30s the existence of polymers was denied by many academic scientists. They maintained that the special properties of natural polymers, like starch and rubber, came from aggregates of ring-like molecules of the same composition. After the pioneering work by Professor Staudinger,[1] major polymer research was initiated in industry and not necessarily fully disclosed in the scientific literature. At Cornell, however, one should have been better informed. In their chemistry department, Professor Flory[2] was teaching 'Principles of Polymer Chemistry' until 1957.

Flory's pioneering book provided a way to interpret my thesis experiments (see page 6-35) and is still on my book shelf. My later work then pointed out the limits of his treatment which was restricted to equilibrium thermodynamics. When I wrote my first major research paper on the thermodynamics of defect polymer crystals, it was rejected. From the comments I received, it seemed clear that Professor Flory or one of his students was the reviewer. Since Professor Flory, as a former Baker Lecturer, visited Cornell occasionally, I managed to discuss the paper with him. His comment after a lengthy, interesting, and friendly discussion was: "This must be absolutely wrong since the defects cannot be described by equilibrium thermodynamics." After I countered that the whole paper dealt only with irreversible thermodynamics, his suggestion was that I would have to state this more clearly and that personally, he knew little about irreversible thermodynamics and was not interested in it. End of the discussion.

Even after I made the suggested improvements to the paper, I still could not convince the editor to reconsider. Finally, I brought this dilemma to the attention to the then current Baker Lecturer, Dr. Dainton from the University of Leeds (see Footnote page 7-3). He had done extensive

---

[1] The understanding of flexible, linear macromolecules is based on the research of Hermann Staudinger (1881–1965), working at the University of Freiburg, Germany (Nobel Prize in Chemistry 1953). He also suggested a division of small and macromolecules at the limit of 1,000 atoms.

[2] Professor Paul J. Flory (1910–1985) was one of the most well-known polymer chemists. His first research in polymers was at the DuPont Company under W. H. Carothers, the Harvard professor who after 1928 led the DuPont effort to develop artificial materials (Nylon, synthetic rubber, etc.). Flory was a Baker Lecturer at Cornell in 1948 and stayed as a faculty member until 1957 when he left for the Mellon Institute in Pittsburgh, and later, Stanford University (1961). For the work on macromolecules at Cornell, he was awarded the Nobel Prize in Chemistry for 1974. Most famous are his books *Principles of Polymer Chemistry*, Cornell University Press, 1953, and *Statistical Mechanics of Chain Molecules*, Interscience, New York, 1969.

work in calorimetry of polymers. After reading the paper, he thought it certainly was a novel treatment and, perhaps, a better way to understand the subject. He sent it to the British journal *Polymer*—he was on its editorial board. In this way, the paper was published. Its follow-up had no problem (see Appendix A, papers #16 and 28). This incident illustrates the importance for a young researcher to have easy access to leading scientists with different opinions to test the viability of his or her new ideas. The Baker Lecturers at Cornell offered me this opportunity.

A number of years later, I could offer an answer to the question: *Is polymer physics underdeveloped*? By 1982, I had visited many academic and industrial research laboratories in various countries (see Chapter 8) and found a good resonance for my different approach to the description of non-equilibrium melting and crystallization of polymers. In this year I was invited to give a plenary lecture at the German Physical Society Meeting in Regensburg with the title of this paragraph: "Ist die Polymerphysik unterentwickelt?" My conclusion was, 'Yes, because it is done by chemists.' The explanation is that most research on polymer physics is done by chemists and not published in the style and journals physicists appreciate (like the Journal of Polymer Science, Part B: Polymer Physics). The new developments in the fields of polymers were interdisciplinary. Progress would have been faster, more focused, and connected better to the fundamentals, if experts of both fields would have tackled the problems and have published in the same journals.

At the Humboldt University, the designation of my field of learning was *the study of natural science, "philosophia rerum naturalis,"* (see page 4-8). Prior to the 17th century it incorporated all natural science disciplines presently known, such as astronomy, biology, chemistry, earth science, physics, and the increasing number of cross-disciplines like biochemistry, physical chemistry, etc. Chemistry as a science (disregarding the medieval 'alchemy') began early in the 19th century with the development of the understanding of chemical elements and the structure of molecules. Initially only the *small molecules* from inorganic sources could be identified. Molecules found in or produced by the living world were then called organic, splitting the field into inorganic and organic chemistry. It did not take long and the first 'organic' molecule could be made in the laboratory and 50 millions followed thereafter (counted by the ACS, 2009). This negated the need of organic chemistry as a separate discipline. Instead of recombining inorganic and organic chemistry, a third branch, biochemistry was created for the 'really organic' molecules which defied synthesis. By now many of the 'biomolecules' can also be synthesized or modified in vitro and it is fully realized that fundamentally, all these molecules follow the same structure and bonding principles.

At the beginning of the 20th century it was discovered that many of the solid materials are made of huge molecules and their structure, if crystalline, could be analyzed in detail by X-ray diffraction, a technique largely discovered and applied by physicists, so that a new branch of science was created, solid state physics. It deals with what today one should call *rigid macromolecules.* Certainly, small and large molecules have different properties. Only in the mid-20th century was it accepted that longer molecules may be flexible and are then, again, are different in nature. When sufficiently long, they are called *flexible macromolecules.* Molecules of this type are found among the biomolecules, as well as synthetic (organic and inorganic) polymers. This summary shows the difficulties if one retains a historically developed classification rather than updating it when it proves flawed. In retrospect, it is simpler to divide chemistry, if at all, into three sub-specialties: Those of rigid macromolecules (existing only as solids), flexible macromolecules (may be solid or liquid), and small molecules (may be solid, liquid, or gaseous without losing their molecular identity). This division became obvious from our research (see also the pages 8-66–67) and is described in detail in Appendix A, publication #473, and #500, as well as on pages B-11–13 of Appendix B.

For 1960–63 I had been promoted to the rank of an assistant professor after the three one-year contracts as instructor, at Northwestern (57/58) and Cornell (58/60). This meant, according to the rules of academic careers, with satisfactory performance, in 1963, I could expect a second three-year term as assistant professor. In 1966, promotion to an associate professor with tenure could be applied for, and later, promotion to 'full professor,' expected after another five to 10 years.

In early spring 1962, when looking at the odometer of our car, I saw 60,000 mi on our car. In 1959, I had considered three years the most reliable portion in the life of a car (see the pages 7-10–11). Our car, pictured in the countryside in Figure 27, should be replaced in 1962. Also, we could afford a new car, did not need a station wagon anymore. The AMC had just developed the first upscale compact car. Based on these arguments we bought a Rambler Ambassador 400 with a 5.4 L OHV V8 engine of 250 hp (190 kW), with vinyl bucket seats, head rests, and two-color paint (black with a white roof, to keep the car cool in summer). Figure 28 shows the front end, with a name changed from Rambler to 'Arabell.' The original letters were attached with single, easily removable screws. By buying a replacement 'L' and 'A,' the car could be renamed to be uniquely ours.

Figure 27 *An evening near Ithaca with our Rambler.*

Everyone had an input into options for this new car. Heidel coordinated the inside, outside, and carpet colors, I chose the model, all agreed with the new name. Caryn had just learned in school that seatbelts were essential for safety. She convinced Brent to insist with her on four seatbelts. The car came with two front belts, just mandated in New York State, although their use was voluntary. We gladly agreed with the children, and ordered the car with four seat belts. After the first use, both Caryn and Brent complained about using the restraint. They now had to sit still. But they had insisted on buying the seatbelts. From then on, this gave us the leverage to enforce their continued use in our car.

Figure 28 *Our 1962 AMC Rambler 'Ambassador.'*

In the year before, Heidel had passed her drivers test on her first try. This was also a good grade for me, her driving instructor. Ultimately, I taught all our family members how to drive, and in addition, also a good number of my postdoctoral students.

In our private life, mainly through the efforts of Heidel, we had an increasing number of acquaintances in the community besides coworkers and colleagues. As usual, we retained connections to many friends and relatives from Germany, Hastings, and Chicago, and from around the world.

<u>*Major travel projects in 1962*</u> were to the First Biennial Symposium of the Polymer Division of the American Chemical Society in East Lansing, MI and the Fall Meeting of the American Chemical Society in Atlantic City, NJ (listed in Appendix C). Figure 29 shows us on the way to East Lansing, in June 1962. On these longer trips, frequent stops were made with vigorous physical exercises, to break the monotony.

Figure 29 *Exercise while traveling to East Lansing.*

The Biennial Symposium was held on the grounds of Michigan State University. It was the first of the later well known meetings where polymer scientists could get together and discuss the progress of the last two years. In 1962 I gave my first invited lecture describing the new way of looking at macromolecules and their transitions, as mentioned on pages 7-19–20. To the proceedings, I had submitted optical color micrographs (paper #21 in Appendix A). Printing in color was quite expensive ($500 per page) and the cost had to be born by the author.

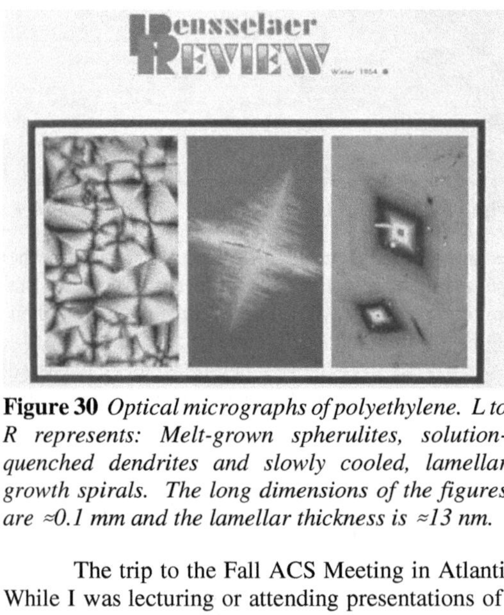

Figure 30 *Optical micrographs of polyethylene. L to R represents: Melt-grown spherulites, solution-quenched dendrites and slowly cooled, lamellar growth spirals. The long dimensions of the figures are ≈0.1 mm and the lamellar thickness is ≈13 nm.*

Fortunately I knew Milt Nelson, the manager of the Cornhusker Press in Hastings, NE, (see page 6-14). He had the color separations made and printed the copies for the journal for a much lower price. The separate pages were then bound into the issue without further charge and were the first and only color reproductions in the journal for a long time. Figure 30 is a copy of the color separation representing the three main morphologies of folded-chain polymer crystals as seen by polarizing (left two figures) and interference microscopy (right figure). After I had joined Rensselaer Polytechnic Institute in 1963, a special effort was made to promote this work and the Figure 30 was the front page of their quarterly alumni publication which reprinted a Scientific American article of mine (see paper #31 in Appendix A).

The trip to the Fall ACS Meeting in Atlantic City was also done with the whole family. While I was lecturing or attending presentations of my colleagues, the family could play on the beach, as is illustrated with Figure 31. Brent was by then old enough to travel with us.

The year before, I attended my first international meeting, held in Montréal Canada, by the IUPAC (the <u>I</u>nternational <u>U</u>nion of <u>P</u>ure and <u>A</u>pplied <u>C</u>hemistry). It was their first Symposium on Macromolecules. At that time we had left Brent back in Ithaca with distant relatives of Heidel's

Figure 31 *On the beach in Atlantic City during the Fall ACS Meeting.*

(Horst and Gisela Wilde), who had come to Ithaca for a one year experience in the US. Horst had taken a job as technician in the Physics Department of Cornell, and Gisela was glad to look after Brent. Figure 32 shows that Heidel and Caryn were enjoying their spare time on the balcony of the hotel. Brent, was not happy. On our return, he was mad for leaving him behind. Only after he was assured that we would not leave him again, things returned to normal, and he was glad to join us in Atlantic City in 1962.

Figure 32 *With Heidel and Caryn in Montréal (IUPAC, in July 1961).*

On the way to Montréal, we made a stop at the Thousand Islands in the St. Lawrence River, and before our return from Montréal, we had a chance to get to know the foreign visitors during a conference-sponsored trip into the Laurentian Mountains,. In this way we combined vacations with business travel. Whenever we could travel by car and the children could have fun during the trip, we all went. If we thought, it was not to their liking, or it was too expensive, I traveled alone.

Figure 33 shows that I needed some play time to satisfy Caryn and Brent after my travels. Other rewards on my return were visits to the animal farm of Cornell University.

Only twice did we need a live-in babysitter for larger trips. One was to Japan [in 1966 with Ms. Rank from Chicago (see page 6-37) staying in our house in Troy, NY, see pages 8-25–30], the other was to Poland [in 1972 with Vati and Mutti stopping off at our house on their trailer-travel through the US (Vati was retired by then and Joachim and Manfred were independent), see pages 8-62]. After Heidel and I were by ourselves in Knoxville, TN, Heidel would always accompany me to meetings longer than one or two days.

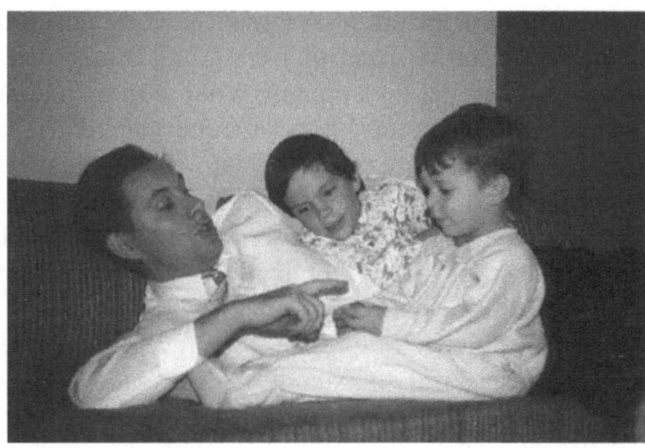

Figure 33 *Play time after coming home from work.*

When the year 1962 ended, we prepared again for our annual travel to Hastings, NE, for the Christmas Holidays. The trip was, as usual, long, but now, with a more powerful car, it also was enjoyable. We arrived a few days early and had a

Figure 34 *An abandoned 19th century Nebraska farm.*

chance to inspect the countryside. Figure 34 is a reminder of an excursion to an abandoned farm, recalling early life in Nebraska. The all-important Christmas eve celebration with the decoration and lighting of the Christmas tree and exchange of gifts is captured in Figure 35. Joachim and Mutti guide Caryn and Brent into the living room to recite their Christmas poems.

Back home in Ithaca, life continued as usual. The special occasions within the Chemistry Department were celebrated in the Statler Hotel on campus, run by the School of Hotel Administration. On one of these occasions, Professor Debye (see page 6-36), whom by then I knew well, sat with Heidel and me at the same table. As we knew from his visits in our house, he was (as I am) a great fan of soups. In this case the conversation around the table was to his liking and absentmindedly he took his coffee cup, added salt and pepper and after 'eating' the first spoon full commented they really do not know any more how to make a good soup.

The stories about Professor Debye were many. Let me stick to the ones I can vouch for. He was a pleasant guest and eagerly participated in the conversations, not only when they dealt with science. If, however, the topic was for some time not in his field of interest, he became quiet and soon would get up and collected Mrs. Debye to go home.

Figure 35 *Christmas Eve 1962 in Hastings.*

During a lecture of Professor F. C. Frank from the University of Bristol, England, of solid state physics and polymer physics fame, Professor Debye attended. After a few minutes, he fell asleep. At the end, the applause woke him up and he immediately dominated the discussion and touched on all the points he had just (not) heard. He was very familiar with all Professor Frank had done. I also remember that this was the seminar where I approached Professor Frank after the discussions and asked him if he had ever heard of a superheating of crystals beyond the melting temperature, as we had just observed in polymers. He was very interested in this observation and pointed out several early publications on the superheating of silicates, dating back to 1905.

A final experience I had with Professor Debye concerned a lecture he gave on "Macromolecules in Solution" to inaugurate a new lecture series of the Cornell Materials Science Center. It was a truly outstanding lecture and the Center wanted to have a transcript of it. A tape recorder was placed on the desk and a secretary was to type it. But, she had not only difficulties with the scientific terms, but also with the German (not Dutch) accent of Professor Debye. So she left out many portions of the text and asked me if I could help her filling-in the blanks. I listened carefully to the tape and managed the task. When the secretary proudly gave him the final manuscript, he came back after looking it over and commented: "This is not what I said ... ."

Smoking was at that time still unrestricted at the universities. For me, having given up smoking almost on the day I started (see page 2-13) this was quite bothersome. In particular, the faculty meetings were a problem. Most of my colleagues were smoking, often cigars or pipes. Every time I came home after one of these meetings, Heidel needed to air my clothes for a day or more to get rid of the adhering smoke. One day, the janitor, who always provided me with the newest gossip which I did not want to hear, came to my office and said: "This is what you must see." He led me to one of the older colleague's office, a few doors from mine. There he had tried to clean a dirty window that had bothered him for a long time. (Maybe he was not that good a janitor.) Over the years, a thick, brown layer of tar was deposited, which he was unable to remove after all the other dirt was gone. How must the colleague's lungs have looked?

This was the time when I was a strong advocate of forbidding smoking in public and increasing smokers life and health insurance rates. I did not succeed, and by the time forbidding to smoke was fashionable, I had other things to feel strongly about. At the end of the century, my problems with smoking were restricted to visits of countries outside the US. By now, even Europe is making good progress, although restaurants are still not all smoke free. How can one spoil the great taste one expects with such terrible smell? The Asian countries have to learn, even now.

Another problem I had recognized long before general awareness of it, was the harmful effect of the ultraviolet irradiation from the sun. Every chemist knows that the energy of light in this color range is enough to break any bond of the compounds that make up the skin. At that time, again, comments of mine about it were not appreciated by most (friends and family). By now, again, this is a non-issue for me. It is self-understood that we cut the play in the sun to a minimum and are good customers of the manufacturers of sunblockers.

Figure 36 *Our last Chemistry Department picnic, catered 1962 by the Cornell Agriculture Department.*

Figure 36 shows a photograph taken at the last annual picnic of the Chemistry Department we attended. These picnics were catered by the Ag Department of Cornell. The Ag-students used a barbeque method and sauce developed for a PhD thesis on the subject by one of their students. We got a copy of the recipe and are still using it today (see Chapter 8, Figures 23D and 63).

My search for a more secure job began late in 1962. As summarized on page 7-21, I expected, and was later in the year given, a second three-year contract at Cornell as an assistant professor. By 1962, however, the newly hired faculty was not employed as instructors, rather assumed immediately the rank as assistant professor. This meant, I needed three more years for promotion than they. I thought, this was not fair! Could I do anything about it?

On November 14 and 20, 1962, I had two extensive discussions of this point with the new department chairman, Professor Harold A. Scheraga, a respected biophysical chemist. It was my opinion that tenure proceedings should be considered before 1966. Over the years, I had attended Professor Scheraga's literature seminars with his students. His field was closest to mine, and I could learn much from his research methods. He could, perhaps, judge my accomplishments best.

I had built up an active research group. Figure 37 shows the topics worked out at the third major internal research accounting of my group. Such summaries were the basis of meetings of one hour or more, held once or twice a year with the whole group participating. These major conferences were held in addition to shorter, weekly research meetings. The hand-written notes were made at the next meeting, toward the end of 1962, when everyone's progress had been assessed. I have kept 34 similar pages, covering the time from 1961 until 1985. Appendix A documents that by 1962, I had about 20 papers produced or in the works which originated at Cornell. Furthermore, my external lectures and consulting proved a reasonable acceptance of my work (see Appendix C).

According to my notes, the following answer was given to my request: "I can assure you that in 1963 you will be offered a second three-year assistant-professor contract. The request of early tenure, however, was not agreed with." My answer was that in this case, I would want to start looking for another university with a different promotion policy. He offered help in finding a different job and promised to work on getting permission to move all of my equipment in case I should leave for a new location. Also, if the need should arise, he offered that I could use only one year of my next

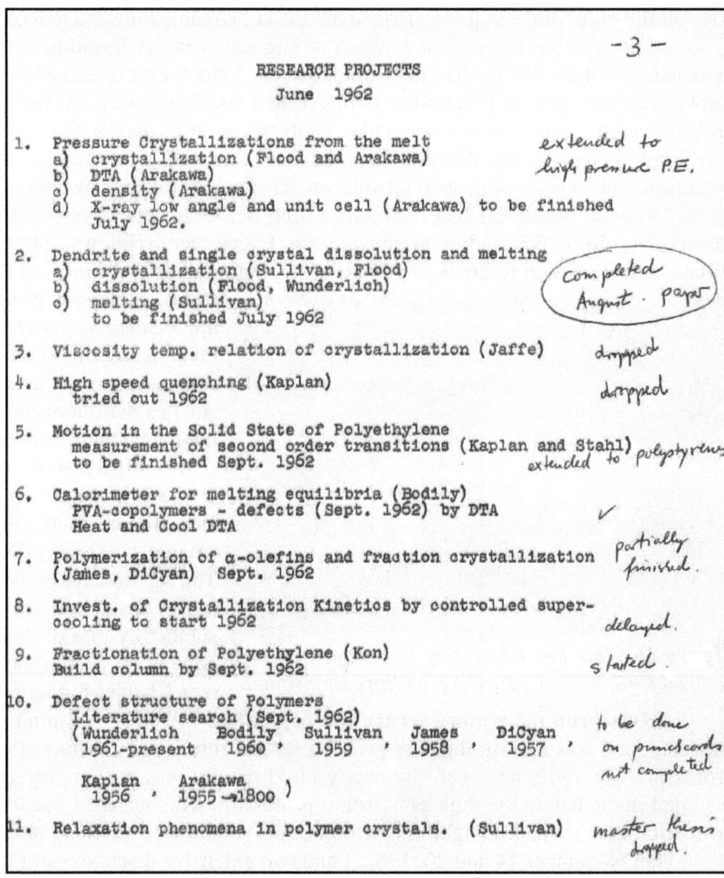

-3-

RESEARCH PROJECTS

June 1962

1. Pressure Crystallizations from the melt — *extended to high pressure P.E.*
   a) crystallization (Flood and Arakawa)
   b) DTA (Arakawa)
   c) density (Arakawa)
   d) X-ray low angle and unit cell (Arakawa) to be finished July 1962.

2. Dendrite and single crystal dissolution and melting — *completed August. paper*
   a) crystallization (Sullivan, Flood)
   b) dissolution (Flood, Wunderlich)
   c) melting (Sullivan)
   to be finished July 1962

3. Viscosity temp. relation of crystallization (Jaffe) — *dropped*

4. High speed quenching (Kaplan) — *dropped*
   tried out 1962

5. Motion in the Solid State of Polyethylene
   measurement of second order transitions (Kaplan and Stahl) — *extended to polystyrene*
   to be finished Sept. 1962

6. Calorimeter for melting equilibria (Bodily) — ✓
   PVA-copolymers - defects (Sept. 1962) by DTA
   Heat and Cool DTA

7. Polymerization of α-olefins and fraction crystallization (James, DiCyan) Sept. 1962 — *partially finished.*

8. Invest. of Crystallization Kinetics by controlled super-cooling to start 1962 — *delayed.*

9. Fractionation of Polyethylene (Kon) — *started.*
   Build column by Sept. 1962

10. Defect structure of Polymers — *to be done on punchcards. not completed*
    Literature search (Sept. 1962)
    (Wunderlich 1961-present, Bodily 1960, Sullivan 1959, James 1958, DiCyan 1957, Kaplan 1956, Arakawa 1955→1800)

11. Relaxation phenomena in polymer crystals. (Sullivan) — *master thesis dropped.*

Figure 37 *Research topics, responsibilities, and my later assessment.*

contract. It would help him, to fill my position, should the need arise. This sounded like my future at Cornell was not sufficiently secure. I got to work to find a new position as soon as possible.

Discussing my situation on November 15 with John Burry, Executive Director of the Chemistry Department, he commented: "A few tenured professors have personal reasons not to support you. It is not that your qualifications are insufficient." He suggested, I should apply to industrial laboratories, but this would be a distant second choice for me. I still could stay three more years! John had told me already in 1960 that I might not survive at Cornell, not everyone in the department was happy with my being there. I had, however, paid little attention to this opinion.

Later, Prof. M. J. Sienko, one of my colleagues, when I asked for advice about new positions on 1/7/63, volunteered that the department made a wrong choice, although he could not discuss the details. Finally, Prof. R. F. Porter, another colleague, offered a similar opinion. He helped me later with the assessment of possible future positions, particularly my contact with Purdue University.

As obvious from this summary, I kept careful notes about the job search, as I did when trying to get into the Humboldt University in 1949, described on pages 4-3–8. On November 15 and 16, 1962, I contacted either in person, or via letters all influential researchers I knew for possible support. These included, at Cornell: Professors P. Debye (Chemistry, emeritus), H. Sack (School of Applied and Engineering Physics), R. L. Sproull (Director of the Solid State Physics Laboratory), and the Baker Lecturers: M. Eigen, P. J. Flory, and F. S. Dainton; my former research Director M. Dole (Northwestern); and my professional acquaintances Prof. H. F. Mark (Brooklyn Polytechnic Institute), Prof. W. H. Stockmayer (Dartmouth College) and Dr. Billmeyer (my contact at the DuPont, Wilmington). All expressed their surprise at my leaving Cornell, and offered help.

A total of 14 openings were finally identified. I was introduced by Dr. Billmeyer to Rensselaer Polytechnic Institute in Troy, NY, Purdue University in West Lafayette, IN, and Ohio State University in Columbus, OH. Professor Dole initiated connections with Stanford University in Palo Alto, CA, Northwestern's Engineering Department in Evanston, IL, and Brooklyn Polytechnic Institute. Professor Scheraga wrote to the University of Texas in Austin, TX. Professor Sack arranged for letters to North Carolina State University in Raleigh, NC, the University of North Carolina in Chapel Hill, NC, and Pennsylvania State University in State College, PA. By myself, through advertisements or other contacts, I found possible openings at the University of Illinois at Urbana-Champaign, IL, Washington University in St. Louis, MO, Illinois Institute of Technology in Chicago, IL, and Princeton University, in Princeton, NJ. Six of these schools had either no opening, or I did not like the description of the available position, its level, or the make-up of the department. Finally, a few contacts were later abandoned because of the earlier success.

On November 23, we visited the Haags. Werner worked at ExxonMobil, by then in Princeton, NJ (see page 7-12 and Chapter 8, Figure 38). Werner and I spent some time at the university. The polymer program in chemistry was directed by Prof. A. V. Tobolsky, whom I knew well, but, he was out of town and I could see only two of his postdoctoral fellows. Later in the day, I also visited Prof. J. Wei in the Engineering Department. Both departments were considering new faculty. Professor Tobolsky wrote later that he was sorry to have missed my visit, but would like to stay in contact. On November 29, a connection was made with the Chemistry Department of Rensselaer Polytechnic Institute, RPI. The Chairman, Prof. George J. Janz, called me and requested vitae, list of publications, three references, and arranged a visit and lecture for December 14. After several mail and phone contacts with Northwestern University, a visit and lecture were arranged there for December 17 by Prof. D. H. Whitmore at the Materials Science Department.

The visit and lecture at Rensselaer Polytechnic Institute, RPI, went well. The Department Chairman, Prof. Janz, and the Dean of the School of Science, Prof. Bauer, and I got on splendidly. It helped that the three of us were dressed alike, in suits with vests, not common attire, even at that time. Also, we all had European backgrounds. Professor Bauer was, in addition, formerly doing research in polymers. He had dealt with viscosity under elevated pressure, not far from my interests (see Figure 4). His plan was to build up a polymer program at RPI. Toward this goal, a Materials Research Center, MRC, was under construction, supported by NASA. They suggested, I could get a large new laboratory, still changeable to my specifications. On my inquiry, they proudly described their teaching program in freshman chemistry (≈1,000 students). In recitation periods, students had contact with the senior faculty in groups of 30, and I could likely work out some lectures given to 500 students at a time. Having never heard of RPI before, I was impressed.

Next, our whole family stopped in Evanston on the way to Hastings for the Christmas Holidays (see page 7-23–24). In addition to giving a lecture and having the interviews in the Materials Science Department, I visited Professor Dole and some of our friends in Chicago. The interview went well, although the question arose wether I could teach engineering classes. When I broke off the discussion with Northwestern after I had decided to go to RPI, they contacted me in 1964 and offered an associate professor position for September 1965 in both, chemical engineering and materials science. It was tempting, but it was too late, as seen from Figure 24 of Chapter 8.

The first solid offer came to Hastings, NE on December 20, 1962. Professor Janz from RPI phoned and offered a position as associate professor with tenure after a short time, a salary of $10,500, and $8–10,000 for new equipment, with details of space in the new building, teaching, the transfer of my Cornell students, research equipment, and the applications for and transfer of research grants to be discussed in January. For me, this was a great Christmas present.

This almost ideal offer led in early January, after our return to Ithaca, to a series of conversations with my closer friends about the various merits of the other still open possibilities. It seemed that little could be improved by pursuing these. Thus, the offer of RPI was finalized on January 18, 1963 during another visit. On January 21, I mailed a summary of the answers I had been given orally to my questions by Dr. Janz. They concerned a possible starting date of June 15, the support of my new students, the cash-transfer ($650) for equipment owned by Cornell (mainly the Leitz microscope and the pressure equipment), the transfer of my research grants to RPI and the move of equipment bought with these, details of new equipment, the research space before and after completion of the MRC, moving expenses, the courses to be taught, and a review for tenure in 1964. On January 28, written agreement was given to all these verbal answers of January 18.

At the Winter Gordon Conference in Santa Barbara, CA, on Polymers, February 4–8, which I attended, I checked with many colleagues on the virtues of this and the other possibilities. After my return to Ithaca, I made a number of wrap-up phone calls to all other still open university positions (February 12, 1963). This was followed on February 13 with my acceptance of the RPI offer. A later photo of the campus is shown in Figure 38.

Figure 38 *Overview of RPI, A photograph taken about 1985.*

CHAPTER-08

Rensselaer Polytechnic Institute, Troy, NY, 1963–1988

The first half of 1963 was used to wind-up work in Ithaca and to arrange for life in Troy. *This involved the planning of the research for the spring term at Cornell and the summer and fall at RPI.* It included the thesis work for the students finishing their degrees in Ithaca and the initial work to be started in Troy at Rensselaer Polytechnic Institute, RPI.[1] As indicated on pages 7-6–7 and Figure 37 of Chapter 7, two PhD students (Arakawa and Bodily) remained at Cornell and finished their degrees in 1964 under my supervision; one, finished his MS (Sullivan), and one his BA (Jaffe). In June 1963, the latter two moved with me to RPI as graduate students.

At RPI, by being able to keep up the level of support, I could hire two new Postdoctoral Fellows from Germany, Drs. Eckhard Hellmuth and Herbert Baur. Two new graduate students, Mr. T. Davidson and Ms. C. Cormier, joined the group later in 1963 from the regular RPI student body, so that by early 1964, the research group was up to eight active students, allowing to continue the research without interruptions of ideas, support, and workforce.

After completion of the high-pressure work for Dr. Arakawa's thesis, the equipment of Figure 4 in Chapter 7 was moved and set up by Mr. Davidson in the Walker Building, using a temporary laboratory in the basement. The differential scanning analysis, DTA, that was started by Dr. Bodily, was accelerated and improved by acquiring a first commercial DTA of high precision, the DuPont DTA 900, purchased with set-up funding from RPI. Dr. Hellmuth had prior experience in microcalorimetry in the laboratory of Professor Müller of the University of Marburg, Germany.

---

[1] The 'Rensselaer School' was established in Troy, New York, in 1824 by Stephen Van Rensselaer "for the purpose of instructing persons ... in the application of science to the common purposes of life." It is "...the first school of science and school of civil engineering, which has had a continuous existence, to be established in any English-speaking country" [P. C. Ricketts, *History of Rensselaer Polytechnic Institute* (1914)]. In 1833 the school became the 'Rensselaer Institute,' and in 1861 the 'Rensselaer Polytechnic Institute.' RPI is a nonsectarian, private technological research university comprising five schools: Architecture, Engineering, Humanities and Social Sciences, Management, and Science. In the 1960s, RPI had ≈5,000 students and 500 faculty. The freshman chemistry lectures of ≈1,000 were taught by close to 30 members of the Department of Chemistry. In 2006, RPI was ranked among the top 50 universities, as listed by US News and World Report.

He first evaluated and calibrated the new DTA, shown in Figure 1. It also was located in the Walker Building, as was my first office. His main research topic became the analysis of fast, irreversible melting, using heating rates with the DTA of up to 100 K min^{-1}.

The structure analysis of the analyzed polymers was concentrating in these early days at RPI on optical, hot-stage microscopy. It later developed into different forms of interference microscopy. Initially, I did the microscopy. Later, it was taken over by Peter Sullivan and Mike Jaffe, who quickly became experts in the field. With a specially designed hot stage, heating rates of up to 3,000 K min^{-1} were possible. The basic research was irreversible thermodynamics of polymer melting and the connection between heat capacity and latent heats, as before.

Figure 1 *The first commercial differential thermal analyzer at RPI in the basement of Walker with Dr. Hellmuth as operator. Special temperature-gradient crystallization equipment is shown in the back, 1963.*

Dr. Baur had worked earlier on the theory of equilibrium and time-dependent, non-equilibrium thermodynamics in the research laboratory of the "Badische Anilin und Sodafabriken, BASF" (see also page 5-4). The results seen on crystal melting, fast heating, and superheating led to an exciting number of new ideas and observations (see Appendix B, Sect. B). We had succeeded in getting experimental and theoretical information on equilibrium crystals, as well as found methods to test metastability of semicrystalline and glassy samples.

Several visits in Troy were made to discuss laboratory space, equipment, and housing. Before the research plans could become reality, we had to find new living quarters in Troy. An opportunity arose when on March 14–15 1963, I was invited to lecture on "Polymer Crystallization at Elevated Pressure" at a general meeting about 'Pressure Effects' held at the Watervliet Arsenal. This is the facility of the US Navy developing, testing, and manufacturing large cannons. It is located across the Hudson River from Troy. The whole family went, staying in a motel on Route 7 at the edge of the town.

Figure 2 *211 Winter St. Ext., Troy, NY, March 1963.*

With the help of Prof. G. Janz, the Department Chairman at RPI, we had contacted a realtor, Mr. David W. Heer, who had lined up several houses for us. One, on 211 Winter St. Extension, was built in 1928 and is shown in Figure 2. Our wish list said, a new house would be best. But, this old house made a good impression. It was outside the city, which at this time was in a sorry state of decline. Almost all its industry had left, and the downtown was dilapidated.

In March, there was still snow on the ground, and Figure 2 shows a muddy drive up to the house. Not to get stuck, I had to move across the edge of the lawn. The house came with 2½ acres of woods and lawn, bordered by mature sugar-maple trees (see Figure 77).

The next house we saw was closer to our wishes. It was less than 10 years old. Important for this time of the Cold War, it also had a fully functioning A-bomb shelter underneath the basement. Figure 3 shows the appearance of the house just outside of Troy on Route 2, the route to Massachusetts. A busy highway. All was modern and ready to move in. The intercom piped pleasant music through the well-heated house. There were sufficient grounds, but all was bare, as can be seen from Figure 3. A third house we looked at was similarly new and just across from RPI, also close to the local schools, but on a small city lot.

Figure 3 *The new house with A-bomb shelter in Troy.*

In the afternoon, I had the meeting to attend and an interesting discussion, chaired by Prof. H. Eyring from the University of Utah. He was one of the originators of the 'hole theory' of liquids, a theory we had used earlier to describe the glass transition of polystyrene (paper #23, Appendix A).

Into the following day's schedule we could squeeze an interview with a local banker to check if we could afford any of the houses. It was a friendly chat with one of the Vice Presidents of the Troy Savings Bank. As a future member of the RPI faculty, I was a preferred customer. With an expected cash return of the money I had paid into the pension plan of the Cornell retirement system, we had a sufficient down payment for any of the three houses. The interest rate, at that time, was fixed at 5½% for 25 years. The banker knew two of the houses we were interested in, the one's depicted in Figures 2 and 3. The third, we had already decided against. We expected too much traffic from school children in front and back. The Troy Savings Bank held the mortgage on the house of Figure 3 and we were told that the present owner had lived for less than five years in the house, and the price he asked was excessive, unless he had done major improvements (which he had not). The house of Figure 2 belonged to the widow of a respected jeweler in town. She had died a few months before and the estate was eager to settle. He thought, the house was built to a high standard of quality, but needed updating. His suggestion was to take a second look in a week or two. Then, he said, we should get an informal appraisal from Mr. Tenczyk, a reliable local contractor he knew. We gladly accepted the advice and arranged a return visit with Mr. Heer.

On this second visit, things went quickly. Mr. Heer told us the heirs of the house in Figure 2 had reduced the price by $4,000, to $25,000. The house in Figure 3 had no change in price, and on a second inspection, there was no music greeting us, it was heated only in few rooms, and the bomb-shelter was dark, not as impressive as it was before. In the afternoon, we met Mr. Tenczyk, who turned out to be a congenial, hands-on builder. He spent time with me in the basement and pointed out all potential problems. For example, when he opened the electric fuse box, a big gush of water came out of the wall! I would have turned and run from the house, but he said, I expected this, I have seen the power connection. In the back of the house it goes underground and the protecting pipes are corroded where they enter the ground. There, the melting snow has leaked, but you need more powerful connection and modern fusing anyway. The heating system is old, but has been updated, you can wait a few years to install full air conditioning and electronic air filtering. The whole house needs redecoration and the floor, sanding and new varnishing. The vintage kitchen needs to be redone. Later, the driveway should be blacktopped and in back, a patio could be built. Otherwise, things are in good order. You could never build such a solid house today for the price asked. When I asked him for his expenses, he answered: "I do not charge for my time, but I certainly would like to do the remodeling." On my question of how much would that be? He answered: "$5,000, and I have it done by the time you move in June." We made a first offer of a low $21,500, which was accepted on March 27, and we had bought a house. The cost of all our special wishes for the initial remodeling was included in the mortgage.

Figure 4A–D illustrates two views of the interior before and after the remodeling. Figure 5 shows the new drive, which was finished two years later with Heidel's first new car, a Volkswagen, VW, beetle, imported from Germany (see Figure 26).

We certainly enjoyed our modern 'old' house and had given up our wish to live in a 'new' house (until we moved to Knoxville, see on pages 8-129–130). The move was done by a mover from Ithaca who, after he arrived in Troy, hired the additionally help for setting up the furniture. Overall, this was a one-day job and much less stress than the move to Ithaca, described on pages 6-38 and 7-1.

Figure 4A,B *Living ro om., kitchen & pantry, before.*

4C *The remodeled living and sun-room.*

Figure 5 *Heidel's new car on the blacktopped drive.*

4D *The remodeled kitchen and pantry.*

Everyone had improved quarters. My office is shown in Figure 6A with most of the bookshelves still empty. Brent's and Caryn's upstairs bedrooms are displayed in Figures 6B and C. Mr Tenczyk's work, which also included the shelving in Figure 6A, was so good that he was asked to do all future improvements which included the driveway and setting up of the patio behind the house with fountain and rock garden. Later work was a new rear entrance, an updating of the well and septic system, and in 1969, a heating and cooling system via a heat pump. It worked well by using oil heat for the days with temperatures below zero °F. By fall, everyone was well established in the new surroundings and Caryn also had a new school. In Figure 7 she shows Brent the one-room school, responsible for the initial grades. She still had to attend it the first months before transferring to the elementary school in our village of Wynantskill.

Figure 6A *My office with empty book shelves, 1963.*

Figure 6B *Brent's room, ready for play in 1963.*

Figure 6C *Caryns's room for studying.*

Figure 7 *The Little Red Schoolhouse in 1963.*

Figure 8 *Overview of the city of Troy. Dr. Heinze, from the BASF, Ludwigshafen, on a visit to RPI, 1966.*

When we had arrived in Ithaca in 1958, the beautiful countryside in the immediate vicinity of the city was overwhelming (see the descriptions on the pages 7-2, 10, 11). *Arriving in Troy, a much different impression was obtained*. The city of Troy was an eye-sore which was better not viewed from close up. Figure 8 gives a much better view from the hills lining the Hudson. As one moves east, the countryside is more beautiful and directed our weekend travels toward the New England states. The city of Bennington, VT, is only ≈30 mi away.

Figures 8 and 9 show two *visitors from Germany*, Dr. Heinze from the BASF, and Professor Müller of the University of Marburg. These and other visitors kept up my European connections, initiated largely at the1961 IUPAC Macromolecular Symposium (see pages 7-22–23). Dr. Heinze was later the laboratory director of Dr. Baur, whom we welcomed at the airport in New York City in the first months after our move to Troy. Figure 10 shows the Baur family with Heidel and Brent at a stop on the way to Troy. Dr. Hellmuth and family had arrived a few weeks earlier. He was a graduate of Professor Müller's laboratory. Figure 9 shows Professor Müller in the Green Mountain National Forrest. We had tackled Mount Snow (3500' high)—with help of a ski lift.

As became the custom, Heidel was always involved in hosting such visits. For the postdoctoral fellows and newly arriving students, she also arranged for suitable housing. Many interesting stories are connected with these activities. Especially *coworkers from Japan, sometimes brought surprises*. The Japanese custom was not to divulge the size

Figure 9 *With Professor Müller in Vermont, in the summer of 1963.*

Figure 10 *Stop on the trip from Kennedy Airport to Troy with Dr. Baur, wife and baby, Heidel and Brent, 1963.*

of their family, or even mention that there was a wife. It was never clear how many seats were needed for the car(s) to pick them up at Albany Airport. Sometimes, the arranged apartment had to be changed quickly. The biggest surprise was when a Japanese family arrived with three children and a dog. One boy was accepted by the couple from a friend who wanted his son to gather experience about life in the US. It took a few days before suitable quarters could be found.

My frequent duty was to teach the newcomers how to drive. Once, when I had not yet appreciated that Japanese do not like to say 'no,' I picked out a car with my student (I hope he got what he wanted, but he had said 'yes'). He got his learner's permit, and when we picked up the car, I asked: "Can you drive the car home?" He answered: 'Yes.' We entered the highway, and in the middle, I could just get him to stop before he ran into the traffic coming from the right. We quickly switched seats and it took some time until he was safe for driving on the right-hand side.

On another occasion, we had the new coworker with his wife for dinner. His English was very poor and we could not exchange much information. Even writing down questions and answers did no help much. His wife said practically nothing. We thought, she knew even less. Wrong! Her English was much better than his. But, she did not want to embarrass her husband by divulging it.

Some other time, Heidel was told by a wife of a postdoctoral fellow that when he would bring home his salary in cash, as was common, she had to greet him by bowing, and telling him: 'You are a good husband and provider.' Only then, would she get the money. To the satisfaction of the wife, this custom disappeared with the event of direct bank deposits. But, in contrast to our household, where bigger purchases are always discussed jointly, the Japanese housewife is expected to decide most domestic matters with little or no input from the husband, including buying a house or car. We and our students learned much about customs in different countries.

At home it was my initial job to tame the 2½ acres of wilderness which, probably, had not been touched in more than 20 years. Figure 11A shows Brent and Caryn on one of our discovery safaris. After a few weeks, I managed to clear a nice little area in the midst of the woods, seen in Figure 11B. There we could arrange a picnic area for hot summer days. The only air conditioner in the window of our bedroom, dated

Figure 11A *In the wilderness in Troy.*

Figure 11B *A picnic in the depth of the woods in 1963.*

from our Chicago time (see page 6-30) and was augmented in 1966 with a small unit in the kitchen. Summers in Ithaca were not as hot as in Chicago or Hastings, but being able to sleep well and work efficiently in the kitchen during the few days above 90°F was worth the expense. In time, we developed plans to air-condition the house (page 8-4) and consider cars with air conditioning.

Our house was thought out well. The windows had copper-clad, steel frames and roll-up screens hidden in the walls, so that the netting showed only when used. A second set of windows existed, to be put up in winter. Double glazing was not invented by 1928. The front door had a protection against draft with an automatically lowering contact bar. The outside walls consisted of two layers of brick, enclosing an insulating air space to be energy-efficient. The basement had an old, coal-fired stove to produce hot meals during possible power outages (see page 8-131).

When exploring the house after moving in, we found pages from an unidentified architecture magazine with drawings of the '1928 house of the year.' Our house was close to identical. Finally, we found out that the house was erected by a builder from New York City who was involved in construction at the Rockefeller Estate at Pocantico Hills on the Hudson River. He had designed the house for himself, to live far from the 'Big City.' After a number of years, he sold it. His wife missed the life in town. Some of the information about the early days of the house came from the son of the builder. He visited us on a nostalgic visit to see where he had lived as a boy.

As mentioned on page 8-6, the citiy of Troy was past its prime (as well as the adjoining cities of Albany and Schenectady, both on the other side of the Hudson River). *To appreciate the beauty of the area, one had to go into the surrounding countryside*. Turning north, toward the Adirondack Mountains, or south along the Hudson valley provides both a historically important and scenically beautiful environment. Figure 12 shows an overview of the 1909 restored 18th century Fort Ticonderoga[1] in the north. It was to protect the British-controlled Hudson River Valley from the French influence in the St. Lawrence River region. It is located less than 100 mi from Troy at the eastern edge of the impressive Adirondack State Park and near the northern edge of Lake George on the southern Lake Champlain Narrows, just a 'cannon shot' from Vermont which begins with its eastern shore.

Figure 12 *Fort Ticonderoga on Lake Champlain.*

Going south, one finds many historical mansions along the beautiful Hudson valley. On the right-side of the Hudson River, in the foothills of the Catskill

[1] Built in 1755 by the French as Fort Carillon, initially out of logs fortified by dried mud. Later it was rebuilt with stones from a nearby quarry. In 1759, it was given up to the British who repaired and renamed it Fort Ticonderoga. In 1775 the fort was captured in a surprise raid in the first battle of the American Revolution (by Ethan Allen and Benedict Arnold who led a band of 83 'Green Mountain Boys'). In 1909, the Fort Museum was opened after extensive restorations.

Mountains, we found Roland Lindemann's Catskill Game Farm. A delightful, private zoo of 1,000 acres with about 2,000 animals of over 150 species. It opened in 1933 and saw its peak when we frequented it with Caryn and Brent in the 1960–70s. Unfortunately, it closed in 2006 and all its animals were auctioned off. On our first visit in 1963 we had asked Mr. Lindemann about possibly keeping a few fallow deer. Getting his expert advice, we gave up the idea. It would have meant to fence not only the property, but also each tree, to preserve the bark, edible to the deer. Also, the daily maintenance would have been costly and time consuming. It was better to spend the time visiting the game farm, which was about 50 mi from Troy. Figure 13 is an illustration of such a visit.

Figure 13 *Caryn and Brent in the petting zoo, Catskill Game Farm in 1965. This is how the fallow deer would also have looked in our back yard in Troy.*

New York City, about 150 mi south of Troy, is an attraction for a visit, but not to live there. Smaller cities offer a more pleasant life, particularly when a bigger city is not too far away. In the next two years, the 1964/65 World's Fair at New York City's Flushing Meadows made us frequent visitors to the city. As one would expect, our move to Troy and the World's Fair got many relatives and friends to arrange visits to us. In time, we were good guides for the Fair and the countryside.

Completing the picture of the surroundings of Troy toward the west, one follows the old Erie Canal in the direction of the expansion by the settlers toward Chicago. This was followed by the heavy industry, leaving the textile mills in Troy. These were lost later in the move south to benefit from cheaper operation costs.

We slowly felt at home in the not so attractive city. It had an ideal location and we could see no reason ever to leave the city where Uncle Sam was buried.[1]

Having described the initial steps of our move to Troy and giving a picture of the interesting assets of the area, it must be said that **the neighbors, colleagues, and coworkers were part of a pleasant support** for our new life. We noted already in Ithaca that going from the Midwest to the East, the behavior became more 'European,' more formal and reserved (see page 7-18–19). Perhaps this was based on the larger population density. The same was true in Troy, although at the university, I had a feeling of being much more accepted and appreciated, and also, our newly gained acquaintances and friends, once we got to know them, made us feel quite at home. First, I want to comment on our neighbors. We were the 'junior family on the block.' All of the three neighbors we had in the 1960s and 70s are

[1] Uncle Sam was named after Samuel Wilson, born in Arlington, Mass. on September 13, 1766. In 1789, he and his brother Ebenezer moved to Troy, New York. During the War of 1812, Wilson was in the slaughtering business and packed meat. He provided large shipments to the US Army in barrels that were stamped with the initials 'U.S.' Supposedly, someone took these initials for 'Uncle Sam' Wilson's. This led to the idea that Uncle Sam symbolized the federal government. Samuel Wilson died in 1854. His grave is in the Oakwood Cemetery in Troy.

by now deceased. Our neighbors on the right, Bill and Shirley Geddis and son Mark greeted us friendly and we always got along well. Bill told us he was in the Air Force during WW II, and it took a while to freely talk about the terrible happenings during the war (see page 2-23).

These reservations, resulting from our past, were even more serious with our neighbors across the road, Dr. Adolf and Mrs. Lilo Glaser. It took several months until we got acquainted. We slowly gained contact after Heidel went on Halloween 1963 with the children to 'trick or treat' at their house across the street. Dr. Glaser was a well-established Jewish surgeon in Chemnitz, Germany and was lucky to flee in the late 1930s after the first confrontations with the SA ("Sturmabteilung," Nazi assault group, stormtroop) tried to close his practice. After they left Germany, Dr. Glaser was unsuccessful finding employment in the neighboring countries, traveling as far as Morocco. Finally, with help of relatives, they came to the US in 1940. With hard work, he learned English and passed the needed exams to open a medical practice again.

As time went on, they became our best friends and we learned a lot from them. Dr. Glaser's education was based on the German "humanistische Gymnasium" (humanistic high school) typical for the 19th and early 20th century Germany, much different from my "Realgymnasium" (high school for mathematics and natural sciences, see page 3-13). His foreign language training was in French, Latin, and Greek, while mine was in English with only a small amount of Latin and almost no Russian. His main subjects were literature, philosophy, and history, while mine were mathematics, chemistry, and physics. He also had a strong interest in art, and Mrs. Glaser was an accomplished painter of water colors. We still display several of her exquisite works in our home. Since up to then I had to narrow my effort to the professional development in chemistry, it now was high time to broaden my outlook. Having the Glaser's as friends helped in this endeavor, which subsequently continued to my enjoyment. By now, I wonder if there should not have been a broader mix of subjects in high school. Particularly for me, the time spent in school could have been longer since I lost much time due to illness, the effects of the war, and the customary half-day instruction.

Figure 14 shows the Glasers. Dr. Glaser never used his first name, Adolf, which recalled for him the terrible past in Germany, but he also did not consider changing it. In appearance and behavior, he was a typical German aristocrat. We were often met in our houses and also had memorable parties with friends, family, and in our case, also with students. There were many parallel interests and common behaviors. Figure 15A shows a Sunday afternoon in the Glasers' back yard with friends of theirs, and 15B depicts a scene of a more exuberant New Year's party in our house. For the last, we had also invited Dr. Medicus, originating from Switzerland, and his wife Hildegard, who was from Germany. Dr. Medicus was a professor of physics at RPI and headed a student exchange between RPI and the ETH in Zürich.

Figure 14 *The Glaser's inspecting travel photos with Heidel, during one of their visits in our house.*

As time went on, one by one, *Dr. Glaser introduced us to his Jewish-German friends*. Each time we felt the existence of the barrier created by the holocaust. After a while,

Figure 15A *In Dr. Glaser's back yard with friends, debating questions of common interest. The ladies were shown changes in the garden.* **B.** *A New Year's Eve Party in our house.*

however, it melted away. Many of his friends, all having victims of the concentration camps in their family, became also our friends.[1] One cannot run away from one's past, but on a person-to-person level, there does not have to be hate based on religion or race, or political persuasion. Such hate needs indoctrination by living in closed society groups where such behavior is taught. Despite their horrible past with Germany, we found that most of these Jews still liked German literature, music, and customs. Also, most took every opportunity to speak German.

Dr. Glaser was enthusiastic about glamorous cars, but could not get himself to buy a Mercedes because of it being built in Germany. He, however, gladly rode in mine. We noted a similar resistance in some of our American friends to buy anything Japanese. Also, I noted that Japanese scientists were up to the 1970s only hesitantly invited to professional meetings in England. The ferocity of the fighting and treatment of the POWs and interned civilians in Southeast Asia was too much ingrained in the memory. Unfortunately, such memories of the past can last in some parts of the world for hundreds of years. They can only be defused by frequent positive personal contacts. Otherwise, in the long run, the once created prejudices may cause more problems than the initial evil deeds. More about causes and effects of such hatred should be included in our education.

To close out the short list of our nearest neighbors, Mrs. Lilian Bennett owned the house on our left. She was an elderly widow and was a good friend to our children. She always had an eye on our house when we were traveling. Later, she rented a room to a succession of my foreign postdoctoral fellows toward whom she felt like a (grand)mother. After she left in 1977, to live with her children on Shelter Island, the new neighbors were the Youker family with four boys. Two of them were closer in age to Brent, so he gained friends with whom he is still in contact. We also are still exchanging Christmas letters with Sandy and John Youker.

Another essential: *We had to find the supplies for German cooking and baking* which were not available in the standard supermarkets. There was a German butcher in Wynantskill, Herr Jacob C. Würtemberger, who worked in a meat packing plant. After hours, and on weekends, on order, he sold any cut of meat out of his store, built into his garage. In Albany, across the Hudson, and in

[1] Mabel & Dr. Heinz Bieringer, Hedda & Dr. Hans Schein, Esther & Sol Boxer, Inge & Dr. Milt Schiffman, Drs. Alena & Carel Polesny, Mrs. Irene & Dr. Freeman, and Mrs. Hanna Gumpel.

downtown Troy there were bigger German style butcher stores where one also could buy special sausages and imported goods. These were the places where recent and earlier immigrants not only shopped, but also met and connected. Soon Heidel was part of a German "Kaffeeklatsch" group which met periodically. We, however, resisted becoming members of the "Deutsche Klub." The German connection was only a part of our life, although the language within our family remained German (see page 6-24–25) and many of the new German friends have become 'old friends' and are on the list of 150, or so, relatives, friends and former students we annually have contact with.

Going to work at RPI, one of the first persons I met, was the secretary, Mrs White. She was long past retirement, but just irreplaceable. She looked at me and classed me at the bottom of the list as a junior person who needed lots of help. A good accent helps giving this impression. We heard of cases where people having difficulties in spoken English were also assumed not to be able to read or write. It, however, did not take long and I was Mrs. White's favorite. She loved to have everything neat and proper. When I gave her a manuscript to type, I got it back with dozens of little remarks where I could make improvements. When, for example, she thought I had misspelled a name, she would go to the library to look up the references I had listed. When she found out, how much I appreciated her help, my work was always finished first. I was sorry when she finally retired after 2–3 years. Coming the long way from Ithaca, my next secretary was Mrs. Helen Carroll. By now I had enough business with my research group and was working on several book chapters, and later books, so that 'Helen of Troy' was my sole secretary in the MRC building.

At RPI, as at Northwestern and also at Cornell, there was an annual get-together of the departmental professors, postdoctoral fellows, students and staff. At RPI, this took the form of a Christmas party, catered by the faculty spouses. It was an affair well attended and liked, held in one of the decorated dining halls. Besides this, at least once a year, we had my students at our house. Invited in groups, or separately on special occasions. On these dinners, it was of interest to hear about the background and aspirations of my students and their spouses. Figure 16 shows the first of these group dinners in 1963. To manage the group, the postdoctoral students (see Figures 1 and 10) were invited separately. Six more chairs for the two postdocs and their families would not have fitted around the table.

Figure 16 *Student dinner, ready for dessert.*

This brings me to my colleagues. On top of the scientific scale ranked Professor Paul Harteck,[1] who headed research on

---

[1] Paul Harteck (1902–85), Distinguished Research Professor at RPI since 1951. Immigrant from Austria/Germany; PhD in 1926, Berlin; Professor of Physical Chemistry, and subsequently Rector of the University of Hamburg (1934–50). Until the end of the war, he worked in Germany on nuclear reactions, artificial radioactivity, and the separation of hydrogen isotopes by diffusion. On April 24, 1939, Harteck and Groth send a letter to the German war office stating that the country that first exercised the use of atomic energy "has an unsurpassed advantage over the others."

radiation chemistry, the atmosphere of the earth, and the other planets. His teaching involved occasional special courses. On the bottom was Mr. Rackster, MS in Chemistry. He expertly handled the teaching laboratories in the Walker Building, shown in Figure 17. Everyone else had a certain ratio of duties for research to teaching.

Figure 17 *Walker Laboratory and the path to West Hall with the large lecture-hall for teaching freshmen.*

In a short time, Heidel and I got to know every one of my colleagues. We both were invited to most of their houses, alone or for larger parties. All contributed their best to a well functioning chemistry department. The goal was to give a first-rate education to undergraduate and graduate students, and to accomplish an increase in quality and volume of research. Most of the ≈30 professors handled from one to three of the freshman recitation classes of 25–30 students. This meant, many contributed to the teaching of this course and had an input to its success. Despite the large number, each student had at least once a week close contact with one of the professors. No teaching assistants were used for this duty, assuring quality of instruction even though the lectures were given to large groups. A textbook "Principles of Chemistry" was developed between 1960 and 1963 by five of the main lecturers (L. G. Bassett, S. C. Bunce, A. E. Carter, H. M. Clark, and H. H. Hollinger) and then published by Prentice Hall, Englewood Cliffs, NJ in several editions. It was designed for the interests of engineers as well as scientists. All students had to take these lectures (and buy the book) in their first year (see Footnote page 8-1).

At RPI, thus, **I had a chance to get involved in teaching of freshmen**. Initially, I taught two of the recitation periods, i.e., I had to answer questions and elaborate on the material taught in the main two lectures each week, as well as give and grade quizzes and exams with the help of teaching assistants. Soon, I took over 3–5 of the main lectures. The lecture room in Walker Laboratory (see Figure 17) could hold about 200 students. It had the proper steeply sloping rows of seats to have the students see the experiments which traditionally illustrate and liven the introductory chemistry lectures. By the 1960s, the number of students at RPI had increased to such a volume that it became impossible to instruct the 1,000 freshmen in the Walker Building. Five parallel lectures in the chemistry lecture hall were unthinkable. The still rather new and untried television had to come to the rescue.

The biggest room on campus was the Auditorium in West Hall. It could hold 500 students in a setting with a sizeable stage and gently sloping seating. Not even from the first row could one observe the experiments. The solution was to display the demonstrations on television monitors. Four cameras were employed, handled by two operators. One, to always follow the lecturer, one to show the experiments, and one to allow reading of three blackboards placed on the stage. The fourth was a stationary, overhead camera for slides, movie clips, and overlays. It was operated by the lecturer himself. The picture displayed on the monitors was also recorded, together with the sound. It was then available for review and as a means for the students to study missed lectures. The lectures with experiments were given in the morning and repeated in the afternoon.

As soon as sufficient other rooms were fitted with monitors, the lectures were broadcast across campus, and the secondary rooms were supervised by recitation instructors. All students could then be instructed simultaneously. Surprisingly, although the lecturer on stage was quite remote from his audience even in West Hall Auditorium, the students overwhelmingly preferred this to watching in the satellite rooms. This is an indication, that a movie about chemistry must be of much better quality than locally produced lectures in order to compete in pedagogical value.

My experience with experiment-illustrated lectures went back to the high school days (see Figure 14 of Chapter 3 and page 3-20) and were brought to perfection by watching and copying several years of Professor Thilo's outstanding presentations in Berlin (see page 4-11). The technique of handling of large-volume student groups, I picked up at Northwestern (see page 6-32). With a little bit of showmanship, introduced to me by a short acting career (see the pages 3-26–27), I was well prepared to contribute to this major teaching effort at RPI.

To produce perfect lectures, I spent about 40 hours in preparing each lecture. Then, I added plenty of well-rehearsed and tried experiments, used two of the three blackboards as memory devices for initial information and a final summary. The third was for writing of new material and derivations of equations. By writing during the lectures, I made sure that the pace was proper for the student's to make complete notes. An outline script of the lecture was prepared for the camera crew and director and was briefly rehearsed. The result was satisfying. Students were there on time, and more importantly, stayed to the last minute of the lecture. Finally, they often gave spontaneous applause and the recitation periods were full of exciting questions and comments. My colleagues made sure that I was named one of the 'outstanding educators of the United States of 1971.'

The challenge of how one can make sure that every new student at a university can have the experience of outstanding instruction led me to *further analyze and experiment with teaching*. The analysis had to answer a number of questions: Why did chemistry attract more interest int the past than it does now? Was the subject matter treated in chemistry not as interesting anymore? Why is the number of schools so small where really first-rate chemistry is taught? What can be done, to improve the present state of affairs? As with any of the major problems in this world, ranging from poverty on earth to global warming, the problems slowly become obvious to everyone, the reasons and solutions, however, do not. The simple solution like we need more chemistry teachers, we need to share the wealth, and we need to reduce the CO_2 content in the atmosphere, are certainly not only insufficient, but often outright wrong.

Answering some of the questions raised in the last paragraph, one must remember that from approximately the middle of the 19th to the middle of the 20th century, chemistry was the science which contributed most to the growth of knowledge about the nature of matter, its analysis, and the synthesis of old and new materials of interest to man. Many of the results could be applied immediately to practice and were noticeably increasing the standard of living. An incomplete and random listing of new or better products includes fertilizers, dyes, explosives, elastomers, plastics, fuels, solvents, disinfectants, preservatives, pharmaceuticals, metals, ceramics, detergents, adhesives, textiles, films, and packaging materials. The misuse or excessive use of these products is a topic for a different, but also important discussion (see page 8-53).

Initially, demonstrating brilliant color changes, melting of iron with the spectacular thermite reaction (ignition of powdered aluminum and iron oxide), or showing the harmless burning of hydrogen gas changing to a violent explosion by altering ratio to air or oxygen, were not only drawing students into university lectures, but were even fairground attractions. Since the knowledge behind such experiments is relatively simple, the interest in learning about chemistry was high. The experiments based on current chemical research are not as spectacular and are more difficult to understand without prior training. In addition, we have developed a less enthusiastic attitude toward new or better products by being more informed about possible problems in their use. Finally, movies and computer games can conjure up much more impressive, 'out-of-this-world' pictures which happen not to be real, but can excite the mind more than the real chemistry experiments.

The decrease in quality of teaching also has many roots. First, the number of students to be instructed has increased, and thus, more teachers are needed. This makes it difficult to have top quality instruction everywhere. The instructors themselves may not have been exposed to outstanding teaching. At universities, one finds an additional change. Instruction of first-year students gives less recognition. Conducting successful research attracts funding from outside and lightens budgetary restrictions of the school. Teaching freshman chemistry, as a result, is often put into the hands of professors less active in research. For exciting teaching, it is necessary to excel in both, teaching *and* research. Just as I noted with high school teachers, the interest in 'how to teach' must not overcome the 'need to understand science' (see page 7-5 and also 9-86). It surprised me that, as I became more successful in research, I felt less appreciated by the fellow teachers. They had developed their fixed method of teaching and resisted the changes I tried to introduce.

Support of my research activities by RPI was strong. On arrival at RPI, I was not knowledgeable in college sports. So, I appreciated it only later what was meant being greeted by the Provost with the comment they had lost their ice hockey coach to Cornell this year, but my coming was as a bigger gain.[1] He certainly made sure of the quick transfer my old Cornell research grants and the generation of new ones at RPI. The editor of the *Rensselaer Review*,[2] Mr. Daniel J. Priscu,[3] arranged through his connections as a journalist to publish a paper of mine in *Scientific American* and later, he reprinted an abbreviated version in his publication (see Appendix A, #31, and Chapter 7, Figure 30). Also, he was helpful in my being named to the editorial board of the ACS journal *Chemistry* (1965–1969). There, I wrote a general article aimed at freshman teaching: "How Big Molecules Crystallize and Melt" (Appendix A, #46). Having joined the American Chemical Socierty, ACS (see page 6-33), I also was involved in teaching an eight-lecture ACS short course (see Appendix A, book #37) and became a Tour Lecturer for local sections of the society (1967–1985, 32 lectures, see Appendix C). In 1966, my election to the New York Academy of Science was supported and I attended several of their meetings and lectures in New York City.

---

[1] Rensselaer boasts one of the oldest intercollegiate hockey programs in the East, playing its first game in 1902. In 1954 and again in 1961, the 'Rensselaer Engineers' won the national championship and remained competitive in the NCAA (National Collegiate Athletic Association). In 1963, after three NCAA appearances, their legendary coach since 1950, Ned Harkness, left RPI to successfully coach at Cornell until 1970. Then he became the first college coach joining the NHL (professional National Hockey League) as a (not so successful) coach of the Detroit Red Wings.

[2] Now available on-line under the web address: http://www.rpi.edu/research/magazine/

[3] Heidel and I knew Mr. Priscu from Ithaca through his wife, Ruthild, another German immigrant. The Priscus also came in 1963 from Cornell to RPI, somewhat later in the year than we.

Figure 18 *North Hall, RPI, 1963.*

Our research began in Walker Laboratory and North Hall where some additional space was available. The old high pressure apparatus, shown in Figure 4 of Chapter 7, and the new differential thermal analysis equipment, DTA, illustrated in Figure 1, were placed in the Walker Laboratory, depicted in Figure 17. The microscopy was moved into North Hall, depicted in Figure 18, an old dormitory converted into the first building of MRC, the Interdisciplinary Materials Research Center. At that time, the new MRC building was already under construction. It is shown in Figure 19. After completion in 1965, we could move all our activities into a four-room laboratory suite, 2,000 square-feet large and custom-built. All was on the third floor, together with my new office.

Despite leaving Cornell, or possibly because of the move, this period from spring of 1963 to spring of 1965 was especially productive. Not only in the new aspects of teaching, but also in research. We could use the discovery of equilibrium crystals of polyethylene by crystallization under high pressure and their time-dependent melting to expand the description of the solid state of flexible, linear macromolecules using a balance of both, equilibrium and irreversible thermodynamics. The details are given in Section B of Appendix B, pages B-3–7.

In 1964, invitations to the Gordon Conference[1] on Polymer Crystallization, held in New London, NH, and to the General Electric Professors Conference in Saratoga, NY, discussing 'Progress on Crystallization of Macromolecules,' created national and international interest in our work.

At this time, the Gordon Conferences were held mainly in small colleges during the summer vacations in New Hampshire. The ≈100 attendees stayed in dormitories. In spring, a smaller number of such Gordon

Figure 19 *The new Materials Research Center, MRC, under construction. Moved into by 1965.*

[1] The Gordon Research Conferences, a non-profit organization, are managed by and for the benefit of the scientific community. They were founded by Neil Gordon in 1931 in Baltimore ('Summer Research Conferences on Chemical Physics,' held at Johns Hopkins University). By now the conferences provide an international forum for the presentation and discussion of frontier research in the biological, chemical, and physical sciences, and their related technologies. For further information see http://www.frontiersofscience.org/ and http://www.grc.org/

Figure 20 *Gordon Conference at Colby Sawyer College in New London, NH. Photo in summer 1964.*

Conferences were held in California. In 1963, I had already attended the one on Polymer Melting (see page 7-28). At the Gordon Conferences, the attendees had a whole week to discuss a limited number of lectures. The conferences in California broke the long winter in New York, however, the conferences in New London were, at that time, often closer to our research. Figure 20 illustrates the relaxed atmosphere, enhanced by the comfortable chairs collected from all over the campus. Over the years, I attended ≈20 Gordon Conferences (see Appendix C).

The New Hampshire countryside was attractive and could easily be reached by automobile. The 1964 Gordon Conference, I attended by myself. Thereafter, we made it a family affair and combined work with a vacation. As long as Brent was under 10 years old, he could not stay at the dormitories, so we rented a cottage. To have enough vacation time, we usually 'extended' the trip into a second week (see page 8-50 and Figure 58). Not only was the countryside worth exploring, the antique shops and old book barns were also of interest, and it did not take long to discover a number of exquisite restaurants. The stay at the Gordon Conferences as well as other meetings in the north-east of the US were occasionally also extended with sightseeing trips into Vermont, New Hampshire, Maine, or along the Atlantic coastline. Figures 21A and B are snapshots from visits to Mount Washington (1963 and 1965).

These conferences allowed also for additional international connections and led me to realize that it was time to attend research meetings abroad and to visit the important research laboratories in the world. A plan took shape, to visit Europe in 1965.

Figure 21A *The 1869 cog railway up to Mount Washington. (Old Peppersass).* **B** *Caryn having reached the top of Mount Washington.*

Our first plane-ride as a family occurred also in 1964. It was a vacation trip to the Rocky Mountain National Park in Colorado. We parked our car at Kennedy Airport and flew with TWA to Denver, where Vati und Mutti picked us up, to drive to Grand Lake. They had rented a cabin there for two weeks on the shore of the lake (see also page 8-50). Figures 22A–C illustrate the first big vacation of ours with Vati, Mutti, Manfred, and Joachim. It was quite an impressive two weeks. The only way I had experienced the Rocky Mountains before, was by looking out of the window when flying to San Francisco or Los Angeles. Even more exciting was to get up onto the Alpine Visitors Center at 11,796 ft., and seeing fresh snow falling in July.

Figure 22A *Vacation at Grand Lake (left).* **B** *Evening at the lake (right, top). Vati, Brent and Caryn in front of the cabin feeding a chipmunk.* **C** *(Right bottom) Manfred and I, speeding on the lake.*

The biggest public event in 1964/65 was the World's Fair in New York City, less than three hours away by car (see page 8-9). Figures 23A–B give excerpts of one of these days. The upper left starts with an overview of the site from the observation tower. The next snapshot was made when inspecting the Belgian Village, for us, one of the main attractions. In fact, this was the time and place for the introduction of the "Belgische Wafels" to the US. Another main event, mainly for Vati and me was the Norwegian buffet. A selection of fresh fish was offered as it was hard to get in Troy. Such exhibitions allowed a 'travel around the world' within a single, exhausting day. It took some time until we could do this in reality, but not in one day. (see the pages 10-13–23).

Figures 23A–D *World's Fair time, 1964/65.* **A** *Upper L: The symbol of the fair.* **B** *Upper R: Mutti, Vati. Caryn, Brent, and Heidel in the Belgian Village.* **C** *Lower L: Back from the fair, on the finished patio with the new 'Barracuda' in back.* **D** *Lower R: Backyard barbeque, Cornell style.*

The two bottom pictures, Figures 23C–D, are of our finished patio, fountain, and rock garden, completed during 1964/65 taken after one of the drives to the World's Fair. As already pointed out on pages 7-10–11, with the large amount of driving of about 20,000 to 30,000 mi a year, it seemed advantageous to get a new car every three years. The left lower picture shows my new, sporty, 8-cylinder, 1964 Barracuda. With the introduction of the 'pony cars' at the World's fair by Ford with their Mustang model, Chrysler had countered with the Barracuda, built on the Plymouth Challenger chassis. It was released two weeks before the start of the World's Fair.

Heidel initially kept the much heavier Ambassador (see page 7-21), but by 1965 she changed to the small VW beetle, seen in the figure and also in Figure 5, above. Buying a car delivered in Germany and shipped back to the US dealer in Troy was cheaper than buying it from the show room in the US. Finally, Figure 23D illustrates the Cornell-style chicken barbeque (see page 7-2) as done on the end of the patio on the occasion of one of Mutti and Vati's visits.

In 1964 I did not get tenure as associate professor despite having the feeling that all went well. What was wrong? I had been promised tenure after one year (see page 7-28). With all the activity, I was not very concerned, but talked a few months later at an opportune moment to Dr. Janz. When mentioning my understanding, I saw he was taken aback. He had forgotten to get the ball rolling for the tenure hearings 1964. He promised he would get to work immediately.

At the end of March 1965, I received the letter reproduced in Figure 37. It conferred not only tenure, but presented the promotion to a full professor. I should have expected this only during a second five-year term as an associate professor, as discussed on page 7-21. What a surprise, satisfaction, and reason to celebrate. Much later, Heidel heard over the grape vine of the Faculty Women's Club that the additional promotion was helped considerably by my having actively pursued permission and funding for the trips to Europe in 1965. It was thought that there was a chance that I might want to stay there.

Indeed, this question was brought up at the University of Mainz during my first sabbatic stay in Germany, but at that time the opportunities in the US for finding support for research were superior. In addition, it was also necessary to consider that the standard of living in the US was much higher, and the intensity of the Cold War was still increasing. The Berlin Wall had been built just a few years before, as discussed on pages 7-7–8.

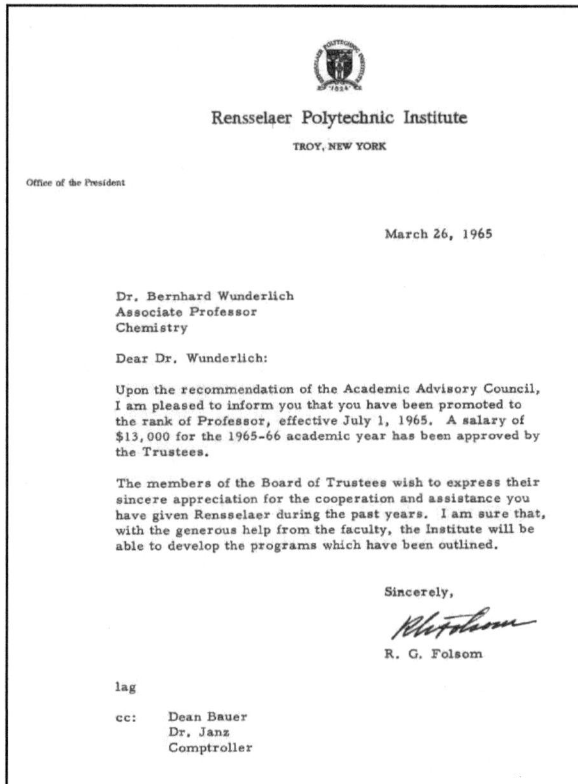

Rensselaer Polytechnic Institute

TROY, NEW YORK

Office of the President

March 26, 1965

Dr. Bernhard Wunderlich
Associate Professor
Chemistry

Dear Dr. Wunderlich:

Upon the recommendation of the Academic Advisory Council, I am pleased to inform you that you have been promoted to the rank of Professor, effective July 1, 1965. A salary of $13,000 for the 1965-66 academic year has been approved by the Trustees.

The members of the Board of Trustees wish to express their sincere appreciation for the cooperation and assistance you have given Rensselaer during the past years. I am sure that, with the generous help from the faculty, the Institute will be able to develop the programs which have been outlined.

Sincerely,

R. G. Folsom

lag

cc: Dean Bauer
Dr. Janz
Comptroller

Figure 24 *Appointment to Professor of Chemistry at RPI.*

Is the promotion in 1965 to a full professor the end or the beginning of my career? This was the question I had to tackle after the initial euphoria about the promotion had passed. Since my beginning of university study, 16 years had passed. In this time, Heidel and I got married, escaped the GDR, continued study in the FRG, and made a new life in the US. Finally, I completed my PhD in 1957, and now, in eight years, I had reached the top of the academic ladder. This was the end of academic apprenticeship. I decided, this was only the beginning of my career. Other possibilities, to work toward becoming a department chairman, dean, or take any other administrative position did not appeal to me. I could see the value these offices have in the overall functioning of a university, but personally, I did not feel that there would be as much joy in seeing an organization prosper through my coaxing others to better performances. My goal was to understand what I had set out to explore. My chosen research into the solid state of macromolecules was still far from complete (see page 7-19). In fact, as I write this paragraph, more than 40 years later, there still is much to be done.

Another path that could be followed was to *directly apply research to industrial uses*. A first opportunity arose when finding the extended-chain crystals (see Appendix B, page B-4). They had very different properties. "Why not patent this finding?" I was urged by colleagues at the DuPont. Growing up in a business household, I knew what it takes to shepherd an idea into a successful enterprise. Doing this, in addition to research, was not to my liking. To be successful in bringing research results to application, it would also be vital to look for other inventions as well. The time needed to combine all knowledge for a new application would be lost from my main interests.

This does not mean that looking for practical uses in my work was avoided. Quite to the contrary, often my research was governed by questions arising from industry, such as mentioned on page 6-32 when indicating the start of the high pressure work. Also, teaching industrial researchers new insight and its possible use was important to me. Frequently I checked the production facilities to resolve problems which were brought up during consultant visits. This bypassed the verbal descriptions which often pointed in the wrong direction when based on 'old' knowledge.

With respect to missing out on potential monetary rewards, I felt sufficiently compensated. Also, I felt that the government support of my research should not be used to generate additional personal income. Always, I am appalled when hearing what enormous fees are asked by politicians when presenting public lectures during their tenure and after leaving government offices. The information transmitted in these lectures was enabled by the taxpayer and should, just as the results of my research, be available without undue further compensation.

On a personal level, life continued to improve from the zero level it had reached in 1945 described on page 3-5. By now, Heidel and I headed a well-adjusted family with two rapidly growing children, were owner of a house in a suburb of Troy, and our two late-model cars made us independently mobile. Brent had started kindergarten in August 1964. The building was the 'Little Red Schoolhouse' of Figure 7. Figure 25 shows Brent with the traditional "Schultüte." Caryn, three years ahead of Brent, was making progress in the Wynantskill school.

Since I had never imagined that my occupation would involve so much travel, the idea of buying or becoming part-owner of an airplane crossed my mind. It certainly would be fun. On checking not only the expenses, but also technical details, it quickly became obvious that it would involve considerable training and cost when trying to go faster than 150 mph and to be able to fly not only in perfect weather. My most frequent travel was to Wilmington, DE, and points south on consulting trips. From Wilmington, it was possible after quitting time to get home with a tiring five-hour automobile trip. To get to Wilmington, I needed to go the day before and spend a night at the DuPont Hotel. By air, it involved changing planes in New York City and flying into Philadelphia, followed by a lengthy limousine ride. Flying myself with driving to the airport, getting the plane ready, filing a flight plan, and, in Wilmington Airport, calling a limousine from town to get

Figure 25 *Brent on his first day of school in 1964.*

to work would also not have been possible in the morning, i.e., the excuse I need a private plane to save travel time was flawed. The distance range of 250–500 mi, where an hour or two could be saved, was too narrow. Shorter distances were more convenient by car. Beyond this distance, commercial flights were faster and more economical, not even considering the uncertainty caused by the weather. The consequence was, that I decided, rather to spend the saved resources on even better automobiles, as done during my first sabbatic in 1967/68, described on page 8-38–48.

The promotion to a full professor gave a big boost for my first return trip to Germany in May 1965. I was invited to lecture at the "Kunststofftagung in Lübeck," followed by seminars at the Universities of Mainz and Marburg, the German Kunststoff Institute in Darmstadt, and the Research Laboratory of the BASF in Ludwigshafen (see Appendix C). The whole family came along on this first trip back to Germany after 11 years. Many more were to follow. The meeting was close to the start of the school vacation and we promised to sent Caryn to school in Germany. Shortly before, Papa and Mama had moved from Brandenburg in the GDR to Horrem, near Cologne ("Köln") in the FRG (see page 7-15). Figure 26 illustrates all of us touring Köln and our new VW on a trip to the relatives which were located the FRG. (Yes, I was the driver of the car and had to fit into it, too.) For us, we could see how much further the "Westen" had recovered from the effects of the war.

Figure 26 *Left: Inspecting the Rhein in Cologne and right: On the road; Heidel's new VW "Käfer, 1965.*

Since Caryn had to go to school, Brent, wanted to go also. Papa accompanied them daily. Brent had not started first grade yet in the US and had to struggle, and for Caryn it also was not as easy as she was used to in Troy. Both saw, it was not a foible of their parents to speak German at home. There was a whole country where this was done. We ended our stay with a short trip by plane to Berlin to see friends and to look over the 'Wall,' as shown in Figure 27.

Figure 27 *A quick look across the Wall in Berlin.*

The next trip to Europe was late in August of the same year, 1965. This time I had to go by myself. As usual, I did not know if it was possible to get the expenses reimbursed out of any research grant. This meant, if I would not have taken the risk, I almost never could have gone abroad. On my return, however, there always was a way to cover the expenses. Often, the visited companies helped out. This second European trip started just after a Gordon Conference (see pages 8-16–17), i.e., the travel calender began to fill rapidly.

The first stop of this trip to Europe was for an invited lecture at the Polymer Institute of Hungary in Budapest. Just after starting the lecture, I was interrupted and asked: "We know, you are originally from Germany. Could you do us a favor and speak in German?" I could, but it was not easy to switch language when the preparation was done in another. I had to spend more time in the future to relearn in German what was studied only in English. But, the Hungarians were not experts in scientific German either, so it went well.

On the weekend, I was introduced to Hungarian history, the beauty of Budapest, and the special cuisine using loads of paprika. A final problem arose. My hosts took care of all meals, but to get the visa, I had to purchase food coupons in New York in US dollars. What to do? They were not reimbursable. I found the best restaurant near the Danube. They had great gipsy music and most guests looked too well dressed to be the general public. I found a friendly waiter who translated the enticing menu. I told him my problem with the coupons and offered him all, if he could serve me the best meal I ever had. He did, and also served outstanding wines. He confided to me that one either had to be a high government official or a western tourist to eat here. I barely made it back to my "Hotel Duna," but had, indeed, the best meal in a restaurant up to that time.

Figure 28 *Symposium opening ceremony in Prague.*

The trip continued to the IUPAC Macromolecular Symposium in Prague, Czechoslovakia (note the earlier IUPAC Symposium, described on pages 7-22–23, Montreal 1961). Figure 28 is a view of the opening ceremony. Overall, I attended 24 IUPAC Meetings (see Appendix C).

In Prague, I found out that the Polymer Division of the ACS had sponsored a first charter flight to IUPAC meetings including some free days of sightseeing. The trip was arranged by Dr. H. Z. Friedlander. I took this opportunity to get information for the next flight to the IUPAC Meeting in Tokyo and Kyoto in 1966. Heidel and I ultimately went there when our neighbor from Chicago (see pages 6-37–38) agreed to babysit Caryn and Brent in Troy.

From Prague, I continued via a weekend in London, to Aberdeen, Scotland, to attend the First International Conference in Thermal Analysis, ICTA (see page 4-21 for a meeting with a GDR friend at the meeting). At this ICTA a new society was founded, the International Confederation for Thermal Analysis (later changed to ICTAC by adding 'and Calorimetry'). The second ICTA was in Worcester, MA, 1968. I attended this, as well as most later ICTA(C) meetings (see Appendix C).

Research after 1965 was located in the new MRC building (see Figure 19). By now, there were 8–10 coworkers with me, mainly graduate students. In 1968 Dick Bopp joined us as a capable technician from the local industry. He had not finished his bachelor's degree. Over the years, he worked diligently to improve himself as an evening student and completed his BA, then worked with me to complete an MS degree in 1975, and took a job at the General Electric Research Laboratory, but worked with me part time, to finally receive a PhD from RPI in 1993 as an off-campus student. This was a long haul, but it certainly was worth it. We still occasionally cooperate on mutual research. Between 1969 and 2004, we published **11** papers together. Even more publications arose from the work of Dr. Umesh Gaur(**32**), who came in 1978 from India to complete his MS degree, then stayed to get his PhD with me and spent two years as a postdoctoral research associate until 1981. His main job was to shepherd the initial ATHAS Data Bank (Advanced Thermal Analysis System). The all-time champion, however, was Dr. Marek Pyda, who joined me in 1994 as a postdoctoral associate, was promoted to Research Director of ATHAS and Research Professor. Until 2006, when I closed the laboratory, he participated in **62** publications. Other *very productive students and postdoctoral associates* with over 20 publications were: Dr. S. Z. D Cheng (**35**, from 1986–1999); Dr. B. G. Sumpter (**32**, from 1989–1996); Dr. Y. Jin (**25**, from 1990–1999); Dr. J. Grebowicz (**24**, from 1981–2001), Dr. R. Androsch (**24**, from 1998–2007), Dr. A. Xenopoulos (**24**, from 1990–1996), Dr. A. Boller (**23**, from 1995–2000), Dr. J. Cheng (**22**, from 1991–1999), and Dr. W. Chen (**21**, from 1994–2000). For more details consult Appendix A.

By the end of 1965, the high pressure research, calorimetry, and microscopy were moved to separate, fittingly furnished rooms in the new MRC building of Figure 19. One other large room was used for two new directions of activity, accelerating at this time. They were our own synthesis and preliminary structure analysis of new materials to document different classes of molecules and structures. The vacuum rack for synthesis of extended-chain crystals of poly(tetrafluoroethylene) and later poly(p-xylylene) can be seen in its early stages in Figure 29. A discussion in my office with a new student, Bruce Prime, is shown in Figure 30. In his thesis work he was doing the second generation work on 'Equilibrium Melting of Homopolymers' by combining molar mass, calorimetry and structure data. For this structure work, the second accelerating activity, *we collected a basic set of research tools*. At that time we could acquire all for less than $50,000. It included, besides the DTA, dilatometry, and optical microscopy, new low and wide-angle X-ray diffraction with photographic Polaroid image recording, and electron microscopy. The small electron microscope, was acquired for $5,000. It is illustrated in Figure 31.

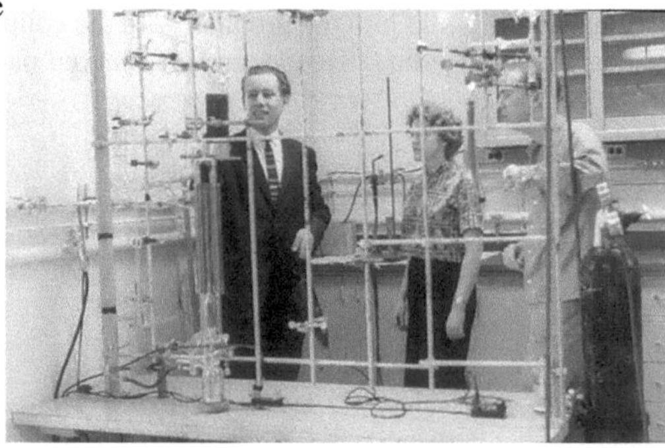

Figure 29 *Vacuum rack for polymer synthesis from the gas phase, instructing Ms. C. M. Cormier, 1965.*

Figure 30 *Research discussion with Bruce Prime, PhD student, 1965–1968, in my new office at MRC.*

From the experience gained from the series of easy-to-execute, semi-quantitative analyses, it was a simple matter to select the samples and conditions for quantitative, high-quality calorimetry, X-ray diffraction, and electron microscopy. Any one of these latter measurements required state-of-the-art equipment, needing investments of over $100,000 each. Such electron microscopy and X-ray diffraction equipment was available for the research groups specializing in this work at RPI and was set up in common access facilities. When the need arose, we cooperated with the professor-in-charge to interpret the more precise crystal structures and morphologies. Later, a much further broadening of the characterization of our samples was possible after joining Oak Ridge National Laboratory in 1988 (see Chapter 9).

Quantitative calorimetry became possible in our laboratory after we purchased one of the first Perkin Elmer differential scanning calorimeters, DSCs, in 1964 (see Appendix A, publication #38). For the high-cost adiabatic calorimetry at low temperature, we first relied on literature data. After 1990 we cooperated with Professor B. V. Lebedev of the Institute of Chemistry, University of Nizhnii Novgorod, Russia.

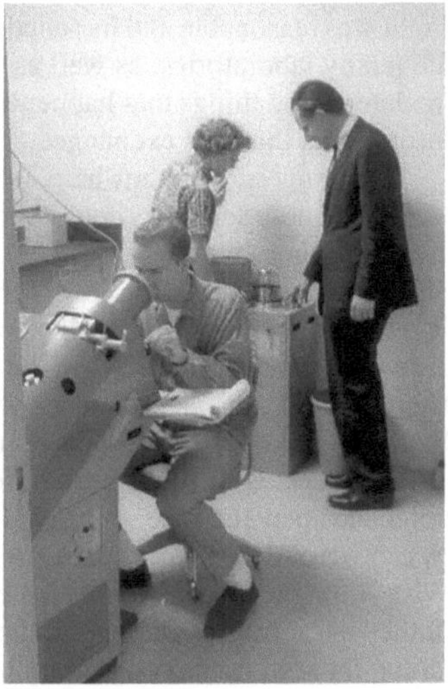

Figure 31 *JEOL JEM 30 Electron microscope with 5 nm resolution.*

To conclude the discussion of the semi-quantitative analyses, I would like to stress that these first-round data were of great importance. We could check a large volume of samples with this broader range of techniques. This usually cannot be accomplished with elaborate, high-precision instrumentation. Naturally, the semi-quantitative work is of limited value, if not supported strategically by quantitative evaluations. Few laboratories have the chance to do such two-tier analysis, others limit their work to the low resolution analysis with qualitative or semi-quantitative findings. Without the high precision follow-up, however, such data are of lesser value for the scientific literature.

On a much different note, I can report that the electron microscope in Figure 31 came in a large, well-built, wooden crate that benefitted Caryn and Brent. Figure 32 shows that it was simple for me to fashion a roof, an open window and a door to convert the crate into a hideaway in our backyard. It served them for a good number of years, until both outgrew it. Perhaps this caused them to also be eager to work later in my laboratory when they were students at RPI. We even published papers together, as is described below on the pages 8-79–81.

Figure 32 *Caryn and Brent in their hideaway, the electron microscope crate in our back woods in 1965.*

While I have no intention to comment on all domestic and foreign travels listed in Appendix C, some were so important to us, that I have to give a brief description. The IUPAC Meeting in Japan, 1966 was the first experience of the Orient for us, and can be counted as the first of what I like to call our '*three* trips of a lifetime.' As mentioned on page 8-23, it was possible for Heidel to come along because we found a babysitter to whom we could trust our children and house. Furthermore, the cost of the ACS charter flight was reasonable and its schedule allowed a generous amount of time for visiting company and university laboratories, as well as doing some site-seeing. There is not enough space to relate all the interesting things that happened on this trip—in addition to the enormous amount of scientific information that was exchanged. A few more memorable moments of the trip are given next.

To my surprise, my help was often needed after I once chaired a discussion session in Tokyo. Many Japanese scientists could 'read' their lecture well. No problem for anyone. A Japanese in the audience, then, would ask a question in English. No answer, even after repeating the question. Since I had learned during my own period of learning English to guess what was said from only a few recognizable words, I helped out by repeating the answer in simple English, but taking care to pronounce every letter. Now the speaker understood, and answered, also in atrocious English. Helping out again, since I understood his prior lecture, I could easily repeat the answer in distinct English. The questioner understands. Everyone laughed, and I had a job as a 'translator.'

The charter flight started from New York, Kennedy Airport, with a delay of 5 h which was not made up by the time we arrived at Haneda, at that time Tokyo's airport. We were overwhelmed by crowds and the heavy traffic. With the airport limousine we made it easily to the railroad station. With slight problems only, we bought tickets to Kyoto, our first stop, and found the assigned seats in the fastest train in the world, the Shinkansen,[1] inaugurated in 1964. The train was fast, indeed. Going through tunnels, the sharp tunnel boom was felt, rattling interior doors and passengers.

We saw the difficulties mount when approaching Kyoto. First, it was not easy in 1966 to even know at what station one was at. All names were written in indecipherable Japanese Kanji. And then, at the station, how to get to the Okazaki Hotel? We knew Japanese taxi drivers do not understand spoken or written English, and experienced later that if they could not understand or know where it was you wanted to go, they just left. It appeared rude, but proved the best solution.

In Kyoto, the train arrived on time, so that we easily could get out at the right station. Looking up and down the platform, to our surprise, we saw a familiar face, my recent student Tamio Arakawa (see page 7-6). All problems were solved, and we noted that he also was happy having found us. How did he know when we would arrive? He did not. He only knew when the plane was supposed to arrive in Tokyo. He then assumed, we would immediately take the Shinkansen to Kyoto, calculated the arrival time and waited for the five hours the plane was delayed. Several trains had arrived without us, and Kyoto station had two exits between which he had to run to scan all passengers getting off. Fortunately, standing one head above all Japanese, I was easy to spot.

Arriving by taxi at the hotel we learned two things. Taxi-drivers and waiters were not tipped. When the driver did not have the minor amount of change, we did not wait, but asked Tamio to tell him, not to bother. Checking into our mini-suite of two rooms, we could wear our shoes in the Western-style room, but in the Japanese-style bedroom with tatami mats, no shoes! Shoes were also not permitted in Japanese houses, temples, laboratories, and even lecture halls. For visitors, slippers were always kept, but for my quadruple-E-width feet none fit (see page 1-18), so that I often lost my

---

[1] Shinkansen means nothing else but 'New Trunk Line.' It uses standard gauge rails instead of the narrower gauge of the old lines and has many tunnels and viaducts to straighten the track. Its average speed was 100 mph (maximum speed of 130 and a record of 277 mph). This compares to the California Zephyr's 60 mph and maximum of 85 mph (see page 6-3). Today's 900 series of the Shinkansen has top speeds above 200 mph and its bulbous streamlining reduces the tunnel boom.

slippers to the general amusement of host and audience. After we had settled in the hotel, the surprise came, the taxi driver located our room and brought the minimal change!
With Figures 33A–F, chosen from 250 slides of my travel archive, our experiences are illustrated:

Figure 33A *Heidel, exploring the side streets in Kyoto. The main streets are as modern and busy with automobiles as in any big city in the world.*

B *The Arakawas, Tamio, Katsu and Masahiko are hosting a tea ceremony in one of the temples.*

C *Typical nature path with lantern in a garden.*

D *The most beautiful bridge in the Shūgaku-in.*

E *In Professor Takayanagi's home in Fukuoka, after dinner, when planning the next day travel.*

F *Kobayashi's Meiko girl party with us visitors.*

The first week we stayed in Kyoto. It was great fun to branch off the main streets and walk through the quaint, narrow streets (Figure 33A). Next came a full program of seeing the sights in Kyoto. It started with one of several tea ceremonies arranged for us. The one in Figure 33B was arranged by Tamio. From our pained expressions, one can see that we had to suffer being on our knees. Still, I liked the tea served, and from this day on, preferred Japanese tea to others. This liking was later transferred to Chinese green tea which is my every morning delight. I am always kept in good supply by my Chinese students. The tea ceremony was fascinating. The tea bowls matched our feel of beauty, so that we brought two of them home, along with a whole set of lacquerware and a couple of netsukes, the small ivory carving used to hold a cord to attach articles to the sash of a kimono. These antique netsukes started my collection, just that by now the better ones have become frightfully expensive. What was then few dollars in the hotel lobby, can now be $1,000 in antique shops in the big cities in the US, and even more so in Japan.

The gardens of Kyoto kept us busy for a number of days. Figure 33C relays their intimate, close-to-nature feeling. Soon after returning home, I bought a Japanese stone sculpted lantern as in Figure 33C which over the years has acquired the needed signs of age, an overgrowth of moss and lichen. The bridge in the garden of the 'Shūgaku-in Imperial Villa,' Figure 33D, fits masterfully into the landscape. I photographed it again in 1996 as Figure 59, Chapter 9. I have seen nothing nicer.

A short excursion went to Osaka to visit Professor H. Tadokoro at the University, one of the foremost experts in polymer X-ray crystallography and infrared analysis. Another was to the Teijin Research Laboratory, where Tamio was employed. Here, the research was organized along the path from raw material to the finished textile, all in one huge, many-storeyed building. The General Manager, as a special recognition, invited Tamio and family with us to his home for a tea ceremony. This was probably the first recognition of Tamio by the company (see page 7-6).

In the manager's house we admired not so much the extremely expensive museum-quality tea bowls we drank out of, but the completely undisciplined behavior of Masahiko, Tamio's son (see Figure 33B). There was great danger that the beautiful home would be wrecked, but none of the adults corrected the unruly child. A surprising observation, explained later by Tamio. Before going to school and being able to comprehend the 'honor and reputation' of their family, children have an unrestrained childhood. Even more so than in the US. As soon as these children go to school, this changes drastically. Their responsibility toward the family (and state) was (at that time) taught. Once it was learned and accepted, the child and later adult, lived accordingly. The big crowds of (uniformed) school children we saw at the tourist attractions in Nikko and Nara behaved perfectly.

My second Japanese PhD student in Troy, Mr. Shigeo Kubo (1968–1971), was an undergraduate of Kyoto University. Professor Kobayashi of the same university (see page 8-30) visited us in Troy and we had invited him for dinner. I picked him up at the downtown motel at 5:00 p.m. At about 8:00 p.m., Mr. Kubo telephoned. He was waiting at the motel. He was asked by Professor Kobayashi, who thought he would be back at 7:00 p.m. to discuss Shigeo's research. I passed the phone. A short discussion in Japanese followed. At 10:00 p.m., another call. Shigeo was still waiting. Now Professor Kobayashi was rather abrupt. Our dinner and later discussion, was rather animated, and when I delivered Professor Kobayashi close to 1:00 a.m., Shigeo was still waiting. It was his *duty*, thus the harsh comment of Professor Kobayashi.

Next, some comments about Japanese food. It was outstanding, and we quickly learned how to handle chopsticks. The raw fish we appreciated, and even today we pay a weekly visit to our local Japanese restaurant for a miso soup, sashimi, and sushi. But, there was also *live* fish. When visiting Professor Takayanagi of Kyushu University, he had a surprise for us. For lunch he had reserved a room in a very fine fish restaurant. The food was served with sufficient ceremony. It was a huge plate with a large living lobster in the center, surrounded by sashimi and an outer ring of moving, live shrimp. The lobster had its tail shell removed and the meat cut in small cubes, for us to pick up with chop sticks and eat while the lobster was still moving. It took a little bit of "gute Kinderstube" (good upbringing) to overcome the initial shock. Surprisingly, it tasted very good, better than the dead, cooked 'live lobster from Maine.' It got somewhat more difficult to eat the live shrimp. You had to peel the shell and eat quickly so that the shrimp neither died nor bit your lip. At home, I prefer sweet (raw) shrimp to live ones.

All would have been an enjoyable learning experience if we were not trapped into a well planed game Professor Takayanagi played with many of his foreign guests. As soon as the fish was served, his students got up with their cameras and were hoping to catch a terrified look on our faces. To the defense of Professor Takayanagi, I must add that he had ordered a full Western-style meal in case we did not pass his test. We were told most of his US visitors did not.

When he visited us in Troy, some years later, we had a chance to pay him back. We had found out on this visit that Japanese do not like raw meat, not even a rare steak. Heidel served raw beef (tartare, "Schabefleisch") as appetizer. Naturally, he behaved as we did. He ate, and seemed to like it. His comment was 'interesting taste," but he enquired about the details of the preparation.

Some years later he, in turn, could repay us. His students had arranged a symposium on the occasion of his having received the Ford Prize of the American Physical Society (APS) in 1983. Since Heidel and I were accidentally in Fukuoka, traveling on a Fellowship of the Japanese Society for the Promotion of Science (see page 8-97–99), I was invited to contribute a lecture. As a great honor, Heidel and I were invited to the dinner with all of his present and former students. When we arrived at the restaurant, it had as its main attraction a huge basin full of lively squid. Heidel looked at me and said: "He doesn't dare, does he?" I knew he would. After we all were seated (on the floor) the squid was served on a big platter in form of freshly cut rings with tentacles arranged around the rim, still moving furiously. A horrible sight. Again, the "gute Kinderstube" required that we had at least to take a little taste. Surprise, surprise—it tasted much better than the rubbery stuff one gets in most Japanese restaurants. After half an hour, the spectacle was over, and the good taste was disappearing. The squid did not move any more. The still half-full plate was then removed and after a little while brought back, but now cooked, and again, it tasted much better than what we knew as calamari. After the meal, we were graciously excused and brought back to our hotel, so that the all-male student-professor group could have their traditional Japanese celebration.

Figure 33E was taken the evening after the live seafood lunch in the house of Professor Takayanagi (seated on the left). The dinner Mrs. Takayanagi served was much less 'exciting', but still much to our liking. This get-together was to plan our next day's touring of Mount Kagami which overlooks the coastline of Kyushu (see Figure 45, below). Our host and driver for this trip was Miss Sato, daughter of a friend of the family (seated next to Heidel). She was a very modern, highly-educated young girl, just returned from her study in the US. Now she was waiting for her parents to pick a husband for her!

The role of women was changing, but women still had no part in their husband's professional life. Mrs. Takayanagi, for example, was cooking and serving the food on this evening, but she did not join us to eat, she had her meal in the kitchen. In this respect, Heidel taught our hosts that this does not have to be so. When Heidel came along on the trip in 1983, she asked if the wives could not join a large reception held for Professor Takayanagi, so that Heidel did not have to be the only woman in the room. Mrs. Takayanagi obliged, but hardly any of the other's wives did, and Mrs. Takayanagi obviously felt out of place and tried to help the serving staff (see also page 8-7).

Back in Kyoto, we met Professor Kobayashi from Kyoto University. He had just completed building his new, high-voltage electron microscope, valuable for analyzing polymer crystals and biological samples imparting only minimal decomposition. He also was one of the first observers of folded-chain single crystals of linear macromolecules. Figure 33F shows a traditional dinner for which he had hired Meiko girls to have mixed company. As a break with tradition, he brought his wife (left) along, to show how modern he was, and to teach Heidel the value of Eastern etiquette. He did not succeed on this and other occasions, although I must say that the service from the Meiko girl on my left was faster than from the other side on which Heidel was sitting (Heidel took the picture, sitting between Professor Fische from the University of mainz, Germany, and myself).

During the larger gatherings catered buffet-style, I would have starved, would I not have been accompanied by Heidel. I was always surrounded by younger colleagues and also students who wanted to try their English on a 'native' speaker. With all this attention, I often did not make it to the buffet. Heidel, seeing the problem, made sure that I was not neglected. This quickly became her permanent job in similar situations at home and world wide.

More could be said about the gardens, temples, palaces, and museums which we visited during this and later travels in Japan (see also the pages 8-97–99, 9-55–57 and 10-9). It may suffice to mention that the historic centers of Kyoto, Uji, Otsu, Nara, and Nikko, all visited by us, were later declared UNESCO World Heritage Sites (1994 to 1999).

Other companies and university laboratories were visited on this trip, so that I had a broad, impressive view of Japanese research. Of central importance was naturally the IUPAC Symposium. New conference centers were built in Tokyo and Kyoto and shown off by the split-location meeting. Figure 34A shows the Kyoto site. Figures 34B and C were taken with Professors A. Kovacs from the NCR (Strasbourg, France) and Professor A. Keller (University of Bristol, England), both émigrés from Hungary. Some of the other European researchers in the field were from Germany [Professors Kern, Fischer, and Meyerhoff (University of Mainz); Cantow (University of Freiburg); and Müller (University of Marburg)] and Czechoslovakia [Professor Wichterle (University of Prague)]. Overall, I would estimate that <u>*more than 50% of the top academics in the field were attending, and most of all, I, as the youngest, got to know them all*</u> and was well accepted.

Figure 34A *The Conference Center in Kyoto, called the Palace of Nations. IUPAC, 1966.* **B** *With Prof. Kovacs.* **C** *With Prof. A. Keller.*

Figure 35 *Route to the IUPAC Meeting in Budapest, 1969; (- - -) driven by rental car.*

Of the later overseas conferences, *two more IUPAC Macromolecular Symposia I attended were served by ACS charter flights*. The first went from New York via Amsterdam and Vienna to the 1969 Meeting in Budapest, Hungary. Heidel could again travel with me, while Caryn and Brent stayed with the Haags in New Jersey. The second charter flight, I had to take alone. It went to the 1974 Meeting in Rio de Janeiro, Brazil.

The flight to the *Meeting in Budapest* we exited in Amsterdam. Our separate itinerary is depicted in Figure 35. Figure 36A shows the arrival in Amsterdam on August 10, 1969. A rental car was picked up at the airport. It was French, with a Dutch license. When asked, based on this 'dual' nationality of our car about our origin and destination, we had to admit that we were born in Germany, but US citizens, traveling to Hungary. This even surprised Europeans.

There were four goals for our trip: To visit relatives and friends, illustrated by Figures 36B and C; to do some sightseeing since we had traveled little when living in Germany before (see Chapters 1–5); to lecture at a number of industrial laboratories and universities (see Appendix C, page C-10–11); and finally, to attend the IUPAC Meeting. Mama and Papa, who had by then moved into an apartment in Bergheim, near Cologne, are shown in Figure 36B with the family of Heidel's Onkel Hubert (see page 11-3). This stop was followed by travel through the FRG and Austria with an unforgettable, sleepless night in the "Gasthof Krimmlerfälle" (Hotel Krimml Waterfalls). We were tired and had another long day of travel ahead of us, but the songs of a "Heimatabend" kept us awake until morning. In Budapest, we had the usual clandestine meeting with Hans and Inge Gottschalk, shown in Figure 36C with their Hungarian friend Mr. Takacz.

Figure 36A *Amsterdam, 1969.* **B** *Visiting family in Cologne.* **C** *The meeting with the Gottschalks.*

How could we meet clandestinely in Budapest? Simple, since our name was on the Gottschalks' approved list of receiving mail from a western country (see pages 7-7–8), we had written a long time before the actual date that we planned to attend the IUPAC meeting in Hungary at a given hotel. A few months later, Inge would write that they had a vacation planned for this year in Budapest to visit their friends, the Takaczs. No date or details.

On this meeting, we saw again how right our leaving was. Walking on the Margaret Island, we noted that whenever we touched on a political problem, Hans would turn around and look for possible informants! Once, at another IUPAC Meeting, he even got out on the balcony of our room, to check for signs of listening devices. Such behavior always brought back the terrible times until 1953 which we had so quickly been able to put behind us. Still, with these clandestine meetings we kept in-touch and found out in what ways we could help, as described on page 4-28.

The IUPAC Meeting went well. At such meetings in communist countries, we also could contact colleagues from the East in addition to the ones from the Free World. The eastern European meetings were heavily subsidized by the communist Governments to show off their supposedly high standard of living. I remember a lavish conference dinner in Budapest were fancy food was offered never seen in East Germany. A first-class Gipsy Orchestra played and at the end was forced to play encore after encore with never decreasing applause.

The last day we spent sightseeing in Vienna, as shown in Figure 37. On the arrival in New York, we had to collect Caryn and Brent from the Haags in Trenton, NJ, illustrated with Figure 38. The Haags repaid us for our babysitting their children in Chicago when Werner and Hanna were traveling to inspect the job and living conditions offered by ExxonMobil in 1958 (see page 7-12).

The final trip by ACS charter flight went to Rio de Janeiro in 1974, Brazil. On this flight, in July/August, I was by myself. It was a specially well designed charter flight with an eye on tourist attractions. The first stop was in Brasilia, the long-planned "Distrito Federal" in the center of the country. Its initial construction was completed from 1956–60, designed by the famous internationally recognized Brasilian architect Oscar Niemeyer. Now, the city has over 2.5 million inhabitants. It is the only city built in the 20th century declared a UNESCO World Heritage Site (1987). Figure 39 is a picture of the buildings of the National Congress at the head of the 'Monumental Axis.' The left dome is the Assembly Hall of the Senate, the right dish, the Chamber of Deputies. Between them, are two office towers, and to the right is the huge flag pole with the national flag of Brasil, illustrated separately in Figure 39B, with the motto 'Order and Progress.'

Figure 37 *Heidel, sightseeing in the formal garden behind the Schloß Schönbrunn in Wien.*

Figure 38 *Ready to pick up Caryn and Brent from the Haags, were they enjoyed (?) the big family.*

Figure 39A *Center of the Monumental Axis in Brasilia.* **B** *Flag: "Ordem E Progresso," 1974.*

The IUPAC meeting was in a Holiday Inn near the Avenida Niemeyer, the coastal road out of Rio de Janeiro. As usual, the conference was outstanding with its opportunities to learn from others, and to exchange new ideas. By now, I knew many of the top scientists of the world from prior meetings and visits during my travels.

Figure 40 is a snapshot of Copacabana Beach, the famous, 3 mi long beach in Rio. When I was there, the beach was not overrun, but beautiful to see with its bordering Portugese pavement. In contrast to life in Rio, the vibrant metropolis of Brazil, one must look at Figure 41 which illustrates life at an open-air market in a small city, high in the Peruvian Andes. We visited there only one week later. What a difference!

Although I had little chance to explore the city of Rio, the merchants were well informed about our presence in the suburb. They offered a free round trip by taxi for anyone who would visit their show room. The most interesting of the jewelers was H. Stern.[1] I accepted his offer of a free ride, not only to check out the store, but also to have a walk through the city. Besides the sight recorded in Figure 40, I also admired a unique, two-door, sporty fastback Volkswagen, named the

Figure 40 *Copacabana Beach in Rio de Janeiro.*

Figure 41 *Open air market, Chincheros, Peru.*

[1] Hans Stern, a German emigrant. In Rio de Janeiro he founded in 1945 a small gemstone trading firm. By now the name H. Stern is synonymous with 'beauty, style, and good taste in jewelry.' His stores are found worldwide. None of the major airports is without a store of his.

SP2 (acronym for São Paulo). It was built only in small numbers in Brazil from 1972–1976 and never exported to the US. Back in Troy, the VW dealer had never heard about this car.

Two other interesting things happened in Rio. In the store of H. Stern, my eye fell on a amethyst bracelet which I thought might be of interest to Heidel, but I had no check or credit card along, only enough cash to get me safely home. The sales lady told me to appreciate the beauty of the many stones which were all absolutely identical in shade and depth of color, supposedly by being cut from a single, large, raw stone. I felt this was not so, three of the 16 stones were different to my color-blind eyes which are very acute in seeing intensities (see page 1-18). The sales lady called the manager, the manager called the gem cutter, who agreed with me and got the records for the bracelet. Indeed, the three stones I had identified were of a different find. He offered to replace these stones within a day. By then, I really liked to buy the bracelet and a matching ring, but I had no money. Now the manager got to work: "We take a check." My answer: "I rarely write checks, so I don't have the check book along." Next try: "What is the address of your bank, we may be able to make the needed arrangements?" I did not know. Finally, he agreed to an IOU (I owe you) and a guarantee that if for any reason Heidel would not be delighted, their New York City store would take the bracelet and refund 100% of the cost (but none of the import duty which I did not pay). Next day, the corrected bracelet was delivered to the hotel, I wrote a note to Heidel to make sure our account had enough money in it and the bank would accept the IOU. When Heidel saw the bracelet, she liked it, but would not wear it. The IOU had not yet been accepted by our bank! Later, we got a bill from H. Stern in New York with a request to send a proper check, which we gladly did.

The second interesting observation occurred when the organizers of the meeting offered a free visit to horse racing at the "Hipódromo da Gávea," the Jockey Club Brasileiro in Rio de Janeiro. After the first few horse races, all looked alike to us. The members of the conference slowly drifted into discussing science. Then, as a special attraction, more like an incentive to increase our interest in betting. One of the races was declared the 'IUPAC Race.' Almost nobody bet on it. I am sure, a meeting of scientists will never be invited to the horse races in Rio again.

After the meeting, we had a two-day stopover in Lima with visits to the historic center which is a UNESCO World Heritage Site with its Torre Tagle Palace, the Cathedral with Pizarro's tomb, and the Museum of Archeology and Anthropology with its exquisite collection of Mochica and Inca artifacts of pottery and gold. Another day was spent at the pre-Inca Sun Pyramid at Pachacamac in the coastal desert. Then *the charter flight went on to Cuzco with its great Inca buildings*. Later, in 1983, the city became a UNESCO World heritage site. Figure 42 shows an Inca base, erected without mortar, topped with a colonial building. Only the base is safe from the frequent earth quakes. Neither the two conferees nor the two Peruvians in the picture seem to be interested in the close-fitting boulders, cut without metal tools!

Figure 42 *Street scene in Cuzco.*

The next day was scheduled to involve a visit of an open air market, but a bridge had collapsed during earlier rains. Not to miss this interesting view, taxis were hired and we drove high into the mountains to see the market of Chincheros, shown in Figure 41. The food items offered, were not appealing, and the textiles I was interested in, I had already bought before. After looking around for a while, I discovered a table with a few old, colonial silver coins. By that time, I had begun to collect Mexican colonial coins (see page 8-49). Here an interesting old, dime-sized Peruvian silver coin was offered. After a long period of haggling, I was shown a print-out of the prices in New York City in a coin publication, dated the week before. The dealer said, he would not go lower than that. Even in this remote area, he knew what his coins were worth on the world market. I bought it. The coin was one grade better than the price asked. After this interesting experience, I looked around and saw a few other tables with upscale goods which the natives would probably not buy. It seemed our coming was announced, and these dealers had arrived just ahead of us from Cuzco. On the way back, we stopped at the Sacsayhuamán Inca fortress overlooking the city. Figure 43 is a photo I took of the unbelievable craftsmanship it displays.

Obviously all sightseeing in Cuzco is only the prelude to the train ride to Aguas Calientes in the Urubamba River Valley, followed by a precarious bus ride up to Machu Picchu, the lost city of the Incas. It was discovered only early in the 20th century. The countryside can be judged from Figure 44A. While being at the altitude of about 10,000 feet in Cuzco, and 2–3,000 feet higher in Chincheros, the coca tea served at the hotel did not fully alleviate a slowly building headache of mine. Also, trying to run up the steps to the cathedral in Cuzco, I found I had to sit down for a moment, not to faint. As the train got into the Urubamba valley, all this altitude sickness disappeared and I had a nice time viewing the ruins of Machu Picchu.

Figure 44B depicts the jewel of Machu Picchu, the Sun Temple. It is by now closed to inspection by visitors to preserve the intricate, not fully understood interior structure. Some 33 years later, I could fulfill a wish created on the first visit, to return and show the wonders to Heidel (see pages 10-14 and 15 and Figures 18 and 19 of Chapter 10).

Figure 43 *The Inca wall at Sacsayhuamán fortress.*

Figure 44A *In the valley, one can (barely) see the Urubamba River (it flows to the Amazon).* **B** *The Temple of the Sun, presumably used for astronomical observations. As seen in 1974.*

Let me return to 1965, before the IUPAC Meeting in Japan (pages 8-25–30). At that time I was *taking stock of my situation at RPI*. The pattern of research, teaching, travel, and family life was set. The questions to be asked were as follows: Should I continue on along the same path? How could I broaden the scope of research? What new ideas could be found for teaching? In what way could travel and family life be combined for maximum effectiveness? The answers seemed to be that the research, funded for 6–12 students, was effective, both at Cornell and RPI. The path chosen seemed to work, but needed to be focused to a new, more specific direction.[1] It had led to the successful exploration of the three states of macromolecules: Glasses, equilibrium crystals, and non-equilibrium crystals. An initial inroad was made into the theory describing the states. In the field of teaching, a difficulty had developed to instruct my students in our basic research interest, the thermal analysis. They had learned about the traditional thermodynamics, but this deals with equilibrium only. For thermal analysis, one needs to have knowledge of equilibrium *and* non-equilibrium, irreversible thermodynamics. A new lecture course needed to be developed.

The experience gained in travel was that together with Heidel we made a more effective team and achieved more than I could do alone. Her special influence was not only the greater enjoyment we had by our traveling together, as can be seen from Figure 45, and the increased comfort for me, allowing more concentration on the science I had to absorb. Heidel also took it upon herself to organize the travel and finding the most opportune sightseeing events. Finally, she was the one who had much more talent than I to establish and maintain personal connections with the other attendees during the receptions, lunches, dinners, and excursions. It slowly developed, that all non-scientific letters were handled by Heidel, including my friends from high school whom she all knew from having been to the same school. In this way, we made a family affair out of a major part of my job as professor of chemistry.

Figure 45 *On top of Mount Kagami near Fukuoka. The picture was taken by Ms. Sato, (see page 8-29).*

This also meant, Caryn and Brent had to be fitted into this frame so that they could have some benefit from the travel. Whenever possible, they were taken on the trips, for fun and to learn about the rest of the world. The family involvement needed to be paid out of our private budget, but particularly in retrospect, it certainly was worth it. By then, however, Caryn and Brent were of school age, and soon, it would not be advantageous to compromise their formal schooling

---

[1] When later asked by my students, I would always advise, that initially, it makes little difference in what direction you start research. Then, you must broaden as much as possible. Once you reach such a breadth that you cannot follow all leads and it becomes difficult to oversee all results, you must focus, based on the prior results. You should choose the direction most promising, and most advantageous for your abilities. After a while, again, the new direction will have broadened and then, a new focus must be found based on the prior work, etc. In this fashion, you will reach a much larger horizon than if one never deviates from the initial direction.

with undue interruptions due to travel. Our detailed planning, thus, was limited in its timing. This led to an early plan for **my first sabbatic[1] leave in Germany, 1967/68**.

The seventh year of my going through the succession of professorial ranks was 1967/68. Considering also the schooling of Caryn and Brent, this was an ideal time for a sabbatic, to make the adjustments discussed on page 8-36. The problem was, RPI did not allow an official sabbatic. There was only a 'leave policy.' Such leaves could be granted every seventh year with external support and were to be approved only for professors with a minimum of six years of service to RPI. This summary of the policy is taken from the Official Bulletin, Vol. 66, No. 39, May 11, 1966. A sabbatic policy with Institute support was introduced in 1972 (Vol. 72, No. 34). This led to the question: "*How can one get a sabbatic if the university does not allow one*?"

Even before the trip to Japan, I had started to circumvent Bulletin 39. First, on May 17, 1966, I wrote to Professor Stuart of the University of Mainz in Germany, asking for a possible sabbatic in his department for the fall term 1967/68. Professor Stuart was known to me from the first major treatise on polymer physics,[2] and from my visit in his department in 1965. This visit had reinforced my assessment that, at this time, Mainz was the top German university in macromolecular science, followed by the Universities of Freiburg and Marburg. The research topic was to center about my proposition: If a randomly coiled polymer molecule always folds on crystallization, it should be possible to make extended-chain crystals by simultaneous '*Crystallization During Polymerization*' and, thus, avoid the randomly coiled state. During the sabbatic, I hoped to discuss this question with as many scientists working on polymer crystals as possible (Appendix B, Part C), and go through the literature for any results which earlier were not recognized in their importance. In addition, I offered to contribute to teaching 'thermal analysis.' By the middle of June, I had the answer. All colleagues in Mainz would be delighted and they would try to find an official appointment as visiting professor that could cover the living expenses in Germany.

On June 21, I applied for a six-month leave with external support. At the same time, I applied for a Fellowship from the National Science Foundation (regrets, 12/2/66) and the John Simon Guggenheim Memorial Foundation (regrets 3/3/67). In the meantime, after discussions with Professor Fischer, who also was in Japan (see page 8-30), the arrangement was finalized with a letter from October 27, 1966 with an offer of a stipend of 9,000 DM-West (≈US $2,100). This first step was followed by innovative juggling of my research budgets and many phone calls to the supporting agencies to assure agreement. Finally, the following external salary support was managed: NASA (33⅓%), NSF (15%), ONR (18⅓%), and begrudgingly, RPI committed the missing 33⅓% from the ≈50% overhead the school collected from my research grants. It was a generous help from all. In the budget sheets I found in my notes about the sabbatic, the I reached the conclusion that the stipend from the University of Mainz would approximately cover the cost of the plane travel and the rent of the apartment, this meant with full RPI salary, we did break even.

---

[1] From the Greek "σαββατικος" (of the sabbath, hiatus, rest from work). In the Bible there is already a commandment to desist from working the fields in the seventh year. Today, sabbaticals are taken by professors, pastors, and some other professionals. Universities have specific sabbatic policies, usually allowing a sabbatic every 7th year, with ½ year at full pay, or a full year with ½ pay.

[2] H. A. Stuart, "Die Physik der Hochpolymeren." Vols. 1–4, Springer, Berlin, 1952–56.

The next arrangements were easier. To get permission to take Caryn and Brent out of school, we only had to promise that in Germany we would enrol them in a German school (6th and 3rd grade). The tickets were arranged in Troy by Copeland Travel for the whole family and a single, separate two-week round trip for me in December, to look after the graduate students, the empty house in Troy, and also satisfy my consulting obligations for six months at the DuPont Company. This called for a special round-trip ticket of no-less than 14 and no-more than 21 days.

A final question was the transportation in Germany. To take my Barracuda (see page 8-19 and Figure 23D) or another new American car was not practical. Transport to and from Germany would have to be arranged privately at great effort and cost. A German car could easily be bought in the US for delivery abroad and, after touring the country, shipment back to the US. The price was less than for a straight sale from the show room in the US. Since the shipping would take a few weeks, we decided to buy a German car delivered in Germany, and would ship it back early, to be delivered at Kennedy Airport to await our arrival. For the last travels, we would then buy a replacement VW in Germany for Heidel, and have it arrive in Troy three weeks later. This was complicated and, ultimately, it did not work. When we arrived back in New York on March 29, 1968, the longshoremen were on strike and our car was rocking on the ship in the harbor. It took a quick extra plane ticket to Troy and a wait for almost two weeks until I could pick up the car.

But what car to buy? Only three makes had service in the US: Porsche, Mercedes, and Volkswagen. The Porsche was my dream car ever since I had seen it at the "Automobilausstellung am Funkturm" in Berlin in May 1952. But by now, the whole family had to fit into the car, including Mama and Papa, who lived about 125 mi from Mainz. This excluded Porsche and also the VW which was barely sufficient in 1965 (see Figure 26). So we were 'forced' to invest in a Mercedes 250S (≈$5,000, at that time equal in price to a Cadillac in the US). After our arrival, we found that our apartment was too far from the University, the shopping area, and also the school for Caryn and Brent to easily manage with only one car. No real problem, we just went and bought the VW on the next day. Anyway, it was in the budget for later. Figure 46 displays Heidel's VW, on the left, and my 'big car' on the right after we had successfully brought them home in 1968.

All decisions about the sabbatic made, things returned to normal. One bigger assignment was still on the calendar for 1967, my first tryout to be *a tour speaker at local ACS Meetings*. The 17-day tour started January 10 in Raleigh, NC, with a rental car. The schedule was similar every day. At 7:00 p.m. dinner with the local ACS section, followed by the lecture and a discussion at 8:00 p.m. The menu also was surprisingly uniform, usually chicken, which I am not particularly fond of, unless it is our special barbeque (see Figure 23). But, I do not like to eat much before a lecture anyway. Back to the motel at about 10:30 p.m. Next morning, after an extensive breakfast to make up for the slim dinner, on the road again, for the next location. Distance permitting, there was time for sightseeing, and Sundays were free of lectures. The first nine lectures were in North and South Carolina and Virginia. Then I went to Florida for the last six, and

Figure 46 *Both our cars safe at home in Troy.*

ended the tour in Miami (see Appendix C). Florida was interesting state with its often flat and bare countryside, except for the orange groves and the beautiful gardens and beaches. Visiting some of the smaller colleges made me appreciate the outstanding education they offered.

At this time, the NASA Cape Canaveral rocket launching facility was gearing-up for the moon shot. I spent a Sunday there, inspecting buildings and hardware. It turned out to be a few days before the tragic fire in the Apollo 1 capsule, January 27, causing the first three astronaut fatalities.

Private travel in 1967 included a trip to Hastings, NE during the summer vacations to visit family, and for a trip to Yellowstone National Park with Vati's car. The Park is by now a UNESCO World heritage Site (since 1978). After seeing the spectacular mountains and geysers, we observed a number of bears near the road, and Figure 47 shows our encounter with the tamer wild donkeys.

Figure 47 *With Vati's car on an excursion to the Yellowstone Park, visiting with wild donkeys on the way back to Hastings in 1967.*

The sabbatic began with the Meeting of the German Physical Society in Garmisch in September 1967. On the 25th I left from Troy to go to Stuttgart, the home of Daimler Benz, to inspect the Mercedes assembly plant and to pick up my new 250S (Figure 46). After getting all instructions, on my question: "What do I do if anything goes wrong with the car when I am on the road?" I got the reassuring answer: "You bought a Mercedes, nothing will go wrong." Indeed, nothing did. Late in the evening, I arrived in Garmisch-Partenkirchen at the rim of the Alps. European meetings started at that time on Tuesdays, so that the Monday could be used for travel, not as in the US, where one was obliged to travel on a Sunday. The meeting was different, not only was the lecture published, the discussion was also. The questioner and lecturer would get a numbered sheet to write down question and answer as soon after the lecture as possible. When the galley proofs of lecture and discussion were ready, both questioner and lecturer could correct the discussion remarks, and if necessary, argue over the way things were said. This made for a very special issue of the "Kolloid Zeitschrift and Zeitschrift für Polymere" for attendees, as well as for others who could not attend (Appendix A, #58). In Figure 48 I document the continuation of the discussions on extended walks through the beautiful Alps, a pleasant diversion.

Figure 48 *Science discussions in Garmisch in 1967.*

After the meeting, I slowly drove to Frankfurt to pick up Mama and Papa who had come from Cologne to meet Heidel, Caryn, and Brent on their arrival. My car still had to be 'broken in.' It should initially not be driven above 40 mph. Next, the speed could, in stages, be increased to 80 mph. The instructions then said, that after 1,000 mi, one should run the car at maximum speed for at least half an hour to properly condition the motor. This time came a few weeks later. I found a new stretch of Autobahn and stepped on the gas. At about 120 mph, the speedometer stopped climbing. The car was beautifully steady and quiet, it felt like going 80 mph with my old 'Barracuda.' But, when I tried to brake, I found that the distance to get down to a manageable speed was too long. Even a gentle curve could hide an obstruction so that I would not be able to stop in time to avoid it. I always tried to be a good driver, not only avoiding accidents, but also trying to anticipate possible accidents caused by others. Going over 100 mph, was one of those accidents waiting to happen. When I bought the next Mercedes in 1975 ($9,900, compare to page 8-38), there was no break-in period anymore. This car, a 300D, had a five-cylinder diesel engine, running 90 mph at top speed, but it, and the faster 6-cylinder 300SD (bought in 1988, see page 9-10) fulfilled my early dreams in motors (see page 3-12). I still dream about the next improvement, a Diesel-hybrid Mercedes not only to increase efficiency in city traffic by shutting off the main engine, but also, to give an extra boost for peak acceleration on high-speed passing.

The year before, Professor Fischer had taken over Professor Stuart's position as institute director. He had rented a very comfortable apartment in Mainz-Kastell, across the Rhein River from the city of Mainz and the University. Our address was "Am Pionierübungsplatz 7" (the exercise yard for combat engineers). It was a short side-street going from the bank of the "Rhein" to the four-lane main street, the "Wiesbadener Straße." Figure 49A shows the view across the river toward Mainz. The Wiesbadener Straße led SE to the "Rheinbrücke" into Mainz, and NW to "Wiesbaden."

Figure 49B shows the front entrance into our apartment on the occasion of a school friend of Heidel's visiting from Berlin (Eva Lehmann, née Sternsdorf and husband). In the back of the house, from our living room, there was a patio with fountain and pool with gold fish, all set into a small garden. The owners of the house lived upstairs with an entrance from the side. A number of American military families were living in the area since the former German military barracks across the Wiesbadener Straße were occupied by American soldiers. Brent and Caryn, thus, could find children to play among the German and American families.

October 2 and 3 were used to get settled, and October 4 and 5 were filled with discussions at the University with six of the American and British conferees from the Garmisch Meeting. All aspects of polymer crystallization were covered with Drs. J. D. Hoffman, J. I. Lauritzen, Jr., A. Keller, P. H. Lindenmeyer, H. D. Noether, and H. D. Keith, all well-known experts in the field.

Figure 49A *The Rhein in the morning mist with the city of Mainz on the other side of the river.*

B *Am Pionierübungsplatz 7, bei Sieberts, 1967.*

The following, Monday work started in earnest. Most of my time was used to search the literature and starting to write the first review of the topic of the sabbatic: 'Crystallization During Polymerization' (see Appendix A, paper #61). It produced the new focus of our research for many years I was looking for, and is described in more detail in Part C of Appendix B. Later, after sufficient broadening had occurred again, it was used as the basis for the second sabbatic in Germany in 1986/87 which further defined and focussed our research direction (see page 8-36).

The discussions and lectures away from Mainz were also part of the sabbatic. They are listed in Appendix C on pages C-8–9. The teaching I could do during the sabbatic involved a lecture course with the heading "Differentialthermoanalyse von Polymeren." (DTA of Polymers). Fifteen "Vorlesungen" (lectures) were given to the research community at the University. Seventeen of the senior graduate students in physics and chemistry attended, and frequent visits by Professor Fischer and other faculty members enlivened the discussions.

Later, these lectures became the base of a course given at RPI to about 10 students and then led to the development of new teaching tools, as can be seen from the entries in Appendix A as follows: First, there was a book chapter (#53), followed by two editions of an audio course (#109 and 184), a second book chapter (#104), then computer courses (#201, 421, 464, and 560), and a succession of two books (#290 and 500) of which the last appeared in 2005. Several entries in encyclopedias and treatises rounded out the publication program of this new field of 'Differential Thermal Analysis of Polymers' (#141, 193, 276, 311, 410, 567, and 582). Also, there was an extended discussion with Professor Fischer in January to possibly write a multi-volume treatise, later published as 'Macromolecular Physics' (#95, 102, and 123 in Appendix A).

Despite the busy schedule, there was time during the weekends and holidays to discover the area around Mainz and Wiesbaden, as exemplified by two typical scenes along the "Rhein" in Figure 50, as well as comparing, for example, mediaeval dwellings in Germany with those in England, which I could see on my excursion into Great Britain (see below). The result is shown in Figures 51 and 52. Naturally there were also many new portions of the FRG we had not seen before and a somewhat nostalgic trip was to Berlin, to be commented on below. Even during meetings, my colleagues were often kind enough to make sure that we could see more than just hotels, meeting halls, and laboratories. All travel was done by the Mercedes, except for the excursion to Great Britain which was by plane. At the end of the sabbatic, our family car had logged 10,000 mi.

Figure 50A *Typical vineyard, close to the Rhein near the 1200 year old city of Oppenheim, about 15 mi to the south of our temporary home, Mainz.* **B** *The left bank of the Rhein with a view of Bacharach, and above the town, the Burg Staleck, ≈40 mi WNW from the city of Mainz.*

Figure 51 *Caryn, Heidel, and Brent explore the old city in Mainz with its half-timbered houses.*

Figure 52 *View from the excursion to England (a house with a thatched roof behind a flint wall).*

Being involved in the teaching of students at the University of Mainz allowed a look into what *changes had occurred since I was a student*. The Goethe University in Frankfurt was only 25 mi away. As poor students we were looking at subcompact cars and discussed how much more efficient our study could be with such a car. By now, a number of students did have automobiles. Particularly the simple French 2CV Citroën ("Ente," duck, built from 1949–1990) was well liked by the students because of its exceptional performance and economy. Unfortunately, these cars were not used to study more diligently, as fantasized on page 5-18. Also, the first Saturday morning in Mainz, when I wanted to go to my office, I found the door to the building locked. Since I did not yet have a key to the building, I waited. After about one hour, it dawned on me, that by now, Saturday was part of the weekend for German university students.

For my graduate students, I always made a simple calculation: If you miss all Saturdays, an average graduate study of four years with six-day work weeks will quickly turn into five years when working only five days each week. The monetary loss due to the extra year is not to be counted by the extra stipend needed, but by the one-year loss of professional income. Over your lifetime, this is to be taken when reaching maximum income. Gaining this one year can easily pay for the whole four-year graduate study. The same problem arises for a student taking off a year, or having to serve in the military, concluded also to be a bad practice on page 7-9 for other than monetary reasons.

During my lectures in Mainz, I found that the students always sat toward the back of the room, and the professors and assistants who attended my lectures assembled in front. My two-hour lectures were designed to allow for extensive question and answer periods. The students were very hesitant to ask questions with their teachers in front. Since I was used to the lively questioning in the recitation periods at RPI (see pages 8-13–14), after answering a number of questions from the front row, I made the comment that the lectures were mainly for the students, and deliberately marched to the back of the room to make sure that the students were not intimidated by having to get up and talk above the heads of their professors. It worked, and it was interesting to note that the student questions were directed more to understand the basics of the problems, and as such, more essential. The more advanced researchers in front, asked more about the influence of the new material on the old, more established thought. After a few lectures, questions came all across the room, and everyone enjoyed the discussions which did not lack a sprinkling of humor. One such comment: A student, is someone who does not yet know all (Greek "ἰδιώτης," idiot, layperson, in contrast to expert) and has the duty to ask many questions. A professor, in contrast, is an expert, should know many answers. And naturally, there are fewer stupid questions than answers.

Another observation I made in Germany was that the colleagues who were within a few years of my age were quite aggressive in the discussion after my lectures. Not that I minded, I like to look into all possible aspects of an incompletely understood problem and enjoy a spirited discussion. After it became clear that I was not interested to compete for a professorial appointment in Germany (see page 8-20) we all got along splendidly. One day, one of the senior professors at another university talked to me before my lecture about one of his assistants. I was not to be too critical in my discussion with the questions of this coworker. He might bring up his own description of polymer melting, somewhat different from mine. Several professors in the audience were members of his "Habilitationskomitee," and I might jeopardize his chances for his upcoming hearing.

An academic career in Germany is different from that in the US, as it also is in most European and Asian countries. After one's PhD, one has to work as an assistant to a professor (not an assistant professor, as in the US system). The professor dictates much of the research work to be done and assigns his teaching and supervision duties of students. He also decides whether to encourage the assistant to work towards a "Habilitation" which qualifies one to accept a professorial position. The "Habilitation" requires a written research document beyond the PhD thesis, the "Habilitationsschrift," to be defended in front of the "Habilitationskomitee." A successful "Habilitation" results in the "venia legendi," (Latin for permission for lecturing) or the "ius docendi" (the right of teaching for life) in a specific academic field at universities. This process takes approximately as long as the US system allows for getting tenure, but was less regulated, so that sometimes, it took longer than 10 years. Only in the 21st century has there been a major effort to reform the German system. In my opinion, what made the German system less desirable was not only the long time to complete the apprenticeship, but the restriction of the candidate in his or her creative abilities. While I was free to do my own, independent research starting with the PhD, this was not so in Germany, i.e., I had a substantial advantage in following my own ideas, direct my own research, and handling my own students, as described in Chapter 7. Now, I was in Germany on my first sabbatic, at a time most German PhDs of my age had not finished their "Habilitation."

Since most radically new ideas are usually developed early in life, it is of importance to be able to express and test these ideas early in one's career. This is a time when one is not fully ingrained in the existing picture of a science. One extreme case of living in the shadow of the professor, I saw with a Japanese postdoctoral fellow of mine. As assistant, he was chosen to be the successor to an only two-years older professor. This meant, with the strict Japanese rules of professorial retirement at 60, his truly independent research was limited to two years!

Having commented on some negatives in education, I have to relate an occurrence which dramatizes *the burden of the German bureaucracy*. Similar problems arose on the second sabbatic in Germany in 1986/87, and while efforts are being made to address the academic career hurdles, the bureaucracy seems not to have made much progress. By now, retelling this story, it is almost comical. On the first payday in Mainz, I was told, yes, your money is here, but you first need a permission to work in Germany. You get it at the "Stadtverwaltung" (city administration). It was late afternoon, and by the time I arrived there, they were just closing. I was told: "Yes you are at the right place to get a working permit for foreigners. Come back tomorrow at 8:00 a.m." Next morning, after waiting my turn, the official checked my passport and asked for my address, and then said, Mainz-Kastell is not in my jurisdiction. Mainz is in the province ("Land") of "Rheinland-Pfalz," Mainz-Kastell is in Hessen, and the proper office is in Wiesbaden, a 12 mi trip. Fortunately, I had a car. Arriving in Wiesbaden, and after the proper waiting, telling my wish to work as a 'foreigner' at the University of Mainz, it was found that my papers were incomplete.

First, I needed to register at the "Einwohnermeldeamt" (registration office at the local police station, see also page 7-12). This meant, back to Mainz-Kastell (9 mi). By then, it was noon. I asked to be registered. The answer was another question: "Where is your filled-in registration form?" I didn't have one. I first had to purchase the form at the "Schreibwarenladen" (stationary shop). "Where is the next shop?" I got directions. By the time I arrived there, it was closed for the usual, long midday break (1–3:00 p.m.). I went home for lunch, then bought the form, and made it back to the police station, but it was too late to go back to Wiesbaden.

Next morning I waited at the Wiesbaden office, hoping to spent the afternoon at the university, but it was not to be. Now both, Heidel and I needed a medical exam at the Mainz University Clinic. Arriving there in mid-morning meant adding our names to a long list of patients. We asked when will it be our turn? About 3:00 p.m. Can we leave in between? "No. If we call you and you are not here, you have to register new." We left anyway and returned, for safety's sake at 2:30 p.m. We were done by 5:00 p.m. Two days later could get my permit at the "Ausländeramt."

When I arrived again in Wiesbaden, there was a large crowd of Turks, the most common foreign worker in Germany at that time. I sat down in the waiting room. When the next was called, the official spotted me and said: "You are next." When I protested that I just had come, he said no, *you* are next. I wore a suit and tie, the others were in their work attire. In Germany it counted, to be well dressed.[1] I got my working permit without further problems, but when I presented it proudly at the cashier at the university, he said, you are not considered a 'worker' and do not need the working permit. The information you got last week was not correct! This example was not given to mean that there is no bureaucracy in the US, but I am sure, that on contacting the first office, the clerk would have written a detailed itinerary and might have pointed out the pitfalls for foreigners.

The standard of living in Germany had increased considerably. A commonly cited statistic suggests the real wages had increased by a factor of two since 1950. Most of the destruction from the war was removed and the acute shortage of housing overcome. We could see then, and also in the following visits in one to three-year intervals how one thing after the other was improved.[2] Most interesting was the concentration of the country on one of the 'finer things of life' after the other. It started with the "Freßwelle" (eating wave) in the late 1950s, then acquisition of fancy electronics and cameras, went on to high fashion, automobiles, to large scale restoration of old city centers, historical buildings and churches to their original appearance, to replacing the initial blacktopping of the streets in the old cities to the historically correct cobblestones (with parallel redirection of the traffic). Ultimately, by the end of the 20th century, Germany looked more beautiful than ever before.

---

[1] At that time, when flying Lufthansa, we would make sure that our ticket read *Dr. and Mrs.* Wunderlich and I wore suit and tie. When asking for better seats, or even an upgrade to business class on check-in, having a title and being well-dressed, considerably increased the success rate.

[2] Including my first trips back to Europe in 1965, by the year 2000 I had made about 30 round trips across the Atlantic (see Appendix C), each of them requiring 15–20 hours in the air, all together as long as a single round trip of the MS Gripsholm in 1954 (see pages 5-19–22).

A few other exciting happenings during the sabbatic need to be mentioned. Perhaps the most different entertainment from what we were used to around the area of Berlin was the "Fasching" (carnival) as it was celebrated in Mainz. In 1968, it was on February 25–27. In these days it was impossible to work, so I gave up and we joined the crowd, as is shown in Figure 53. The spirit of the festivities can be glimpsed from these photos. The floats of the main parade had themes ranging from humor to political satire. The title of the 1968 Mainz Carnival is seen on Figure 53A: "BABBLE NIT, KOMM MACH MIT," in local dialect this means 'don't blabber, come on, take part.' As the floats passed, sweets were thrown to the children, a special fun for Caryn and Brent.

The carnival organizations start preparing for "Fasching," on November 11, three months before. We attended only one of their "Karneval Sitzungen," with plenty of special songs created for the event, humorous poems and lectures, and naturally, the wine of the region. It was a most enjoyable evening. The police were, however, on the lookout to catch anyone who might get into his car with an excess of wine consumed. The drunk-driving laws were much stricter in Europe than at home, with some of the eastern countries having a zero tolerance.

Figure 53A *Head of a parade.* **B** *Crowds in front of the Dom.* **C** *Detail of a float in the parade.*

D *A satire of DeGaul as he leads the rest of Europe.* **E** *Brent and Caryn on their lunch break.*

Another few minutes of excitement were experienced during the first winter travel on the Autobahn. When winter approached, I looked for advice for snow tires. In the 1960s, studded snow tires had just become highly acclaimed. Each tire had some 100 steel spikes, and the Mercedes dealer recommended to have four of these mounted on a second set of wheels.[1] In light of the harsh winter in Troy, this is what I did. The first test was a big success. We were to drive from Mainz to Cologne for one of our visits to Mama and Papa. Over night, snow had fallen. A good test for the new tires. Traffic on the autobahn was light, and considering the snow, we drove about 60 mph (slow). Then came the hills. The road became icy. Still, the studs performed ideally, but one had to keep up the speed since it was not possible to accelerate up-hill because of underlying ice.

Then suddenly, just before we reached the crest of a somewhat higher hill, a policeman jumped into the street and frantically waved us to stop. This could not be done! Touching the brakes made the car slide sideways despite the snow tires. A little acceleration straightened us out. We were down to about 30 mph when we passed the policeman who surely noted our predicament. When we crested the hill, we saw the reason for his action. There were dozens of cars and trucks haphazardly parked at all angles down the road, into the valley, and up again on the next hill. To stop, my choices seemed to be to drive into the ditch or to run into one of the parked vehicles.

But no! I saw an unbroken zig-zag path though all the confusion. Trying to slow as much as possible without slipping, I gingerly turned the steering wheel to follow this path. The tires performed perfectly as long as they could be kept moving. Just before reaching the bottom of the valley, I had gained enough confidence and saw the uphill stretch clearly enough, that I now speeded up and just barely made it up the next hill without stalling. Reaching the crest, I could have halted, but by that time, I certainly did not want to stop anymore (or talk to a policeman).

Beyond, all was clear and we made it to Köln without problems. Overall, we had 90% luck, and 10% courage counted also. The next winter, back home in Troy, I did not have much opportunity to test the studded tires. Traffic was sufficiently dense that it stopped as soon as the worst driver and car could not proceed. Passing was rarely possible. Some years later, the spikes had to be pulled out of the tires, the city had forbidden their use to preserve the roads.

At the time of the Carnival in Mainz, we had just returned from *an excursion to Berlin* by automobile. Caryn and Brent had to go to school and were left under the watchful eyes of Mama and Papa, staying in Mainz. During this trip we had planned another courageous outing. I had accepted an invitation by the Academy of Science of the GDR to lecture at their Fiber Institute near Berlin, in the GDR proper. The lecture was arranged by Dr. C. Ruscher and I had asked him to investigate if I could get a guarantee for this visit. Our effort to make our exit from East Berlin as nonpolitical as possible, paid off (see page 5-13). After extensive checking, he assured a safe trip.

We had to pick up visa at the "Polizeipräsidium" (police headquarters) in the "Neue Königstraße" near the "Alexanderplatz." I had ridden the bus past this imposing, infamous gray building every day from 1949 to 1953. This building was already in use before WW II with the same notoriety. Surprisingly, the "Bienenkorb" (bee hive), a well-known café of ill repute, was

[1] In the posting http://www.consumersearch.com/www/automotive/snow-tires/review.html of March 2008, one can read: "Studded tires have traditionally performed better on ice than non-studded tires, but some new studless tires perform just as well in recent tests. Studded tires are noisier as well. ... Be aware, however, that many states and municipalities limit or ban the use of studded tires because they contribute to road damage. ... Get four matching winter tires."

nestled in a corner on the opposite side of the street and seemed to thrive. Being once in East Berlin, we took the time to drive to the old "Chemische Institut" in the "Hessische Straße" (see Figure 5, Chapter 4). The experience there was already related on pages 4-34–35.

Shopping for specialty items in East Berlin, we tried the showroom of the "Meissen Porzellanmanufaktur" in their prestigious location "Unter den Linden." We had heard from Dr. Glaser (see page 8-10), that there was a re-issue of the sculpture "schlafendes Bauernpaar," made by the famous sculptor Barlach.[1] When we walked into the showroom and inquired, we were told, yes it is available. "Could we take a look at it?" "Yes, I will show it to you." It was gorgeous, as can be seen from Figure 54. So after admiring it for a few minutes, we decided, yes, we would like to buy it (for 900 DM-Ost, see page 3-28). The saleslady made a somewhat sorry face and said: "I will put down your name, it will be about two to three years until you can pick it up." When we grinned and said, sorry, but by then we are not here anymore, she got the idea that we may not be the run-of-the-mill GDR citizen. She asked: "In what currency will you pay?" "In ours—US $," was the answer. "Then it is no problem." She removed the name of the person, who had waited for two years for his sculpture, wrapped it, and gave us the export papers. We hoped it was a high communist party official whose sculpture we took.

Figure 54 *Barlach's "Russisches Bauernpaar" Meissner Böttgersteinzeug, 1968 (11.5×16×11.5").[1]*

Early in the morning of February 20 we went to Teltow for the lecture, going through the American Checkpoint Charley in the Friedrichstraße. Instead of just driving through, as usual, I stopped, went into the little gatehouse, seen in Figure 23 of Chapter 4, and asked to be registered to go into the GDR to Teltow, to return late the next day. If we would not report back on time, they should start an inquiry. This done, I had to stop at the more imposing control point of the GDR with concrete barriers and heavy steel gates. Again, I had to buy East German license plates as described on page 4-35. Then we were on our way with our car, distinguished as seemingly belonging to a high ranking GDR official. When we arrived at the "Faserinstitut," there was already an East German car parked. We knew, that Hans and Inge Gottschalk, our friends from Brandenburg, wanted to meet us for the first time since our New Year's Eve party in 1952 (see page 4-28). Before we stopped the car, I had told Heidel, please be careful, no big hug and ceremony when you meet. People will be watching. People were watching, but Heidel and Inge had forgotten all caution. It was clear to everyone this was one of those rare reunions of people separated by the Wall.

---

[1] "Schlafendes Bauernpaar" (sleeping peasant couple) von Ernst Barlach (1870–1938, he lived from 1910 in Güstrow, Mecklenburg, where there is a museum of his work, see page 9-34). It was offered to Meissen in 1922 under the name "Russisches Bauernpaar". Five copies were made in Böttgersteinzeug on September 3, 1923. In the 1960s, additional copies were made in the GDR.

When we entered the building to arrange the days plan, Herr Ruscher had the situation well in hand. I received a fee of 100 DM-Ost. This money could not be exchanged and had to be spent before we left. No problem. Heidel and the Gottschalks planned a visit in nearby Potsdam, and I would join them later in the evening in our fancy new "Potsdam Interhotel." The hotel, and the dinner for all of us, was paid, courtesy of the "Akademie der Wissenschaften." Of the delicacies of the evening, I remember two items. A Russian soljanka, a soup which was immediately included in our list of best soups of the world, and the homegrown "Blutwurst" (blood sausage) which outside the area of Berlin does not taste as good because different spices are being used elsewhere. The next meeting with Hans and Inge was already arranged for 1969 (see the pages 8-31–32).

The following day we were back on our way to Mainz. Since the "Autobahn" went past Brandenburg, we thought, having had permission to go to Teltow and Potsdam in the GDR, we might just as well take an hour to inspect our hometown. So we took the road into Brandenburg shown in Figure 3 of Chapter 3 and drove once through the city from one end to the other. It had not changed very much, except for a number of new, bland, apartment blocks. After about an hour we were back on the "Autobahn."

At the border town Helmstedt to the FRG, the GDR officer asked: "What took you so long?" I felt no reason to make up an excuse and said that we drove through Brandenburg since we had visa for the GDR. He looked at our passports and said: "No, your visa was voided yesterday. Please wait, we have to find out what to do." After half an hour of anxious waiting, a higher officer appeared and asked for a detailed account of our visit in Teltow, ending with: "What do you think as a US scientist about the work done in Teltow? Does it measure up to international standards?" When I assured him that I was impressed by their work, he thanked me, and we could continue.

One never knew how border guards would react. They certainly could cause a lot of trouble. So, I made sure that I did not say what came to my mind when he asked about the Teltow Institute: "I would be glad to have Dr. Ruscher as a coworker in Troy." This would have been construed as an invitation to flee the GDR, a serious crime. Years before I had heard a story from a friend of Vati's, who in the 1950s was driving a truck from West Berlin to the FRG. On the same border checkpoint, he took a rest, got acquainted with one of the border guards, and when he left, was waving to him from the cab and said as a friendly goodbye: "Nice if you could come along!" His truck was immediately stopped again and he was jailed for over a week until with help from officials in West Germany he was freed and could continue his trip. He never took that route again.

As can be seen from Appendix C, many other trips were taken during the 1968/69 sabbatic. Toward the end, I also made a round trip through the southern part of Great Britain without family. Enough colleagues and friends lived in this area to arrange the trip with a nice mixture of lecturing, discussions, and sightseeing. Especially impressive was a drive through the countryside to Oxford. The nature and architecture of British universities were new to me, as well as the narrow country lanes and historic buildings. Figure 52, above, is one of the pictures of a house of special interest. I thought its thatched roof looked beautiful, and I knew from Vati, that he was qualified to build roofs like that. My idea was why not make the thatch out of a weather-resistant plastic? This would give the roof a longer life, and probably would not invite mice and other animals to take residence, as usually happens in thatched roofs. But, I did not want to change my occupation at this stage of my career, as already explained on page 8-21.

In Bristol, many interesting discussions were held with Professor F. C. Frank, who wanted to know how our work on superheating was progressing (see page 7-24) and Professor A. Keller (see

Figure 34C) who was interested in our extended-chain crystals and their properties. He was the first to identify the folded-chain crystal morphology in 1957. A very interesting walk with him was to the John Cabot ("Giovanni Caboto") Memorial, built in 1898 for the 400th anniversary of Cabot's reaching the North American continent in the name of the British Crown. Cabot sailed from Bristol. It is known with certainty that Cabot discovered North America[1] and his Italian origin is not in doubt. Columbus ("Cristóbal Colón" in Spanish), in contrast, never reached North America and his Italian heritage, although born in Genoa, is not certain. But the world seems to have embraced him as their favorite Italian discoverer, forgetting about Cabot. May be, in the light of this, it is not so unfair that the new continent was named after Amerigo Vespucci, who was only involved in discovery trips to South America, but popularized the idea of a new continent, different from Asia.

Let me conclude this section by mentioning *acquisitions that also influenced our future life*. One was that while discussing chemistry with Dr. Hans Wilski (see Appendix C) I was also introduced to coin collecting. Dr. Wilski was an avid coin collector of Ottoman coins. He also showed me his German collection of small change which included the first coins I ever saw, the "Sechser" (see page 1-16). This was so exciting that, when I found out that they were easy to obtain and not expensive, I bought them in Mainz. Next, I found a beautiful 16th century "Thaler" from Saxony in a coin shop in Worchester, MA, the coin that later shaped the US Dollar. It is reproduced in Figure 55A, and I made sure that I had all "Thaler", covering the time to 1918. This was followed by a type set of US coins. At that time a gold Double Eagle ($20) was worth $58. In 2008, its value approached $2,000. My copy is shown in Figure 55B. Some ancient Greek coins were beautifully sculpted and so reasonably priced that I could afford them, as is illustrated in Figure 55C. Finally, the 16–19th-century colonial coins of Mexico were of interest to me after several visits there. Among them are a number of early cob coins. They were no more than an irregular, weighed peaces of gold or silver with a hammered-on royal seal and date. Later Mexican 8 "Reales" as reproduced as Figure 55D, were used in the US until the first dollars were minted in 1794.

A second acquisition arose out of a trip to Freiburg and the nearby "Schwarzwald" (Black Forest). In the future, the Glottertal, a small Black-Forest village near Freiburg, became for a number of years *our favorite vacation spot*. Figure 56 illustrates its beauty. We always lived in the

Figure 55 *From left to right:* **A** *Silver Thaler, Saxony, 1632.* **B** *Gold US $20, 1911.* **C** *Greek silver Tetradrachms and a Didrachm from 250–500 BC.* **D** *Silver Mexican eight Reales of 1751.*

[1] Omitting reference to the Vikings which colonized "Vinland" (Newfoundland) from their settlements in Greenland (Leif Ericson, early 11th century). They continued to visit North America for ≈300 years to import lumber and possibly iron. When the Greenland colonies became extinct because of the climate turning colder in the 15th century, the discovery of America was forgotten and the surviving sagas were not believed, until finding definitive archeological evidence in the 1960s.

"Pension" of Mrs. Gabriel, in the house shown in Figure 56C. Relaxing walks in the beautifully wooded hills and a number of first class restaurants and the good wine of the area made for ideal vacations. By 1985, this oasis was spoiled by being overrun by tourists due to a very successful German television series playing in the Glottertal ("Die Schwarzwaldklinik"). From then on, we

Figure 56A,B *Heidel above the Glottertal.* **C** *Our vacation apartment.* **C** *The Schwarzwald.*

replaced the Glottertal with the small town of Kössen in the Tirolian Alps in Austria, just two driving hours from the airport of München (Munich). We stay there always in a small farm house with a most breathtaking view of the "Kaisergebirge," as shown in Figure 57. In 2007, we and our hosts, the Schlechters, were honored by the Mayor of Kössen with a Champagne brunch since we had vacationed at that time for the 25th time in his town.

Figure 57 *A 180° view from 'our' balcony in Kössen. L to R: Unterberg, Wilder Kaiser (in the clouds) and Zahmer Kaiser in the distance, pear tree, Staffenberg and Feilenberg. 180° Panorama.*

As a family tradition, vacationing usually means to go to the same, once found place. It is much more relaxing to know where to go and to anticipate the familiar countryside. Note that Vati's first such vacation spot was Lautenthal in the Harz Mountains (see pages 1-8 and 23 as well as Appendix C). After getting settled in Hastings, NE, he chose Grand Lake in Colorado, where we once also went all together as is described on page 8-18.

Our vacation-spot in the US was Chase Pond, near New London, NH. There we stayed a number of years in connection with the Gordon Conferences (see the pages 8-16–17). Typical views are illustrated in Figure 58. The entertainment there was mushroom hunting for chanterelles which we knew from Germany and Colorado, swimming in Chase Pond, and for Brent, fishing. The mountains in the vicinity were pleasant challenges and offered great views. Travels to other locations were considered explorations of places which we felt needed to be seen, not vacations.

Figure 58 *From left to right:* **A** *Heidel in front of our favorite cabin.* **B** *Caryn and Brent rest from mountain climbing.* **C** *Beach and boats behind the cabin.* **D** *Climbing Mount Kearsarge (2937 ft).*

Figure 59 *The completed living room in Troy, 1970.*

Back home in Troy, by 1970, we had completed the remodeling of our house by adding heating/air conditioning, modernizing the water pump and sewage system, and adding new ceilings in the bedrooms. The living room was completed with a modern cabinet, stretching along the wall, as shown in Figure 59. It also accommodated our new music system which was acquired in steps. The furnishing was completed by adding a number of Oriental carpets, bought from our favorite source, Kermani's in Schenectady. Sometimes, when we could not make up our mind in the store, Mr. Kermani gave us all the carpets in question to take them home. To his and our benefit, this led to more than one being bought.

Naturally a number of novel technical gadgets had also to be tried. We were the first in the Albany area to purchased a microwave oven in 1968. Also, our old black and white TV had been replaced by a new color TV.

Finally, Figure 60 is a scan of the first of our collection of original engraving of German city views from the 17th century. This engraving

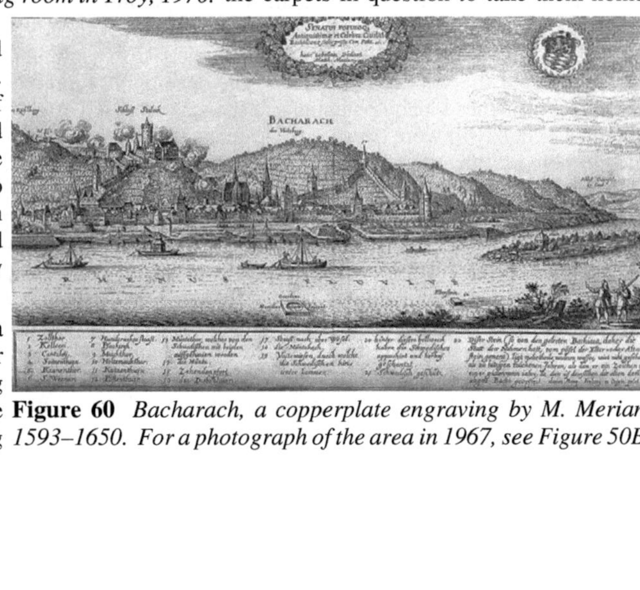

Figure 60 *Bacharach, a copperplate engraving by M. Merian, 1593–1650. For a photograph of the area in 1967, see Figure 50B.*

was a memento of our stay in Mainz, just 40 mi away. At that time, prints of overviews of old German cities were in vogue and appealed to us. We discovered that the originals could be bought in antique shops for only two or three times as much as the reproduction. This, we thought, was a better deal, although the large cities, such as Mainz, were hard to find. Still, over the years, our collection grew to include many cities and villages, all with some connection to our travel.

By Matthæus Merian, my favorite engraver, we collected five more prints: Königstein (the place of our honeymoon, see pages 5-15–16), Köln (the second home of Mama and Papa, which we visited often, see pages 7-15 and 8-22), Freiburg and Ulm, the cities of our second sabbatic in Germany (pages 8-109–116 and 121–128), and Leipzig, close to the city where Vati was born (page 8-125, we also visited Leipzig after the "Wende," see the page 10-4). Others are from the 18–19th century. One from Stockholm, where we spent a week, to lecture in 1992 (see page 9-39), and four are hand-colored from Brandenburg (our hometown, see Chapters 1–4), the Glottertal (the first of our vacation spots in Germany, see Figure 56), Reit im Winkl (a town in Germany in walking distance from Kössen, our second vacation spot, see Figure 57), and a view of the shore of the Phraya River near Bangkok, Thailand, printed in Berlin. We visited there in 2003 (see page 10-9). The 20th century is represented by 15 additional engravings and small paintings of various spots we saw. All of these are displayed at home and recall special memories and supplement the photographs which are stored as digital pictures and in the many albums, created by Heidel.

By the end of the 1970s, I had reached half of my stay at RPI, although I did not know it then. It was clear by this time that the 'Golden Age,' described on page 7-9, had already begun to fade. **What caused the decline of the 'Golden Age of America' in the sciences?** This needs to be addressed next. As mentioned on pages 8-14–15, there usually are many, seemingly unrelated, causes to any such major changes. Otherwise, it would be easy to foresee a developing problem and, perhaps, even anticipate and prevent the development. The quick rise of the 'Golden Age' provided one of the causes itself. The increase occurred exponentially, and every exponential rise must level off, or lead to an explosion. Had the budget of the NSF increased with the same acceleration as documented for 1957–67 on page 7-9, it would by 2007 have been bigger than all federal, state, and local taxes collected. Another reason for the decline was the decrease of the public interest in technological advances after the superiority of the US in the space race with the Soviet Union was re-established with the moon landing of NASA's Apollo 11 in 1969.

The decline in teaching efficiency and quality was already outlined on pages 8-14–15. This also led to a decreasing interest in science. By the 1970s, a number of changes in American society further affected the science education. The developing hippie counterculture during the 1960s negatively influenced many potential science students. All established morals, discipline, values, and rules of behavior were questioned, tested, and often changed or discarded. This development was furthered by the unending debate over the involvement of the US in the continuing Second Indochina War (1959–75) between North Vietnam (supported by its communist allies) and South Vietnam (supported by the US and a few other nations). This hot focus of the Cold War dragged on longer than any other major war in the 20th century. From our own experience in Europe, it was clear that the communist goal was to create their style of dictatorship in all Indochina. The outcome was just that: The creation of the Socialist Republic of Vietnam (1976), the Lao People's Republic (1975), and Cambodia changed to the People's Republic of Kampuchea (1979). These were tragedies that led to untold refugees and deaths, but did not bother the fiercely anti-war protesters.

My observations about changes in students were that their dress and grooming became more informal. When early for a lecture, the students would not stand and wait, but sit on the floor to read, do homework, or eat, blocking corridors and leaving a mess of discarded things. Perhaps, these signs of lacking discipline seem unimportant, but the spilled over into more careless work, a serious flaw for science education. Even more troublesome was the unwillingness of students to participate in discussions of cause and effect. Starting in the 1970s, I had difficulties filling the recitation hours with questions. By now, to keep up grades of the less prepared students, a habit developed among professors to have well-delineated subject areas tested in their exams.[1] This created the most frequent and deadening discussion question: "Will this be on the exam?"

Poor teaching and learning habits were mentioned already on pages 3-33 and 4-17–18. I remember the first 'Earth Day' in my recitation period on April 22, 1970. It was highly publicized among the students and I had prepared discussions to make the period relevant. I did not only want to address what was in vogue, but open questions such as the development of the earth before the appearance of man, the change of it by man, etc. I wanted to cover topics such as the chemical composition of the original atmosphere, the appearance of oxygen and disappearance of carbon dioxide, the need to use, but not to overuse chemical fertilizer and pesticides to support the present population, and the overall changes due to agriculture. No meaningful discussion was developing. The students were only interested in repeating political slogans (at that time about dangers of nuclear power, leaded gasoline, air pollution, and the ill effects of pesticides). The details of any of these problems of the day, and a discussion of possible solutions based on their topic of learning, chemistry, seemed not of interest. Science was separated from public opinion!

This development points to another problem, *the villainizing of chemistry*. The very popular advertising slogan of the DuPont company "Better Living Through Chemistry" had to be abandoned,[2] because this positive message until the 1960s, slowly turned negative. Chemistry is the basic science which strives to understand all substances on earth and all others that can be imagined and made in the laboratory. Next, chemists try to understand possible changes of any compound, including inorganic, organic, and biological molecules (see page 7-20 for a better subdivision of chemistry). How could they be at fault for real and imagined problems of society? The answer to this question is simply the misuse of, or ignorance about the results of chemistry when it comes to large-scale applications. There are many examples where true or imagined misuse of a chemical product led to banning of compounds as a political reaction based on flawed or incomplete information. This is a poor alternative to correcting problems after they are properly understood.[3]

---

[1] In my opinion this is cheating by professors and causes a poor preparation of the student for further advancement. Students will need their newly acquired knowledge only after reaching a professional position, i.e., any short-time learning is a waste of brain effort (see page 3-33).

[2] In 1935 the DuPont company changed its image from a 'powder company' to one that improves daily life with the slogan 'Better Things for Better Living Through Chemistry.' In the 1980 the 'through chemistry' was removed. In 1999 the slogan was changed to: 'The Miracles of Science,' emphasizing a science company and avoiding the negative image of a chemical company.

[3] Example: Overuse of DDT in agriculture. It led in 1962 to Rachel Carson's book 'Silent Spring,' scientifically flawed, but best-selling. The book did not only stop agricultural overuse as a positive action, but it also curtailed use of DDT in the fight against malaria. From 1950–70 more than 500 million lives were saved worldwide by DDT (Natl. Acad. Sci., The Life Sciences, NAS Press, Washington, DC, 1970, pg. 432). Nothing better, safer, and cheaper exists, even to date, so that outlawing DDT may have killed more people than the wars of the 20th century (see page 3-37).

There were *other signs of decreasing interest in chemical research* in the US. During this time, I noted a drastic decrease of the American presence at international meetings. While in the 1960s it made little difference where a meeting was held, the US contingent was the biggest. This changed continually, so that by the end of the 20th century, US scientists were often in the minority, being not only outnumbered by the host country, but also by Europeans and Japanese. In the future, Chinese and Indians may be added to the list. The largest fractions of missing US scientists were from industry. In the US, scientists found it increasingly more difficult to convince companies to pay the expenses for such 'extravagant' travel and to prove the benefits of these meetings.

The annual Gordon research conferences kept attendance records documenting the change of the affiliations of their conferees. These records[1] revealed quantitatively the increase in participation of foreign scientists and the decrease of US industrial scientists. Does this mean, industry is not interested in fundamental science anymore? Seemingly so, and I may also suggest a specific reason: Over these years, much of the management of chemical companies was transferred from the research-trained PhDs in chemistry or chemical engineering to managers with an MBA degree (Master of Business Administration) and little or no training in research. The profit of research and the time needed for its success is difficult to perceive for anyone who has not gone through a research experience himself. On discussing research in economic downturns, I found it difficult to convince administrators, that this was the time to increase research activity, to be in a strong position during the subsequent upturn. To an administrator who is only concerned about the quarterly profits, this makes no sense, but for a researcher, it is perfectly logical. The pay-off for successful research has a longer timescale than most financial crises.

Two other topics contributing to the decline of the sciences in the US are *the change of research support by government agencies, the salary structure of professors, as well as the support of graduate students and postdoctoral associates*. As reported on page 7-9, the research spending leveled off in the 1970s (see also above, page 8-52), while the increase in researchers and students continued somewhat longer, i.e., the success rate of proposals for research support decreased. Also a negative point was that it became customary by the end of the 20th century to judge proposals not only according to their own merits,[2] but also according to the politically preferred broader impact on teaching and society[3] (copied from NSF judging-details). These conditions of selection can restrict fundamental research by favoring topics in accord with the second set of criteria.

---

[1] From 1950 to 2000 the number of Gordon conferences increased almost linearly from 30 to 180 with a total attendance changing from 1,300 to 17,000. The foreign participants increased over the same period from 1.5 to 33%, while the industrial participants decreased from 70% to 13% and the academic attendees made up the difference, increasing from 17 to 77%.

[2] Assessed by: "Importance to advancing knowledge and understanding; qualification of its proposer; creativity and originality; concept and organization; access to the necessary resources."

[3] How well does the proposed research (a) advance discovery and understanding while promoting teaching, training, and learning, broaden the participation of under-represented groups (e.g. gender, ethnicity, disability, geography, etc.), (b) enhance the infrastructure for research and education, such as facilities, instrumentation, networks, and partnerships, (c) get disseminated to enhance scientific and technological understanding, and (d) benefit society?

This selection method also prefers research conceived as important by the general public, i.e., it entices to follow recognized trends to the exclusion of new ideas. For example, at the beginning of the 21st century one notices a turning of more researchers to the field of 'nanophase science' than, perhaps, is warranted by its importance. Many of the 'trend-experts' do not have sufficient time to develop an understanding of the fundamentals before another 'cutting edge' topic emerges. Prior 'trends' I observed (and sometimes helped to deflate), were the unimportant polywater (which was only dirty water), the overemphasized liquid crystalline polymers (which were often conformationally disordered (condis) crystals, and cold fusion (which was not so hot).

The changing salary structure at universities can best be seen from my own experience. Most other universities did not fare better, and even today, this problem has not been resolved. On analyzing the changes in salary, I discovered a letter I wrote to the RPI Administration (by then different from the one of the 1960s). The letter is shown as Figure 61. On reading it after so many years, I am not surprised that I never got an answer, and, that I ultimately left RPI. Too much had changed. My inflation-adjusted salary had reached a maximum in 1971 and dropped thereafter by approximately 15% before leveling off. Between 1979 and 1981 the annual inflation rate had exceeded 10%. Taking the maximum income at 1971 as a base, it turned out that by the time I retired from RPI in 1987, my inflation-adjusted income had not reached the 1971 level. By 2001, it had increased by about 50%, mainly because of an increase in salary with the new position at UTK and the pension I received from RPI. This summary shows how much scientists lag behind other occupations like corporation officers, trial lawyers, physicians, and entertainers (including athletes). Income expectations, naturally, are one of the factors when students make their career choices.

By the 1970s most graduate students in chemistry were supported as teaching assistants by the department or as research assistants by professors with research grants. Both jobs are considered half-time employment. Postdoctoral positions are full-time positions, but because of the learning component, the compensation is still much below that of industrial employment. The problem of funding of these positions was exasperated by the fact that many students from foreign countries were eager to study in the US. They usually came from an environment of lower standard of living and were able to live more easily under lower income conditions (such as Heidel and I did, see Chapter 6). The department's admissions committee or research professors would gladly hire *two* foreign students at the cost that could properly support only *one* American student. In addition, the applying students from foreign countries were often superior in ability and sometimes even had a better undergraduate education. For the country as a whole, this was a benefit since the cost of the earlier education of these students was born by the home country, and more often than not, the students would ultimately stay in the US to make up the shortfall in number of US students.

A quick look at Appendix A reveals, that of my 39 graduate students, about half were US born, and no American student joined my research group after 1988. Of the 57 postdoctoral associates, only four were US born, but half stayed in the US. This trend of foreign students remaining in the US, is decreasing with an increase in standard of living and better employment opportunities in the respective home countries. Similarly, the increase in personal freedom in some parts of the world decreases the number of political refugees. This experience-born discussion may provide part of the answers, some of which never made it into the nationwide discussions. The unfortunate conclusion is that the US are not #1 in graduating scientists, and American scientists are in the minority at international conferences. Clearly, *the 'Golden Age of America' in the sciences has ended and a new effort needs to be started* to keep in front of the rest of the world.

Bernhard Wunderlich

Professor of Chemistry at Rensselaer Polytech. Institute
Troy, N. Y. 12181, Phone (518)-270-6456
Specialization: The Solid State of Polymers

211 Winter Street Ext. Troy, N. Y. 12180
Phone (518)-283-1141

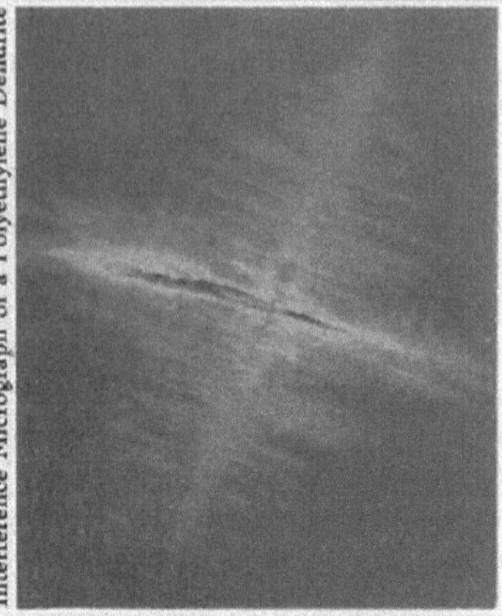

Interference Micrograph of a Polyethylene Dendrite

April 15, 1974

Dr. R. Loewy
Vice President and Provost
RPI

Dear Dr. Loewy:

Two weeks ago I received from you this years salary notification. I would like to let you know my reaction to it. Taking into account a moderate evaluation of the increase in cost of living, my salary for 1972-73 showed no change, that for 1973-74 showed a decrease of 2% and the presently given figure for 1974-75 indicates a decrease of 8%. Two years ago, I had no comment, last year I discussed the continuing trend of lower and finally negative increases with the department chairman, seemingly with no result.

My complaint is not so much the actual decrease in salary, but I think it is an insult to write for now the third year in a row that 'This action of change in salary is based upon an evaluation and review of your activities.' In my case I hope you also realize that the last three years have been the most productive in my life. They resulted in the completion of two textbooks, two monographs, and over 40 research papers. In addition, I developed a completely new approach to teaching which can increase productivity manyfold (Audio Courses), tested a new curriculum for General Chemistry (with Prof. Wiedemeier), and taught 3 new graduate courses. My research group was about 10 post-doctoral and predoctoral students and technicians. All support for my research activities came from outside the school. External recognition was given by invitation to many national and international meetings and the award of the NATAS Prize for Thermal Analysis and the nomination as Outstanding Educator of America.

You can imagine that all this could only be accomplished by using every minute for school activities, neglecting in particular my family who hardly saw any vacation or free weekends. I have, however, difficulties to keep up this pace if at the end of each year recognition in word or deed is not forthcoming from Rensselaer.

I would appreciate it, if this could be rectified in the future. Salary notifications as I have received in the last three years should contain an apology that possible accomplishments could bot be rewarded and an accounting should be made why a particular part of the staff cannot be properly compensated. In case you are dissatisfied with my services, I would appreciate notification so that I can change my activities to suit the goals of the school better.

Sincerely yours,

Copies to
Dr. K. T. Potts, Chairman
Dr. G. H. Handelman, Dean
Dr. R. J. Grosh, President

Figure 61 *Letter to explain problems of RPI's 1974–1975 salary structure and evaluation policy.*

The results of our research and the increasing worldwide travel convinced me that the time was right to write **a monograph about the 'Crystalline Solid State of Linear High Polymers.'** The project was first discussed early in 1968 with Professor Fischer in Mainz (see page 8-41). In the light of my seeing a possible decline of research support in the future, I thought there may be time to undertake such an enormous job. When I sent the detailed proposal of the book to Professor Fischer to get his final commitment, it turned out that he did not think there would be enough time for him to participate. Fortunately, he quit before a major effort was spent by either of us.

Ultimately, it turned out that my research support remained stable due to faithful support, mainly by NSF, NASA, and ONR (see the Footnote on page 7-5), i.e., there was no extra time to write the book. But, was I not always interested in the solid state? (See the pages 4-11–12, 5-2, 6-20, 7-19, 8-20.) *Was this a chance to understand the field even better*? After studying a subject as a student, I always felt well prepared. Only when I tried to lecture about it, did I notice that there was much more to be learned and understood. This teaching of others is, I felt, what make the expert in the field. Now, I experienced that writing a text book, which one hopes the colleagues and their students will use, is one more step of self-improvement. It requires to connect all separate aspects of a subject, link it to the historic development, and then simplify the concepts without losing precision. The simplification helps the next generation of students not to have to follow the many erroneous paths that historically are followed on the way to better understanding.

How could I find the extra time for such a book? Was it possible to get *another sabbatic*? This would allow me to stay home, concentrate on searching and studying the literature, and write. By then, the policy of leaves was relaxed somewhat and now, I truly was in my seventh year at RPI. So, on May 19, 1969, I submitted a proposal for second sabbatic to the Chairman, Prof. Janz.

The six major arguments listed in support of the sabbatic were: (1) In the last six years, our research in the area of macromolecular crystals, their properties, and their mode of formation were successful. For the first time, we could give a detailed description of the crystalline solid state. (2) No other researcher in the field seemed to be undertaking such a monograph. (3) The treatise could unite separate results gained in chemistry, physics, and materials science, all fields of interest at RPI, and also of value to the future teaching at all levels of instruction. (4) My stays abroad had taught me, that the work generated in Europe was different, more physics-oriented, so that I would be uniquely suited for this job. (5) I was in contact with most of the active researchers in the field in the US, Europe, and Japan. Finally, (6), for the Department I suggested to develop an updated Freshman Chemistry course, including the new polymer knowledge.

This proposal sounded convincing to Professor Janz. He forwarded it to the dean and suggested that it was "meritorious and well conceived." It was accepted as a release from active classroom teaching, not a sabbatic. Travel could be covered by a grant from Owens Illinois Co. in the amount of $7,500. It supported three trips to Europe, as can be seen from Appendix C. Most exciting was the trip to Switzerland to the 3rd ICTA (see page 8-23), where I received the Mettler Award of the North American Thermal Analysis Society (NATAS, see Appendix C, page C-13–14).

At the time, it was customary to exchange preprints of research papers, including copies of original photographs. This was a good base for the 500 graphs and photographs I expected to need for the proposed 1,000–1,500 pages of the book (actually 1,420). It was best for me to redraw all graphs to make them uniform and fitting to the text. The permissions to copy photographs were easily obtained (except for two of micro-crystals, published in the *American Scientist*. They requested a licence fee of $20 per photo—which I personally paid).

By June 1971 the first half of the manuscript of Volume 1 was submitted as a first report to the Department. The book had by then been divided into two volumes, to make it more easily printable. This manuscript was also offered to 'Academic Press' for evaluation and led to a contract for publication. The first volume was completed on January 16, 1972 and the manuscript submitted to the Department. Further consideration was given for its completion in reducing my teaching load also in 1971/72. After updating and correction, the book appeared in 1973 with the dust cover shown in Figure 62 as "Macromolecular Physics" with the subtitle 'Crystal Structure, Morphology, Defects.' The new title was chosen by the publisher with the reasoning that 'The Crystalline Solid State of Linear High Polymers,' the title I suggested, does not contain any word fitting in the categories of librarians. Since they are the key people to order books for the library shelves, it was important to put the well-known word 'Physics' in the title. The price was too high to assume a large student purchase. The first volume was followed by Volume 2 (Crystal Nucleation, Growth, Annealing) in 1976, and Volume 3 in 1980 (Crystal Melting) (Appendix A, publications #95, 102, 123, a and b). By now, the books are out of print, but I retrieved the copyright and an updated, free PDF copy is available over the internet (see Appendix A, #95b, 102b, and 123b, listed on page A-4).

Figure 62 *Dust cover, published 1973,76,80.*

Later, I found out that a pirated, Russian edition had appeared, translated by Drs. Yuli K. Godovsky and W. C. Papkov of the Karpov Institute in Moscow. The USSR did not recognize the US copyright. When I finished the next volumes, this had changed, and "Физика Макромолекл," published by "Мир" became properly licensed for translation and printing between 8/79 and 10/84, with $1,750 going to me. By then, I had met Dr. Godovsky at the IUPAC Meeting in Prague in July 1979 (see Appendix C, page C-21). He greeted me by saying: "I have translated your books!"

When we visited China in 1984, I found paperback copies in the apartment of Dr. Miao-yong Cao, at that time my postdoctoral fellow in Troy. His wife, Cui-xia Liu, who had not accompanied him, had hosted an outstanding dinner for 12 in their small apartment in Hefei. We found out later, the university had lent added furniture, the books (?) and also supplied food delicacies to impress us. On inspection, I found, my books were pirated paperback reprints of acceptable quality. Thus, *Soviet and Chinese students had much easier access to my books* than US students.

The hours in the library researching for the book, I found of great interest. Every day I discovered something new. It did not originate so much from the direct questions I searched for in the Chemical Abstracts (or the "Chemische Zentralblatt," the German equivalent, see page 3-19). Rather, the random reading of adjacent abstracts or references in the retrieved papers opened new avenues and ideas. Did I get rich from the sales of the book? No! Counting the hours spent, the pay was far below minimum wages. The value of the gained knowledge, however, was priceless.

In the summer of 1972, Heidel arranged a European trip with the charter club "Schwaben International" to follow an invitation by the Academy of Science of Poland (see Appendix C). This was close to Heidel's original hometown near Beuthen (now Bytom), it became a trip into the past. First, we had to find someone to look after Caryn and Brent. Mutti and Vati agreed to help. Vati had retired in 1969 and was now traveling with Mutti in their Airstream trailer (see page 1-33). They parked their trailer in Troy for the duration, as shown in Figure 63 during a barbeque after our return.

Figure 63 *Vati and Mutti's Airstream trailer. With Dr. Wiedemeier (see page 8-67) and Papa in 1972.*

This trip was typical of combining business with visits, vacation, sightseeing, and nostalgia. First, we spent time with Mama and Papa in Bergheim and I visited companies and universities. In Bergheim, we inspected the open pit mines, the gardens of "Schloss Paffendorf," and shopped in Köln. After a lecture at the University of Marburg, we traveled with a bright-yellow, rented VW 'beetle or bug' (see Figure 67B) to Brandenburg. Hans and Inge (see page 4-28) had arranged a two-day permission for us to stay with them. By then, the GDR felt secure enough to show off their 'progress' to us. It was a surprising feeling to be back again. Little had changed. The appearance of the city was not different, and the continuing oppression became obvious from discussions with friends. Figure 64 shows the area close to that in Figure 1 of Chapter 2. All historic buildings of the city center were gone. The ruins of 1945, of Figures 17, 18 and 24 of Chapter 2, were removed, rather then restored. The city center was a big parking lot for the few available cars. Figure 65 is a snapshot of the narrow-gauge street car of 1910 vintage! (See also Chapter 1, Figure 24.)

Figure 64 *Heidel and Inge in the center of Brandenburg in 1972.*

Figure 65 *The 60-year-old street car must still be used.*

Next were short visits to Potsdam and West Berlin, followed by an invitation of the "Physikalische Gesellschaft der DDR." to the Faserstoff Institut in Teltow (see also pages 8-46–48). Then came the trip to Poland. Via Posnan, we drove to Łódź. There, we were guests for a lecture and many discussions. Our host was Professor Marian Kryszewski of the Center of Molecular and Macromolecular Studies of the Polish Academy of Sciences. He taught at the Technical University in Łódź, and also was a member of the Academy in Warsaw, where he showed me the Academy's new Polymer Institute, while Heidel had a city tour with an expert guide. In the laboratory, all basic equipment was available. Just, that I noted there were hardly any researchers at work, despite the fact that it was a weekday. Also, the discussion was somewhat reserved. No new ideas seemed to be pursued. The effort was, to repeat what was done in the West. I pleaded with them. You must do something new, perhaps, find a superior 'Polish Polymer!' Some seemed to get the idea, but they were not sufficiently free to choose their own research.

Figure 66A *Market place in Warsaw, 1972.* **B** *Rebuilt city gate.*

Prof. Kryszewski, later, allowed one of his coworkers whom we had met, to spend time with me as a postdoctoral fellow (Dr. Janusz Grebowicz, now living in Houston TX, see pages 8-109–110, 9-53–54). Then, at noon time, I joined Heidel for an impressive walk through the old and new towns of Warsaw. The old town is now a UNESCO World Heritage Site (since 1978). It was the first city bombed in WW II and had a sad experience under German occupation. A special effort was made to rebuild it to the state of 1939. Figure 66 illustrates the old town and the city gate between the old and new towns.

We continued our travel, to see the Black Madonna of Częstochowa, and from there we went to Kraków. For me, a return to a city I had seen almost 30 years before. I could remember the beautiful St. Mary's Church and the Royal Castle, the "Wavel" (also a UNESCO World Heritage site since 1978, see Figure 10 of Chapter 2). Much hidden politics was revealed by the guide to the "Wavel." The lady was outright anti-Soviet and gave us a long list of what was confiscated during the Soviet occupation. She knew her "Wavel," well and had keys to many back rooms.

At the hotel, only for tourists, we sat in the dining room with another German couple and had an interesting conversation about problems in the GDR. From the dialect we new that they were from the GDR, but they did not know where we came from. Our Berlin dialect could come from either East Berlin or West Berlin. Initially they considered us as to come from the GDR. As the conversation went on, they slowly got the idea that, instead, we were 'foreigners.' The conversation stayed interesting, but they made sure that they did not reveal their name and hometown, as would otherwise have been customary. Better safe than sorry at these times.

The next step was to inspect "Auschwitz" (Oświęcim), as mentioned already on page 3-39. Then we went to the hometown and school of Heidel in "Mechtal" (Miechowice), a suburb of "Beuthen" (Bytom). With our car, all was reached much faster than Heidel experienced before, on foot. Figure 67 shows the road to "Andreashütte" (Zawadzkie) and our yellow rental car in front of the old city hall, the last place of residence of Heidel and her family before they left in the summer of 1939 for Brandenburg.

The last stage of this trip was to visit the relatives who opted for Poland and stayed after the end of the war in their ancestral farmhouse. Some, by now, had also married into other families living nearby. A picture of one of the reunions is given in Figure 68. It was taken in the garden of Papa's elder brother (Onkel Robert, at the far right) and his wife (Maria, on Heidel's left), cousin Steffi (on Heidel's right with her husband second from the right), and in the center are Heidel's god parents (see page 11-3). Many of the children in front wear clothes which are hand-me-downs from Caryn and Brent. Packages were sent regularly to Poland to support the family left behind.

Figure 67A *County road between Bytom and Opole ("Oppeln"). The main transport on these small roads is still by foot or horse-drawn wagon.*

B *Our VW at the city hall in Zawadzkie, then* "Andreashütte."

Figure 68 *Family reunion in Szczedrzyk in 1972.*

Onkel Robert complained that we did not let him know early enough when we would arrive. The card we sent from Łódź had arrived a few days before, but the butcher had meat only on Wednesdays, so he had to slaughter a chicken for dinner. As the honored guests, we were asked to eat alone in the "gute Stube," (best room) of the house. It took all our convincing power to change their mind. We had come to talk to them, not to be served a meal beyond their means.

Heidel and I had to conclude that these (former) Germans who stayed behind to protect their possessions were the ones who truly lost the war. There was little recovery since the end of the war. This was the state the Allies wanted all Germany to be reduced to (see page 3-29–30). Only the Cold War had protected West and East Germany, from the same fate.

Beginning our way home, we thought to save a considerable distance by driving through Czechoslovakia instead of back through the GDR. Our goal was to reach the Glottertal near Freiburg for a week's vacation. It was suggested to us by Professor Wiedemeier, a colleague who had moved to RPI a few years after us. It was to be a first of many vacations there (see Figure 56).

The travel though the GDR needed no visa for transit without overnight stay, so we thought, the same would be true for Czechoslowakia. Not so! When going through the Polish border station, our visa for Poland were cancelled, and ¼ mile later, the Czech border police checked our passports. "<u>*You have no visa*</u>." My answer: "I think wee do not need one, we are only in transit, traveling for about 200 mi. We did not need visa through the GDR unless we stayed overnight." "No, you need a visa from the consulate in Warsaw to cross the border." "But my Polish visa is not valid any more!" "I cannot help you." And away he went. What could we do? After 10 agonizing minutes, the policeman came back and asked: "Do you have small change?" "How much?" "Twenty dollars." "Yes." "Please wait, I will get you a receipt." We got a receipt for 20 Czech Korun, a minimal value. The paper we obtained had no information outside of name, passport number, and car registration. Now the guard became very friendly and told us what stops we could make to see interesting things and what to buy before leaving Czechoslovakia. We did not stop anywhere.

On the border to the West, we handed in our paper. The border guard acted surprised: "There is no information!" Then he disappeared. In front of us, another car and luggage was practically fully taken apart to look for contraband and possible hidden passengers. It did not look good. Another $20? A higher-ranking guard appeared, he did not look at our passports or anything else but said: "You drive along this free lane. Have a good trip!" How much we could have smuggled! The border guards shared their private income from the other side of the country!

Later on this day we began our first vacation in the Glottertal. When we arrived, all hotels were booked out. But, a waitress who overheard our inquiry, suggested seeing the mother of a friend of hers, Frau Gabriel, who ran a bed and breakfast out of one floor of her house. The pictures of Figure 8-56 were taken on this visit (see the pages 8-49–50). Every time we went to the Glottertal, we could stay with Frau Gabriel and enjoy the outstanding food and wine in the many good restaurants which also hade clientele from Freiburg, which included us during our second sabbatic in Germany in 1986/87 (pages 8-112–114 and 9-21). After recovery in the Glottertal and a lecture in Ulm, we again joined the "Schwabenverein" in Stuttgart to fly back to the US.

Once home, we had to pick up the collection of special wines we bought in Ludwigshafen and had shipped to New York. The BASF operates a large, successful wine cellar, not only to handle the harvest from their experimental vineyards linked to their fertilizer production, but also to compare their wines to those available on the market. Since the wines were liked by their employees and friends, they were for sale. Every time it says in Appendix C that I gave a lecture at the BASF, one can be assured that we had several boxes shipped to New York, to be picked up at the harbor. Shipping and duty added very little to the cost of a good bottle of wine.

How did I find the *time to complete Volumes 2 and 3 of Macromolecular Physics* so quickly after Volume 1? On April 20, 1972, there was a bulletin by the RPI President for a true sabbatical leave policy (see page 8-37 for the bulletins). Since the earlier leaves were declared non-sabbatics, in order not to set a precedence, it was now possible for me, to officially apply for a 'true' sabbatic with pay from RPI. One week later, Professor Janz had my request for a one year's sabbatic on his desk. It was appreciated by me that he was not fed up with my continued requests for leaves. Professor Janz was of great support to me and an outstanding chairman. The department prospered under his leadership. We were told, the first applicants for one of the new sabbatics will be considered first. Also, the one who requests a full year sabbatic at ½ pay has a better chance to succeed. My immediate effort to raise additional funds from the Guggenheim Foundation, NSF, NATO, and the Fulbright Program were, as before, not successful (see page 8-37).

The sabbatic leave was given on May 31, 1973 to be effective from September 1, with a grant of half-salary. It turned out, that this financial penalty was not as hard as it seemed. Of the grant, $300 per month were tax-free, and the lower tax bracket helped also (20% instead of 30% for the combined local, state, and federal taxes). The initial manuscript of Volume 2 was submitted by the end of the sabbatic, and published in 1976. It contained 470 pages of text, 85 pages of tables, 165 figures, and 1,500 references.

For Volume 3, planning was started after the sabbatic, and writing began in January 1975. The writing of this manuscript was easier. Its topic, the melting of crystals, was our active research at this time. Most references were already collected. Volume 3 was published in 1980.

Plans to complete the project with a fourth and fifth volume, covering irreversible and reversible thermodynamics in cooperation with Dr. H. Baur of the BASF (see page 8-1, 2, 6) could not be realized because of the unavailability of Dr. Baur at that time, and my pressing teaching commitment, including the developments of the audio courses. Ultimately, Dr. Baur published the intended book separately through Springer, Berlin, in 1998 (Thermophysics, Part I, Theory, 356 pages). The initially intended Part II, was to be written by me, treating 'Experimental Aspects.' It had to wait for my retirement. It became the 'Thermal Analysis of Polymeric Materials' (see Appendix A, publication # 500), which appeared in 2005, again, too late to be the continuation of Part I. Although not all was completely contiguous, the five books together give a unique, seamless base of 'Macromolecular Physics' based on the microscopic structure and motion, as it is linked to the macroscopic properties, describable with irreversible and reversible thermodynamics.

The Appendices A and B contain the information about **progress of research during the 1970s and 80s, as well as the development of new teaching methods**. The major influence of the travel and leaves, as well as the increasing realization of the changes of the position of science in US society form the backdrop of most of the work in this time period. *The research planning and execution* was least affected, as can be seen from Sections C and D in Appendix B. The new focus was to study the successive or simultaneous polymerization of small to large molecules to bypass the random coil which was identified as the cause of chain-folding on crystallization (see pages B-7–10 and Figure 7 in Appendix B). The increase in precision of differential scanning calorimetry led to new measurements and to an early effort to generate a data collection in 1969 (Appendix A, paper #62). Through the major effort of Dr. Umesh Gaur and other available students, including Brent and Caryn as undergraduate assistants, the first ATHAS data bank was collected, and then published between 1981 and 1983 (see Appendix A, publications #151–53, 159, 171, 177–80).

Figure 69 is the cover of the first an annual report from our laboratory which was now given the heading '<u>A</u>dvanced <u>TH</u>ermal <u>A</u>nalysi<u>S</u>,' ATHAS. The report was shipped to approximately 100 recipients each year. By 1996, the internet became a more efficient tool and all information was displayed there. Appendix B, Section D, summarizes the scientific background of the early ATHAS effort. As our research into thermodynamics and kinetics of polymer crystallization developed, it was presented at important international meetings in England, Belgium, Japan, and at an ACS Meeting in Boston, MA (see Appendix C, pages C-22, 41, 58, and 64, respectively).

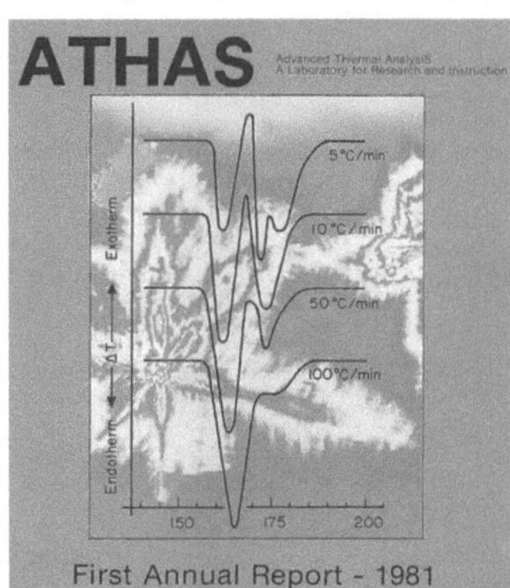

Figure 69 *ATHAS Report, as published until 1995. The shown curves represent the thermal response at different heating rates. The picture shows an interference micrograph of sperulites.*

The more *wide-spread application of computers* simplified the collection of the literature for the ATHAS data bank, as well as the treatment of literature data (assessment, conversion, scaling, plotting, and tabulating). Finally, even operation and calibration of the calorimeter and data calculation were helped by computers. My early interest in calculators went back to Vati's old, black, Swedish, hand-cranked 'Facit' of 1942. It was used to simplify the bookkeeping (see page 2-10). The next step involved the many calculations needed for the adiabatic calorimetry in Professor Dole's laboratory (see page 6-19). These were carried out on electric 'Marchant Figurematic' machines. At the same time, more powerful electronic main-frame computers became accessible as a special university facility. My first independent paper in 1958 made use of the IBM-650 (see page 6-33, for the paper, see Appendix A, #4). At RPI, I tried to reach the ultimate in mechanical calculation with a 'Friden' calculator with mechanical storage and back-transfer into the accumulator after completion of intermediate computations. Quite a mechanical feat. In 1972, 'personal computing' was revolutionized by an electronic, hand-held calculator, the Hewlett Packard HP-35. It could do scientific calculations and not just the basic adding, subtraction, multiplication, and division. Naturally, I had to have this calculator as soon as possible.[1]

In 1974–76, we began to automate our third Perkin-Elmer DSC (2). The first DSC was type 1 (see page 8-25), followed by an upgrade to 1b. The analog response of the DSC was digitized, and the output produced on a punched paper tape. I did not trust immediate computation. By having the tape, one could not lose the measurement. On changing one's mind about the calculation, the original data were still available. The tape was fed into a programmable, electronic calculator with 420 memory registers for the program and also temporary data storage. The output was either a data table or a plot. With this progress we were already many years ahead of commercial equipment. The second generation of an automated calorimeter was completed and tested by 1978. A copy of the block-diagram of its flow chart is reproduced as Figure 70. The upgraded

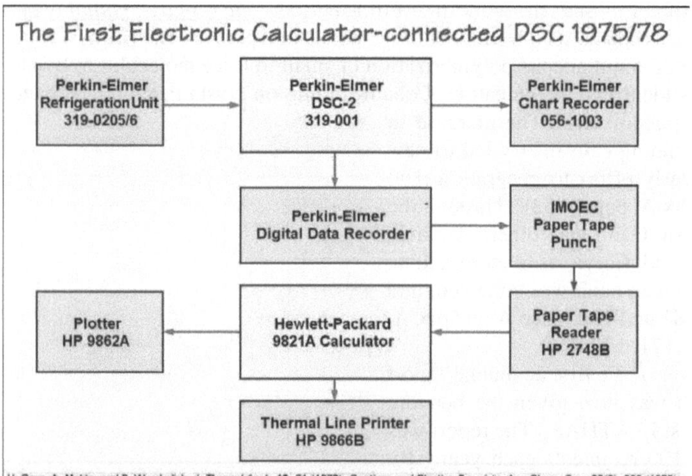

Figure 70

[1] The Hewlett Packard HP-35 scientific calculator was first introduced in 1972 with sales exceeding 300,000 units. The unit weighed only nine ounces. It was named for its 35 keys. Its speed and accuracy surpassed even that of my 5-digit circular, spiral slide rule. Its purchase price was only \$395, much less than my last, most advanced, mechanical Friden calculator (≈\$1,250).

computer had now 929 registers and additional access to 6,000 registers from a digital cassette tape, sufficient to store all needed programs. By optimizing all parameters, an accuracy of ±0.5% could be achieved on measurement with this equipment (for details see Appendix A, #130 and 138).

Progress continued with the introduction of more powerful and faster computers. The computer became part of the calorimeter, able to automatically optimize measurements and carry out subsequent calculations and plotting. Our own research effort in this direction is described in Appendix B, Section D. By 1982 a new programming company (Laboratory MicroSystems) had started operation in Troy, NY with help from RPI. They generated the third and fourth generations of our automated calorimeter by developing the LSM-1000-DSC data station. By then, the intermediate punch tape was eliminated, and in the last version, an earlier Chromemco Z-2 computer was replaced with an IBM microcomputer (Appendix A, #195, 206, and 257).

General discussions about the use of computers and data storage in thermoanalytical practice followed these first successes. With Dr. P Fellner from the Slovak Academy of Science, I chaired a workshop on this topic, held at the 8th ICTA Symposium in Bratislava, Czechoslovakia, in 1985. The main problems discussed were prophetic warnings about the difficulties caused by the use of proprietary software as was supplied by then with commercial calorimeters. The scientist is then not fully informed about the way the measured data have been collected and subsequently treated. Furthermore, with such packaged software, it is often not possible to extract the raw data and analyze them, using more suitable and transparent, or advanced calculation methods. A company with a first-rate calorimeter often does not have an expertly designed computation program.

By the middle of the 1980s, I had *my own complaints about computers*. By keeping up with the rapidly changing development of computer technology, I felt, we often did not save any time when compared to doing the same without computers. The continued designing, relearning, and testing of new programming needed more time than saved by the faster handling with computers. At that time, we just did not do a sufficient volume of routine measurements within our university research to overall save time with any given program. By the end of the 20th century, the development settled down. Once a program was understood, it could often be transferred with little changes to the next generation computers which also did not appear in as quick succession as before. The overriding advantage was now the better accuracy, higher speed, and greater convenience.

Another complaint was becoming more serious. It was harder to contact the programmer who knew the details about more sophisticated software. The information in manuals usually does not teach understanding. Sometimes, if the programmer had left the company, it was even impossible to find out details. For the DSC, these vital questions are: How were the data normalized, averaged, rounded, extrapolated, etc.? What routines are used for integration, baseline correction, lag correction, etc.? What was the number of measured points used per time unit, and at constant heating rates, were integrations made over time or temperature?

An inability to answer such questions brings one back to the Middle Ages. Then one could not fathom the reasons behind the experimental outcome, one had to believe in 'magic.' Typical were the efforts to produce potent medicines or to make precious metals out of base materials. Today's computer programs are created by man, but often they produce 'magical' results because we are not informed or cannot understand the programming. Frequently we assume 'the computer' gave the wrong result! This is rarely so. Either the operator or the programmer made the mistake. Certainly, there is no magic. Even worse is the situation of an 'out of service' computer. Work cannot continue since manual operations, used earlier, are forgotten.

Unforseen possibilities developed with steadily improving personal computers. Figure 71, shows the installation in my office at home (compare to Figure 6A). At RPI, being an engineering school, the utility of the first widely used personal computers was quickly realized. By 1981, the school offered a discounted Atari Computer for students and professors to enhance learning and teaching. The package consisted of the computer, a monitor (a standard TV-set), and a dot-printer.

All at once, a manuscript could be printed and corrected without retyping. Typists were in less demand. The professors had to learn not only how to type, which I still do only poorly, but also had to learn how to set up a letter or publication, keep address files, etc. Did this increase our productivity as a teacher and researcher? I certainly do not think so. Especially the other amenities offered by a secretary were lost as well, like having the mail opened and sorted, screening telephone calls, keeping class lists, and relaying messages to the students. These are all services which I still could enjoy at the beginning of my career, but they would soon disappear.

Figure 71 *Working out lectures, manuscripts, and computer courses in my office at home in the 1980s.*

In addition, earlier, there were janitors who would support the professor by cleaning the blackboard, seeing that chalk and other tools were in their proper place, putting the lecture room in order before each lecture, and keeping my office and laboratories clean. It is not only that many (lowly paid) secretaries and janitors lost their jobs, but these jobs were taken over by (not sufficiently highly paid) professors without any training in these jobs. I would estimate that in this way I lost 1–2 hours each day. The computers do not always increase efficiency.

Since I am complaining about the amenities lost in the life of professor's over the years, I might as well mention that in my first years at RPI, it was strictly forbidden to collect garbage or do any noisy maintenance during the lecture hours. If absolutely necessary, these activities had to be scheduled for the intermissions between lecture hours and during lunch periods. Would it not be nice to have today the same quiet environment for teaching and studying at a university?

Having discussed the new *tools for research*, I need to turn to *the new tools for teaching*. The experimentation with in-room TV for live experimentation was described already on pages 8-13–14. Not only was it possible to teach using TV, it was also learned how it had to be done without degrading the quality of teaching. Giving a poor lecture to a small class may cause some complaints by some students, teaching poorly to an audience of all 1,000 freshmen is a disaster!

With the broadening of the field of chemistry, advanced studies needed to be increasingly split into sub-specialties to manage the larger volume of information (see page 7-20). I tried to address the problems arising from this development for teaching in two ways. One was to *re-visit and simplify the basic knowledge* in the light of new information. The other was to find economical ways to teach small groups of students and even individuals.

An especially interesting example of *simplification* arose out of my effort to combine enough material to describe the progression from atoms to molecules to phases. The teaching can start with the often neglected observation that atoms have only a limited range in size, estimated from their diameters (0.16–0.67 nm, using H and C atoms as limits). Next, the atoms are connected by bonds of which there are only four limiting types (ionic, covalent, metallic, and weak). All, including the intermediate bonds, fit into the Grimm tetrahedron.[1] Only covalent bonding allows flexibility. This scheme of bonding can next be connected to the three possible types of molecules described on page 7-20. The rather diffuse grouping into inorganic, organic, and biochemistry becomes then unnecessary. The final step after this progression from atoms to molecules continues to macroscopic 'phases.' A phase is a uniform assembly of atoms or molecules and should not have noticeable fluctuations in chemical composition and overall physical state.

Historically, the phases can be linked to the four elements of antiquity of earth, water, air, and fire (more modern, solid, liquid, gas, and heat, see Appendix A, #500, Section 2.1.1). To understand phases, one must not only judge their structure, but also identify their molecular motion. Ten phase types were first proposed by us in 1984 (Appendix A, publication #200). A diagram and further details are given on page 10-4 (and also in Figure 13 of Appendix B). A last addition to the description of phases is their division into macro- and microphases at the micrometer dimension, and the more recent introduction of nanophases (see page 9-46). This overall concept can be taught easily, as is illustrated with the textbook and computer course of Appendix A (#500 and #560).

A number of topics of interest to engineering students, the main constituent of the RPI student body, were collected for a pilot section to restructure freshman chemistry as part of my special leave (see page 8-57). Experiments and applied examples were designed for teaching in the updated audio-visual facility, to be completed 1973/74. It was a joint assignment with Professor Heribert Wiedemeier, who came to RPI in 1964 from the University of Kansas in Lawrence, KS (PhD of the University of Münster, Germany). His main interest was solid state chemistry, specifically the growth of crystals of chalcogenides, a topic which led him to experimentation on several of the space shuttle missions of NASA to study crystal growth at zero gravity. Our results on teaching were reported at a symposium of the American Chemical Society on "Teaching Chemistry Through Macromolecules" (Appendix C, page C-15, Appendix A, #101).

The teaching of freshman chemistry continued with contributions of mine, but now more frequently on polymers. In addition there was the handling of recitation periods and supervision of laboratory sections. By 1987, I completed a workbook which includes the new ideas of teaching freshman chemistry (see Appendix A, #255). The efforts to shift the topics taught in chemistry toward the interests of engineers were not accepted by the department. Ultimately, the engineers were sufficiently disinterested in the courses taught by chemists to develop their own introductory chemistry. This, in turn, led to a reduction of the number of professors needed in chemistry.

In time, my lecturing was increasingly in graduate education. Here the problem was to find enough students to justify teaching of new courses. To be productive, RPI required at least 10 students in a lecture (compare to he HU, page 4-11). In 1972, I started to record my senior and graduate courses on audio tape and added soft-cover books as 'blackboard material' and notes.

---

[1] The 'Grimm tetrahedron' made it into my notes of Professor Thilo's lectures (1949–50, see page 4-11). Next, I used the Grimm tetrahedron in my lectures in Freshman Chemistry, starting in 1963 (see pages 8-13–14 and Appendix A: Page 63 of publication #255, Figure 1.6 of publication #290, and Figure 1.5 of publication #500). W. B. Jenson (Prof. Chem., Univ. of Cincinnati) traced such representation of bonding in his detailed article in the *J. Chem. Ed.*, **75**, 817–828 (1998) with a full set of references. [See also M. Laing in *Education in Chemistry*, **30**(6) 160–163 (1993).]

Based on the initial 'live lectures,' the teaching was changed to *self-study with audio courses*. The courses were first published in 1973–75. Their titles were 'Crystals of Linear Macromolecules' (Appendix A, publication #108), 'Thermal Analysis of Linear Macromolecules' (#109a,b) and 'Crystallization of Linear Macromolecules' (#124). As a part of my 1980/81 sabbatic (Appendix C, page C-23), the thermal analysis course was changed to include all types of molecules (#184).

To check the efficiency and cost of the audio courses, I kept detailed records of the time to teach the live courses compared to supervising the students studying with audio courses. For the audio courses, I needed an average of six hours of personal discussion during office hours or by special appointment (average over several years). The students were required to write preliminary term papers. After my corrections and suggestions, these papers were resubmitted for grading. An oral final exam of ½ h concluded the course. The students would register for the course in the Division of Continuing Education. When the number of students exceeded 10, the course was to change to additional live lectures. This never happened, except for off-campus teaching. The tuition for the course was divided into three parts. One, I could use for the production of the course (tapes, packaging, printing, and purchase of suitable tape players), one part, RPI kept as overhead, and the third, was paid to me for teaching at an agreed-upon hourly rate. Figure 72 illustrates the offering.

Figure 72A *The three audio courses in final form, as offered until 1988.*

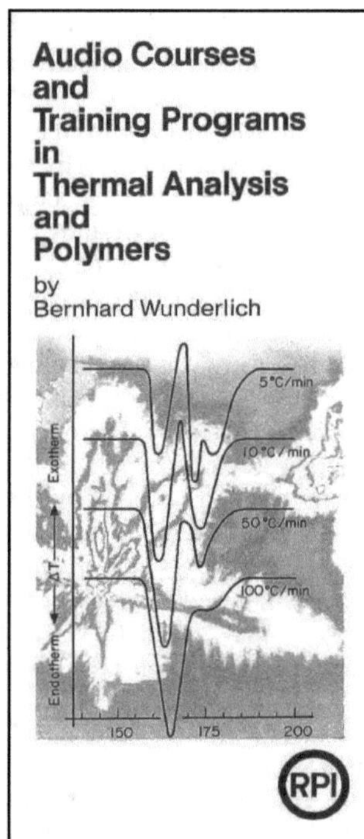

B *Cover page of the mailer of RPI's Continuing Studies, 1973.*

The experience with these audio courses allowed much more flexibility in teaching. The 'Crystals' course was also offered by the ACS, the American Chemical Society, and industrial companies could suggest these courses to their researchers to increase their educational level, and even as a means to earn graduate credits as an off-campus student. The discussions were then handled by visits of mine at the beginning and end of several weeks of study with the tapes. Two of these sessions with 15 and 20 students were arranged by Dr. Muñoz-Escalona in Caracas, Venezuela at the IVIC (in 1977 and 1978, see page C-20 in Appendix C).

Overall, about 200 students were taught with these audio courses. Occasionally, when the distance a student would have to travel to Troy was too great and I would find myself close to the town of the student, I would arrange a day and time for the exam at a nearby airport. These experiences were discussed in appropriate publications (see Appendix A, #119, 134). Although the production of an audio course would take about three times as long as preparing for teaching a live lecture, it offered a number of advantages. It increased the possible offerings by the department. In fact, if everyone in a typical department of 20–30 Professors would maintain 3–5 audio courses, the listing of specialized courses of instruction would increase easily by 100! Industrial researchers with enough experience, and professors of other universities, actively employed or in retirement, could offer such courses at nearby universities (see also the proposal on page 8-118).

Experience taught that the audio lectures could not be the recording of a live lecture. They are best shortened by about 50% and should be recorded in a studio from a script, checked for errors. The students appreciated the condensed lectures. For difficult passages, they had an instant replay. Easier or known material could be fast-forwarded. The printed 'blackboard material' gave precise schematics, images of equipment and experiments, and equations with their derivations. When expanded with notes by the student, review of the material was easy. Finally, audio courses were always available and could be studied for any length of time, a few weeks in the summer or stretched over several semesters. If there was no need for academic credit, the study could be open-ended.

In the assessment of the courses in publication #134 of Appendix A, I concluded: "For the first time since the discovery of printing, I see a possibility of a major improvement in teaching. The chore of repetitive derivation of basic information can be reduced for the teacher. There is no need to unite teacher and student at a fixed time and location. Discussions with students involved with the audio course are at a consistently higher level and the teaching of the audio courses extends the reach beyond the location of the university. The audio courses offer a third alternative for study, the others being represented by the learning from books or lectures."

The next stage of new teaching techniques was *the introduction of computer courses*. It was based on the experience gained by automation of the research, described on pages 8-64–66. The first experiments with computer assessment of learning were made from 1975 to 1976. A reader for pencil-marked IBM cards was purchased (see page 5-3 for IBM card readers). Either questions with multiple choice or numerical answers were prepared beforehand or formulated at the spur of the moment. The students were at the appropriate time asked to quickly answer a question. The results were printed and displayed with the student name codes within 20 min of my reaching the office after the lecture. The method was tested in recitation periods with freshmen, and in1980 for a course on 'Introduction of Polymer Science,' with about 50 juniors and seniors. At the end of the semester, I gave the standardized American Chemical Society exam. My students scored 12 percentage points higher than the ACS average and had no negative comments in their year-end report.

Full scale development of computer courses was begun after the introduction of the Atari computer to RPI students and faculty in 1981 (see Figure 71). This computer had a number of advantages, although by today's standards it had a rather limited memory for data and programming (32 kB). Its roots in computer games made it easy to simulate thermal analysis curves. The student could then examine these and enter his or her results for immediate evaluation. Instead of flying missiles, molecule trajectories could be discussed, etc. Another great advantage was that for data and program storage, it used a standard, two-track audio cassette tape. One of the channels could be used for the replay of sound. At the same time, illustrations and programs could be loaded synchronously with the lecture and displayed on the screen. In addition, it could easily be programmed. A first computer course and text: 'Introduction to Differential Thermal Analysis,' were produced by 1984 as shown in Figure 73 (see Appendix A, publications #201–204, and Appendix C, page C-26, for a seminar given at RPI in 1983). This computer course was also available until the 1990s to my graduate students to sharpen their skills in handling of and learning about differential scanning calorimetry. The Atari, was an antique by then.

The more comprehensive 'Thermal Analysis of Materials' was developed at UTK, the University of Tennessee at Knoxville. Its last version of 2007 is available on the internet, see Appendix A, publication #560, and for a description see #478 and 545. Figure 74 is the first screen of the course. Although an enormous amount of resource material is collected in this course, it certainly is not yet the ultimate in computer courses. There is presently no universal program to combine sound and figures with the ability to jump conveniently between screens within the common operating-systems.

Unfortunately, every time there was a new update of the course, it also required additional changes in the operating system. Ultimately, Eberhard, my nephew, at that time a graduate student at the University of Pittsburgh, was able to write the program necessary for the course to become self-contained on the Java, Sun Microsystems platform. Microsoft, is the main culprit of non-compatibility. They do not recognize even their own, earlier software. At present, they have not approved my course. The course runs well, so I do not plan to oblige them to pay for this approval service.

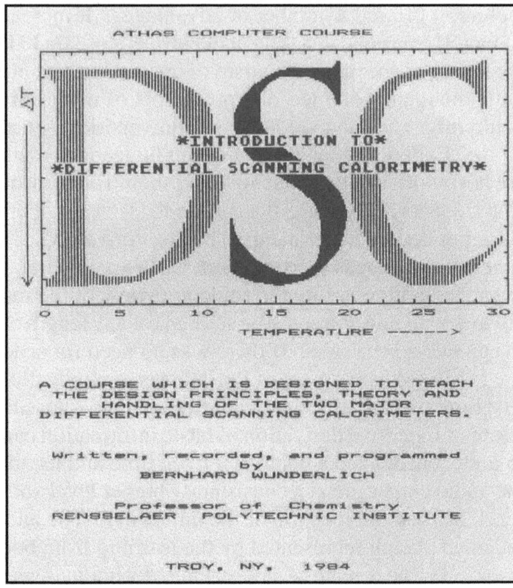

Figure 73 *Cover of the text to the first computer course. Note, the only fonts available on the Atari were* `Courier`. *For all headings my own font was developed. Besides the text, four cassettes were needed for the storage of program and audio tracks.*

Figure 74:

My poor experience has been that after I finished the first several hundred DSC traces and graphs, in MSP (Microsoft Paint) used for my 1990 book 'Thermal Analysis' (publication #290), they were soon outdated. Microsoft had changed the Paint program to the PCX format (Zsoft Paint). By now, it has been for a number of years BMP (Windows Bitmap). The most economical for simple color presentations, however, is TIF (Tagged Image File Format) or GIF (CompuServe) which I used in Figure 74. The older formats cannot be converted to the newer ones with the operating system by Microsoft. I had to buy other graphics programs to do this (easy) task. The same is true for Microsoft's own 'Word' text files. As a result, I will not change from the better 'WordPerfect' word processor which was required earlier by the university to allow easy exchange between offices. With it, I can still convert my WordPerfect files from the 1980s to 2005. To interface to'Microsoft Word,' I use the 'Open Office' by Sun Microsystems which can be downloaded free of charge.

By now it may be useful to have **a look into the future of teaching and learning**. After leaving RPI, I had no further connection with freshman chemistry. In fact, my teaching schedule was drastically cut. At UTK, I was employed to conduct research and, supposedly, taught only for ¼ of my time. As already mentioned on page 8-15, the group of professors acting as 'experts in teaching freshman chemistry' are usually not in favor of someone who might upset their established routine, so that all my teaching concerned graduate courses, often using the just described computer course which was expanded, developed to its current state and given the companion book listed in Appendix A as publication #500.

The ancient form of transmittal of knowledge and experience was verbal, and manual techniques were also taught by example. For us, the achievements of this early period in precise transmission of knowledge, religious beliefs, genealogies, sagas, and literature over many generations are quite amazing. What took later libraries to document, was memorized with no more tools than verses and song for formatting, which also made the presentations more exciting.

After writing and assembly of larger manuscripts was invented, learning was simpler and teaching was helped by written texts and instructions. After forming academies and universities to assemble teacher and student communities,[1] the teaching and learning were further advanced by the ability to print books. Now, it was possible to learn without the constant presence of instructors. On hand of a sufficient library, one could become self-taught, an autodidact. In the 19th century, large-volume printing became cheap. Now, one can even assemble one's own library.

Still, most students required support by teachers, as is exemplified by our present system of education. After reaching a high school diploma after 12–13 years of class-room instruction, this is followed by studying for 3–10 years at a university to complete a professional degree. In the 20th century, additional learning tools were developed and tried, as described above, including video and audio recordings, as well as using computers. It is now possible to relegate a larger part of teaching and learning to the new media and reduce (a) the effort of live teaching and (b) the restriction of student and teacher to live as a separate university community. In the 21st century the opportunities of learning with modern media became more pervasive, especially in the basic levels of education. As expected, in subjects of general interest, private enterprises lead the way since it can generate profit from the initial investment. Universities are still hesitant to embrace these developments.

---

[1] Early universities provided up to three services: Education, research, and the conferring academic degrees. The word itself derives from the Latin "universitas magistrorum et scholarium," with the approximate meaning 'community of teachers and students.' Such institutions were first established in the 9th–13th centuries (Arabic Madrasahs and European, medieval Universities). The even earlier institutions were limited to teaching and learning only, as exemplified by Plato's academy, established in the 4th century BC) and similarly old institutions in China, India, and Persia.

Before addressing problems of the future of teaching, one must analyze the goals that should be accomplished. There seem to be six, not only the two or three identified in antiquity (see the Footnote on page 8-71). First, one must collect the proper scholars (students). After the basic student preparation, a selection is required of who is able benefit from higher education. Some observations about elementary and high school education, based on my own experience, were made already on pages 1-14–19, 3-6, 3-33, and 4-17–18. Present day practice and advice to students is not always in accord with these observations. Hopefully, progress will be made in the future.

As goals 1 and 2, one knows that the millennia-old premise that *education consists of two vital parts, the teaching and the learning* must be kept. The problem of decreased interest in chemistry was already touched upon on pages 8-52–53. It is a quite serious problem since it will channel the needed top students into different fields of study. This can only be corrected by reminding the general public that many problems troublesome to man (and, as is presently fashionable to say, 'to the environment') must be solved using the 'scientific method' of chemistry by doing the appropriate research. The present preoccupation to find the blame in chemistry and searching for political solutions based on incomplete, preliminary investigations, is often outright wrong, and certainly more expensive in following its erroneous path than proper research.

Teaching of science and engineering early in life through toys can show amazing things possible in 'reality' science. Such toys should reappear on the market and replace the too frequent playthings based on 'fairytales' and 'unrealistic gadgets.' Superman, Spiderman, James Bond, and similar imaginary heroes cannot save the world. But any student can, by tackling the problems with common sense and a good education in the natural sciences. Excitement in chemistry should be ignited by teaching up to date knowledge and pointing to the mistakes of the past and present. Should we not object to the enormous fleets of windmills built and planned to be built? Are windmills effective enough? Do they not destroy the beauty of the countryside? Don't we know other solutions that could be made effective and environmentally friendly?

The importance of the two goals of education is often not fully appreciated. (1) The main goal of *teaching* is to create excitement in the student about a given topic to make him or her want to learn the subject and master the laboratory techniques necessary for definitive experiments. (2) The major component of *learning* is the solitary work to anchor the various facts in one's memory and connect them to prior experience. Any discrepancy between different bits of information must lead to further learning, engender discussion, and lead to actual research in case the question raised cannot be solved with the present knowledge. The actual learning involves the remembering and reviewing of information gained by the study of appropriate learning implements, supported by personal notes. The internet quickly gains on importance as a learning tool. It procures vast volumes of information, but it still lacks much of the critical evaluation of the stored facts which has taken place in the scientific literature and reviews, as well as textbooks, and encyclopedias.

The learning process is initiated by teaching. The amount of teaching needed, however, varies widely for different students. The teachers may be the parents, friends, siblings, colleagues, school teachers, or university professors. Important is that they bring about the initial excitement and contribute to the learning by detailed discussions, arguments, and demonstrations.

The observed decrease in eagerness of students to contribute to discussions, mentioned on page 8-53, is a significant loss to education. The threats of impending examinations and the method of cramming to get by, is a poor way of furthering actual learning (also pointed out on page 8-53). This was brought home to me during my involvement in teaching of selected freshmen at RPI (page 8-67). These students had taken the standard exam during their last high school year and passed with a grade sufficiently above average to be granted 'advanced placement.' To get to know the ability of my 75 students, I gave them the *identical* exam again during the first hour of my lecture. The performance by then did not exceed the average high school result anymore. In less than one half year, all 'advanced' knowledge had been forgotten!

Considering the points raised about teaching and learning, I offer an opinion, based mainly on my own experience: *The students in the US may be the best taught in the world, but they certainly are not the best learned students in the world* (also see the pages 9-36–37). Improving teaching and learning is necessary and must be a continuing process. The basic frame, represented for chemistry by the many texts of freshman chemistry does well for learning the fundamentals, as does the information transmitted by most professors. What often is missing is the relevancy to today's interests and the continued updating of older information.

Finally, a short comment on when it is time to end formal education at a university and to start the lifelong, continuing process of independent learning. The end of the university education is graduation with the ultimate degree. In chemistry, this usually is the PhD. How long should it take to complete this last degree from the prerequisite undergraduate degree (BA or BS)? Most universities have a minimum residency of three years. This is a reasonable time to learn how to independently solve a well-defined, scientific problem. How is it possible that sometimes 5–10 years are used? The simplest answer may be that the candidate was unsuited for the job, if this is true, it should have been noted earlier. Frequently, however, the candidate is qualified! After having proven his ability to produce publishable work, he is required to solve additional problems. This is not fair! The competent, lowly paid assistant, is by then employed beyond his learning stage for the PhD. He or she should have become a postdoctoral research associate and be paid accordingly, or be permitted to continue the professional career at full compensation.

Next, goal 3 is that the *research by the professor joined by his students should question established knowledge and expand into new areas* in a fundamental manner. The first question concerns what research should be done at universities? One hopes that much of the academic research contributes to the fundamental questions neglected by commercial and government laboratories. The level of competence, however, should be high, and the use of the most advanced equipment must be accessible, to be able to research and teach at the highest level. This does not mean that academic research should be oblivious to the problems of industry, government, and society (see the pages 8-54–55), but one should be careful to avoid research which could better be done elsewhere. Many academically interesting problems arise from applied research, as I experienced throughout my career. Being a consultant for DuPont (Appendix C on page C-4) helped me to identify basic research problems which could solve applied questions. I feel, however, that applied work and patents should not be pursued at a university (see page 8-21).

The worst scenario is when a company supports and makes use of university research and does this at the expense and decline of its own effort. In certain cases, this may bring cheaper and, perhaps, better research to a company, but it would ultimately cause the loss of the job the student was hoping for after graduation. Still, the university professor should be informed about applied problems, and the research carried out at universities should also be disseminated by the applied laboratories. Advances in the fundamentals should be applied in short order by keeping close contact and consultation. The teaching ability of the university should, furthermore, be made use of for periodic upgrading of outdated PhDs and to help when a change in expertise in a laboratory is necessary. Modern off-campus teaching media could further these goals (pages 8-68–70) and establish closer connection between fundamental and applied knowledge.

Another common observation is that researchers away from their field for many years are often out of touch with the latest developments. Accordingly, university professors who concentrate on teaching and neglect research may have difficulties to excite and properly direct students. What I found so valuable during my early student years at the Humboldt University was that my early teachers were also top, up-to-dare researchers (see Chapter 4). In the US, in contrast, as my research was yielding results, I was increasingly discouraged to contribute to teach beginning students. This was specially bothersome, since my field of macromolecules is for many students, unknown to them, the area of their later employment. Most universities in their early courses do not give proper recognition to this field. How can one then attract graduate students, if the possible candidates are not informed about the importance of the field they are to choose?

As goals 4 and 5, it should be recognized that *conferring an academic degree has the dual purpose, to evaluate past performance and to discourage unsuited candidates* from entering the profession. The goal 4 is to evaluate the proper learning during the study. Its connection to teaching was illuminated above. One should only remember that today the teaching can take many forms and does not have to be by the classical lecture and laboratory instruction.

Presently, the examination, goal 5, is done to a large degree by the same faculty that handles the teaching, leading to the problems mentioned on page 8-53. Only in PhD and MS examining committees have I experienced additional faculty members of related fields of the same department, and occasionally even a different department. The chairman, however, with the main vote, has always been the director of the thesis work. In addition, the prior graduate learning was usually assessed by the grades from the written exams of narrowly chosen subjects. Little testing of wide-area learning and the ability to apply the gained knowledge is done. The thesis work itself, again, is usually limited in scope. One should make use of a *separate faculty* for examinations, to get an independent evaluation long before the PhD level is reached! Both teaching and examination faculties should work together on an up to date syllabus of knowledge for the student, but the examiners should have no prior contact with the students. From what is written above, naturally, I also would opt for fewer examinations, less written, and more oral testing.

Separate exams from university education are well known in the US from the Bar Exams in the law profession (administered by the judicial branch of the given state) and the Board Certification in the medical profession (a voluntary certification by the American Board of Medical Specialties, ABMS). Perhaps the oldest example for the use of separate examinations from teaching is the development of the ancient Imperial Chinese Examination system for civil service positions (Keju). Student learning was available through private and later government supported schools. With changes, this examination system lasted for more than 1,300 years. Regardless of parentage or age, all (male) Chinese were initially eligible to take these exams. The system reached its greatest success in the Ming Dynasty (1368–1644). The final decay occurred during the Qing dynasty (to 1911) when the system became corrupt and positions (exams) could be purchased.

Having described some of the problems and possible solutions for the improvement of the educational system as I see it, one should suggest as goal 6 solutions to the even more thorny question of *how to pay for the education*. Perhaps, I was lucky by having experienced the two widely developed systems of education. One is the socialized education where in one form or another, tax money is extracted from the population and used for an education with little cost to students and parents. The main problems I had with this system are clear from Chapters 4–5. They ranged from unbearable government interference in education (Chapters 2–4) to a seemingly acceptable system with minimal interference and reasonable levels of student support (Chapter 5).

The other educational system is more or less a free enterprise system which I experienced as described in Chapter 6 and beyond. Tuition and fees are required to be paid by the student to cover a major part of the cost of education. In private schools, it is attempted to get the (substantial) missing funds from gifts of former students and other fund-raising schemes. The time period of my study and the prior level of education put me into the fortunate position that with Heidel's added effort in the US we could manage to finish our education without financial help from family or loans, ending up with a close to zero balance, but without debt (see page 6-32). Today, such a result would be more difficult to achieve. The state-schools in the US, I experienced as a teacher only, but I can attest that their management is less flexible than that of private schools (see Chapters 9 and 8). The tuition in state schools is usually much lower, the bigger part of the budget is covered by state taxes—and the state has a much bigger voice in how things are to be done and paid for.

Recognizing the greater degree of freedom that can be achieved when staying away from tax funding, but also avoiding subsidizing the education of the few students who could afford the full cost, I suggest the following: Each school must annually charge the actual expenses of education to their students. At the same time, the student must be given the possibility to pay for education and subsistence by receiving a loan, independent of the will of state and even family. Using the various teaching media available, it should be possible for every school to experiment with more efficient teaching and examination. Indeed, I would expect that then, the overall cost of education would be divided into the suggested major categories of, teaching, examination, and research.

It is not within my experience to suggest ways to pay back loans. But why can top athletes, educated on sport scholarships (an oxymoron in itself), receive a signing bonus from professional sports organizations which goes into the megadollars? I calculated, based on record keeping on some of my PhD students, that at that time (1980–90) a full academic education in chemistry of about eight years, would cost between 1 and 2 million dollars. Why can the employers of PhD students not have a similar signing bonus for a top scholars? For all others, it may be possible to attach a tax surcharge on their life-time earnings which satisfies the cost which arose during the education. It could be compensated by raising the salaries commensurate with the value of their education to society. This system would give an incentive for the student to go to the university which can offer the best education for the least money, and it would also offer a motivation to the family of the student and the student himself to reduce the loan as much as possible to increase the net take-home pay later in life. It must not be forgotten, that at present, the cost of education is covered, although I think it may be uneconomical. Would it not be an appealing result when everyone can say, I (and my family) paid *fully* for my education? As was stated on page 3-6: 'Something is wrong with education if the consumer does not value the goods.'

After this excursion into teaching, **I need to return to the changes in our family** to cover the next two decades. After our first sabbatic in Germany in 1967/68 (see pages 8-39–49), Caryn and Brent were less flexible in their schooling so that major travel of the whole family was restricted to summer vacations. Frequently, we took a vacation in connection with my attending a Gordon summer conference in New Hampshire (see page 8-16–17 and Appendix C). *The cottages at Chase Pond near New London* were then often our place to stay (see page 8-50 and Figure 58). Figure 75A is a snapshot when searching for chanterelle mushrooms along the road.

There were enough cottages that we could also invite Vati and Mutti, our nephew Eberhard when he continued his study at MIT, and occasional also other friends from Troy. Later, Brent with Anna and Stefan could also join us there. Grandson Stefan was especially intrigued by the different life style. When he noted that there was a goodly layer of sand in our cabin after being barefoot on the beach, he went 'home' to Anna and Brent's cabin and declared: "Oma's cabin is dirty!" An expression that stayed in the family. In the summer of 1971, the whole family went to the 8th IUPAC Microsymposium on Macromolecules, in Prague, and visited with Gabi and Rolf Bögel, old friends from the GDR (see pages 4-16, 17, 30). Figure 75B is from this trip to Czechoslovakia.

Figure 75A *Mushroom hunting on the road to Romanov's Chase Pond cottages, New London.*

B *Caryn and Brent on the Charles Bridge in Prague, Czechoslovakia in 1971.*

The typical <u>*fun in winter time in Troy*</u> is illustrated with the Figures 75C–E. There was always plenty of snow. Our slightly sloping property was ideal for simple skiing, sledding, and skating. In the back of the house we had set up a small ice rink for a number of years, ready for fun after school. One year, it was raining onto the frozen ground to cover the whole driveway with ¼" of solid, black ice. The path to the road by car was one power-glide until the wheels could grip the salted road, but in the evening, I had a beautiful, slightly sloping, iced drive to skate on.

Figure 75C *Winter weather for skiing in Troy.*

Figure 75D *Caryn and Brent, Winter Street Ext.*

E *Efforts to skate on the mini rink on the patio.*

Figure 76 *Using the snow blower on the patio in deep snow.*

The car needed to be parked at the bottom since it was not possible to get to the top (not even with spikes, see page 8-46).

Our down-sloping driveway had a big advantage. We could always get out of the drive in the morning and leave the car at the bottom on return. I did not miss a single day at RPI due to snow or ice. Sometimes, it was difficult to cut a path into the snow. The John Deere snow blower made it possible to connect the house with the garage, as photographed in Figure 76. To produce a sufficiently wide path to reach the road or mail box from the garage, however, was not as easy.

It did not take long to recognize *the precious possessions we had in our sugar maples*. In spring, we would boil-down the sap collected in plastic bottles through spouts drilled into the maples. For good running of the sap, the nights had to be below freezing, and the days, above. To get perfect maple syrup, the rough boiling was done outside, on the side of a steel barrel, as illustrated in Figure 77, using fire wood, collected from the woods. The fine adjustment, was done in Heidel's kitchen.

Figure 77 *Boiling maple syrup (you can see the collection-bottles on the trees). The yield ≈1 gallon per year (from ≈40 gallons of sap).*

Other entertainments for Brent and Caryn were the children's parties. Examples are shown in Figure 78. It looks like Caryn was happier among the small boys and girls (left picture) than Brent among the older girls (right picture). After a short while, the Little Red Schoolhouse (Figure 7) was augmented by the new elementary school in the village of Wynantskill. The junior high (Gardner Dickinson School) was built new, not far behind our house, on the other side of the hill. Caryn, and later Brent, could walk there in good weather.

While Brent was eager to play Little League baseball, Caryn asked to take piano lessons. This meant to invest in a piano. To make sure that this was not a passing interest, the next time we were in Hastings, NE, we asked Barbara (cousin Jim's wife, see page 11-2, bottom center and page 6-12) to give Caryn piano lessons. Caryn passed her scrutiny, and we bought a used upright piano for Caryn's room in 1967. It was difficult to move it upstairs, but shortly thereafter, she started private lessons with Ms. Alma Deep. Ms. Deep stayed Caryn's piano teacher until she went to medical school. It was never necessary to remind Caryn to practice, and quickly it was a pleasure to listen to her efforts.

Figure 78A *A birthday party for Brent in 1964.* **B** *Caryn's 9th birthday party in Troy, 1965.*

For both Caryn and Brent it was quite clear that they would ultimately try to enter RPI. A major benefit of the faculty of RPI was that their children did not have to pay tuition. Even at that time, this was a big financial help. For us it meant to slowly help them to identify their interests.

Caryn had for some time been talking about becoming a forensic scientist, and Brent liked machinery and collecting stamps and coins. In the summer of 1970, we undertook a longer trip by car, first going to inspect Williamsburg and Jamestown to learn about the early history of the settling of North America. Finally, we stopped in Washington, to tour our capital and study the running of the government. To make sure everyone could satisfy their interests, the evening before we made a plan. Heidel wanted to go to the art museum. Caryn wanted to take a tour through the FBI building. Brent wanted to go to the treasury building, and I wanted to take everyone in the afternoon to the Smithsonian. Since Brent was still too small to go on his own, the two of us went together. Everyone was to meet with Heidel at the art museum for lunch and a quick look at the museum. Figure 79 illustrates this meeting for lunch. Initially all went well. I dropped off Heidel, showed Caryn the two-block walk and delivered her to the FBI for the tour, and then went with Brent to the treasury. A big sign in front said 'closed today.' What to do? This was Washington. The White House is just across the road. Brent wanted to go. When we looked at the entrance, however, there was a queue more than a block long. We might not have that much time! So, I took Brent to the head of the queue to ask how long it might take for us to get in. The doorman said: "Wait, I just have to open the gate and then I will answer your question." When the door was opened, everyone was swept into the building. We did not intend to crash the gate, but by then, we did not really have a question anymore. Since everyone had lost their place, nobody complained, and we had a nice tour of the White House. Everyone was happy and at lunch we had a story to tell.

Figure 79 *The family gathering for lunch at the National Gallery of Art in Washington, DC, 1970.*

On graduation from junior high school, in 1970, Caryn was the valedictorian. To make sure, she was challenged enough in high school, *we enrolled Caryn as day student in the Emma Willard Shool*[1] instead of Troy High School. The Emma Willard School I had to pass every day when going to RPI, and obviously, we were well informed about the unique qualifications of the school. Caryn liked the idea, and after getting to know her classmates, she also agreed that it was a good idea not to be a boarder in the school. In fact, she frequently brought some of her friends to our house. Since the school had many electives, Caryn had to make a few choices. What language? Her choice (without asking us) was Russian, her justification, I know enough German to get by. Indeed, she took four years of Russian at Emma Willard, and when she was an exchange student in Switzerland, she went to St. Petersburg (Leningrad at that time) to take a summer course in Russian. We probably would not have chosen Russian because of our still not forgotten experiences of 1945–53, but rather suggested Spanish (she thought it is too easy) or Japanese (too difficult, she thought). Another typical response of Caryn was not to study a musical instrument in school, but to continue lessons from Ms. Deep. Nobody knew of her piano playing. When later, one of her friends could not find an accompanist for a flute recital, she offered to everyone's surprise her services and did an outstanding job. A picture of her graduation in the quadrangle of Emma Willard is given in Figure 80.

The next European trip for the family was in 1974, just after Caryn's graduation from high school. We all went for a 10 days vacation to the Glottertal (see Figure 56). From there, we continued to the 4th ICTA in Budapest (see Appendix C). We again met with Hans and Inge from Brandenburg (pages 8-31–32). They had brought their youngest son Uwe along, so that the three 'children' could have their own party. By this time, Caryn was old enough to travel alone. She left us in Vienna to go to St. Moritz in Switzerland. There, we had arranged two months of summer work for her at 'Randolin's Hostel.' on recommendation of the Haags, Then she started study at RPI as a chemistry major with interest in medicine.

Yes, the time went fast. Before Caryn left home to go to medical school, she did two noteworthy jobs with me. They were to help in my laboratory and preparing a first translation

Figure 80 *Caryn's graduation.*

[1] Since 1814, the Emma Willard School has been one of the nation's leading college preparatory boarding and day schools for women. It is known for its academic rigor, the promotion of intellectual curiosity and disciplined study habits through a challenging curriculum distinguished by a wide array of advanced placement courses and electives. About 300 students and 60 teachers, school motto: "Gaudet Patientia Duris" (Patience Rejoices in Adversity).

The Quadrangle of Emma Willard in Troy.

of an essay by Gottfried Leibniz about the German language (see Appendix A, publications #135 and 149). Leibniz was a founder of the modern science (1646–1716). The booklet was edited by Dr. Calinger, a professor in the History Department of RPI. The book was a concise account of the life of Leibniz's universal learning and was dedicated to Professor E. B. Allen (1898–1974), formerly of the Department of Mathematical Sciences at RPI. When I was approached to undertake the job, I needed a coworker to correct my English without destroying the link to the old German of Leibniz. Caryn was the right person. Emma Willard had prepared her well for the job.

Three years later, the education of Brent had to be directed after his graduation from junior high school in 1973. His grades were good, but we could not find a suitable private high school in the Troy/Albany/Schenectady area. La Salle Institute in Troy was close. It was a Catholic school, established in 1850 by the Brothers of the Christian Schools and was run for grades 6–12. Their mission was listed as 'Builders of Boys, Makers of Man.' When looking into the school, we noted two negatives. We did not appreciate the uniformed and military-like demeanor of the students, and also not the overly religious fervor. Most of the other schools of the 'Capital Region Independent Schools Association' also did not fully fit our ideas of education, or were difficult to provide transportation to. We certainly did not want Brent to leave home as yet. So, *Brent went to Troy High*. To further his interests, we enrolled him for the 1976 summer school in the Phillipps Exeter Academy in New Hampshire.[1] As with Emma Willard, we think this was a worthwhile investment. Besides learning to appreciate life at home, which he could expect to last also through his college years at RPI, he improved markedly in his ability in English writing, and found a strong interest in astronomy. Thereafter, he spent many a night in the dark, to find and observe the stars with his telescope. A hobby he still has not fully lost today, although it is taking a distant second place relative to his interest in philately.

The close connection between Caryn and Brent, despite minor occasional quarrels at home, came to light when we got a request to pay for a phone call from Brent while he was at Exeter. When Heidel answered: "Yes we accept the call, it is from our son," the operator replied: "I am sorry, the call is not for you, it is to go to Caryn Wunderlich in Pennsylvania." Caryn was at a summer job with Aldrich Chemicals. Brent rather liked to talk to her about whatever problem he had at the moment. We gladly paid for the conversation, but were not informed about the problem.

[1] The Academy was established in 1781 by Dr. John Phillips and wife Elizabeth. It focuses on intellectually, physically, and spiritually enriching the 'whole' person. Exeter is noted for its Harkness education, a system based on a conference format of teacher and student interaction, similar to the Aristotelian method of learning through asking questions and creating discussions. It is a coeducational, independent, private boarding school for grades 9–12 in Exeter, NH, fifty miles north of Boston. The most famous early alumnus is Daniel Webster (class of 1796). The enrollment is about 1,000 students for a faculty of 200. Each summer, Phillips Exeter hosts over 700 students for the five-week program of academic study Brent attended. The three mottos on the Academy's seal are: "Non Sibi," not for oneself; "Finis origine pendet," the end depends on the beginning; and "Χαριτι Θεου" by the grace of God.

On graduation from Troy High, Brent also entered RPI as a freshman with interest in chemistry and engineering. While Caryn's work was more connected to research, Brent did an outstanding job as an assistant in my laboratory, contributing to the ATHAS Data Bank (see Appendix A, publications #156,177–179).

Soon after Caryn and Brent had reached the age to get a driver's permit, they took driving instructions offered by the school. Naturally our help was needed to make sure both could drive with a manual shift. At every suitable occasion, we now had chauffeurs (with a licenced driver in the next seat). Caryn could make her driver's licence relatively early, but Brent had to wait until graduation from RPI. Our insurance would have increased by more than 100% if we would have a newly licenced, male driver under 25 years of age in the family. By the time Brent left with a BS in Management and Industrial Engineering for his job with Bausch and Lomb in Rochester, NY, he had saved enough money from student loans (which he did not spend since he lived at home) to buy the car, shown in Figure 81.

Figure 81 *Brent's first car in 1981, a used American Motors 'Pacer.' By now a much appreciated antique.*

Brent's second job, starting in 1983, was with Carrier Corporation in Syracuse, the 'World Leader in Air-conditioning' and a part of the United Technologies Co. When established in this job, Brent went to Syracuse University as an evening student and received his master's degree in 1987. We could arrange to attend his graduation by traveling for the event from Germany where we were on a sabbatic (see page 8-112–128). Figure 82 shows the convocation in the arena of the university. Figure 83 adds two more snapshots of the subsequent festivities. Brent and Caryn are shown in Figure 83A working on hood and gown. Figure 83B is a picture of the family and friend. Note that by then our family circle had expanded (see page 11-4).

Brent met Anna Kosar, a neighbor shortly after his move to Syracuse. They were married on November 24, 1984, in Utica, NY, as summarized in the Figures 84. One of Anna's two brothers,

Figure 82 *Ceremony for Brent's graduation from Syracuse University with an M.B.A. in May 1987.*

Figure 83B *Brent, MBA, with family and friend.*
← **Figure 83A** *Brent being readied by Caryn.*

John, and Brents friend Venkat(eswara) Rao from Troy High are shown on the left of Figure 83B. Caryn and Anna's sister Rosie underneath, Anna is in the center, next to Brent and Heidel. After this graduation, we returned to Germany to the sabbatic in Ulm (1987, see the pages 8-121–128).

Brent and Anna's wedding in 1984 was the occasion to bring friends and relatives together. We never had such a big family affair again (between 200 and 250 guests). Most of the guests were from the bride's side because the whole congregation of the Volodymyr Ukrainian Church was always invited for the ceremony and dinner when one of their midst married.

Figure 84A is a picture of the "Polterabend" (Rehearsal Dinner). We wanted to combine German and Ukrainian customs. Heidel had collected a sizable pile of old and broken china. It was now wrapped as a wedding gift together with two smaller packages. First, the small packages were given to be unwrapped by Anna and Brent. They contained, surprisingly, a small broom and shovel. Next, Heidel picked up the big package, and promptly dropped it in front of the table. To the shock of all present, this shattered all that could still be broken. Then the explanation was given. In Germany, all friends and neighbors would come the evening before a wedding and break china in the street in front of the house ("poltern"), as a wish of good luck. The bride and groom have to

Figure 84A *Rehearsal Dinner, on November 23, 1984, at the Sheridan Hotel.* ***From left to right:*** *Shahla and Eberhard W., Rosalie and John K., Tina (friend of Anna)and Venkat (friend of Brent);*
Anna and Brent
Vera (friend of Anna) Daniel K., Maria K. (mother of the bride), Heidel, Papa, and Caryn.

Figure 84B *Official Wedding Photo 11/24/84.*

Figure 84C *The head-table for the celebration.*

sweep up the shards, and reward the guests with a piece of cake or similar. We had slightly altered the custom. By having informed the service staff, it went without problems and, naturally the broom and shovel came in handy for the clean up by the bride and groom. Figure 84B shows the official wedding picture with Rev. Bohdan Smyk, who officiated. Figure 84C is a view of the head table in the basement of the church where the dinner for all was served, and the festivities with a band, skits, etc. took place. A memorable occasion which lasted until early in the morning.

To have at least a small counterweight to the large number of Ukrainians, *we had invited all our friends besides the few relatives*. We finally accounted for 34 guests (only). There were the *Haags*, we had met in Chicago and with whom we frequently exchanged visits (see page 6-24 and Figure 38). They drove up from New Jersey. Next, were three colleagues from RPI, Professor *Wiedemeier* (see page 8-67, full professor since 1971) with his friend of many years, Carleen *Goeckel*. Although Carleen has moved on, we are still in contact with her, at least once a year. The other colleagues were Professor Jim and Lotte *Moore* (see page 8-85). He had spent a year as a postdoctoral fellow at the University of Mainz 1967/68 where he met his wife. This year overlapped with our sabbatic leave in Mainz, but we did not meet there. Physical and organic chemistry of polymers were seemingly too far apart. The third colleague was Professor Heinrich *Medicus* and his wife Hildegard from the physics department, already introduced on page 8-10. Hildegard was an orthodontist in Troy. Of my research group, we had invited two Chinese graduate students, Stephen *Cheng* and Miao-yong *Cao* (see Appendix A), and my secretary Beth *McGraw*, who would later look after our house during the next sabbatic. The Chinese were to get a look at the different marriage ceremony in the western world. My former student and postdoctoral fellow Dr. Umesh *Gaur*, who got to know Brent during their common work in the laboratory (see Appendix A) came from New Jersey. Our neighbors Sandy and John *Youker* were there (see page 8-11). There are also the friends from Troy we still have frequent contact with, Susi and Ivo *Buhac*. Ivo was a physician specializing in gastroenterology and two years later was crucial in detecting my kidney cancer early enough, so that I survived. The Buhacs were part of our (then still) Yugoslavian connection. We got introduced to them by my former student Nico Gjaja (see Appendix A) who, on his name day, had an annual party of all Yugoslavs. With the later break-up of Yugoslavia, this group also broke up into Croats, Slovenes, and Serbs. From the same connection were Zlata and Phil *Phillips* and Angelica and Miran *Milkovic*. Angelica was also the official photographer at the wedding. Elli and

Henry *Maag* were there, another Swiss connection, who also owned property in the Adirondack near ours, as will be described below. Next, Inge and Milt *Schiffman* are to be mentioned (see page 8-11, Footnote). Milt was our Optometrist. And last, another couple we had close contact with was DeForest *Pitt* and Charlotte M. *Blandy*. 'De' was a supreme court justice of New York, and Charlotte M., a graduate from Emma Willard, was his long time companion and from 1986 on, his wife. Charlotte M. was a postal employee in Wynantskill, where she got to know Heidel from her frequent need to send packages, letters, and postcards to overseas. To this list, the relatives and friends of Brent, listed in Figure 84A need to be added to describe the full lopsided party where the main language was Ukrainian with small sprinkles of German. Our immediate family could not make it to Brent's wedding as complete as we had hoped. Vati had a heart infarct before, and was not sufficiently recovered that he and Mutti could undertake the trip. Joachim and Manfred, living still in Nebraska (see the pages 7-16–17), and Frank, who had by then moved to Columbia, SC, (see pages 6-31–32) could not arrange their schedules for the long travel to Utica, NY.

Brent's marriage was not the first broadening of our German-American family. Our nephew Eberhard was married in 1978 to Shahla Mozzafari (see Figure 84A) who came from Iran. Both were studying at MIT in Boston for their PhD and were often guests of ours in Troy. Shahla had studied at the American University in Beirut, Lebanon, before coming to MIT. Naturally, these were not to remain the only different heritages to be joined to our family. While Frank was married already before coming to the US, Joachim and Manfred married American born wives. Caryn was married in 2006 to the Irish-born Brian (see Figure 29 in Chapter 11 and bottom left of page 11-4). Eberhard's son, Roland, as the following generation, was married in 2008 to Sherry Wu, a Chinese-born girl. We will have to see what the future will bring.

The *prior larger family gatherings were the golden wedding anniversaries*. The first was was Vati and Mutti's in 1975 in Hastings, NE, shown in Figure 85A. In the front row left is Joachim with Ian, and Manfred with David. Second from the right is Frank with Eberhard (above). Brent and Caryn are left and right from Eberhard. The four daughters-in-law of Mutti and Vati are in the top row, left (Carol, Dagmar, Heidel and Sharon). Mutti and Tante Lieschen are in the middle.

The golden wedding anniversary of Heidel's parents in Bergheim, Germany, was three years later than Vati and Mutti's. I could not make this trip, so Heidel had to represent our family. Figure 85B is a snapshot in front of the restaurant for the dinner. Above Heidel, to the left are Irmchen with husband Henning, continuing in the back row are Onkel Sepp with Tante Gertrud and daughter Ria. On the far right is Onkel Hubert with Tante Grete II, below. Next to Mama and Papa is Tante Luzie,

Figure 85A *Golden Anniversary in Hastings, NE, 1975.*

Figure 85B *Glolden Anniversary, 1978.*

diagonal above are Tante Marie and Tante Käthe (a cousin of Mama's) with her son Werner Bückner on her left. On her right is Charlotte, wife of cousin Herbert who (most likely) took the picture. For the connections with the relations in Figures 85A and B, see the pages 11-2 and 3.

In Germany, the 50th anniversary is more widely recognized than in the US. The "Bürgermeister" (Mayor), "Oberkreisdirektor Erftkreis" (District Director), and the "Regierungspräsident Köln" (District President of Cologne and surrounding area) sent each a separate congratulatory letter and a monetary gift or flowers. A nice and appreciated gesture.

By the time the 1980s were reached, one could note a distinct **decrease of interest in polymer chemistry at RPI**. The strongest supporters in building up the study of macromolecules had been our chairman, Professor Janz and the Dean, Professor Bauer. This can be seen from the help I had in my research and teaching, as well as in obtaining the time needed to complete the three-volume treatise on Macromolecular Physics, described on pages 8-57–58, 62–63.

In my early years at RPI, an effort in polymer science existed already in Chemical Engineering through Profs. *Sternstein* and *Hansen*, and in Mechanical Engineering through Prof. *Birnboim*. Somewhat later, Professor *Chung* joined RPI in Materials Science and Engineering . He became my neighbor in the MRC. To cover a full program in polymer science, it was usually assumed that up to 10 professors were needed at one university to cover a full range of polymer expertise. The polymer effort at RPI was initiated by Dean *Bauer* (PhD 1929, University of Wisconsin), who before his deanship had worked on boron hydrides as rocket fuel and the rheology of polymers. My joining in 1963 added thermal properties and the solid state physical chemistry of macromolecules (see page 7-19). To expand this group, in 1965, Professor *Billmeyer* was added as another senior member of the Chemistry Department. He came from the Central Research of the DuPont Co., see page 7-27 (PhD, Cornell University, where he worked with Profs. Flory and Debye). His 'Textbook of Polymer Science' was fist published in 1961 and then saw two further editions in 1970 and 1984 while he was at RPI (Wiley-Interscience, New York). It is still in use in many places. After arriving at RPI, he quickly set up a first-class polymer characterization laboratory and developed his other specialty, color science and technology. Next, Professor Sonja *Krause* joined us in 1967 as assistant professor (PhD from the University of California, Berkeley). Her research dealt with the dilute polymer solution, block copolymers, polymer compatibility, and transient electric birefringence. All three of us were housed in the MRC building. The expertise in polymer synthesis was provided by Professor J. *Moore*, who was hired in 1969 (see also page 8-83) he had earned a PhD from the Polytechnic Institute of Brooklyn, and before coming to RPI was at the University of Michigan, Ann Arbor, MI, as a research associate with Professor C. G. Overberger, a well-known organic polymer chemist.

We slowly approached the breadth necessary as a major polymer center. The others in the US were the Polytechnic Institute of Brooklyn (now Polytechnic Institute of New York University), the University of Massachusetts in Amherst, MA, Case-Western Reserve University in Cleveland, OH, and the University of Akron in Akron, OH. None of the better-known universities have sizeable polymer groups. Efforts were made at RPI to hire a leader for our polymer group, who could try to raise funds for a major 'Center of Polymer Research.' Various researchers were invited to the campus in support of the polymer work and to be looked at as possible hires. In fall 1968, Dr. Richard Bonart, whom I had met during my first sabbatic in Germany (see Appendix C), was a one-semester Visiting Professor in Chemistry from the Bayer AG in Leverkusen. He lectured on 'Low Angle X-ray Scattering in the Field of Polymers.' In spring 1970, Professor R. H. Marchessault,

Chairman, Department of Chemistry of the University of Montréal, gave lectures on 'Solids and Solutions of Polymers and Biopolymers.' Other, shorter visits were by Drs. P. Lindenmeyer and Ron Eby, both well known polymer scientists with possibly leadership potential.

But, slowly, the interest of the Department in creating a major polymer effort seemed to decline. In retrospect, the start of this change coincided with the retirement of Dean Bauer in 1972 and the end of the two 5-year terms of Professor Janz as Chairman in 1973. The new department leadership, Professor Kevin T. Potts, followed in 1980 by Professor James P. Ferris, was more indifferent toward polymers. The new chemistry building, Cogswell Laboratory, was dedicated in 1972 and was already called the 'Chemistry and Biology Research Building.' It unified the Department, except for the polymer research group. We were now split between the synthesis, done in Cogswell, and the properties research in MRC. More problematic was that no serious attempt at further growth of the 'Polymer Group' was supported later on.

The major planning and *fund-raising effort of RPI in 1980 for 'Rensselaer 2000'* still lists in chemistry the three 'key' areas of growth (in order) as: polymer, solid state, and bio-organic. An outline to adjust the teaching of chemistry in this direction, the 'Chemistry 2000,' on which Dr. Wiedemeier and I initially worked, was ultimately not accepted (see page 8-67 and Appendix A, #255). In 1980, the Chemistry Department counted 31 active professors of which four were designated as polymer scientists, today (2008) it lists 20 active chemistry professors, of which seven are designated as biochemists. Only two are having an interest in polymers, and none declare solid-state chemistry as their field of interest. Three additional professors of the Biology Department are listed in the chemistry roster (raising the total to 23). It underlines the major biochemical interest of the department to the exclusion of all others.

The discussions in the early 1980s were of interest, but when checking which of my suggestions were taken up, or at least considered, I came to the conclusion: None! I certainly could not afford such a waste of time, and thereafter, I hardly attended any further faculty meetings. My basic motto in these meetings had been that we must recognize that we are #2 in the race to the top spot among institutes of technology in the North-east (MIT being #1, although RPI has a slight seniority). To overtake #1, one cannot mimic its development—it has been there before. One must find something new and then build on this new strength. It is a risky undertaking, but one may have a better chance of success. Teaching better fundamentals by building biochemistry based on its true roots, the macromolecules, was my suggestion, but these 'roots' were cast aside.

The de-emphasis on polymers became clear when I now re-read the annual reports by the Department of Chemistry which I have kept in my files. Until the end of the leadership of Professor Janz, they were issued each year and were based on the input from the faculty with little editing. Counting the length of the description of the faculty activities in 1973 (by number of lines printed), 25% were contributed by the polymer group, being 13% of the faculty. In addition, 40% of the publications originated from professors of the polymer group. In 1983, the reports were severely edited to achieve equal input for each professor. Without changes in number, the 13% of the polymer faculty provided the permitted 13% of activities, but still listed 28% of the publications. Extra effort, did not count anymore.

On my return to duty after the third sabbatic leave, mentioned on page 8-68, I found that my teaching duties for the spring term 1981 were unduly increased. My complaint to the chairman, Dr. Ferris, did not lead to any change. Because the memo contains quantitative information for a typical semester's work of mine, it is recreated on the next page as it was transmitted to Dr. Ferris, based on my file-copy (onion-skin paper). Some added explanations are enclosed in brackets [ ].

Inter-Departmental Memorandum

To J. P. Ferris, Chairman, Department of Chemistry — Date 1/13/81
From Bernhard Wunderlich
Subject: Teaching for Spring 1981

Last week I received my teaching assignment for the Spring term which listed:

1. A total of eight teaching hours which included one 400-level course with 70 students, two recitation hours (usually 60 students) and one laboratory.
2. A discussion with Dr. Hepfinger [the Associate Chairman] revealed that there may be a good chance that little or no help will be available for grading [through teaching assistants].
3. The memo sent with the assignment of the teaching duty contained a graph suggesting that if 40% charge-out [from research grants] could be obtained, the teaching duty should be about three contact hours each semester, linearly increasing to 10 h for 0% charge-out.

I would like to discuss the following points:

A. As you may know, I have kept over the last 10 years half-hourly records of my actual time spent on various duties. Averaging this statistically, I find that for Fall and Spring terms under normal conditions (excluding sabbatic leaves) I spent an average of:

18.5 h each week on teaching (5.05 credit hours per term)
9.0 h each week on graduate student supervision (4–6 graduate students, 2–3 postdocs, 0–3 technicians, undergraduates and evening students).
6.5 h each week on mail, review of proposals and papers, and writing of proposals, etc.
13.3 h each week on scholarly work (books, papers, preparation of lectures for seminars, meetings, etc.).
11.7 h each week on Departmental and Institute business.
1.7 h each week with visitors and at seminars at RPI.

Total: 60.7 h each week.

Any time during the semester (of 90 working days) I spent away from campus is made up on Saturdays and Sundays. As you can see, there is no way to increase the total working hours. In fact, I have been asked to decrease these hours toward more reasonable levels [by my family].

This semester's increase of average teaching load from 5 to 8 h represents an increase of 60%, or increases the weekly teaching load from 18.5 hours by more than 10, to 29 h. This calls for drastic cut-backs in:

1. Attention to graduate students,
2. Proposal writing and reviewing,
3. Scholarly work.

B. I would also like to know if the suggested reduction in teaching hours with charge-out will be permanent departmental policy from now on. Last time this question came up, I discussed this with the NSF monitor. I was told that total funding would not change. At that time, the then department chairman [Dr. Janz] decided rather to use the money to support students than to increase charge-out. In my case, it would mean, if this is future department policy, I would suggest to support myself through NSF for 40% (dropping support for two graduate students or one postdoctoral fellow). This would permit me to set up an appropriate weekly schedule of 12 h teaching, 8 h graduate student supervision, 6.5 h mail, etc., 13 h scholarly work and 13.4 h Departmental business and visitors. This makes an acceptable 53 h weekly work-load.

Please let me know your response, so that updated renewal proposals can be written appropriately.

BW:bm

The written reply of Dr. Ferris was totally unacceptable. He said in essence: This will be departmental policy from now on. I will be happy to discuss it with you further. If you want to increase the charge-out for the academic year, let me know soon. The further discussion was not very friendly on either side. For the present semester, I promised fewer proposals, and all NATAS activities to be dropped. I suggested, no further lectures outside RPI, and no promise of support to teaching assistants who chose to do research with me. (I always found support for teaching assistants when they joined my group, many of the other professors did not.) For the next year, I would cancel one postdoctoral appointment, and initiate the needed changes in my NSF grant.

I did not tell the truth! All I did, was work more hours, go to no further faculty meetings, which I already had decided before (see page 8-86), and I refused most other work for the Department. Next, I worked to find support for a sabbatic in 1986/87, and, as I know now, became receptive to leaving RPI. This happened without my actively seeking it, as will be described below.

The decline of polymer science at RPI was not as obvious to me as it became when writing about it now. It was a gradual process. Each step was initially seen as a bothersome, isolated development. As time went on, the unpleasant trend accelerated. When Professor Billmeyer retired in 1984, he was not replaced. A final polymer hire, was Professor G. Wnek in 1987 (PhD from the University of Massachusetts, 1980, followed by spending some time at the Department of Materials Science and Engineering at MIT). I left shortly after he arrived, and he stayed only until 1996.

In the meantime, there were **more pleasant developments in Troy** to be commented on. Had we not felt at home in Troy? Had I not reported that Troy is an ideal location (see page 8-9), such that we could see no reason ever to leave the city where Uncle Sam was buried! One of the special attractions not yet mentioned is the *Sterling and Francine Clark*[1] *Art Institute*, built in 1955. Figure 86A shows our family in front of the building, photographed by one of the several visitors we took through the collection. It is located just mi 30 mi east of Troy in Williamstown, MA, a beautiful, small town which is also home to Williams College. Our interest in the museum's collection focused on the European and American paintings and sculptures. The greatest strengths of the collection are the French impressionists.

A game we always play when visiting an interesting museum is to suggest which one item would we take along if we could? My favorite picture from the Institute was a Claude Monet (1840–1926), The Duck Pond, and Heidel's favorite was a small painting by Camille Pissarro (1830–1903), depicting a country road on a rainy day. They are reproduced in Figures 86B and C.

Figure 86A *The Clark Art Institute in Williamstown.*

---

[1] Robert Sterling Clark was born in New York City, 1877 (son of Alfred Corning and Elizabeth Scriven Clark), an heir to the sewing machine fortune (his grandfather had been Isaac Singer's business partner). He also inherited his father's interest in art collecting. In 1910 he settled in Paris and began collecting. Together with his wife Francine, they created a remarkable collection of paintings, silver, sculptures, porcelain, drawings, and prints, fully according to their own taste.

Figure 86B *Monet's* *and* **C**, *Pissarro's painting.*

In the meantime, our family in Troy had gotten smaller. First, Caryn had left home for her 'Junior Year Abroad' at the ETH, the "Eidgenössische Technische Hochschule" in Zürich, Switzerland. This student exchange between RPI and the ETH was arranged annually by Professor Medicus (see page 8-10). The Swiss portion of the exchange was relatively easy to arrange. The state paid the tuition for all students. Only board and room plus transportation needed to be paid by the parents. The tuition at RPI, in contrast, was rather high, and there were no stipends available. To raise the needed money, Dr. Medicus had approached a Swiss importer. He donated a sizeable amount of Swiss cheese each year with all proceeds from sales going to the tuition of the Swiss students at RPI. The past, and future exchange students were involved in these annual sales. We certainly bought our share of the delicious cheeses.

For the senior year, Caryn was back at RPI and I could personally present her the bachelor's degree in chemistry at the ceremony of the Department. Next, *Caryn had to find a medical school and arrange for the finances*. In contrast to graduate study in chemistry, which most students can finance with teaching or research assistantships, the substantial medical school tuition needed to be paid by the student. The best solution was to share the financial burden equally by raising the money by refinancing the mortgage on our house, and to get a student loan for Caryn.

There were a number of medical school interviews to travel to. At the end, Caryn had several good possibilities. She chose the Pritzker School of Medicine, a part of the University of Chicago. She earned her MD on June 10, 1982 and we could join the celebration at the Palmer House in Chicago and watch her taking 'The Oath of Hippocrates.' Figure 87 shows the celebrants. The following day, Brent had to return to his job in Rochester, while the three of us went on to Hastings, NE, to visit the rest of the family.

Figure 87 *Doctor Wunderlich, Brent and Heidel.*

By 1981, Brent had left Troy after his graduation from RPI, as shown in Figure 81, above. This means, we should have had an empty nest. We had, however, as Heidel called it, purchased a new baby. *In May 1977, we bought 100 acres of the Adirondack State Park,*[1] near Stony Creek, NY for $10,500. To enjoy the wilderness,

[1] The Adirondack Park is the largest park in the contiguous US and the largest area protected by any US state. It covers some 6.1 million acres (the area of the combined Yellowstone, Yosemite, Grand Canyon, Glacier, and Great Smoky Mountains National Parks). More than half of the land within the Park is privately owned, including several villages, hamlets, and our tract of land.

we were required to construct a new entrance road. The existing road was too steep, prone to erosion, and only poorly maneuverable by car. Next, we set up a simple gazebo, as shown in Figure 88A. To build a second home had to wait until we would have some extra income. For the clearing of the underbrush, straightening the logging roads, and 'maintaining' the beaver dams, I bought a Ford tractor. It was equipped with four-wheel drive, detachable front loader, rear scraper, and a power pump. Figure 88B shows it in operation, handling the snow. The road from the main spur out of Stony Creek was plowed diligently by the county (after an occasional bottle of brandy), so I only needed to handle the entrance up to the high area.

When the second home was ultimately built, it looked as shown in Figure 88C. We named it *the 'Hideaway Beaver Lodge*.' Much of the interior and exterior finishing we did ourselves, so that many weekend visits were spent working instead of relaxing. Still, it was a different experience. Often I would leave already on Friday evening, and Heidel would come on Saturday with Brent, and all of us would return on Sunday afternoon with dinner on the road. The trip was about 1½ hours, and we had made sure to have all modern amenities, ranging from a Swedish water-less toilet, to a hot shower, a kitchen with a (propane) gas stove and oven, a telephone, and an oversized hot water tank, fed with creek water by trucking the water and pumping it into the system with 200 psi.

Figure 88A *Heidel, ready to work.*

Figure 88B *The four-wheel-drive Ford in the snow.*

Figure 88C *The completed Hideaway Beaver Lodge.*

Electricity we had in form of a big-sized battery, charged with solar cells. Not quite living in the wilderness!

Of the interesting happenings there, let me relay three, dealing with the animals we shared the property with.

First there were the beavers. We had three large beaver ponds of perhaps one to five acres each, and about two to ten feet depth. One of these, I would have liked to drain. With the tractor, I went into the shallow lake and, bit by bit, pulled the dam aside. A hard job, they had built the dam well. The lake was then draining down a small creek. Next morning, the water was largely gone and I began to clean the lake bottom. Next week, the eager beavers had rebuilt the dam and made it even higher. On the second attempt, I broke the steering gear and could barely make it out of the lake. The drive back to the lodge was by driving backwards and steering like a tank by braking one wheel or the other. The beavers had won, I gave up my fight.

The next excitement was caused by a bear. Not that I ever saw one, but one day, when I was alone in the woods cleaning up one of our walks, I heard barking of dogs, yelling, and shooting. Seemingly this was on our property. After a while, it stopped. When I got back to the lodge, Heidel told me that someone had wounded a bear with bow and arrow, but could not track him, so he had called the game warden who brought the dogs. Finally, they found and killed the wounded bear. There was no restricted hunting with bow and arrow! Yes, indeed, it was on our property, and I could easily have met the wounded bear. Unfortunately, Heidel did not ask for a share of the kill so that we could try a bear steak or the paws which supposedly taste particularly good.

The third animal was a porcupine. In this case, at first I noted something had gnawed on the seat of my tractor. Each week a little more plastic was gone. Then the insulation of the exposed wires disappeared. Asking around, I was told that porcupines like this kind of 'spicy' food. Indeed, one day, Brent discovered a porcupine sitting on the branches of a tree close by. I told Brent to scare him away with a rock. He did so from quite a distance, but he hit the porcupine squarely in the head. From this time on, the plastic and rubber parts of my tractor were safe.

The Hideaway Beaver Lodge was not only our hideaway, but also my study away from home. I could look out of the front window from my desk, so that several papers were written there within a pleasant environment and without interruptions. This was the time, when hand-written manuscripts were delivered to the secretary for typing, i.e., no word processor was needed. On frequent occasions, we also would invite friends of ours for coffee and long walks.

For Yuri Cheban from Moscow, one of my postdoctoral fellows, a stay at the lodge was particularly appreciated. He was in Troy on *an exchange program supported by the US State Department*. He was selected by the Soviet equivalent, and well trained politically. In fact, although relatively young, he was the Vice Chairman of the Physics Department of the Lumumba People's Friendship University, a school founded to educate students from Asia, Africa, and Latin America. While my coworkers wanted to tell him how much better life is in the US, I had sufficient experience not to try to convince him with words. I showed him the huge construction project of new highways and state government buildings in Albany, under construction at that time. He was impressed. His comment: "We have this also in Moscow." Certainly you have, I said, but this is only Albany, the capital of New York State! After a while in the US, when he was shown a specially big selection of goods in a department store, he said: "We also have all this in the Soviet Union—but not so much." Finally, closer to the time of his return to the Soviet Union, when he was shown a big, new main-frame computer, he said with envy: "We do not have this in the USSR—but we will have it soon." He slowly was learning, without realizing it.

Perhaps this information was not good for Yuri's survival. We only received once two identical, clandestine Christmas cards mailed from different locations, but never could make contact with him again, despite having friends of ours look for him in Moscow after perestroika (see Footnote on page 9-23). We only heard unconfirmed rumors that he was transferred to Ghana.

This was also the time when the American students in Moscow were restricted by the Soviet Government to 30 mi from their residence. The US State Department would counter this by putting the same restrictions on our visitors, and I, as the host of Yuri Cheban, was told on a visit by the local FBI agent to report any infraction. Naturally, I refused. My comment was, do not restrict our visitors. Rather, expand the freedom of the Russian students and support them with books they cannot get in the USSR! When Yuri is with me at meetings, or in the Adirondacks there is no danger. If you want to know where Yuri is at other times, you have enough means to look for him.

From his college time, Yuri was an accomplished wrestler, and I always admired when he was putting huge logs into place for the bridges I was building for my tractor to go over the various creeks. Yuri also enjoyed the solitude and safety of private conversation out there. He told me that he did not like his present job in Moscow. "Too much administration and too much politics." He hoped with what he had learned in our laboratory to get back into research. His knowledge in computer programming helped us to set up the first ATHAS calculations (see Appendix A, # 191).

The extra income needed to build the log home in Figure 88C came from a job for the DuPont as an expert witness in Federal Court in Wilmington, DE, from November 1977 to May 1978 in _a civil suit about the patent rights to polypropylene_. Over the years, I was asked occasionally to act as a technical expert in law suits by various departments (see Appendix C, page C-4) . Usually, after a few visits, the problem got settled. The polypropylene case was different.[1]

The basic synthesis was discovered by Prof. Dr. Karl Ziegler in 1953 by using a new catalyst to polymerize ethylene (CH_2=CH_2) to linear polyethylene. Ziegler was director of the Max Planck Institute for Coal Research in Mühlheim (Ruhr), and received the Nobel prize in 1963 (with Prof. Guilio Natta) for the development of the catalysts. To polymerize propylene was more difficult. Propylene has a similar double bond (CH_2=CH–CH_3), but the resulting polymer has an extra methyl group (–CH_3) which can be located on either side of the chain. Only highly regular (isotactic) polypropylene can crystallize and give a useful plastic, thus, Ziegler turned only slowly to synthesize polypropylene. Natta, working for Montecatini Chemical Company of Milan, was aware of Ziegler's work and secretly filed a polypropylene patent 10 days before Ziegler.

In 1958, the US Patent Office observed 'interference' involving five applicants for polypropylene patents and asked the companies to present evidence that they had invented a "normally solid polypropylene, consisting essentially of recurring propylene units, having a substantial crystalline polypropylene content." Other companies sued over other aspects of the process. It was a heaven for attorneys. For a scientist, this phrase describing the invention is not precise! What means 'normally solid,' 'essentially recurring,' or 'substantially crystalline'?

It took until 1971 when the patent interference seemed settled after 18,000 pages of testimony given by 106 witnesses. The Patent Office awarded priority and a patent for the invention to Montedison on February 6, 1973 (formerly Montecatini, patent of Natta and coworkers). But! The DuPont, Phillips, and Standard Oil companies appealed, resulting in the 1977/78 retrial in Wilmington, mentioned above, presided over by Judge C. M. Wright. There, I could contribute to the additional 15,000 pages of testimony, which finally overturned the earlier decision.

---

[1] A more detailed description of the polypropylene discovery and the 278 polypropylene patents around the world is given in a general article of 4,400 words by D. B. Sicilia called: 'A Most Invented Invention.' Published in Invention & Technology Magazine, Volume **6**/1, 44 (1990). [It can be read on line at: *www.americanheritage.com/articles/magazine/it/1990/1/1990_1_44.shtml*].

On November 28, 1977, I was given a separate contract as a consultant and expert witness for the civil litigation that had begun on September 19. The Legal Department paid much better. My expertise was to explain: "What is a solid state?" "What is substantially crystalline?" And to a lesser degree: "How much do impurities, included in the macromolecule, change the crystalline melting point?" Not only did I have to testify about these matters, but also read all the testimony of the other experts and check where they were not up to date. Since dozens of experts were invited from all over the world and we knew each other, we had our own scientific meeting, dealing with the more interesting scientific problems of the day. Naturally, we were not permitted to be present inside the court room while others testified, nor were we to discuss the details of the case.

Overall, I think I made a good witness and could teach the judge what is a solid and that, at the time of the patents, substantially crystalline may have meant less than 50%. Also, I added that all equilibrium-derived conclusions by the other famous witnesses are suspect. Semicrystalline polymers are not in equilibrium!

After all the testimony, Dr. Roger A. Hines, attorney for the DuPont, whom I knew well by then, issued his post-trial brief as a bound book of 192 pages (September 8, 1978). The three other parties in the suit, I expect, filed similar briefs. In his decision of 200 pages, Judge Wright awarded the priority discovery to Phillips (1/11/1980). Again, the losers appealed, but the Federal Appeals Court upheld the decision, and the US Supreme Court refused to hear the case. Finally, 30 years after the discovery, Hogan and Banks of Phillips Petroleum Co. were granted the patent. My testimony did not help the DuPont, but it helped to establish 'what is a solid,' a topic I first discussed in 1969 for my ACS Short Course (Appendix A, #37)—and it paid for the log cabin of Figure 88C.

Of the many national and international lectures outside of RPI, listed in Appendix C, I would like to detail **several interesting travel destinations in the 1970s and 1980s**. The trip into our past to Europe in 1972 is described already on pages 8-59–62, and the IUPAC trip in 1974 to South America, on pages 8-32–35. In the first half of the 1970s the airlines had a special promotion. When flying from coast to coast, one could interrupt the flight without additional cost. On flying from New York to Los Angeles, I could include a stop inside the US or also in Mexico City!

The first time I made use of this offer, was after attending a Gordon Conference in Santa Barbara, CA in January of 1971. The conference was held at the Miramar Motel, bordering the Pacific Ocean. Figure 89 is a view out of one of the blocks of guest rooms. As usual, the Gordon Conference was a very relaxed week, filled with discussions on nucleation, growth and properties of macromolecular crystals (see pages 8-16–17). On the way to the airport in Los Angeles, I stopped for a visit at the California Institute of Technology in Pasadena, CA. In addition to the lecture, an afternoon's walk through the interesting Huntington Gardens, nearby, was arranged.

Figure 89 View from the *Motel in Santa Barbara.*

Following this conference travel, I made use of the airline promotion and interrupted my flight to Troy in Mexico City. The extra time was possible due to my special leave (see Appendix C, page C-13).

<u>*From Mexico City I flew to Tampico*</u> and joined the Wally Byam Caravan to travel for two weeks with Vati and Mutti in their trailer (see pages 1-33, 8-59, and Figure 63). In 1969, Vati had retired and went each January from Texas, where he stayed in Brownsville through the winter, to Mexico in his Airstream trailer. Figure 90A shows the caravan of more than 100 trailers parked in a sports arena. During the evening meeting shown in the figure, the program for the next day was discussed, consisting of sightseeing in the area by bus and evening entertainment by the local talent. Travel days were planned in detail, so that the Caravan could assemble over a period of an hour or two in the next, prepared location with police protection. Usually, a field outside the city or a sufficiently large sports arena was rented for the stays of two to five days.

Figure 90A *Drivers meeting of the Caravan, 1971.*

This travel was an anthropological excursion. It went from Tampico (in Tamaulipas) to Veracruz (in Campeche) to Merida (in Yucatán) with intermediate stops. There I left the caravan, which drove back to the US, while I flew to Troy via Mexico City.

The Olmec civilization of 1400–400 BC is represented by Figure 90B. Next, we visited the ruins of El Tajin (1st to 12th century) and Palenque from the early classic period of the Mayas ≈400–800 AD (UNESCO World Heritage Sites since 1992 and 1987, respectively). Then we stayed in Merida with bus trips to Chichén Itzá, Uxmal (UNESCO World Heritage Sites since 1988 and1996, respectively), and Kabah, shown in Figure 90C. The last three towns are from the end of the Maya

B *One of Olmec monumental sculptures photographed in the Museum Park of Villa Hermosa (excavated in La Venta).*

C *Vati and Mutti, at the entrance to the unrestored ruins of Kabah, a site near Chichén Itzá, visited 1971.*

civilization (≈1200 AD). The ruins in Figure 90C are from the 7th–11th century. All archeological sites were of interest to me. When Professor Allen died in 1974 (see page 8-80), his widow gave me his collection of vintage books about the Mayas, which I have expanded with modern books and DVD recordings to document the substantial recent increase in knowledge about the Maya culture.

In 1973, I attended the APS Meeting in San Diego, CA, and gave lectures at IBM in San José. During this trip, I flew to Mexico City and stayed at the Holiday Inn at the airport for the weekend. The first day, after renting a car, I drove to Teotihuacan, the pre-Aztec site which reached its zenith between 150 and 450 AD (UNESCO World heritage site since 1987). Figure 91A is an overview of the 'Avenue of the Dead.' The next day, I traveled to Cholula. Its pyramid is the biggest in the world. It beats the Egyptian pyramids in volume (but not in age or in height). Cholula was contemporary with Teotihuacan, but survived to the time of the Aztecs. Figure 91 B shows a picture from 1973 when the restoration had just started.

The Aztec empire (1325–1521), with its capital Tenochtitlan, collapsed with the arrival of Hernán Cortés. Mexico City (of which the historic center is a UNESCO World heritage site since 1987) was built on top of Tenochtitlan. Its Museo Nacional de Antropología is the center of Aztec artifacts. In 1985, while at a meeting (see Appendix C), Heidel and I had a chance spend time there. The colonial period of Mexico also caught my interest when collecting coins (see Figure 55D).

Figure 91A *Teotihuacan, near Mexico City. View from the pyramid of the moon with a fish-eye lens. The pyramide of the sun is in the back, left of center.*

Figure 91B *The pyramid in Cholula On conquering of the Aztecs by Cortés, the cathedral was built on top of the pyramid.*

Two more stops on coast to coast flights allowed me to have weekend vacations. One was in July 1973 in connection with lectures at the University of Utah, Salt Lake City, UT. The stop led to 2½ days in Mesa Verde National Park (a UNESCO World Heritage Site since 1978) and study the buildings of the ancient, cliff dwelling Pueblo Indians (classic period 1100–1275, not connected to the Mesoamerican Indians, but to me, showing great similarities). Another APS Meeting in San Diego permitted a weekend in the Grand Canyon (another UNESCO World heritage Site since 1979). There I had an unforgettable plane ride. I had explored the canyon from the rim, and on the next day took to the air. The up and down drifts were so strong, that I was constantly pushed either towards the ceiling or the floor of the small plane. I wished that they shortened the flight. During the rough flight, I could not even change the film in my camera, so that no documentation exists.

Two Asian trips, to India and Japan, were supported by fellowships. The first, was arranged by the NSF in 1980. It supported travel to a symposium entitled: "New Frontiers in Polymer Science and Polymer Applications." The meeting was held in Madras (now Chennai) on the east coast to the Indian Ocean. Much effort was expended to make us welcome. The meeting was a success. A special effort was made to further discussions with the Indian scientists and induce us to visit top institutions like the Indian Institute of Technology in Madras (see also Appendix C).

Since Chennai was a largely 'dry' city and the local cuisine was mainly vegetarian, extra arrangements were made for the visitors. After it was established that there was no religious taboo for us, alcohol passes were issued. Alcoholic beverages could be purchased with the pass as long as the permitted quota was not exceeded. Without the occasional cold beer, served at a separate bar in the hotel, located behind a hidden door, the hot, humid climate would have been harder to bear. Interestingly, the students who took care of this arrangement collected the passes before we left. They could use the remaining quota for their final party (without religious permission?). Since much of the quota was left on most passes, I guess, it must have been a grand party.

A meat-line was arranged for lunch in the cafeteria. I tried it only once before going back to the much better tasting vegetarian food. Even the veggie burger was good!

The contrast of life in India, seen on a walk through town, is shown in Figures 92A and B. As usual, the special tourist attractions were also shown to us. Near "Chennai" we visited the temple town "Kancheepuram."

Figure 92A,B *Madras (Chennai) snapshots, 1980.*

Figure 92 C *The Taj Mahal, without tourists!*

A visit at the University of Delhi allowed me to make a bus tour to Agra and see the "Taj Mahal" (Figure 92C) and also Fort Agra. It took until 2007 to see these exquisite buildings again, which since 1983 are UNESCO World Heritage Sites of the Islamic Mughal architecture (summarized on pages 10-19–20 from our Trip Around the World). As a finishing touch, I interrupted my return flight in Athens for two days to see the Acropolis, also a UNESCO World Heritage Site (since 1987).

A second fellowship was by the 'Japanese Society for the Promotion of Science.' It covered my as well as Heidel's expenses. On May 16, 1983, we arrived at Tokyo's new Narita Airport, and visited 11 universities, 6 companies, and 3 government laboratories in 12 cities (see Appendix C). Just as the visit in 1966, summarized on pages 8-25–30, it was an exceptional trip.

Figure 93A *View of the Imperial Garden and Palace with a clear view of Mount Fuji on May 18, 1983.*

The first morning when we went for breakfast at the top floor of the Government Hotel, 'Takebashi Kaikan,' we had a clear view of Mount Fuji, ≈60 mi away, as captured in Figure 93A. For all Tokyoites, a rare occasion and a sign of good luck. The wooded area in front is the Imperial Garden with the hidden buildings of the Palace.

In typical Japanese style, we were picked up after breakfast and driven to the Ministry of Education and then greeted by a high official. He ceremoniously gave me ½ million Yen (¥) in a huge bundle of cash which I was afraid to count before signing for it ($1.00 was ≈240 ¥). It easily paid for all of our travel expenses. The US–Japan flight was separately arranged by Japan Airlines. A typical envelope of cash, given after a lecture by the companies, paid for some extra mementos.

Our second stop was Kyoto, reached by Shinkansen (see page 8-26). There, most of my former students resided. Again, we stayed at the Okazaki Hotel, now called 'Sunflower Kyoto.' Figure 93B is from a reunion with students in a sushi bar (Note, wives were not invited, an exception was made for Heidel). A company visited nearby was Toyobo in Katata, with Heidel spending the day exploring Mount Hiei and Enryakuji Temple. Their generous gift was used to acquire a beautiful, spherical vase by one of Japans 'living treasures,' the potter Takaya Kato. It is displayed proudly above our fireplace (see Chapter 9, Figure 15).

Figure 93B *Heidel and I with students (from L to R: Nobuhiro Toyota, Reikichi Iwamoto, and Fumiyuki Hamada, at a reunion on my visit in Kyoto, 1983.*

The next two days were used to visit and revisit gardens (Figure 33C,D). To enter the Saihiji Temple and the Moss Garden Kokedera, depicted in Figure 93C, we now had to join the rites of kitōu, skakyo, and zazen.[1] The rules were posted in English at the entrance. We obliged with difficulty, particularly for the last two items. The sutra was preprinted lightly on the paper. Mrs. Hamada who accompanied us,

[1] "(1) listening to the reading of sutras for happiness and ancestry by priests, (2) writing of a sutra by hand, and (3) sitting in the Zen fashion as in the practice of Zen meditation."

Figure 93C *Picture in the Moss Garden.*

discovered that I was tracing the text upside down. For the zazen we concealed our legs sideways to survive the half hour torture. At the end of the ceremony, we could write a wish on the paper of the completed sutra which was to be fulfilled. I thought that gods could read all. Having copied the sutra in Kanji, I now wrote the wish in "Sütterlin Schrift," an example of which is shown on page 1-19. I forgot if the wish was granted. Probably!

Back in Tokyo we had a few days of academic lectures with an exciting birthday celebration. We were both picked up from the 'Takebashi Kaikan,' where we stayed again, to go to the Tokyo University of Agriculture and Technology, outside the city. During the hour-long subway ride, our host told us, it was his birthday. But it was also mine! After a nice day, I told him he did not have to escort us back, it would cut two hours of his birthday celebration. If he would guide us to the right train and write the station name for the hotel in Rōmaji (English letters) and Kanji, we would find our way. He did—but it was the wrong train! I became suspicious when the above-ground section seemed to take much longer. Then, after the crowd in the train reached a maximum, it thinned out. Had we passed the city center? Checking with the neighbor, he looked at our destination, looked at us, and gave the typical Japanese sign of trouble: He sucked his breath between his teeth. With help of another neighbor who knew a little English, we were told to go back for two stations, and then change trains and go for three more stations. Fortunately, we got the idea that the train back would leave from the next platform and got help to change trains at the station. It took only about 15 minutes longer. By now we looked forward to a good birthday dinner at the hotel. It also was also not to be! The dining room was reserved for a wedding, interesting to watch, but we could not crash the dinner reception. To discover a good restaurant nearby also failed. All we could get was a fast food meal in the cafeteria. I had, of all soups, a Russian borscht. In Japan, it was not as good as later in Russia.

From Tokyo we flew to Sapporo, on Hokkaido, and thereafter back to Osaka. In Osaka we visited Dr. Iwamoto at his employment, the Government Industrial Research Institute and also a product display of the Matsushita Electric Industrial Company, demonstrating an early version of a CD. This was followed by travel to Okayama, Kurashiki, Hiroshima, and Iwakuni. Parties were arranged after each lecture, similar to the one seen in Figure 33F. Often it took all effort of Heidel to support my wanting to get some rest and skip the later private, late-evening geisha parties which the hosts were eager to arrange. I would not have survived four weeks of too long partying and not enough sleep, if I had followed these invitations. Figure 93D illustrates the Kintaikyo Bridge in Iwakuni (not so easy to walk or bicycle across, and no cars!).

Figure 93D *The 300-year-old Kintaikyo Bridge with cormorant fishing boats in front.*

Figure 93E *Murakami's pictures of us in Hiroshima in front of the Castle and* **F,** *the ruin of City Hall (a UNESCO World Heritage Site since 1996), kept in the state of 1945, as part of the Memorial Museum, 1983.*

A side trip was to Miyajima Island National Park (see page 9-56), where we went after the lectures in Hiroshima. But, no one can visit Hiroshima today without being shown Hiroshima Castle, built in 1589, one of Japan's treasures, destroyed by the atomic bomb[1] (August 6, 1945, see also page 2-30) and reconstructed in 1958. The other 'must see' item is the Hiroshima Peace Memorial Museum, picturing these horrors of war. In typical Japanese fashion, both visits were captured in photographs by our host, Professor Ichiro Murakami in Figure 93E,F.

The visit to Japan was continued with stops in Fukuoka on Kyushu, where I could meet again with Professor Takayanagi, a visit mentioned already on pages 8-29–30. Final lectures of the round trip were given in Tokyo and in Tsukuba. After the trip, an eleven-page, overall positive report was submitted to the Japanese Society for the Promotion of Science. It detailed our experiences and observations. For a not so positive quote from the report on the retirement practices in Japan, see page 9-80.

The following year my student Stephen Cheng arranged <u>*a four-week stay in China*</u> for Heidel and me. Was it possible after such a perfect and well supported trip to the rich country of Japan to have a successful stay in the relatively poor China? It turned out, that we were received even more enthusiastically in China and everyone went to the limit to make the visit enjoyable.

Stephen Cheng had planned an academic career in China, but was interrupted in his university study by the Cultural Revolution (1966–1976).[2] Before he could complete his degree,

---

[1] Truman and other allied leaders issued on July 26 the 'Potsdam Declaration,' an ultimatum outlining the terms of surrender for Japan. It stated that without a surrender, the Allies would attack Japan, resulting in "the inevitable and complete destruction of the Japanese armed forces and just as inevitably the utter devastation of the Japanese homeland," but the atomic bomb was not mentioned. On July 28, Japanese papers reported that the declaration had been rejected by the Japanese government. Emperor Hirohito, who was waiting for a Soviet reply to Japanese peace feelers, made clear that the Imperial Regalia of Japan had to be defended at all cost (see also page 3-9 for the Communiqué of the Potsdam Conference concerning Germany after its capitulation).

[2] In 1981, the Central Committee of the Communist Party unanimously passed the following Resolution on Certain Questions in the History of Our Party since the Founding of the <u>P</u>eople's <u>R</u>epublic of <u>C</u>hina, [PRC, "Zhōonghuá Rénmín Gònghéguó," founded in 1949] which states: "Chief responsibility for the grave 'Left' error of the 'Cultural Revolution,' an error comprehensive in magnitude and protracted in duration, does indeed lie with Comrade Mao Zedong. ... Under the mistaken leadership of Mao Zedong, which was manipulated by the counterrevolutionary groups of Lin Biao and Jiang Qing, [it] brought serious disaster and turmoil to the Communist Party and the Chinese people." (*http://en.wikipedia.org/wiki/Cultural_Revolution#Official_Historical_Assessment*)

he was sent to the Mongolian border for a Maoist re-education by working under most primitive and poor conditions in the countryside. When in 1978 scholarly exchanges began between China and the US, and in 1979 relations between the two countries were normalized, Stephen was able to come to the US in 1982. He was supported largely by his family, so that he had some freedom of where to go and what to do. At RPI, he interviewed the professors working in the polymer field and decided to work for me, supported by my research grant. His aim was to get a sufficient education to teach at his home institution, the East China Institute of Textile Science and Technology in Shanghai. Stephen was not only working hard, he also quickly understood his PhD project and the functioning of ATHAS (Appendices A and B).

Stephen became one of my most productive and innovative students (see page 8-24). Shortly after his arrival, he suggested to arrange a trip to China. Initially, I did not take this seriously, but in time detiled arrangements were made. The conditions were that I would pay our airfare and his Institute and the Ministry of Textile Industry of China would assume expenses in China for Heidel and me. In return, I would give lectures at the institute in Shanghai and allied universities.

Stephen was leaving earlier from Troy to make final arrangements, to see his wife, and to get acquainted with 'Wendy,' his daughter, who was born after he came for the US. Figure 94 is a reproduction of a note, Stephen gave to me before he left for China. He must have felt as we did, when first going back to the GDR (see pages 8-46–48 and 59). He wanted to make sure that nothing went wrong. The difference between the two systems became obvious as soon as we arrived in China. While in the GDR there was a prominent presence of police and army in the streets, this was not so in China. Germans were afraid of the "Stasi" informants (see pages 3-23 and 9-26), in China, *everyone* was obliged to tell on your conversations and thoughts, an even more sinister system.

Looking at the second paragraph of the Memo of Figure 94, it turned out that I was only asked about Stephen's performance in my laboratory by his senior Professor Qian Bao-jun, who was very proud of his student. About the third paragraph, the criticism of Chinese society, he had to be more afraid of himself than of us inadvertently expressing subversive opinions. More than once would Stephen show us and discuss the flaws of the system, so that *we* had to caution him. As far as I know, all went well. In January 1985, his wife Sue could join him in Troy, but the daughter had to stay behind with her grandparents, as assurance of Stephen and Sue's return. Wendy arrived in October 1987, shortly before Stephen started his first job in the US at the University of Akron.

With respect to the last paragraph, the professors at various universities must have been getting similar instructions. Everyone was eager to show me their work and discuss the science involved in it. But, no one made direct overtures to join my research group or send a student. Possible cooperation was only brought up with the East China Institute. On my last day, there was a formal, scheduled discussion session for this topic.

To Stephen's duties, expressed at the end of the Figure 94: He fulfilled them superbly. He certainly took care of all our needs and made sure that we were exposed to many aspects of life in China and saw a good number of its treasures. He was not only an excellent guide, but also an outstanding translator. I wondered often why I was needed at all. Only 10–20% of the audience understood some English. Stephen managed to let me make a detailed point by talking several minutes without interruption, still, he never lost his continuity. His lecturing technique was also perfect. I noticed that he succeeded to get laughs of the whole audience at the same places in the lecture where I could excite the small group of English speakers. With his help we could even start real discussions after the lecture, something rarely done in China at that time.

Figure 94

THE MEMORANDUM OF VISITING CHINA

To: Dr. and Mrs. Wunderlich
From: Stephen Cheng
Date: April 5, 1984

The visiting date is coming soon, and I sincerely hope that you will have a very nice trip, and it will make a deep impression on you. Also, I am feeling that I have to remind you of something which will possibly happen in China. I have classified them into three respects: about me; about the society system of China; and about relation between you and some Chinese professors.

About myself, as one of your students, some questions will be associated with me when you visit China. Would you please do not mention anything about my political opinion about the society system of China and my real feeling to U.S.A. which I often talk with you, as well as my future, except for my study and research? Otherwise, a lot of trouble will come to me soon. I will say nothing in China too. The only way I will show off to my family and my friends is [through] many slides which have been taken for last two years. The judgement will be from them: which one is much better and how can China do?

As you have fully understood, please do not criticize the society system of China very much during your visit even if it should be exposed somewhere. I really know that your opinion which we discussed here is extremely all right, unfortunately, I do not think it will be accepted due to many reasons by people who still live in China.

I believe that you will be welcome very warmly and friendly since it is a Chinese traditional habit. Sometimes, probably, the welcome will be too warm to be accepted. Some Chinese professors must have some requirements, such as they want to send someone to your group, etc., in time. I would like to say that please keep me away from those things. The decision should be from you fully. The only suggestion from me is that such things [left] undecided would be better. I will try my best to be a good interpreter and guide of yours. Making you happy during your visiting is my sole duty.

Again, have a very nice trip and see you at Shanghai!

P.S. Please do not bring it [the Memorandum] to China.

Of the trip itself, an approximate itinerary and summary can be seen in Appendix C. My estimate is that about 1,400 students and faculty members attended the various lectures. Four factories have been toured. Seven universities and technological institutes under the direction of the Academy of Science or the Ministry of Education were visited. A total of 19 fancy dinners and lunches, with 8–20 guests each, let us appreciate the Chinese cuisine. A nine-page report was written for the trip, as for the trip to Japan the year before. This report about our trip to China contains a section dealing with 'Impressions.' I have quoted a major portion of this section of the report on pages 8-103–106 to generate an image of China as we saw it in 1984. Added explanations and references to the figures are indicated by [brackets]. Left-out portions are indicated by '...'.

An enormous change was noted during our second visit to China in 2000 (see Chapter 9 pages 9-62–66). My picture collection of the 1984 visit consists of 192 slides, made with my little half-frame camera. In addition, Heidel produced an album of our pictures, those contributed by others, and mementos, descriptions, postcards, etc. Figures95A–E and 96A–D give a small selection of relevant images. The trip was the second of our once-in-a-lifetime travels, of which, fortunately, we could experience three. The others are described, beginning on pages 8-25 and 10-13.

Figure 95A *Our hosts, the Polymer Faculty. In the center, Professor Qian Bao-yun, President of the Man-made Fiber Institute.*

B *Entrance of the guest house of the institute with the official car, ready for us.*

C *Jing-shan Chemical Company, photo of the two head managers.*

D *Dinner with the Cheng family in Shanghai.*

E *Our hosts in Suzhou, the paradise, on 5/13/84.*

B *Our walk along the Chinese Wall, June 3, 1984.*

C *Temple of Heaven.*

Figure 96A *One-time Flower in Winter 1984*
Wu Sun Yong

Chemistry Fiber Institute Gift to Professor Wunderlich June 5, 1984, to Commemorate Lecture on Visit to China.

Figure 96D *A visit to the Truck Factory in Nanjing on May 31, 1984.*

'Impressions'

The impressions such a trip gives, are many. Let me start with the research laboratory and teaching facilities. In practically all cases the physical facilities were outdated and too little was spent on upkeep. The lecture rooms and laboratories had a 1930-like appearance with missing modern facilities. Although automatic slide projectors were always available for my lectures, screens and particularly darkening of the room were often make-shift, leading to the need to cut-down on the number of half-tone slides shown in the lectures outside of Shanghai. [Figure 95A shows our Shanghai hosts, Fan-ting Li, Bao-yun Qian, Sun Tong, Zong-quan Wu, and Li-hang Wu(?). On our arrival at the Shanghai Hongqiao airport in the evening of May 10, after 25 hours of travel ($\approx$15 h air time), all greeted us on the tarmac and brought us, by minibus, to the guesthouse shown in Figure 95B. There we had a two-room apartment with a full bathroom (specially installed for guests). We found out soon that the rest of the floor was empty (for safety?). The room one floor down facing the stairs was always open and had a guard sitting at a table to check anyone wanting to visit us. Still, many of the professors chanced a visit and brought gifts, to welcome us.]

Up to date, high quality equipment such as FTIR, NMR, X-ray, TEM, SEM, and DSC seems to have arrived in all institutes visited, but in most cases only in the last few years, often with help of World Bank funds. All of it is imported from Japan, US, and Europe with the usual problems of service. Only very basic research equipment seems to be manufactured in China. Its functionality is outdated. The Chinese-made DTA, for example, though workable, is typical 1950 vintage. Much of the less sophisticated, everyday equipment such as high quality balances, pH-meters, pumps, optical devices in the range of $500–3,000 are missing. Mainframe computers are commonly poorly accessible and often even less powerful than the integrated computation facilities of the modern instrumentation. A typical situation which arose was that our graduate-student guide in Beijing could not accompany us to functions also important to him, so that he would not lose his 20 minute weekly allotments at the terminal of the [not very powerful] institute-computer.

Most important are, however, the impressions gathered on the scientists, their research and teaching ability. Many of the senior scientists had their training in the Soviet Union and are only now making a major effort to connect to the science published in English. All suffered a devastating interruption during the Cultural Revolution. For the younger scientists, this has lead to weaknesses in the basic knowledge which may never be recoverable and has its strong effect on their teaching ability for the next generation. Since all teaching is done from Chinese texts, it will take some time until up to date texts and teachers using these texts will be available. Although modern teaching techniques are often listed as strong points of institutes, I saw little evidence of wide-spread use. In fact, the showplace Children's Palace in Shanghai also recently shown to Mrs. Reagan [on her first visit to China] seemed better equipped with science teaching tools than some universities. The research topics introduced to me were of modern character and seemed to be oriented to catch-up with the forefront of development. Hard data, however, were few, and interpretations were frequently not offered with confidence. To support the often imported modern plastics and fiber industry and, in addition, develop basic research capability on a broad front, the numbers of researchers seem too small by almost an order of magnitude. A large number of the researchers made a highly intelligent and eager-to-work impression, but seemed unsure of themselves. The discussion of 'new developments in the western world·' was to them of more importance than the discussion of their own research. I found it often necessary to remind them that in basic research it is important to follow one's own ideas, rather than to follow those of others too closely. [An interesting topic came up when we were discussing the selection criteria for students to study chemistry in China. I was told that under no circumstances were color blind students admitted. It was thought they could not see the color changes of the indicators in titrating acids and bases. I could not have studied chemistry[1] in China! (See also the pages 1-18 and 8-34.)]

---

[1] Note that today such titrations are rarely performed with color-change indicators. But even as a student, I could outperform my fellow students in precision by finding the changes in intensity instead. Using reference solutions when determining the endpoints, I even could identify red-green color changes . The ability to see intensities better than a person not color blind helped also to increase precision to a fraction of a nanometer when measuring the thickness of lamellar crystals by interference microscopy (see Chapter 7, Figure 30). Only my student Mike Jaffe could later match this feat (he was also color blind!). Remaining problems are color markings on electrical wiring, some traffic-signs, and identifying the colors that are now used in computer-graphics.

The technology observed at the Jing-shan Chemical Complex [2 h bus ride from Shanghai] was that of a modern plastics and fiber plant with an eye also on environmental improvement. The newly built city was somewhat austere in our eyes, but improved over living conditions in other cities. An official description of the project from about 1980 announces living quarters of more than 300,000 square meters for a population of 50,000. Many of the production lines are in operation 24 h a day and the down-time seems to be low. Of 25 PE blowing machines producing film up to widths of 14 m, I noticed, for example, only one not in operation. [But, note the austere dress of the managers of the plant in Figure 95C and our escort in Suzhou in Figure 95E. This official 'Mao' austerity in dressing was still widespread in 1984, except for children. For them, parents dared to be colorful.] On the other hand, a truck plant we visited in Nanjing with about 10,000 employees seemed to be left-over from the 1940s with little evidence of modernization and much indication of idleness. [See Figure 96D for the assembly line, which was idle for several hours each day because the power station could not produce enough electricity. When we were visiting, all workers were chatting, reading, or doing needle work! To avoid these problems in Jing-shan, they were fully self contained, they had their own power station, trucks for transportation, harbor, etc.] The quoted production figures [of the truck plant] were 20,000 light trucks annually [not very efficient, half a person-year's salary had to be added per truck produced!]. A third picture was offered by the carpet factory (hand knotted and sculptured) [in Shanghai], the Silk Embroidery Institute in Suzhou [doing the famous two-sided embroidery, each side in different colors], and a jade and ivory carving factory [in Hefei]. Here, little modern equipment was needed and one could admire the expertise of the workers and the outstanding quality and, for us, reasonable price of the precious products. [To our surprise, however, some of these products could be less expensively bought at Macy's department store in Albany, NY, after our return to the US during a sale on Chinese carpets and porcelain!].

A few words to the tourist-side of our trip. The many historical and anthropological sites make a trip to China a unique experience. The government seems to become increasingly aware of their value and the need for upkeep and renovation. [Note that I do not choose pictures of most tourist sites. We saw many of them, but there is little need to reproduce views which are well known. Exceptions were those of special, personal importance, such as Figure 96B of the Chinese Wall, which we were told would guarantee long life, and Figure 96C of the Temple of Heaven in Beijing, which, indeed, was heavenly.] Also, new tourist hotels are under construction and improvements in travel conveniences are visible. It is unfortunate that a double standard exists for things like comfort in travel (soft class in trains is not permitted for common Chinese), accommodations (special dormitories and guest houses and hotels for foreign visitors) [for Stephen, accommodations and meals were always different from ours], currency (special money and [better stocked] stores for foreigners), prices (from railroad and plane tickets to hotel rooms and meals, foreigners are charged a multiple of the cost Chinese have to pay when permitted to purchase the same). This is the same in most communist countries. Only if these double standards disappear and the continuous control over people-to-people contact is relaxed, can one speak of a new China.

This leads to the last, and perhaps most important impression, that of the people we had contact with. Without exception, we found great friendliness and perfect hospitality. All Chinese scientists and engineers are poor and most of them have had to go through a terrible, hopeless period during the Cultural Revolution. The initial great hope of improvement is, perhaps, just wearing off to some degree. Improvement is slower in coming then expected. Our visit has shown to us that besides the relaxation of suppression, there must be increasing productivity and a participation of a greater fraction of the population in the fruits of production. With pride, we were told at the PE plant [in Jing-shan] that per man-year, 100,000 Yuan surplus was generated and this at an average annual salary in China of, at best, 1,000 Yuan! The first, small-scale experiments in free enterprise are tried in the free markets, seemingly with success. Much more must be done to increase productivity and in decreasing the hours spent on producing. The lack of productivity is particularly obvious when observing scores of peasants tending the fields almost completely without machinery. As everywhere, there are complaints of student performance. A contributing factor in China seems that once having overcome the difficult hurdle of admission to the university, students relax since they are now guaranteed a job and a salary on which performance can change little. This largely fixed income presents a constant source of complaint which is aggravated by rising expectations and slow inflation. ...

We made many new friends during our trip and hope we can develop some continuous contacts and cooperation. Possible research projects and specific coworkers have been identified under the aspect of mutual benefit. I cannot end this paragraph without saying that in all our travels to other countries we have not ever been met with so much hospitality. I do not know anyone in the US, for example, who would arrange for a prior [to him] unknown, visiting colleague a reception and dinner costing more than a month's salary, as was done by many of the Chinese professors. Virtually all professors of the polymer group at the Institute were available during all of our visit to listen to the lectures, discuss research, and host sightseeing and shopping expeditions. Also, the officer of the president's office for foreign affairs of the institute, Mr. Tang, and his Assistant, Mr. Lee, and our driver turned out to be friendly, helpful, and interested far beyond the call of duty.

Finally I must acknowledge the generous financial support of this travel by the East China Institute of Textile Science and Technology. [On the last day of our stay we received the gift reproduced in Figure 96A, by the famous artist listed in the legend who specially dedicated the painting to allow us to take it out of the country.] My student Stephen Cheng must be given special thanks [again], he gave up almost half of his first visit home after 2 years to make many of the local arrangements, accompanying us on our trips out of Shanghai, and acting as expert translator. [His family also was one of our hosts, as shown in Figure 95D. The wife of Mr. Miao-yong Cao, the other Chinese student of mine at that time, was our host for a dinner in Hefei (see page 8-58).] All in all, I must say that this trip was satisfying in all aspects. We could see first hand how the largest fraction of the population lives in this world, learned some of their problems, could bring them some of our scientific and other ideas, plant some seeds for later mutually beneficial cooperation, and made a number of new friends we will not forget.

Report completed. 6/12 1984

Many further stories of interest can be told about our trip to China. For example, the amazement of the student who accompanied us to the Friendship Store, the store where goods for foreigners were for sale. Chinese were only admitted when guiding foreigners. He was so surprised about what all was available and the high prices that we were embarrassed to buy any larger items. In particular, we were interested in a small, sculptured carpet, but we thought it not right to buy it at the spur of the moment. It would have cost more money he had to live on for a whole year. A carpet just like the one we had passed up in the Friendship Store hangs now in my home office. It was bought shortly after our return to the US at Macy's, as mentioned on page 8-105.

Our outings with the university car, usually involved six people. Mr. Tang and Mr. Lee for security, the driver, the two of us, and an English translator, usually Stephen or another professor. Mr. Tang told us, there had to be two security guides. In case of an emergency, one had to stay with us, while the other would have to get help. Heidel reported, that the same crew was accompanying her to the hair salon. Probably there was never a lady with such a large male escort.

The driver explained the difficulty of driving in Shanghai. There were too many bicycles and pedestrians on the road. He also had to save gasoline. He would speed up to ≈45 mph, then turn off the engine and coast, before letting out the clutch to restart the engine and speed up again (just as modern, fuel efficient cars do today). At night, one drove without lights. With only few street lights, this was truly a ghost-ride between the bicycles and pedestrians.

For the 11-h ride to Hefei, we sat in an almost empty 'soft class' compartment. Otherwise, the train was packed. We were brought green (jasmine-flavored) tea. For the rest of the trip, from time to time the tea was refreshed with hot water. On arrival, there still was a faint taste of jasmine left. At one station, a higher army officer entered, distinguished by the larger number of pockets on his coat. Not far from us, he sat down. After settling in, he saw us, and left. It was not safe for him. We might have talked to him, and then he would have to write a report about it.

In the 'Friendship Hotel,' the best in Beijing, our building was #5. It was used for foreigners only and had a quiet air conditioning. What a luxury. Stephen was not permitted to enter with us into the complex of buildings. I suggested, he should wear a suit and tie instead of the conservative 'Mao' uniform he usually wore on traveling days. It worked, he was now accepted as an 'overseas' Chinese and could pass without needing identification.

The next day, a car was ordered. Stephen was to join us at the main entrance of the hotel. It got late. No car came. Finally, a speeding car appeared, picked us out from others also waiting and took off. When the driver did not turn to the front entrance to pick up Stephen, we rebelled and pointed to the main entrance. Stephen could quickly resolve the problem. This was the wrong car! The driver spoke no English! He was late and had to bring two Japanese(!) to the airport. We looked most foreign and anxious to go. Ultimately, we all got to my lecture. But this was Beijing Institute of Technology! In their eyes the East China Institute was second rank. They had their *own* translator. After five minutes, the translating lady broke down, it did not work. She was not familiar with the scientific terminology. Stephen took over. He and his institute gained considerable respect.

Finally, I must praise Chinese medicine. Heidel got a cold on our trip to Hangzhou, another 'paradise' near Shanghai. Its special attraction is the beautiful West Lake (see Chapter 9, Figure 66D). Since there was no 'western' doctor, Stephen got some traditional medicine from the pharmacy. It had several unidentifiable ingredients and included some toad venom. The cold was cured. It probably would have also cured many other illnesses.

The year 1985 foreshadowed major changes in our life. Both, our personal and professional life was to undergo major alterations. The year started as usual. Key was *the March Meeting of the APS in Baltimore* (see Appendix C). This was also attended by a car-load of my students, giving lectures and posters. The APS rule was one lecture per person, i.e., several students had to attend to present our broad spectrum of work. Also, students should be introduced to professional meetings. The first step for a student was to give a poster, and next, a 12 minute lecture after the presentations in our group meetings were acceptable. At this meeting we met with Dr. Don(ald) W. Noid of Oak Ridge National Laboratories, ORNL and discussed possible cooperation in the field of computer simulation of assemblies of molecules. Don had access to the most advanced computers at ORNL (Oak Ridge National Laboratory) and was an excellent theoretical physicist. He invited me to visit ORNL for a lecture and discussions.

At that time ,Vati had serious health problems after his earlier heart infarct (see page 8-84). On May 26–27, 1985, I flew to Hastings, before continuing to Oak Ridge for the lecture at ORNL. Vati was not in good shape, still, we had two good days. After this visit, his heart had increasing difficulty. *On July 29, Vati passed away and the family assembled for a sad funeral in Hastings*. Figure 97A and B are the reminders of this sad occasion. Just three month's later, Mutti and Vati would have had their 60th wedding anniversary. The 50th anniversary was the last happy gathering of the whole family (see Figure 85A).

Figure 97A *Mutti, Heidel, Frank (and I) at Vati's grave site.* **B** *Flower arrangement, 1985.*

Figure 98 *Joachim, Manfred, Bernhard, and Frank.*

Before we all left Hastings, Mutti wanted to have a picture to remember all of her four sons. It is shown as Figure 98, taken in the back of the house in Hastings. She treasured the picture, and indeed, all four of us were never seen together again.

To find a period of rest, Heidel and I started our next trip to Europe earlier. Its purpose was to attend the IUPAC meeting in Bratislava (where we met with Hans and Inge Gottschalk, as always). The extra two weeks, however, we spend for the first time in our new vacation spot in Kössen in the Alps (see page 8-50, Figure 57).

In the fall, Mutti found the courage to travel to Troy for a period of recovery. In her life it was the first travel alone. Together with Papa, who lived with us since 1979, all had a time of pleasant memories and nice excursions in the vicinity of Troy. Figure 99 was taken in Williamstown, visited on one of these trips. Both Mutti and Papa lived into their 90s.

Figure 99 *Mutti, Heidel, Papa.*

In 1985 it was also high time to plan **another sabbatic leave in Germany for 1986/87**. The feeling grew that without another major effort to narrow the research direction, progress would be diluted (see page 8-36). We had completed thermal analyses of a large number of samples, ranging from small molecules to macromolecules of one, two, and three dimensions. Now it became obvious that *the various phases were in need of a better understanding*. Ultimately, I came up with the classification summarized in Figure 13 of Appendix B (see also page 10-4 and Appendix A, #200). A new mesophase for polymers had been identified, the conformationally disordered state or 'condis state'. In addition to understanding the structure of the phases, it was necessary to have an operational definition where a given structure becomes a solid (see page 8-92–93). To understand solidity is of importance to identify the mechanical behavior, which, in turn, is based on molecular motion.

The first time I spent more time on these problems was in 1969 (Appendix A, book #37). For glasses, there is a glass transition to identify the transition from a liquid to a solid or vice versa. At the glass transition temperature, every glass has a similarly high viscosity of $\approx 10^{12}$ Pa s. Accordingly, the glass transition has also been called the brittle point. Crystals, in turn, do not have to be 'very solid.' Most metals are ductile, and some of them, like alkali metals, are outright soft. Similarly, many crystals of organic molecules, like paraffins, one would be difficult to be called a solid. Working with substances of intermediate order, the mesophases, one notes that one type of these has the rather awkward name 'liquid crystal.' They are liquids, and it was long known that they have a typical glass transition on cooling and then become a solid. They became a fashionable direction of polymer research in the 1970s and 1980s (see page 8-55). We found that not only liquid crystals, but all mesophases have a glass transition on cooling. Finally, we observed that even some crystals have a glass transition (see page 8-116), so that an all-inclusive definition of solids was possible (see Appendix B, page B-39).

The key coauthor of the early work on mesophases was Dr. Janusz Grebowicz, who joined my group in 1980 from Poland (see also page 8-24). In Poland, this was the time of "Solidarność," (Solidarity), the name of an independent, self-governing Trade Union.[1] At that time, Janusz could have received political asylum in the US because of the repression against anyone who supported this movement. But Janusz thought things had changed. When he was promised a good position in his home university in Łódź, he left Troy in April 1985, and went back with his family, with a

[1] Founded in September 1980 at the Gdańsk Shipyard and led by Lech Wałęsa. It was the first non-communist East European trade union. The government attempted to destroy the union, but in the end, it had talks with the Solidarity-led opposition. This led to semi-free elections in 1989. Note the difference to the earlier uprising in the GDR in on June 17, 1953 (see Preface and page 4-33), the 1956 Hungarian uprising (see page 7-7), and the 1968 Prague Spring (see page 7-8). All these earlier attempts were subdued with military intervention, led by the Soviet Army. Compare also to the end of the USSR (page 9-23) and the "Wende" in the GDR (pages 9-23–26).

short stop in Mainz, FRG, to visit the Max Planck Institute. His return to Poland was despite all our warning that he had no written commitment.

When he arrived in Łódź, his passport was revoked, so he could not leave again. His job did not materialize. After a while, I received a desperate phone call from Łódź. Janusz wanted to ask me to arrange a visit to the US to complete an aspect of his work during the earlier stay. Before he finished one quarter of the request, the censor had grasped that this was an illegal call, and the connection was broken. Next day, another call, starting where the prior one had stopped. Again, it took a few minutes and we were interrupted. In two more installments, I had gotten the gist of the problem. I managed to find a native speaker to fill in the forms from the Polish Consulate in New York City. He received his passport back, and promptly left with his family for Mainz, where he could start a postdoctoral position in September with Professor Fischer (see page 8-40–41). During this stay in the FRG, we could work together during my sabbatic in Freiburg. Finally, he reached the US again, and worked later at Shell Development Co., and the University of Houston, TX.

The discussion for a new sabbatic leave in Germany for 1986/87 was already started before leaving for the travel to China. After Caryn and Brent had left Troy we were practically free to go without worries. Papa would gladly stay with us for a year in Germany. He had lived with us in Troy since 1979. He had emigrated to the US with 80 years of age after Mama died of a stroke. In May 1985, Dr. Martin Möller, from the Laboratory of Professor H.-J. Cantow of the University of Freiburg, visited in Troy and thought it would be possible to find support for me for a stay at the university in the frame of their new graduate teaching program. He also was the person most interested in our different outlook on phase structures. He had made and analyzed a good number of compounds which would fall into the mesophase category. We concluded, that both structure, as well as motion measurements were needed, and Freiburg had very good electron microscopy, X-ray diffraction, and solid-state nuclear magnetic resonance facilities, NMR. During my absence, Stephen Cheng could supervise my research with occasional return visits by myself. Stephen had finished his PhD in April 1985, and continued working with me as a postdoctoral fellow.

On July 17, 1985, after initial discussions with Prof. S. Wiberley, the then Chairman of the Department of Chemistry, I submitted the first request which received a favorable response. In August, I was assured support by Professor Cantow. Details were then worked out, so that on December 13 the final proposal was submitted to RPI. It listed, in addition to the already mentioned goals, "the simulation of macromolecules by using supercomputers and their application to understand crystallization and fusion kinetics." This new line of research was worked out in cooperation with Dr. Noid of ORNL (see pages 8-107–108). Longer visits during the sabbatic were proposed for the University of Ulm and the Max Planck Institute in Stuttgart. Professor Pechhold, head of a polymer group, and later President of the University of Ulm, had worked for many years on a model to represent polymers in the liquid state—the meander model. At the MPI in Stuttgart, Dr. Gmelin was an expert in low temperature calorimetry we had connections with.

Finally, after permission to shorten the interval from the last sabbatic to 11 semesters, instead of the required 12, the sabbatic was approved on February 24, 1986. Difficulties to find support at the University of Ulm were worked on by Professor Pechhold, who also started an effort to nominate me for a Humboldt Award which would resolve all financial problems with its rather generous endowment. The Humboldt Award was later actively pursued by both Universities, and on November 10, 1986, after we had been in Freiburg for a month, the "Humboldt Preis" was awarded to me, as shown on the next page. With its many benefits, no financial problems existed anymore.

ALEXANDER VON HUMBOLDT-STIFTUNG

URKUNDE

Prof. Dr. Bernhard Wunderlich

erhielt für wissenschaftliche Leistungen
in Forschung und Lehre den

HUMBOLDT-PREIS

im Rahmen der Marshall-Dankesspende
der Bundesrepublik Deutschland

Dieser Preis berechtigt den Preisträger Forschungen
eigener Wahl in der Bundesrepublik Deutschland und
in Berlin (West) durchzuführen

Für die Alexander von Humboldt-Stiftung
im Jahre 1986

Prof. Dr.-Ing. Hans Leussink
Bundesminister a. D.

Humboldt Award, *approved on November 10, 1986, supporting an 8-month stay in Germany, starting January 1, 1987. Based on the nomination by Professors W. Pechhold and H.-J. Cantow.*

Early in 1986 the connection with the ORNL in Tennessee was strengthened. In March, I was formally invited to join the Chemistry Division Advisory Committee of ORNL for a three-meeting term. Then, in April, after an invited lecture at the University of Tennessee in Knoxville, TN, UTK (see Appendix C), I was told both institutions would like to consider me for one of their Distinguished Scientist positions. The university shared these with nearby ORNL. At the first Advisory Committee Meetings at ORNL, June 23–25 we also discussed a further topic for cooperation with ORNL besides the research with Dr. Noid (see page 8-110). The additional field was quasi-elastic neutron scattering, the area of expertise of Dr. Narten of ORNL.

All was planned, *but things do not always go as expected*. As is amply documented in Chapters 1–4, my early years were troubled with a long series of illnesses, but since the last hospitalization in 1951 (see page 4-22), I enjoyed rather good health for the next 35 years and rarely saw a physician. Unfortunately, this period would come to an end. In July 1986, we took delight in one of our vacations at Chase Pond (see page 8-75 and Figure 58). On return to Troy, I felt a minor pain on my left side, which lasted several weeks. Heidel brought this to the attention of our friend, Dr. Ivo Buhac (see page 8-83). He initiated a series of thorough medical examinations. *Finally, an X-ray tomogram revealed a sizable cancer* on my left kidney. Immediate, radical surgery was considered to be the best course of action. On August 6, the operation was done at Albany Medical Center. After a week's stay in the hospital, I was sent home with more pain than I had before. But all follow up exams showed that the malignant tissue had been removed along with the kidney, so that no radiation or chemotherapy was necessary. I could slowly recover.

The semester began three weeks later and I could take up a normal schedule, though still somewhat shakily. I had not missed any teaching. The first lecture outside of RPI, scheduled before my hospital stay, was at Pfizer Co., in Groton, CT on September 5. With Heidel's help, I did not have to miss it either. She drove me there and carried the luggage and watched over me through the lecture. This proved, that we did not have to postpone the sabbatic either. Naturally, I felt funny when Heidel moved all our luggage, but it took half a year until I could lift heavy things again.

A flurry of lectures was carried out later in September (see Appendix C). One of these trips should become rather important to our future. Right after the NATAS Meeting in Cincinnati, I flew to Knoxville, TN, for an interview for the Distinguished Scientist position, mentioned above. This visit was connected with lectures at ORNL and at UTK. Discussion of my plans for the future were held at ORNL with its Director, Dr. Postma, the Director of its Chemistry Division, Dr. Poutsma, and the Co-chairman of the Distinguished Scientist Program, Dr. Christophorou. Next day, at UTK I saw the Chancellor, Dr. Reese, the Director of the Science Alliance, Dr. Riedinger, the Dean of the Liberal Arts College, Dr. Ratner, the Head of the Department of Chemistry, Dr. Mamantov and his faculty, and the Provost of UTK, Dr. Wheeler. During this meeting a verbal offer was made which was finalized after a second discussion in January 1987 (see page 8-117).

On October 11, we arrived in Germany to begin a very productive sabbatic. *The first part of the sabbatic was in Freiburg* ("im Breisgau") at the "Institut für Makromolekulare Chemie der Albert Ludwigs Universität." It covered the fall term, until April 14, 1987. Then, we moved to the University of Ulm. The job to coordinate our move, find an apartment, and even buy a car, fell to Dr. Möller of Dr. Cantow's laboratory. He arranged all perfectly. Figures 100A–G show a cross-section through our experiences in Freiburg. I remembered that Vati had been offered a business after we arrived in the FRG as refugees in 1953. He thought he would have difficulties to fit in with the different lifestyle there (see the pages 5-13–14). For us, as visitors, it was not difficult to adjust. The bureaucracy, however, was not so different from the first sabbatic (see the pages 8-43–44).

Figure 100A *Schallstadt, near Freiburg, Im Zinken 2 bei Burggraf. Our apartment is in the top floor. The shed serves for the wine harvest.*

B *View from Burggraf's vineyard with the village of Schallstadt in the foreground and the city of Freiburg in the background, Winter 1987.*

C *Visitors from Albany, arriving in France, not Switzerland on 11/5/1986. The Basel-Mulhouse airport has two exits, a Swiss and a French one.*

D *A reminder of WW II, left-overs from the French fortifications facing Germany, the Maginot Line. It was completed in time, in 1938.*

E *Weekly market in Freiburg.*

F *Winter in the Black Forest.*

G *Visit of Brent and Anna, 1987.*

Registration, work permits, taxes, insurance, etc., all took time. Most of this was easily forgotten when seeing the beautiful cities and countryside. Our much liked vacation area, the Glottertal (see pages 8-49–50) was just in the next valley. Figure 100A illustrates the house with our upstairs apartment, rented from Mr. and Mrs. Burggraf in Schallstadt in the middle of the wine-growing region, about 15 min by car from Freiburg, as can be judged from Figure 100B. "Herr Burggraf" was employed with the village council. In his spare time, he was working the vineyard, inspecting his wine cellar, and participating in the village men's choir. The small farmers in this area were sharing in a winery and delivered the specific grapes on the proper day, to ultimately get their share of bottled wines. Figure 100B is a view from the Burggraf's property. Being for a longer period in Germany gave an opportunity to visit or being visited by friends and relatives in Germany. Some also came from the US, as shown, for example, in Figures 100C and G.

Freiburg is located close to Basel (Switzerland) and even closer to Mulhouse (France). Basel and Mulhouse have a common airport. This caused us to have some excitement during the five-day visit of Charlotte M. and De Pitt, (introduced on page 8-84). They arrived early in November, as is documented in Figure 100C. The Airport Basel-Mulhouse is, perhaps ½ h by car from either city, but on French territory, with an extraterritorial exit to the road to Basel. Since we do not know French, we thought we could be more helpful picking up the visitors from the Swiss exit. On the Autobahn we went to the German-Swiss border at the outskirts of Basel. But we had no "Vignette," the marker on the car documenting that we paid the Swiss "Autobahn" tax. We argued that we needed to go only two miles on the Swiss Autobahn to reach the road to the airport, which was tax free. No, we had to pay a whole year's tax! But it was already November! Our other argument was that we would gladly pay a one-day share or the road tax, or we would pay all for the next year, 1987, when we would do much more traveling in Switzerland. Neither was accepted. The Swiss border officer also would not tell us that there was a side road bypassing the Autobahn. He was set on collecting money. But I can be stubborn too. Having enough time, we turned around and drove back on the German "Autobahn" to get to France and took the French road to the airport.

We made it in time, but how to tell our guests? We asked 'Information' to make an announcement. We did not hear it, and our guests did not either. When we looked through the big plate-glass window, separating the 'two countries,' we could see them arriving and steering for the Swiss exit from which there was no 'border crossing' to our French side of the airport. Since we could not be heard through the glass, we used sign-language to direct them to the French exit ☞. Figure 100C shows that everyone got properly met on the French side.

A much more pleasant excursion with Charlotte M. and De was to Haeberlin's Restaurant L'Auberge de l'Ill in Ribeauvillé, Alsace, a French, three-star restaurant across the border, about an hour's drive from Freiburg. After crossing the Rhein, we could inspect the remnants of the Maginot Line,[1] shown in Figure 100D. This was of interest to De, who was a soldier during WW II, stationed in Morocco. We arrived at the Restaurant at noon and departed in the late afternoon after

---

1 André Maginot (1877–1932, French statesman who championed the erection of the fortifications, built between 1929–38. The fortifications, however, stopped at the Belgian border, so that the German army could reach Paris in 1940 by going through Belgium. Also, the fortifications were not designed to fight the enemy from the rear, which meant they were also not helpful for the Germans against the Allied invasion forces when they entered Germany in 1944/45.

one of the most fabulous meals we ever had. The printed menu is still part of our collection of memorabilia, and the local Alsatian "Trimbach Gewürztraminer" wine from Ribeauvillé which was served is still one of our preferred dinner wines for all but the heaviest meats.

For Thanksgiving, Inge Gottschalk visited from the GDR (!) (see also Figure 36C). By that time, it was possible to get a limited permission to visit the FRG to attend a special family holiday, like an important birthday of a close relative. Additional conditions were, not to be employed in an important job and leaving behind a main member of the family, like a spouse or children. Inge had such an opportunity for an aunt's 80th birthday in Cologne. She left Hans behind as a hostage, changed her train ticket once she had reached the FRG to visit us for a week in Freiburg, then traveled to Cologne for the birthday, and return with the proper train to Brandenburg. She enjoyed the first breath of freedom in 25 years since building of the Wall in 1961 (see page 7-7).

Having such a welcome guest, raised the question how to get our traditional turkey for Thanksgiving. Thanksgiving is not a holiday in Germany, and turkeys were not favorite birds for dinner. With difficulty, we finally found a butcher in the main market who had a huge turkey. Much too big for our gathering which included, besides us two, Papa, and Inge, also the Burggrafs and Dr. Möller. The butcher, anxious to sell the bird, offered to cut it in half. I never thought half a turkey could be prepared so well. The meal with all the trimmings was like at home in the US and was enjoyed by all. The local wine, brought by the Burggraf's, also fitted well with the feast.

The next visit was Caryn's. She spent the week over Christmas with us. By then, snow was covering the Black Forest, making for a nice Christmas mood. Figure 100F shows our VW Jetta on the snow-covered road. The car was arranged by Dr. Möller with US equipment. Ultimately, it was delivered to our VW-dealer in Troy, who traded it for Heidel's old VW 'Golf' and gave a new-car guarantee. In 1977 the 'Golf' had replaced the earlier 'beetle' described on page 8-38.

Freiburg has a rather mild climate and only short periods of snow, but the surrounding mountains are ideal for winter sports. They reach 4,500' at the "Feldberg." The picture of the weekly market in Freiburg, in front of the "Münster" (Cathedral) in Figure 100E, however, was taken earlier in the year. At the time Caryn was visiting, the Christmas market was the main attraction. Not only to shop for gifts and special sweets and baked goods, but mainly to get into the proper "Weihnachtsstimmung" (Christmas spirit) and to enjoy "Bratwurst," "Bockwurst," or "Leberkäs' " on a hard roll. For us, these 'fast foods' eaten on the street are preferred to a hot dog or hamburger. They have different spices and modes of preparation, rarely duplicated elsewhere.

A visit of Brent and Anna could be arranged for early spring. Figure 100G shows us in the "Kaiserstuhl," a hilly area, producing outstanding wines. Having expressed our preferences among fast foods, we were taught what growing up in different parts of the world makes for the different likings. When Brent and Anna returned from an exploration in Freiburg, they were telling us about their greatest discovery of the day, the 'McDonald' store in the center of the city!

On a later visit of Brent with family on the occasion of Papa's 90th birthday which was celebrated in Bergheim, we found a solution for bringing the different tastes together. When driving to nearby Aachen to see the "Dom" (Aachen Cathedral), we found a typical German Café next to a MacDonald Restaurant. Both had the same owner, allowing us to eat together at common tables in the street. Our generation must shrink. In 2006, when we visited Aachen again, MacDonald had grown and the Café had shrunk to a small "Kaffeestube." Maybe, now it does not exist anymore.

More important than the enjoyment of the country side, visits and refreshing our knowledge of the German life style, was *the work that could be accomplished during the sabbatic*. It was summarized in the final report to RPI. Overall, 59 lectures were given. These included the 20 lecture series at the University of Freiburg which formed the basis of the book 'Thermal Analysis,' later published in 1990 by Academic Press (Appendix A, publication #290). The other 39 lectures were given at different meetings, universities, industrial laboratories, etc., in Germany (FRG and GDR), Switzerland, France, The Netherlands, and the US (on three short return visits). Appendix C gives a detailed list. The number of publications generated during the sabbatic was seven, while the work continuing in Troy generated 25 additional papers (see Appendix A).

The main goal of the sabbatic, the development of an understanding of the conformationally disordered mesophase (condis crystal) was supported by measurements and collection of literature on more than 100 examples. The details were written up in a review paper as a separate volume of the Advances of Polymer Science (Appendix A, publication #263c). All figures were drawn new in a common style and represented the last major effort of mine to make use of my talents in technical drawings (see the pages 1-18, 3-12, and 6-22). Thereafter, practically all graphics was done on personal computers. But even then, a knowledge of drawing by hand was of value.

Major contributions were made by Dr. Möller and his students who added the knowledge of substances they synthesized and studied by solid state NMR. The two other two coauthors to this review were formerly associates of mine. Dr. Grebowicz was a research associate from 1980 to 1985 (see pages 8-109–110) and had started the work on condis crystals. He was at that time in Mainz, working as a postdoctoral fellow at the Max Planck Institut. He had continued to search the earlier literature in this area and was of help in the interpretations. Dr. Baur was a research associate of mine from 1963–1964 (see pages 8-2 and 6). He handled the theoretical question of the phase transitions into and out of the condis phase.

The conclusion of this work was that the condis phase is of abundance in linear, flexible macromolecules and flexible small molecules of all origins, organic, biological, or inorganic. It was shown that the new transitions could be gradual or abrupt, depending on the degree of difficulty for the cooperation needed for the specific molecular motion central to the transition. All condis crystals, as well as other mesophases, were shown to undergo a glass transition on cooling whenever there was no further ordering to closer packing. In addition, it was found that on heating, conformational motion may even begin in crystals without a change in structure, as long as the additional conformations can be accommodated without disruption within the crystal symmetry.

The discussion of simulation of macromolecular motion by computers was mainly initiated with Professor Pechhold and his coworkers. The polymers they analyzed were trans-1,4-polybutadiene and poly(diethyl siloxane). From such discussions it was concluded that molecular dynamics simulations with supercomputers could give additional information. Our work in computational chemistry was started in Troy in cooperation with my colleagues Professors Miller and Hollinger, and was slated for continuation with Dr. Noid at ORNL (see page 8-110). The work on liquid heat capacities, the area of Dr. Loufakis in Troy (see Appendix A, #260 and 263), was extended by him later in Freiburg, supported by a Humboldt Stipend. The connection with the Max Planck Institut in Stuttgart was made, but the construction of the planned calorimeter did not come about. Naturally, parallel with the research and discussions of the sabbatic, the work in Troy continued with the group shown in Figure 101, led by Dr. Stephen Cheng.

The first trip back to Troy was from January 11 to 20, 1987. It was an extremely busy period. First, the lectures listed in Appendix C had to be given. The house needed to be checked. It was looked after by my secretary, Ms. Beth McGraw, and all was in order. Just that before we left, we gave away all fish in the aquarium above the fireplace (see Figure 8-59, right corner). But, I discovered one little fish, a 'zebra,' who had escaped us. Seemingly there were enough algae growing to feed him. I replenished his food supply, and he survived the sabbatic, lonely, but without a problem.

Next was a detailed check-up for me at Albany Medical Center. All was satisfactory. Then, with all students, I needed long discussions. Finally, I had a two-day visit at UTK on January 28 and 29. *During this visit, the offer to join The University of Tennessee in Knoxville, UTK, and ORNL was finalized* (see page 8-112). Before traveling to Knoxville, I had a detailed discussion at RPI with Dean Raveché and the Department Chairman, Professor Wiberley. It was noncommittal to my possible future at RPI.

Figure 101 *The ATHAS Research Group, photographed in fall 1986. From left to right: Ms. M. R. Loverso, Mr. A. Xenopoulos, Mr. M. Y. Cao, Prof. B. Wunderlich, Ms. J. Woertman, Ms. W. Aycock, Dr. K. Loufakis, Dr. S. Z. D. Cheng, Mr. J. Wesson, Prof. H. S. Bu, and Prof. Z. Q. Wu.*

The following is a verbatim copy of part of a letter I wrote to Dean Raveché on February 6, 1987 from Freiburg, after being told that RPI could not even try to match the UTK/ORNL offer:

"... As we discussed before my return here, I visited the University of Tennessee in Knoxville and Oak Ridge National Laboratory to receive the formal offer of a position as 'Distinguished Scientist.' I would like to make a decision about my possibly accepting this offer no later than early May 1987. At present, there are still some questions of the equipment funds that must be resolved and the arrangement of offices and laboratories to be discussed in more detail. For my decision, it is also necessary to work-out a suitable plan for my continued presence at Rensselaer. I would appreciate it, if you could let me know whether the suggestions I outlined in our last meeting [on January 21] and which I summarize once more below would be workable. Naturally, I am open for any changes you or Steve [Wiberley] might suggest, and all would depend first on my accepting the UTK offer. As you know, the strength of their offer is the establishment of a new laboratory and a guaranteed annual support at a level of my present funding. Many of the difficulties in maintaining our research activities could be resolved in this way (I enclose my last 5 'Highlights' submitted to the Chairman's office [in the past five years for his annual report]).

Timing: I will return on August 1, 1987 from the sabbatic in Germany and be available for Fall and Spring 1987/88. There would be increased travel between Troy and Knoxville as construction of laboratories, sale and purchase of housing and initiation of research progresses (we will probably keep our year-around vacation home up North). I might have initial coworkers working full time at Oak Ridge as early as Jan. 1, 1988. Cooperation in research with some Oak Ridge researchers has shown already good results and will be reported at the APS meeting in NYC in March 1987. A possible moving date may be early in May 1988. After this time, occasional travel back to Troy is expected, to supervise remaining research.

To maintain a connection with Rensselaer, as we discussed, I would like to early retire after 25 years of service on July 1, 1988. My present research is funded at RPI through July 1989 by NSF, and I would like to keep this funding [at RPI], supporting one or two students and one to two postdoctoral research associates (students to complete thesis work: W. Aycock, M. Y. Cao, A. Xenopoulos, J. Wesson, and R. Pan, postdoctoral research associates Dr. S. Z. D. Cheng and possibly Dr. J. Grebowicz, the last to be appointed newly). A continuation beyond 1989 will need to be discussed later [in this letter].

Teaching: (1.) I would appreciate it, if I could be assigned during the Fall and Spring 1987/88 the completion of the development of the computer and video tape assisted F. Chem. Special course that was taught by me and Dr. Cheng in the last 3 years, including writing of text, problem sets, and laboratory notes (the main writing of the latter two parts to be reserved for the Spring Term) [see Appendix A, #255, see also page 8-86]. (2.) Since this course carries 9 contact hours and represents a considerable overload, especially during the development phase, I would suggest to appoint Dr. Cheng as a half-time teaching postdoctoral fellow for the Fall term to handle the overload. Book, illustrations etc. would then be completed by May 1988. (3.) All five graduate audio and computer courses (total 12 credits, total registered students of 10 to 30 annually) are to be offered as before in every term, Summer, Fall, and Spring. (4.) It would then also be of interest to the polymer program at Rensselaer, as well as to me, to continue these courses at RPI beyond 1988 with occasional updating ('Thermal Analysis' is slated for complete revamping in Fall 1988). These courses could be kept active, perhaps as now through the Continuing Education Division and at the compensation as before. I would suggest to appoint a professor-in-charge of the courses and, depending on the number of students, have occasional visits of mine at RPI for discussions and exams (travel expenses to be born by the Continuing Education Division, present plane fare about $500.–). Similar arrangements of setting up off-campus teaching are at present being discussed with the Universities of Freiburg and Ulm in Germany, and early interest has also been expressed by Fudan University and the China Textile University in Shanghai, China, and a to be established thermal analysis institute in Bombay, India. This shows that this is one of my major present interests in new developments in teaching.

Continued presence of Polymer Solid State Research at Rensselaer: Since our ATHAS laboratory at Rensselaer has been well recognized in the last 10 years, it may be advantageous for all concerned, if this effort be continued beyond 1989, also at RPI. One suggestion would be to hire one of my former students (U. Gaur, T. Davidson, S. Z. D. Cheng, or J. Grebowicz are possible candidates). I am sure that, for example, Stephen Cheng is one of the best young scientists I have contact with (I enclose a copy of a letter of recommendation I wrote recently for him). In case he could be considered by the faculty, naturally along with other candidates, for an appointment (as early as Sept. 1987) and be chosen as the most suitable, the ATHAS laboratory could continue with a minimum of start-up money (perhaps only $50,000 for equipment updating and completion of the long planned close-temperature air conditioning of the laboratory). Stephen would be the ideal person for me to cooperate with a two-location ATHAS laboratory, and naturally, we could be of mutual help in developing research support and keeping contact with European and Oriental researchers. Stephen has also shown during the Fall term 1986 that he is capable to continue my F. Chem. lecture and book effort that will carry in the future both of our names. Such a development would keep my teaching and research connection with the Polymer Program at Rensselaer long beyond 1989.

Please let me know the overall reaction of Rensselaer to these possible developments so that I can make an informed decision soon. My next short visit in Troy is scheduled in early May, but I would like to make the major decisions before that time."

It took some time until an answer was received. It is reprinted on the following page, and severed all connections of mine with RPI. The decrease of interest in polymer chemistry at RPI described on pages 8-85–88 had continued. It was difficult for me to understand how a university could not seriously discuss an offer of expansion of a program that it had furthered strongly for 25 years and now had a good chance to be headed by one of my outstanding students and bring in more funding through overhead than it would need from the university. Since I was offered research support at UTK/ORNL as part of my position, I would have kept the NSF support at RPI. As a retiree, one normally is permitted to continue research. Indeed, I enjoyed continued NSF support at UTK/ORNL for another 20 years until I closed my laboratory, and a further five years after my retirement at the University of Tennessee at Knoxville.

Guessing at the outcome of a proposal, research or otherwise, *in retrospect, I was frequently right* (see page 2-6). Dean Raveché should be admonished for not discussing the suggestions on pages 8-117–118. Stephen Cheng probably would not have done better at RPI than he did at Akron (see page 8-123–124, below), but he would have succeeded and would have been profitable to RPI. Even more, I regretted the off-hand dismissal of my plan to center innovative teaching of graduate subjects at RPI. It would have worked nationally, as well as internationally, to the benefit of RPI. At UTK and ORNL, my work was centered on research, so that the teaching efforts developed less than they could have with the proposed two-university approach. To tackle the exploding volume and specialization of knowledge, it was my thought to develop professorships connected to several universities. For narrow fields of specialization, this would give sufficient access to students for efficient teaching, as discussed on pages 8-68–75. To document my being serious with this approach, one can look at my RPI stationary, reproduced on page 8-56. It puts the university affiliation to the same level as my research specialization (and invited critical comments at RPI).

A special travel experiment from Freiburg was made for *the 1987 March APS Meeting in New York City* (see Appendix C, details listed under the letters A–H). Three lectures (A–C) and five posters (D–H) were submitted. They covered my work in Freiburg (A), our ongoing research in Troy (B, C, E, F, and H), and our new cooperation with ORNL (D, and G). All was scheduled for Wednesday and Thursday of the week of the meeting. To avoid jet-lag, I left Freiburg by train on Tuesday at 8:55 a.m. for Frankfurt, to catch the flight leaving at 1:30 p.m., arriving at 4:00 p.m. in New York City with a time-difference of six hours. I went to bed as soon as possible and got up according to Freiburg time. After last preparations for the lecture, I had an early meeting with my coworkers and went to the morning and afternoon lectures. This was followed by the usual dinner with all students, arranged by Stephen Cheng in China Town. As soon as possible, I went to bed again and got up not much later than Freiburg-time on Thursday, and left after the last lectures for Kennedy Airport at 9:20 p.m., to sleep as much as possible on the plane. At noon on Friday, I was back in my office in Freiburg. Tired, but, as I remember, with only little jet-lag at either end of the trip. Unnecessary to say, that all lectures and posters were well received, all coworkers did a splendid job. Since my Humboldt fellowship had already started on January 1, they also picked up the travel expenses to this and similar meetings in Europe, listed in Appendix C.

School of Science
Office of the Dean
Telephone (518) 276-6305

Rensselaer Polytechnic Institute Troy, New York 12180-3590

April 10, 1987

Dr. Bernhard Wunderlich
Professor of Chemistry
Institut für Makromolekulare Chemie
Der Albert-Ludwigs-Universität
Hermann-Staudinger-Haus
D-7800 Freiburg I. BR.
Stefan-Meier-Straße 31
Federal Republic of Germany

Dear Bernhard:

I am writing in response to your recent letter about your new position at the University of Tennessee and future interactions with RPI. Stephen Wiberley and I have discussed this in detail.

The Chemistry Department requests that you support Cao on a RA next year so that he can finish his thesis. The department plans to have you teach the Freshmen Chemistry II Special for the Fall semester, which entails 9 contact hours, and the 3 hour contact laboratory in the Spring of 88. We do not have the funds to appoint Dr. Cheng to handle any teaching overloads.

Because of the department's future interests and priorities, it cannot support the presence of the ATHAS laboratory beyond the time of your departure to Tennessee. Nor would we plan to continue the audio and computer courses which are delivered via Continuing Studies. As indicated in your letter of February 6, we are planning on your resignation from RPI as of July 1, 1988. It will not be possible for us to maintain any laboratory or office facilities for your use beyond that time.

You have made many fine contributions to Rensselaer and I have every confidence that you will build on your distinguished record of achievement and establish an internationally visible program at Knoxville. We are happy about your success at Tennessee and congratulate you for the recognition. We wish you the very best in your new endeavor.

Sincerely yours,

Hal Raveché

Harold J. Raveché
Dean

<u>The move to Ulm for the second half of the sabbatic</u> occurred on April 15, 1987. It was connected with one additional benefit. For six months, I had the use of a new 500-series BMW. We were now a two-car family again, and Professor Pechhold had arranged through a friend, Mrs. Lusser, the rental of a house in the suburb Arnegg, about 20 minutes from work. At the university I was located in the Department of Applied Physics. Figures 102A–E illustrate the flavor of our stay in Ulm, a city on the Danube at the border of the states "Baden-Württemberg" and "Bayern." The beautiful "Ulmer Münster" has the tallest spire in the world. Ulm is also known as the birthplace of Albert Einstein, and was also the home of Max Eyth, my favorite engineer and poet (see page 3-11). The University of Ulm, shown in Figure 102C, is quite new. It was founded in 1967. Figure 102D illustrates the nice woods around the University of Ulm which I valued each morning when arriving from Arnegg.

Figure 102A *The house in Arnegg near Ulm.*

Figure 102B *View of our house (in the center) and country side in the Arnegg suburb of Ulm.*

C *Part of the University of Ulm. Building complex which houses the "Angewandte Physik"*

Figure 102D *My daily walk from the parking lot to the buildings of the University of Ulm, 1987.*

E *Two views along the river "Blau." The blue tint giving it the name, is seen clearly in both.*

Figure 102E represents views along the small river "Blau" (blue) which has its spring near the small city of "Blaubeuren," ringed with beautiful old buildings and a cloister. The pictures were taken close to the spring, the "Blautopf," where the "Blau" emanates from the ground in final size. In Ulm, about 15 mi east, it empties into the Danube. Besides the attractions in and about Ulm there are many places of interest only a short drive away. When Charlotte M. (see Figure 100C, and introduced on page 8-84) came from Albany, NY for a week to visit Heidel, she was given an amazing itinerary: Day 1, sightseeing in Ulm; day 2, drive to Augsburg; day 3, drive to Rothenburg; day 4, a train-trip to Munich. On days 5 and 6 all three of us made a round-trip by car to the Alps (Oberstdorf) and the island Meinau with an almost Mediterranean climate, allowing exquisite gardens. Meinau is located in the "Bodensee" (Lake Constance, Central Europe's third largest lake).

Yes, in Ulm we also were not lonely. As in Freiburg we had good connection with the colleagues, some of them we are still in contact with. The same was true for our landlady whom we could stay with during the 1988 revisit to Ulm, and Heidel's "Schneiderin" (dressmaker), who for many years thereafter made her dresses out of the silk bought first in China in 1984, and later was received as gifts from one of our hosts in China, Professor Wu (see Figure 95A). By then, Professor Wu was a coworker of mine, as is illustrated in Figure 101. In addition, as in Freiburg, our German friends and relatives, took the opportunity to get together with us in the new location, and we also fitted many visits with friends and relatives within Germany into our travel schedule.

Early in May, Heidel and I made *another short trip back to Troy* to attend the graduation of Brent from Syracuse University (see pages 8-81–82) and to change our winter clothes to summer attire. In addition, it was time to resolve the discussions with RPI (see pages 117–120). The newspaper clipping from the Knoxville News-Sentinel tells the story. Another important job was to consult with my students. Also, I had to give a lecture at the Polytechnic Institute of New York in Brooklyn, NY (and to feed the lonesome fish mentioned on page 116–117). On May 4, I had a final discussion with Professor Wiberley. It brought little additional movement in the position of RPI, which was directed by the dean (see page 120). He allowed, however, my removal of all

A8 The Knoxville News-Sentinel, Friday, May 22, 1987

east tennessee news

Distinguished scientist named

Dr. Bernhard Wunderlich, a chemist renowned for his work in thermal analysis, has been named a distinguished scientist at the University of Tennessee and the Oak Ridge National Laboratory.

He will begin his joint appointment Jan. 1, 1988, joining nine other top researchers named distinguished scientists since the program was begun in 1984.

Wunderlich's work in thermal analysis of polymers is considered particularly important to ORNL, where the chemistry division is examining the structure and dynamics of polymers.

"We are delighted to add to our faculty and the laboratory staff a person of his stature," UT Chancellor Jack Reese said.

Since 1965 Wunderlich has been a professor of physical chemistry at Rensselaer Polytechnic Institute in Troy, N.Y.

A native of Germany, he was educated at Humboldt University and Goethe University. He moved to the United States in 1954, attended Hastings College in Nebraska, then earned his Ph.D. at Northwestern University. He is highly regarded for his innovative computer and video teaching methods.

The Distinguished Scientist Program, funded by the state of Tennessee and the U.S. Department of Energy, was created to attract researchers with national and international stature to the Knoxville-Oak Ridge community.

laboratory equipment, the termination of my NSF Grant at RPI (which was reinstated in 1988 in Knoxville), and, finally, Jan. 1, 1988 as the beginning of my early retirement, i.e., he relaxed the condition of the sabbatic from a stay of one year at RPI after my return, to one semester. Before I returned to Ulm on May 11, I wrote the acceptance letters to the UTK and ORNL, and requested my retirement from RPI. The newspaper clipping shows the reaction in Tennessee. At RPI, Professor Wiberley just sent a short, cordial acknowledgment, and there was no newspaper announcement.

A summary of *the development of the student group for the year 1986/87*, as shown in Figure 101 is given in this paragraph. The postdoctoral coworkers (Loufakis, Cheng, Bu, and Wu) and an undergraduate technician (Woertman) completed their appointment in 1987. Another student (Loverso) was a continuing employee of IBM in a resident study program with a company fellowship. She had finished her core courses and cumulative exams with me, and the company asked her to do her thesis in-house, but seemingly she did not follow their plans. A part-time student (Aycock) ultimately discontinued her study (after all her thesis work was completed and written up for publication, see Appendix A, publications # 232, 246, and 274). An off-campus student from GE, Schenectady (Bopp, see page 8-23), completed his degree in 1993 with me returning to RPI for the exam. The two graduate students (Cao and Wesson) completed their degree under my direction.

Professors Cao (see page 106) and Bu (from Fudan University, Shanghai) had come to RPI as visiting professors with support by the Chinese Government via World Bank funds (see Figure 101) . After a few months, we found out that their stipends were quite unsatisfactory, even when living most frugally. Since I had pledged in this arrangement to pay a second year of their stay, I arranged loans to reduce this suffering. When Professor Cao's research was successful, he changed into a graduate student, to get a PhD at RPI. The Cultural Revolution (see Footnote on page 8-99) had prevented him to do so in China. After things were normalizing, he had reached the age of 40, too old in China to be allowed into a PhD program. He succeeded in his graduate studies, with the exam being given on one of my return visits in 1988. His wife (Cui-xia Liu) and son had also joined him from China (see page 106). At RPI, she was working for Professor Chung.

Professor Bu let us know at one of our group dinners in 1985 that by then, he had gained enough weight that his clothes did not quite fit anymore. For the first time in many years he had gained weight. We are still in contact and cooperated in his research in China (see Appendix A). In 2000, on our second visit to China, he hosted our stay at Fudan University (see Appendix C).

This left Mr. Xenopoulos, who had started in 1986 as a graduate student. He decided to move with me to Knoxville, but still get his degree from RPI. He became the supervisor of our move, and the student in charge of rebuilding the laboratories at UTK and ORNL. His PhD thesis was accepted in 1990 with both of us at RPI for the exam. He then stayed for three more years with me as a postdoctoral fellow at UTK before leaving to work for Millipore Corporation, Bedford, MA.

Special mention must be made of Dr. Cheng (see also pages 8-99–100). He was offered a job by the DuPont Co. shortly after his PhD in 1985. His family was trying to encourage him to take this job with much better pay than that of a postdoctoral fellow. This would have easily allowed him to satisfy his obligations toward his family. But, he wanted to get an academic position. We had many discussions to resolve this problem. Knowing his talents, I encouraged him to try to reach his goal. While he was applying to all available academic openings, I first tried to have him stay at RPI after I left (see page 8-118). When this was not possible (see page 120), I tried to move him with me to Knoxville. Before this was fully planned, however, he received a job offer as assistant professor at the University of Akron, one of the few centers of polymer research (see page 8-85).

Since I was permitted to move all equipment from my lab, but had new equipment ordered for the two research laboratories in Knoxville, I had Stephen take the duplicate calorimeters to Akron, to supplement his start-up funding. For 1988–1991 the ATHAS reports (see Figure 69) covered also his research as a 'branch' of ATHAS. By 1991 he had assembled a group of 12 research students and had a list of 34 papers published or accepted for publication. This was a remarkable beginning. He became quite successful at the University of Akron. By now he is their Robert C. Musson Trustees Professor and Dean of the College of Polymer Science and Polymer Engineering.

Special among the meetings during the 1986/87 sabbatic was **the IUPAC Polymer Symposium in Merseburg, GDR**, held at the "Technische Hochschule Leuna-Merseburg" (THLM, Carl Schorlemmer Technical University, founded in 1954, disbanded in the 1990s). The IUPAC required the host country to give visas to all the attendees who applied, regardless of nationality. Also, the GDR had to guarantee international telephone service. This was a big concession for the visitors from the FRG. Also, many improvements were made around Merseburg to make a good impression on the world. The effort to hold the meeting in the city of Merseburg was spearheaded by the university president, a good friend of Margot Honecker (Minister for National Education and wife of the leader of the GDR, E. Honecker, see page 3-22). Merseburg was a city with a population of about 50,000, located close to the bigger cities of Leipzig and Halle. Leipzig is the city known for its annual international "Messe" (trade show), important for the economy of the GDR. Merseburg was in the center of the chemical industry of the GDR. The meeting was held in the newly renovated lecture rooms of the university. Since the city had insufficient hotels, student dormitories were upgraded and reserved for the attendees.

When on June 29 we drove into the GDR. Even our temporary FRG licence plates were accepted. Before, in 1968 on our earlier visit they were not, as described on pages 8-46–48. Naturally, we had again arranged to meet with Hans and Inge Gottschalk (see page 4-28). They were staying in the "Merseburger Hof," at that time the only hotel in town, located in a run-down street which had seen little upkeep since the war. I took a picture on the street where we parked our little VW Jetta. In Figure 103A, the VW looks big among the four GDR cars of the make "Trabant."

Hans and Inge had arrived, but the Merseburger Hof was forbidden to serve food because of sanitary violations (rats and mice infestation) . To eat, the guests had to walk to the railway station, about two blocks away. We met Hans and Inge in the street on their return from breakfast. Inge wanted to know our impressions. We answered with another question: "How it is possible that the city is so dirty and cared for so little?" This brought tears into Inge's eyes. Yes, she said, nothing is done to improve the appearance of our cities anymore.

When I was asked after my return to the US to characterize the state of the GDR, I reported that it looks like the GDR is becoming an underdeveloped, third world country, characterized by neglected buildings and incomplete improvement projects, i.e.,

Figure 103A *Our Jetta near the "Merseburger Hof."*

the construction sites overgrown with weeds. Whatever was done, was of poor quality. The street to our dormitory was newly blacktopped, but the trucks had already cut new ruts. At the university, flagstones were laid to improve the walkways, but they were uneven because of a missing proper foundation. In both cases, it was not only the material that was poor, it was also the workmanship that was sloppy.

Figure 103B *Heidel with Hans and Inge in one of the "Dornburger Schlösser" (palaces) with special felt shoes to protect (and to polish?) the inlayed floors.*

The overall mood of the people seen in the streets was downtrodden. During the meeting, the factories were largely shut down to improve air quality. Still, the atmosphere was noticeably 'chemical.' I was reminded of such steps in 2008, when seeing such method applied during the Olympics held in Beijing. Despite all these observations, nobody could foresee that the economic and political collapse of the GDR was only two years away (see the pages 9-23–27).

During our spare time, Hans and Inge showed us the surrounding cities of Halle, Leipzig, Naumburg, and Dornburg. The last, is illustrated with Figure 103B. The past was clearly more beautiful than the present.

Hans was happy that, for once, he could host us, instead of always having to rely on our support. When we were in Leipzig, he invited us to the famous "Auerbachs Keller" as it appears in Goethe's "Faust." After gaining Faust's soul, Mephistopheles and Faust started their travels from this restaurant. Hans went ahead to reserve a table, but came back quite disheartened. There was no table available, although, he said, it did not look crowded. So, I decided to also try. I went and asked at the reception for a nice, quiet table. No problem, we got the best table in the house, it still had not changed since 1968. The common people of the GDR were treated as second-class citizens in their own country! Despite this bitter incident, I think we all enjoyed the evening together.

Other occurrences reminded us how little had changed over the years. The cadets of the police academy, located next to the Technical University, normally frequented the restaurant and entertainment facilities for the students at the university. For the duration of the meeting, the whole area was off limits to them, so that they would not be exposed to "Westkontakt" (contact with the Free World). When Heidel and I wandered on our own with Dr. Heinze, the director of the BASF Research Laboratory (see page 8-6) into the bar of the "Haus der Kultur in Merseburg" (house of culture), the social center of the city elite. Two officers of the academy asked for permission to join us at the table, as is common practice in Germany if one does not use all chairs around a table. We had a nice, frank conversation. They had much criticism about the situation in the GDR. Perhaps they did not know of our being attendees of the international meeting.

The opening ceremony in the morning of June 30 was in the famous "Leipziger Gewandhaus." It ended with Beethoven's 3rd Symphony, played by the Halle Philharmonic Orchestra. The organizers brought luxury busses from all over the GDR (as I noted from their license plates). The stretch of Autobahn to Leipzig was closed except for our special VIP busses. All was arranged so that everything looked perfect, but the contact with the local population was minimized. As in earlier meetings in communist countries, the 31st IUPAC Macromolecular Symposium was a success, although most attendees were from Eastern European countries. The biggest foreign delegations were (in order of number of attendees) from Czechoslovakia (71), FRG (67), Poland (56), Japan (53), Hungary (40), USSR (38), Yugoslavia (28), and USA (27).

After the meeting, I had *two invitations for lectures in the GDR*. One by Professor W. Berger from the Technical University of Dresden, the organizers of the special symposium I contributed to. The other was from by Professors Dörfler and Sackmann of the University of Halle, where much work was done on the mesophases of small molecules, instead of polymers. These were interesting to me as model compounds for our macromolecular condis crystals.

After the last day in Merseburg on Saturday, we drove to Dresden to spend some time to tour the city and visit some relatives and friends before my lecture on Tuesday. We stayed at the visitors floor at the university dormitory. Heidel had contacted her cousin Adelheid Erbe (see page 11-3). She had never met her or her husband, Werner. When they arrived, the front office of the dormitory advised them we were not there and are not expected, despite the fact that we checked in just a short time before. Fortunately, they had the feeling something was not correct, they had spotted a car in the parking lot that could have been ours, and it was. So, Werner entered anyway and went from floor to floor of the high-rise building. Heidel also had the feeling that we were to be isolated in our room at the top floor. After a while she tried to look for the Erbes. They met in the corridor and had the proper intuition that they were searching for each other. Contact was made, and when we visited in their apartment (see Figure 104), we heard their sad story.

Werner was a geologist with the Russian Wismuth (bismuth) Company, mining uranium ore in the nearby "Erzgebirge" (Ore Mountains). It was a well-paid job. Werner had an uncle who had emigrated to the US many years before and now lived near Syracuse, NY. When he died, a nephew who lived with him in the US was, according to the will, to receive the inheritance. The executor checked with all other cousins whether they had any claims, and also asked for help in locating some of whom he had no address. Straight forward practice. Werner answered the letter, gave several addresses, and mentioned that he had no claim himself. Briefly thereafter, he was arrested for communicating with the West and giving information. He lost his job, and was forced to become a low-level clerk in a market. Yes, he got paid by the lawyer who thanked him and enclosed a valuable check for US $20. Naturally, Werner was afraid to cash the check. This was about 15 years before our visit. We took the check back to the US, found out the funds had been transferred to be held by the State of New York, and we could claim them for a fee of $20! It was so easy to lose one's future in the GDR.

Figure 104 *The Erbes with son, daughter-in-law, and first grandchild on their balcony in Dresden, 1987.*

On Sunday we were invited by Professor and Mrs. Berger for lunch ("zum Mittagessen") in the surroundings of Dresden. They drove with us to the Moritzburg, a hunting lodge of "August der Starke," renovated in 1723, located about

10 mi north of Dresden. Figure 105 shows our hosts and the lodge which had a fine restaurant. In true GDR style, the entrance sported a big sign 'restaurant closed.' We felt sorry, but Professor Berger was one of the chosen who knew that you just step over the chain barring the entrance, go to the back, and join the crowd in the restaurant were you could have all the local delicacies and wines which were otherwise hard to obtain. We had a good meal, followed by an outing to inspect the Elbe Valley, a UNESCO World Heritage site (until 2009 when building of a modern bridge destroyed the view).

On this occasion we also inquired about wanting to travel to Cottbus to visit an old school friend. Yes, he said, you can do that, but you must be back and deposit the key at the dormitory before midnight. Otherwise, they will call the police to look for you. You have to stay overnight where you are registered. I replied: But in case we get too much to drink, would it not be better not to drive back the 65 mi?" We have not seen Fritz in 25 years, we may have a party! He replied you have an excuse, but no permission. When I remarked that we got a second key in the morning, so we could return the main key, and nobody would know that we did not come back, he answered: "I see, you know the system!"

Figure 105 *The Berger's with the Moritzburg in the background, 1987.*

We visited Fritz Bockmühl and his wife Hannelore (see also Chapter 3, page 3-32–33 and Figures 22, 25 and 28). Figure 106 displays his beautiful house and garden, showing that he still had not lost his passion for growing flowers. We had a wonderful evening and went back the next day after inspecting Cottbus, at that time, one of the better restored towns in the GDR. Nobody had missed us in Dresden. We were, by then, also brought up to date on the difficulties in the life of the Bockmühl's, caused by the political oppression. It was particularly difficult for them to get their daughter into a university to studying medicine. It was not desired in a country of workers and farmers to have successive generations of doctors. Little had changed since I tried to be admitted to the Humboldt University in 1949 (see pages 4-3–8).

Figure 106 *Professor Friz Bockmühl, specialist in ear, nose, and throat medicine in the back of his house, posing with Heidel and wife Hannelore, 1987.*

We also took the time to visit the center of Dresden, the city that was heavily destroyed during WW II (see pages 2-17, 18). The center part of 18th and 19th century buildings are the Zwinger Palace and the

Semper Opera House. Much was rebuilt, starting in the 1960s in a special effort by the citizen of Dresden showing their defiance of the war. Still, at the time of our visit, large parts of the city were open green areas where houses stood before. The impressive "Alte Meister" Art Gallery had many of its treasures returned in the 1960s from the Soviet Union, where much of it had been transported as spoils of war from the safe places they were stored in to survive the bombing. When we visited St. Petersburg in 1996 (see Appendix C), we could see in the Hermitage a large display of art that was looted from Germany and was awaiting return to their museums and private owners.

The art gallery, as well as the "Grüne Gewölbe" with its porcelain collection made for an impressive visit. Only the "Frauenkirche," located next to the Zwinger, was still in a sad state, as seen in Figure 107. It burned during the bombing and collapsed, as shown. No funding was available in the GDR for rebuilding the church. Many of the reusable building blocks were numbered and stored at the site of the ruin, slowly being hidden by the weeds. It took until 2005, sixteen years after the "Wall" fell until the "Frauenkirche" was restored according to the plans of its architect, George Bähr (1666–1738), one of the great architects of the German Baroque.

Figure 107 *Ruins of the "Frauenkirche" in Dresden as they appeared in 1987. Only few pillars were standing.*

After these excursions, I gave my scheduled lecture in the colloquium at the Technical University in Dresden. Its title was: "Der starr-amorphe Zustand von Hochpolymeren." On the same day, we continued our trip to Halle.

The institute there, "Chemie Mühlpforte," was not well maintained. For some time, they were out of city water. A fire hose was used to solve the seemingly old problem. But these problems were of no influence on the science done, quite the same as I had observed in China. The best science is not always done in the best equipped institutes. My lecture: "Condis Mesophasen in Molekülen kleiner Masse," was well received and during the discussion, I learned about many other examples investigated during the extensive research done in Halle. Overall, this visit was one of the most productive of my sabbatic leave.

The following day we traveled back to Ulm. But first, we stopped in Quedlinburg, a historically important medieval and renaissance town. Since 1994 its Collegiate Church, Castle, and Old Town are a UNESCO World Heritage Site. For us, it was a lesson about the founding of the "First German Reich." The Quedlinburg castle complex was initiated by "Heinrich I" (see page 1-1). The old city was restored in the 1980s by Polish experts, trained by rebuilding the old city of Warsaw (see page 8-60). We got a tour of the city with a personal guide who accompanied us from Halle and then returned by train. This ended a quite educational IUPAC Symposium and a final look into communist Germany. Next time we visited Leipzig and Halle, reunification had taken place creating its different problems, to be commented on in Chapter 9 (pages 9-23–27).

The rest of the sabbatic leave past quickly. On the 5th of August we were back in Troy. Before, however, we had a family get together to mark the occasion that *Papa had started his new life in a senior citizen home near Bergheim*. Figure 108 is a reminder of this occasion. Since the passing of Mama in 1979 he had emigrated to the US and stayed with us in Troy (see page 8-109). His pension and health insurance, however was only valid as long as he kept his residence in the FRG (see pages 7-14–15). This meant, in the summers he traveled back each year to Bergheim for a few months to be able to talk to everyone and to renew his travel-health-insurance to the US. Despite all efforts, we could not get any health insurance in the US for him. In Bergheim, he had, in addition, his relatives and friend Else Baum (center of Figure 108). Having reached 88 years of age by then, he said he would not want to undertake this strenuous travel in the future. We also found, that he had difficulties getting used to the new surroundings in Ulm after our moving there from Freiburg. At home then, our big move to Tennessee was to come. It would have been quite a difficult time for him to adjust. Thus, we had looked for a suitable home for Papa, which we found in the neighboring town (" AWO Seniorenzentrum Quadrath-Ichendorf GmbH"). On our trip back to Troy in May (see pages 8-81, 82 and 121) he had a trial stay there which was acceptable to him. Thereafter, either Heidel or both of us arranged a visit to him each year and weekly, there was an extensive phone call to exchange all the news. Papa quickly felt at home in his new environment and had a number of new friends.

Figure 108 *Group of Family and friend with Papa.*

Retirement from RPI and a new beginning in Tennessee was the new agenda when we where back at RPI. The fall teaching schedule of mine was set to 9 h, as can be read from the letter of page 8-118. Consulting and off-campus lectures, as listed in Appendix C, filled the rest of the time. (At UTK, the teaching duties were to be only ≈3 h every *second* semester and I gave up consulting for the DuPont.) The work of the remaining students was carried out as described above.

An invited lecture at the BASF-Enka in Enka, NC, named after NK (pronounced En Ka), the earlier built Dutch "Nederlandsche Kunstziidefabriek," located near Asheville, was scheduled to fit a meeting of the ORNL Chemistry Division Advisory Committee. This trip was also used to introduce Heidel to Knoxville, our new town to live in and to find a new house. We drove to Asheville, where the BASF-Enka had arranged a stay at the famous Grove Park Inn. Next day, I visited at Enka, while Heidel drove on to Oak Ridge. In the evening, I followed by bus, and at the end of the trip, we parked the Mercedes 300D in Don Noid's garage to await our move.

After arrival at the Garden Plaza Hotel in Oak Ridge, Heidel met our realtor, Ms. Susan Shay *to find a suitable house in Knoxville* for us. On my earlier visit in Knoxville in January 1987, Susan had shown me the market by making a round trip through Knoxville. At that time, two areas caught my attention, one was at Cherokee Boulevard, an old established area of fine houses and gardens near UTK. The other, called Fox Den, was a newer, well established development, 25 minutes from

UTK and 20 minutes from ORNL. Cherokee Boulevard was not close enough to walk to UTK, and the main road was too narrow and too busy for cycling. After accepting the position at UTK/ORNL, Susan sent a copy of extensive descriptions of all available houses in the area, with about 50 suitable houses marked for us to consider.

Since the house we would buy had to be liked by both of us, a preselection was accomplished by Heidel and Susan by Friday evening when I arrived from Asheville. Detailed inspection was to follow Saturday and Sunday. It turned out simpler than we thought. When concluding the Friday tour, Heidel discovered a house newly put up for sale in Fox Den. By Saturday, Susan had the details and made an appointment to look at the house. Figure 109 is a recent Google photograph of house and property. We both liked it and made a list of all its positive attributes. There were few negatives.

Figure 109 *Our house at 200 Baltusrol Road facing the Fox Den Golf Course on the lower right corner.*

All other houses were compared to this list. None of the perhaps 10 other houses we saw, measured up. So we decided on Sunday, to make an offer. We had really wanted a nice old house as we had in Troy, but see page 8-2–4 for our thoughts in 1963. Also, we wanted a house that was smaller, and with the master bedroom on the ground floor. For the two of us, we would not need the more common four bedrooms. But the house was to have 'character.' This house was new (4 years old), instead of the 2,000 square feet living area in Troy, it had 3,000. It had two bedrooms and an office, *but* it had character. To fulfill two out of four wishes was not bad. This meant, we found a house within three days, and today we still live in it. Surprisingly, a house near Cherokee Boulevard was found which looked amazingly like our house in Troy and we inspected it carefully. But it had more rooms, and would have needed much renovation and updating.

Figure 110 *The Snowstorm on October 6, 1987, caused major damage. The leaves had not fallen!*

<u>To sell the old house in Troy</u> proved difficult. As periodically, the realty market was somewhat depressed. The first few weeks the house was on the market, nobody bid. Maybe, it was because we did not accept the realtor's suggested price, but got an independent estimate of the value of the house and added $20,000 to have some cushion in case we needed to reduce the price to make it more attractive.

Then disaster struck in form of a snow and ice storm. Figure 110 is a view of our patio and back yard. Despite my effort to shake off the snow and break off the ice as much as I could, all of the maple trees had lost one or more of their key branches out of their crowns. The leaves had not fallen, making the snow and ice accumulate. From the woods, it sounded like machine gun fire from the continuous snapping of branches. It did not take long, and most of the area was without electricity. Help was summoned from all over the state, but it took almost a week until we were back to normal. Our extra coal-fired kitchen stove in the basement came in very handy (see page 8-8), as did the fire place.

We asked a friend who was a building contractor to come and help cutting the fallen branches and cart them away. He came with two of his sons, and in two days, all looked liveable, but one should not look at the crowns of the trees. A costly trimming of the trees was called for. If a prospective buyer would have noticed this, I was ready to allow $5,000 for the tree surgeon.

But, it came much different. Only a few weeks later, a buyer came, and told us: "I loved this house for years and always when I drove past it, I liked to live in it." This was our opportunity. When she looked over the house, there came two deer from the back woods, and she was delighted, as they slowly strolled through the property and went across the road to the open field (see Figure 5). This view we had hoped for when considering to keep fallow deer (see page 8-9). In all the 25 years living in our house we had never seen deer that conspicuously on our property. But the deer helped to sell the house without arguing about the price. Next day, we even got another prospect. But the problem was solved, and the house has been, over the years, our best investment. We liked it, and it had increased its value by 750%, while inflation was only 400% over these years.

Another important event fell into this time period. *On August 30, our grandson Stefan was born*. This made our move to a city 900 mi distant much more difficult, particularly for Heidel. Troy was just a three-hour drive away from Syracuse, Anna and Brent's home. Travel from Knoxville would take two days. Soon Anna, Brent, and Stefan came to visit us in Troy, and shortly after the snowstorm, on October 10, we were in Utica for the Christening of Stefan in the church of Anna and Brent's wedding (see the pages 8-82–83 and Figure 84B). The date was chosen to honor the anniversary of Mutti's wedding, 62 years before. Mutti arrived in Troy just before the snowstorm, so we could make the trip to Utica together, as is shown in Figure 111. On this visit of Mutti's we also made a trip to Shahla and Eberhard who lived at that time in Aberdeen Township, NJ. There, Mutti could also visit with her two other great-grand children, Roland and Zeba.

Next came a period of many good bye parties. Noteworthy was only the absence of any notice taken of my retirement by the Department of Chemistry. The polymer group, consisting of engineers and chemists, however, under the leadership of Professor Chung arranged a retirement party of sorts. It was much appreciated and a snapshot is reproduced as Figure 112. Only recently did I recover the RPI scarf which was one of the gifts and is clearly visible in the picture. But, there are very few cold days in Knoxville, so the scarf got pushed into the back of the drawer and saw little use. At the head of the table is Professor Chung (Engineering), followed by Professors Wnek (chemistry) and Birnboim (Engineering).

Figure 111 *The day of Stefan's Christening in Utica.*

Figure 112 *Retirement party from RPI December 11.*

A final Summary of almost 25 years at RPI 1963–1988 is that the first 15 years were quite successful and satisfying. I felt a continued improvement in teaching and research. Then, slowly, the decline of interest by the Department of Chemistry in further expansion in the area of polymer science became obvious (see the pages 8-85–88) and, although my research continued satisfactorily, further development was hampered. Overall, this did not affect the research of my group of coworkers. We could still do what we were interested in, to improve our knowledge and understanding of the solid state of macromolecules, as summarized in Sections B to D of Appendix B. Overall, we wrote 244 publications which appeared under the credit line of RPI. A total of 84 coworkers of different level worked with me during my stay at RPI (14 technicians, 15 MS-level students, 29 PhD students, and 26 Postdoctoral research associates). This group was very active and its success was felt in many fields of polymer science. Many of our ideas are still gaining increasing importance, not only in the field of synthetic macromolecules, but also for biological polymers.

Moving day was January 2, 1988. Right after Christmas, the movers came with hundreds of boxes and packed everything, as shown in Figure 113. There was no way we could have a New Year's party, but our best friends, the Buhacs, had been the favorite place for these parties before. They complained bitterly about our moving. Little did they know that shortly thereafter, they moved themselves, i.e., this was also their last New Year's party in Albany/Troy. Early in the morning of January 3rd, Heidel took off in her Jetta, and I followed in the Jeep Cherokee. (We had bought the 4-wheel drive Jeep three years earlier for winter driving.) The trip was planned well. Points were picked where we would meet in case we got separated. Fortunately, the back and front lights of the Jetta could easily be distinguished from other cars in the dark. Also, we outran a snowstorm from the coast in Virginia, so that we arrived safely on January 4 in Knoxville, two days ahead of the Bekins Van Lines truck shown in Figure 114.

Figure 113 *All packed, and waiting for the move.*

Figure 114 *The fourteen-wheeler for the move.*

CHAPTER-09

University of Tennessee & Oak Ridge National Lab, 1988–2001

The move to Knoxville coincided with the beginning of the year 1988. On December 31, 1987, all was packed and the moving van was parked in our driveway in Troy, NY. The laboratory equipment was loaded earlier, so that on January2, it could be filled with our household goods (see Chapter 8, Figure 114). The last item was the small John Deere tractor with its snow blower (see Chapter 8, Figure 76). We thought never to need it again in the warm, southern climate, but we were not quite right. The Mercedes was already parked in Knoxville (see page 8-129).

On the 4th of January, Heidel and I arrived with our VW Jetta and Jeep Cherokee in Knoxville. We stayed at a Knights Inn, and waited for the furniture, which arrived on the 6th. In the nearby restaurant, we received our first lesson of life in the south. Taking our order, the waitress absolutely wanted to know more about us: "Where are you from?" So we told her, from Upstate New York. Her answer was immediate: "No, you are not Yankees[1]!" On further interrogation we had to admit that some 33 years ago we immigrated from Germany. That satisfied her. She remarked: "*I knew it! You are foreign Yankees*." We never figured out whether this was better or worse than being classified a plain Yankee. Whatever it meant, we were more amused about her southern accent than she was about our German one (see page 6-25), but we were served well.

From January 6 on, our address was *200 Baltusrol[2] Road, in the development of Fox Den*, near the 4th green of the golf course which runs through the development of almost 400 houses in the Town of Farragut,[3] adjoining the city of Knoxville toward the west. People always have had

---

[1] Derogatory for an inhabitant of New England or one of the northern states in the US.

[2] The streets in Fox Den are preferably named after golf courses. The Baltusrol Golf Club is in Springfield, NJ. The name of the 500 acres, 36-hole combined upper and a lower course (each par 72) derives from the farmer *Baltus Roll* who lived on a hill side, now the location of the clubhouse. He was murdered in 1831 by robbers and the club was named after him when it was formed in 1895. The course has hosted 15 USGA tournaments. By 1993 it had hosted the US Open seven times. In 2005 it was the location of the PGA Championship and it is to host it again in 2016.

[3] D. G. Farragut (1801–1870), first Admiral of the US Navy, remembered for his order in the Civil War Battle of Mobile Bay (1864): "Damn the torpedoes [mines], full speed ahead." His action was key to closing the last harbor of the South during the Civil War. Born at Lowe's Ferry near Campbell's Station, TN, incorporated in 1980 as the town of Farragut (population ≈20,000).

difficulties pronouncing our name, from now on, pronouncing our address was even worse. Despite living on the golf course, we are not members of the club, and have never played golf, either. Perhaps, I do not see the beauty in dropping a little white ball into a small cup from quite a far distance when I could drop it just as easily by walking up to the hole. The golf course, however, serves a great purpose. It provides a large back yard of well-maintained grass and a couple of nice looking ponds for ducks, Canadian geese, and occasionally a blue heron.

Figure 1A *Our house, 200 Baltusrol Road in spring.*

The impression of the house is given by Figures 1A and B. Figure 1A was taken in spring 1988 and Figure 1B was taken on

Figure 1B *Back porch view to the golf course.*

Figure 2 *Heidel, waiting in the empty house for the movers on the morning of January 6.*

our first inspection of the house in September 1987 (see the pages 8-129–130). The view from the back porch in Figure 1B is toward the pond near the 4th green. The placement of the house is seen in Figure 109, Chapter 8. Figure 2 illustrates the big emptiness of our living room (hall) before the movers arrived. This emptiness disappeared in a few hours when everything from the moving van was brought in. About 200 boxes filled the space. Figure 3A demonstrates the disaster in my office and Figure 3B has Heidel working already on the unpacking. As the cartons were emptied, they were piled into the garage. By the evening, there was some breathing space and we could take stock of our possessions.

Nothing seemed to be broken or damaged, just that I missed my collection of about 10 netsukes (see page 8-28) which we had carefully packaged ourselves, so they were not on the inventory list and we could not blame the movers. Heidel made another search of all(!) the empty

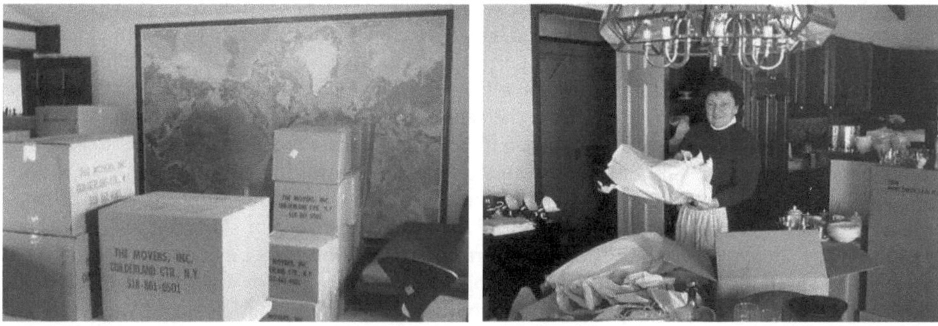

Figure 3A *The books in my office to be shelved.* **B** *Heidel starts to unwrap and place the china.*

boxes in the garage to make sure they would not be taken away next day. No luck, they stayed lost for several years! Then, when I needed my spare Braun razor (see page 3-25). Heidel said, I know were it is, and went to get it. But I knew that it was just underneath the sink, so I got it. In no time, we both each had a small box labeled "Braun Rasierapparat" in our hands. In one was the spare razor, and from the other, I heard a faint cry for help. All the netsukes were there, imprisoned for so long. I had packed them in the empty box of my last new razor, to be moved with other treasures in our car. The movers transported the other box. Each being stored by one of us in a 'safe' place. The joy of recovering the netsukes was big enough to increase the collection from then on for a good number of years. By now, we have a small cabinet to house the collection of 55.

Next day, the moving van was to meet me at UTK to unload the research equipment, the file cabinets filled with paper work, and the books for my office. When I arrived, 1–2" of snow had fallen, and was still accumulating. Fortunately, the truck and driver were from Utica, NY, which meant, a few inches of snow were no bother to them. In fact, it was helpful, since the van could be backed up to the front door, blocking the road to the chemistry building. On campus, nothing was moving. In short order, all goods were inside my laboratory, and by noon, I had temporarily secured and unpacked what was necessary. It was time to go home.

Figure 4 *No, this is not Troy, NY! This is a photo from our back porch in Knoxville, January 7, 1988.*

By now, however, the snow-fall reached officially 8.9" in Knoxville, as also seen in Figure 4, taken in the back of our house. *It was one of the great southern US snowstorms of all time*. From January 5–8, 1988 it blanketed large areas of Oklahoma, Arkansas, Mississippi, Tennessee, Alabama, Georgia, the Carolinas, and also Virginia. Something rarely seen in Knoxville. In most years there is no significant snow on the ground all winter long (but see Figure 48 for an exception). As a result of rarely having so much snow, the city is poorly prepared to handle it. What would caused no problem in

Upstate New York, brought a traffic standstill to Knoxville. What I experienced next, was the most exhilarating drive ever with my Jeep. We had bought it a few years earlier, to protect the Mercedes from the salt and snow in winter. Another purpose was to have a car to get up to the log cabin in the Adirondacks before I had cleared the snow from the road (see Chapter 8. Figure 88B and page 8-90). Heidel and Papa could then get into the cabin and start the fire and did not have to wait at the bottom of our hill. It served its purpose, but the major roads were always cleaned so fast, that until this January day in 1988 I had not driven the Jeep in snow for long distances.

Going home from UTK, I bypassed the Interstate—it might have snow problems. The four-to-five lane Kingston Pike, the main road out of the city with dozens of shopping malls, was not touched by any snow plow. All traffic lights were blinking yellow, and I met no more than 10–20 cars over the 15-mi trip home. With about 30 mph speed, the loose snow was flying left and right, and with four-wheel drive, I had perfect control of the car. With only little delay, I was at home.

There, I had another fun. I could show our neighbors how easy it is to clean a long driveway from 10 inches of snow. It was the last time for my tractor. I soon learned the southern philosophy of snow removal: 'God put it there, He also takes it away.' In spring, I traded the tractor and snow blower to serve in the nearby Smoky Mountains and bought a more effective, wider lawn tractor.

Work began shortly thereafter. As was required in my contract, I made an effort to divide the working hours, as I used to say at that time, into ½ at the University of Tennessee at Knoxville,[1] ½ at Oak Ridge National Laboratory,[2] and ½ in my new office at home, shown in Figure 5.

---

[1] The University of Tennessee at Knoxville, UTK, traces its history to the nonsectarian Blount College, established in Knoxville in 1794, two years before Tennessee became a state. It is the oldest such institution west of the Appalachian Mountains. After the Civil War, in 1869, the university was selected as a federal land-grant institution (see also page 7-2, Footnote), and added agriculture, engineering, and also military science. It was designated as the University of Tennessee, UT, in 1879. By 1988, it had about 20,000 undergraduate and 6,000 graduate students, and a faculty of 1,400.

Logo of the University

The motto in the seal reads: " Veritatem cognoscetis et veritas te liberabit." (You will know the truth and the truth shall set you free). The UT system also has campuses in Martin and Chattanooga. Specialized campuses are the UT Health Science Center, Memphis, and the UT Space Institute, Tullahoma. Since 2000, it manages ORNL in partnership with Battelle Memorial Institute.

[2] Oak Ridge National laboratory, ORNL, as it is now known, was part of the Manhattan Project in 1943. The 'Secret City' of Oak Ridge, the production facilities, and the laboratory were built by the US Army Corps of Engineers. After less than two years, the city housed more than 75,000 residents. The original goal was to separate and produce uranium and plutonium for the use as atomic weapons. After WW II, the research shifted to peacetime goals in medicine, biology, materials, and physics. The first-ever built graphite reactor was then used to produce radioisotopes for medical applications. Today's research agenda is abbreviated by the term 'nano-info-bio.'

Three photographs, two depicting UTK, and one ORNL, are shown as Figures 6 and 7. The bridge in Figure 6A connects buildings of the university for foot traffic across Cumberland Avenue, which ends in downtown. To the right (west), at the end of College Town, Cumberland Avenue becomes Kingston Pike and continues for some 15 mi to Farragut. Figure 6B illustrates the blooming dogwoods in front of Ayers Hall, the landmark building of the university. East of it, across the Circle Drive, is the location of the Department of Chemistry.

Figure 5 *Working at home in my new office in 1988.*

Figure 6A *Cumberland Av. footbridge. To the left is E.*

Figure 6B *Ayer's Hall, UTK in April.*

Figure 7 *View of ORNL in 1988. The road (A) is Bethel Valley Road. The 4500 N & S complex of X10 (B & C) is beyond the parked cars. My laboratory was in 4500 N.*

The view of ORNL of Figure 7 illustrates its appearance when I first saw it. The right-most, tall chimney (D), is next to the building which houses the Graphite Reactor which is now a part of the American Museum of Science and Energy, AMSE, in Oak Ridge, a good source of historical information about the US atomic energy program, its creation during WW II, and its changes to a peace-time enterprise. The AMSE has always been a major attraction for our visitors. Hopefully, ORNL will play a major role in resolving the energy dependence of the US on foreign countries which causes so many problems in the politics of the early 21st century.

At UTK I had a newly furnished office. The first job was to hire a secretary. Ms. Joann Hickson was chosen after interviewing several candidates (see Figures 17, 40, 45, and 46). She was the fastest typist and equipped to handle the new WordPerfect '4.2' word-processor software.

The Department Head, Professor Gleb Mamantov made sure that *the department got its share of the Distinguished Scientists*.[1] Just shortly before I arrived, Professor Georges Guiochon had joined the analytical faculty. On the other hand, Professor Mamantov also made sure that the added faculty would not draw on the departmental budget which, naturally, was (and still is) always too small. This meant, secretarial help was for us to hire, they were not part of the departmental pool (and got paid better in case we thought it advisable). Students and postdoctoral coworkers were a different group of employees. For them, the (rather low) pay was set by the department and could only in rare instances exceed this limit—so that we did not attract excessive numbers of students from the department. Similarly, the cost of all the new office furniture I needed was charged to my account, and occasionally, when the budget of the department was especially tight, we were also asked to help out a little.

Chemistry research thrives via many ORNL/UT collaborations

UT/ORNL Distinguished Scientists Georges Guiochon and Bernhard Wunderlich

The newspaper clipping on the right shows the pride the department took in our being there. We had full freedom to design our work and little or no accounting was needed outside of the annual reports requested of everyone. When after a few years, based on my past practice (see Chapter 8), I brought up the question of a possible sabbatic, Professor Mamantov replied: "This is up to you. You can plan your research and teaching with very little input from me. If you think a stay at another institution for a longer period of time is of value for you, arrange it. You need no permission from the department."

---

[1] The 'Distinguished Scientist Program' was announced in 1983 by Senator Howard H. Baker and Governor Lamar Alexander. It was to attract (up to 30) scientists and engineers of high national and international stature to the Knoxville-Oak Ridge research community (UTK: Jack E. Reese, Chancellor; ORNL: Herman Postma, Director). The initial annual support of each Distinguished Scientist was $210,000, equally shared by UTK (through the Science Alliance, the largest of several Centers of Excellence funded by the State of Tennessee, Dr. Lee Riedinger, Director) and ORNL (which was principally sponsored for its research through the US Department of Energy). Half of the support was used for the 12 months salary, the other half, for the fringe benefits package and for the support of the research group. Administratively, a Distinguished Scientist was placed in a UTK Department and a ORNL Division, to draw upon the resources and facilities of both for research and teaching activities. Additional external grants could be funded either through UTK or ORNL. The financial accounting for the program was through UTK.

As in Chapters 7 and 8, our new location is introduced next. *Let us take a look at Knoxville*. It is a southern city of ≈150,000 population (in 1988). Figure 8 is an overview of downtown toward the west. Knoxville does not only have its climate to boast about,[1] it also is a beautiful river city.

Figure 8 *Downtown Knoxville. On the left is the Tennessee River, UTK is behind the three tall buildings. The sphere is the 'Sunsphere', the right buildings pair is the TVA headquarters. 1990.*

The Tennessee River is formed east of downtown by the confluence of the French Broad and Holston Rivers. The Tennessee flows then for ≈650 mi until it reaches the Ohio in Paducah, KY shortly before joining the Mississippi River. Knoxville is in the center of the eastern US, at the intersection of Interstates I-75 and 40. It is within easy reach of half of the US population. Its airport boasts ≈100 flights daily. Over downtown towers the 200 feet high, 74 feet diameter bronze 'Sunsphere,' one of Tennessee's most unique structures, erected during the 1982 World's Fair.

The city of Knoxville is also the headquarters of the Tennessee Valley Authority, TVA, created by President F. D. Roosevelt in 1933 during the depression to solve the disastrous economy of the Tennessee valley in an integrated fashion. Much of the land had been deforested and farmed too long, eroding and depleting the soil. The TVA, a government corporation with flexibility and initiative of a private enterprise, helped replant forests to better the habitat for wildlife, developed fertilizers, and taught farmers how to improve crop yields. The most dramatic change, however, came from electricity generated by the TVA-built dams. Electric lights and modern appliances made the farms more productive and drew industries into the region, providing desperately needed jobs.

Next is the *Great Smoky Mountains National Park*, about 25 mi south (a UNESCO World heritage Site since 1983). In winter, it is clearly seen from our back porch. In Figure 1B it is still hidden behind its bluish, smoky haze, originally caused only by the transpiration of hydrocarbons[2] and water vapor from its dense pine forest, but now, the haze is intensified by air pollution from

---

[1] The climate in Knoxville still has four distinct seasons. When asked by friends in Troy, NY, I usually explain that fall ends on December 31 and spring starts on January 1. The summer is long and warm. The average high, low, and mean temperatures in °F are: Jan (47,29,38); Feb (52,32,42); Mar (61,39,50); Apr (71,47,59); May (78,56,67); Jun (85,64,75); Jul (88,68,78); Aug (87,67,77); Spt (82,61,71); Oct (71,48,60); Nov (59,38,49); Dec (50,32,41). Annual rainfall ≈50".

[2] Terpenes, $C_{10}H_{16}$, such as α-pinene, found abundantly in conifer oils—turpentine.

such distant cities as Chicago, Detroit, Indianapolis and St. Louis. Figure 9 gives an impression of the ridge upon ridge of forest between North Carolina and Tennessee. The elevations in the park are 875–6,643'. Created in 1934, it is the most visited park with ≈9 million visitors annually. With an age of ≈250 million years, the mountain range is one of the oldest in the world.

Figure 9 *View into the Great Smoky Mountains National Park from the Foothills Parkway, 1990.*

Asheville, NC, about 120 mi south-east from Knoxville, is a most interesting city. The drive to Asheville, a town of about 70,000 population, takes ≈2 h. It is the last town in the Appalachian Mountains, just east of the Great Smoky Mountains National Park. The mountains are crossed along a beautiful stretch of I-40 for ≈40 mi. The city itself was named one of the top seven places to live in the U.S. and was also called one of the world's top 12 'must-see' destinations for 2007 by Frommer's travel guide (a claim one should dispute, I would at least restrict it to the US, east of the Mississippi). Its main attraction is the Biltmore house, shown in Figure 10. At the end of the 19th century, when Biltmore was constructed, it was one of the largest undertakings in the American residential architecture. It was constructed for Georg W. Vanderbilt.[1] Over a six-year period, an entire community of craftsmen worked to build the country's premier home. The estate boasted its own brick factory, a woodworking shop, and a three-mile railway

Figure 10 The *Biltmore House in Asheville, NC.*

[1] Grandson of Cornelius Vanderbilt (1794–1877, the 'Commodore,' shipping and railroad magnate. He held the belief of superiority of free competition over government-granted monopoly).

spur for transporting materials to the site. The celebrated architect Richard Morris Hunt modeled the house on three châteaux built in 16th-century France. It features 4 acres of floor space, 250 rooms, 34 bedrooms, 43 bathrooms, and 65 fireplaces. The basement houses a (now emptied) swimming pool, gymnasium, and changing rooms, a bowling alley, and servants' quarters, kitchens, and utilities. The grounds of the 8,000-acre estate (originally 125,000) were designed by Frederick L. Olmsted, the creator of New York's Central Park. He not only developed acres of gardens and parkland, but in his effort to protect the environment and reclaim over-farmed land, Olmsted established America's first managed forest.

Shortly after we moved to Knoxville, we became seasons' pass holders to visit the Biltmore estate several times each year. After Brent and family moved to Columbia, SC, in 1992 to work for PMSC, now CSC (Computer Sciences Corporation, see page 9-38), it was a nice half-way point either to meet there for a one-day visit, or to stop over and divide the ≈5 h travel time to Columbia, SC.

Figure 11 *Mountain life in Tennessee in the 19th century, as recreated in the Museum of Appalachia near Knoxville, on a visit in 1999.*

What a difference to the 19th century farm pictured in Figure 11, photographed avisits to the 63 acre living history Museum of Appalachia in Clinton, TN, close to Knoxville. This frontier life still existed in Tennessee while close-by greatest luxury and most modern technology were created.

Four other interesting cities within easy driving distance from Knoxville are Gatlinburg (¾ h) Chattanooga (1½ h), Nashville (2½ h), and Atlanta, GA (3 h). All have a lot to offer. *Gatlinburg* is a popular tourist town and vacation resort at the border of the Great Smoky Mountains National Park. *Chattanooga* (Cherokee for 'big catch') has a long history of Indian sites. It is a scenically located city at the foot of Lookout Mountain, (2,400') on Chickamauga and Nickajack Lake, both lakes being part of the Tennessee River. *Nashville* is the capital of Tennessee, located on the Cumberland River and well known as the center of country music. Remarkable buildings and tourist sites are the Country Music Hall of Fame and the Opryland Resort and Convention Center, the largest non casino hotel in the US with several huge, beautifully landscaped atriums, surrounded by the hotel wings. *Atlanta*, finally, is the capital and largest city of Georgia, a modern big city with all its advantages and disadvantages. Its population is over 500,000 with an additional, large surrounding metropolitan area. Its airport is the world's busiest. It is our gateway city, just a hop of ½-h by plane from Knoxville.

Atlanta is also our goal whenever we need to do special shopping. As seen in Figure 2, our house was very spacious. It swallowed up all our furniture and still left much empty room, as seen in Figure 12. As much as possible, we added new pieces from the Knoxville stores, but to fill the living room, we found nothing suitable. We were looking for Danish modern teak furniture. Finally, in early March, we saw an advertisement by the House of Denmark in Atlanta. It offered the perfect style to our liking. On a suitable day we drove to Atlanta and spent a delightful late morning at the House of Denmark. A very stunning, large cocktail table of Carrera Marble caught

Figure 12 *Empty living room after the moving-in of our furniture. To be filled by March 1988.*

Figure 13 *The cars in the garage. Front, Heidel's Jetta, the Jeep, and the Mercedes 300D.*

our fancy (see the photograph of Figure 15, below). To this we added two sofas, chosen to fit the one we brought from Troy, two end tables, a side table, and a desk chair. After lunch, we could pick up the well-packaged imports from the nearby warehouse. To our surprise, almost all fitted into the back of the Jeep Cherokee, shown in Figure 13, parked in our garage. Over lunch, we had met Trautchen (see page 9-33), Heidel's school friend from Brandenburg, living in Atlanta. She helped us by taking the desk chair into her car. Later, she delivered the chair to Knoxville, visiting with her husband Don Hough.

Now the house was well furnished and we could feel at home. But, *the 1975 Mercedes was ailing*. In the last few years in Troy, I annually had to have rust removed and portions of the body refinished. Finally, small holes began to show on the floor boards. Otherwise, the 13-year-old car, 200,000 mi driven, was 'just like new,' however, the excessive salt on the wintery road in Troy had taken its toll. I visited the Mercedes dealer, Knoxville Motor Company, picked out a new car as RPI-retirement gift for myself, and inquired about a reasonable deal on my old car to lower the list price. The salesman inspected my 300D and said to me: "You get more for it, if you sell it on your own." I told him: "But you are the salesman, not I." He would not change his mind and I told him that I will be back as soon as I sold the car, left his office, and tried plan B.

Figure 14A *My classic 300SD Mercedes.*

There was another foreign car dealer in town, US International Motors Inc., USI. They dealt also in restoring antique cars. My 1975 car was almost an antique! I went to USI and asked: "How expensive is it to get my car back into show-room condition?" The answer went as follows: "More than you are willing to pay! But, I can make you a deal, I have a 1986 Mercedes 300SD on the lot that was locally driven for 40,000 mi. I could trade it for less than it cost you to restore your old car. Why don't you take it home and try it for a week?" Heidel and I both liked the looks of the 300SD. Figure 14A is a picture of it. Ultimately, I liked it better than my old car.

The Jeep in Figure 13 was of little further use in the mild climate of Knoxville, so we gave it to Brent when he was in need of a car. Ten years later, in 1998, I bought an additional, used Mercedes, a model 560SEC, driven already some 30,000 mi. It is shown in Figure 14B on the day of delivery at the dealer. Both cars were classic models and I had the idea to keep both until they would increase in price. The 560SEC was a 2-door coupé. It had a big, 5.6 liter, 8-cylinder gasoline engine, and certainly was more luxurious and also more agile than the diesel. Its top speed was limited by a governor to 140 mph

Figure 14B *My classic 560SEC Mercedes.*

Because of the US speed limits, I never drove the car above 100 mph, and I also had no chance to drive the car on the German "Autobahn." I enjoyed both cars, but found that the upkeep of two cars is not economical. Too much maintenance is needed if a car is not driven regularly, and holding insurance on two cars was a nuisance. Unfortunately, I could not convince my insurance agency of over 40 years, State Farm Mutual, that I drive only one car at a time and should not pay for liability and collision insurance for more than one car. It also did not help to mention that in Switzerland one can buy insurance tied to the licence plate. By switching plates, there, one could legally change from one car to another. Just imagine, how many more cars could be sold in the US if one would insure the driver and not the cars! Specially today, with more expensive fuel, one could keep a small, economical car next to the prestigious or fun automobile. Ultimately, I gave up on two cars and traded both for my 'dream car,' a Mercedes CL500 (see Figure 92, below). By then, again, it was a retirement present, but now from UTK and ORNL.

In these first months we also needed to connect with our neighbors and colleagues, find new friends, and reconnect with old friends. Just across from us lived Frank and Alma Conley. He was a retired IRS tax lawyer and WW II veteran, and she had Austrian roots. We got along well and visited frequently. During our travels, the Conleys were always looking after our house. Figure 15 shows one of our get-togethers, and at the same time displays our newly acquired marble table.

Figure 15 *Frank and Alma Conley with Caryn and Heidel in our newly furnished living room in 1988.*

Caryn was visiting at that time from San Antonio, Texas, were she was an intern in radiology. We also got to see San Antonio during the years of her stay there.

Other neighbors were introduced to Heidel, in a get-acquainted coffee arranged to get to know the ladies in the immediate vicinity. Heidel was also quickly introduced to the International Friends Club of Tennessee with about 50 members, most with European background.

For me, the connections to colleagues were through UTK and ORNL. It took some time, however, until we got to know their families. By being only half-time in both places, the bonds were less strong. According to my secretary, it also played a roll that my position was better paid and I could (would) not join the lunch crowds in either place.

Two prior connections were with Professor Ffrancon Williams, by then a long-time faculty member at UTK. He was a postdoctoral coworker of Prof. Dole at NU after I had left. In the Engineering Department, I had prior contact with Professor Ed Clark who, before coming to UTK, was at the Central Research Department of the DuPont Co. where I consulted. We got together with him occasionally and he introduced us to the Wine Society of East Tennessee which we then joined and ever since have been members of. Each month, they offer one dinner with wines and one wine tasting. One pays separately for each event, i.e., one can attend many or few. Typically 30 to 80 members and guests attend these events. Over the years, we learned much about worldwide wines (see also page 8-62). One conclusion was, the more expensive wines are not necessarily better.

As before, earlier friends from Germany, Hastings, Chicago, Ithaca, and Troy were not forgotten. Many of them visited and, in turn, were visited by us. With a good car and easy plane connections, travel was fun. Also, now there were just the two of us to consider in scheduling travels and visitors. Besides friends, there were stays by family and former students.

To remember our guests, *Heidel created a unique guest book*. The type mentioned first by Walter Kempowski in his description of German Society at the beginning of the 20th century ("Aus großer Zeit," published 1978). It is a tablecloth on which every guest signs on the first visit, and Heidel embroiders the signature. Figure 16 illustrates the center of the 'guest book tablecloth,' started in 1988. By now it has over 200 signatures. Our earliest visitors were Frank and Dagmar, driving up from Columbia, SC. Stephen Cheng, now at the University of Akron, could meet with his father in our house. The father, as an engineer in China, was invited by the TVA in Knoxville.

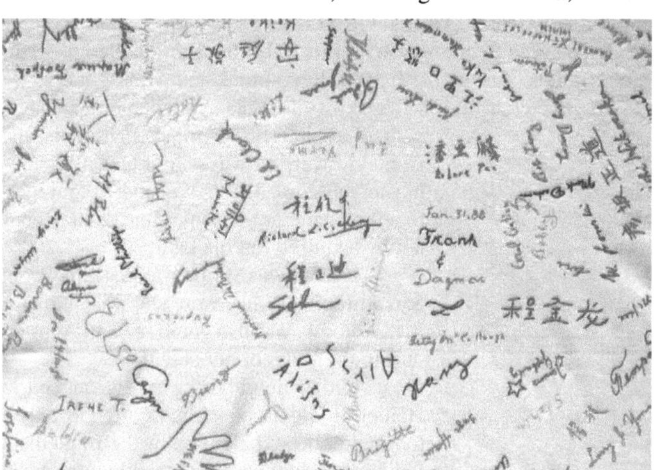

Figure 16 *The center of Heidel's new tablecloth guest book, started in 1988. (Observe the tracing of Stefan's little hand on his first visit.)*

Professor Wiedemeier from RPI visited us on his way to Huntsville, AL, where he had to discuss his space research at NASA. Naturally, Mutti came for a visit from Hastings, Caryn from San Antonio, and Brent and family drove down from Syracuse. Charlotte M. and De started a series of annual visits from Albany, Ivo and Susi Buhac visited from their new location in Worchester, MA. A number of colleagues and students from UTK, ORNL, RPI, and foreign universities are also listed in Figure 16, and many, many more were to follow.

The 'Distinguished Scientist Program' was ideal to do creative research (see page 9-6). There was no need to first write a proposal, to be judged by your peers. Peer reviewers expect that a proposal is a good idea, but not too far fetched, or they might think it could not be done. You yourself, on the other hand, did not want to give away too much of your plans so that others could not 'borrow' the same idea before you had a chance starting to test it. A proposal was accepted when all of the reviewers, usually three to five, and the program director agreed that this was a 'good' idea and it followed the guidelines (discussed for NSF on pages 8-54 and 55). If all went well the first time, it would be at least a year from idea to support and to find the coworker to help in carrying out the experiments. For 'great' ideas, this is much too long. Fortunately, organizations like the NSF still have many scientists as administrators, so that your performance on a given grant is judged more by the success and quality of the research than on your following the letter of the proposal. I never heard that reasonable changes in direction of research were reproved.

One example of *a failed proposal* is the one I sent in 1990 to the Division of Research Grants of the Department of Health and Human Services (NIH, National Institutes of Health). It dealt with 'Thermal Properties of Proteins and Poly(amino Acids).' We proposed that one could unravel the basic thermal properties of many proteins by using only a limited number of new experiments. Proteins have the same basic structure elements as the nylons. The latter we understood, based on research on their thermal properties done in our laboratory by Dr. Xenopoulos. But, the proposal was not supported. The peer reviewers of NIH thought this could not possibly be successful, and if it were, it would not be of interest. The low level of priority for support given the proposal did not encourage us to ever apply again. Lucky for continued progress, the funding as Distinguished Scientist could come to the rescue. Indeed, all that was proposed, was accomplished. Now, some 20 years later, it is recognized as important, fundamental information (Appendix B, Figure 18 and page B-21). Such an example illustrates that a new idea may not be recognized by one's peers. I am sure that the freedom of direction of research as a Distinguished Scientist and the cooperation with the highly qualified staff of ORNL with access to their cutting-edge equipment for experimentation was the key to the success in research detailed in Appendix B, Sections E and F.

Much progress could be made by having similar positions available in many universities. It would allow successful researchers to increase their productivity and creativity. To start such development, the 'NSF CAREER Program[1]' for untenured faculty has been valuable. Its main point is to support the development of a university professor based on teaching the information in a given field and question it. This questioning should then lead to research into the fundamentals and extend and improve the overall knowledge. Such a program must then be followed by continued support, increasingly based on past performance rather than on proposed new research which pleases one's peers. Ultimately, such support should turn into a position as I was lucky to have had.

---

[1] For details on the NSF 'Faculty Early Career Development (CAREER) Program check the appropriate current web site of nsf.gov. The early version of this program, the 'Presidential Faculty Fellows (PFF) Program' was established in 1992 at the request of President George Bush to support the scholarly endeavors of young tenure-track faculty, to "recognize, honor, and promote the integration of high-quality teaching and research in science and engineering fields; foster innovative and far-reaching developments in science and technology;· create the next generation of academic leaders, and improve public understanding of the work of scientists and engineers." The goal of President Bush (1989–93) was to be the 'Education President,' discussed on page 9-37.

A good start of the research was accomplished within the first three years at UTK and ORNL. In Figure 17, the ATHAS group is depicted late in 1988 in our temporary quarters in one of the laboratories in the Chemistry Department. It took several years and an additional temporary move until the very lengthy renovation of 'Buehler Laboratory' was completed and my all new, custom-made research space could be moved into. In addition to the mostly new group of researchers, there were initially still several students at RPI in Troy. They were completing their work or writing their theses, as described in Chapter 8, pages 8-123–124 and Figure 101. They required occasional travels of mine to Troy.

Figure 17 *From left to right: Dr. Y. Jin, Dr. M. Varma-Nair, Mr. J. Cheng, Ms. K. Roles, Mr. A. Xenopoulos, Dr. M. Yasuniwa, Ms. J Hickson (secretary), Dr. B. Sumpter, and Prof. B. Wunderlich.*

The *cooperation between UTK and ORNL was furthered by weekly meetings at 4500 N of X10*, the right wing of the 4500 buildings (B) shown in Figure 7. The meetings were conducted by Dr. Al Narten, the ORNL Polymer Group Leader. In time, everyone briefly discussed his or her past and present research, capabilities, and plans for the future. A later picture of the combined group is reproduced in Figure 40, below. In addition to the meetings at ORNL, the ATHAS group at UTK of Figure 17 met separately each week to detail goals and results, as well as to practice presentations. The research direction was developed on the basis of the new material collected and ideas formed during the 1986/87 sabbatic (pages 8-109–116 and 121–128). The scientific details and the publications can be looked up, as usual, in Appendix A, Part A, and as the specific, science-oriented narrative in Appendix B, Part E.

The ATHAS Data Bank (see page 8-63) was, by then, getting to be close to 10 years old and needed to be updated. Dr. Manika Varma-Nair from New Delhi, India (PhD 1985, Indian Institute of Technology, IIT, in New Delhi) had contact with me since May 1985. She was offered a job to be in charge of this update and other improvements, beginning early in 1988. Manika was highly recommended by Dr. Edith Turi (Allied Corporate Research), a long time friend of mine and a leader in the promotion of thermal analysis worldwide who had interviewed her in India, and also by her former research director in New Delhi. Manika turned out to be an outstanding coworker. The intensive and critical scrutiny of other researchers' data brings many distractions, such as the discovery of omissions and the development of one's own new ideas of possible extensions of the reviewed work. Despite these diversions, she never lost sight of the fact that the sometimes dull collection and analysis of data[1] were ultimately the basis of our success. Her outstanding work led

[1] When I had asked in an early proposal for separate funding for the data bank extension by a critical survey and analysis of the literature, one of the reviewers had written that such "dull and uninspiring work should be left to an untrained secretary and not be given to a highly-qualified postdoctoral fellow." He had not heard about the quip 'garbage in, garbage out.' Naturally, the proposal was denied, and work had to be postponed until I had the Distinguished Scientist funding available (see page 9-13 for the description of another failed proposal, rescued in this fashion).

to the full 1990 Data Bank update (Appendix A, publication #289). In addition, 17 more papers based on her own measurements grew out of her careful research (see Appendix A).

Before coming to the US, however, Manika had to get married (see Figure 54E, below). Indian parents would not let a single lady move out into the world. Surprisingly, all could be achieved. Her husband, Ravi Nair, also entered UTK as a graduate student. After Manika's five years at UTK, she went on to a successful career in industry with ExxonMobil in Clinton, NJ.

The *thermal analysis on new samples of interest* was handled by Dr. Yimin Jin (PhD 1987, Fudan University, Shanghai), a postdoctoral fellow, highly recommended by Professor H.-S. Bu, my earlier coworker (see Figure 101 of Chapter 8). Dr. Jin arrived in September 1988. After his 5-year stay, he moved to the GE in Mt. Vernon, IN, for a career in industry. His first specific job in Knoxville was to improve the DSC toward a one-run measuring technique for heat capacities. He started with a three-position scanning calorimeter (of TA Instruments) which during a single run could measure all three differential temperatures (those between sample, reference, and empty cell, see Appendices A, papers #298, 307, 331, 356 and Appendix B, page B-17). His thermal analysis of substances of interest to the ATHAS group made him a popular coworker (see page 8-24). Substances included, besides mesophases of small and large molar mass, a series of paraffins and perfluoroparaffins, polysilylenes, polysulphones, and the allotopes of carbon (graphite, diamond, and fullerenes). Finally, he initiated our extensive work on temperature-modulated DSC, TMDSC, later continued and expanded by Dr. Andreas Boller (see pages 8-24, 9-46–48 and 50).

Two additional students who spend major time with thermal analysis were Dr. Alexander Xenopoulos who came with me from Troy (1986–1993, PhD in 1990, see Chapter 8 Figure 101 and pages 8-24 and 123) and Dr. Katie Roles (1988–1991, a second-year graduate student who chose our group for her thesis work, see also Figure 40). Alex started major work on nylons and alkylammonium salts (joined with Dr. M. Yasuniwa 1988–1989 and Dr. Q. Wang 1992–1994) and later added X-ray diffraction and neutron scattering (in cooperation with Dr. Narten of ORNL, see page 8-112 and Appendix A, papers # 314 and 315). Katie built on his work and went from nylons to the poly(amino acid)s polymerized from the 20 naturally occurring amino acids and synthetic copolyamino acids. Finally she studied naturally occurring proteins, to prove NIH wrong (see page 9-13, for the results, see Appendix B, pages B-34–35). This work laid the foundation for the study of biological macromolecules through Dr. Ge Zhang (see Figure 48 and pages 9-48–49) and Dr. Pyda (see Figure 47 and pages 8-24, 9-48, 83, and Chapter 10, Figure 4 and pages 10-5–7).

Another experimental technique which was central in my proposal for the sabbatic in Freiburg was *to learn about the solid-state ^{13}C NMR* (see pages 8-110 and 116). It was to teach me and bring the means to measure large-amplitude motion in the condensed state to our laboratory. I found out quickly, that this was a rather difficult undertaking. It would take very long to learn all the theoretical and practical details. Also, the instrumentation was expensive and not easy to attain for a single, small research group. It turned out, that both UTK and ORNL had the proper equipment, but not the needed polymer experts. Another path had to be chosen.

Professor Van Hook of the Chemistry Department of UTK had visited China in 1987 and there, discovered several potential graduate students. Mr. Jinlong Cheng was his first choice. He was employed at the Analytical and Testing Center of Lanzhou University (Ganzu Province, China). His education included a BS degree from Beijing University with a major in magnetic resonance spectroscopy and several years of experience in chemistry and quantum electronics. This was just what I was looking for. In China, Jinlong Cheng was handling a newly established 400 MHZ NMR spectrometer (bought with World Bank funds) and was trained for this job by the manufacturer,

Bruker. He did not have a PhD degree, but had done an outstanding job in setting up and running the facility. When he applied to UTK in fall of 1988 to become a graduate student, I employed him immediately as a research assistant, and later postdoctoral fellow (1988–1993, PhD, 1992, see also page 8-24). With the instrumentation at UTK and ORNL we could solve many questions about conformational disorder and mobility. Dr. Cheng classified a number of samples thought initially to be liquid crystals or even crystals as being conformationally disordered crystals (condis crystals). These can be above or below their glass transition temperature, an observation which later helped in the understanding of the different phases via solid–liquid glass transitions and order–disorder transitions (see page 8-66 and 67, Chapter 10 Figure 4, Appendix B Figure 13).

Starting in 1988, major *cooperation between ORNL researchers and the ATHAS group* developed. It concentrated in five areas: 1.) molecular dynamics simulations (with Dr. Noid, see pages 8-107–108, 110 and 112) using the Cray supercomputers of ORNL and UTK (see Footnote on page 6-33), 2.) full-pattern X-ray analysis by the Rietvelt method (with Dr. Busing), 3.) separation of amorphous and crystalline X-ray diffraction intensities as a function of temperature (with Dr. Habenschuss), 4.) quasi-elastic neutron scattering (with Dr. Narten, already mentioned on pages 8-112, and 9-15), and 5.) somewhat later, atomic force microscopy, AFM (in cooperation with Dr. Annis). All senior researchers who worked with us are represented in Figure 40, below, in a 1990 photograph of the ATHAS/ORNL polymer group. Their help was invaluable.

The *molecular dynamics simulations* with Dr. Noid were already planned in 1985 (see pages 8-107–108). In fact, these initial discussion lead to my move to Knoxville. Before, we had calculated the low-energy conformations for mesophases with Profs. Miller and Hollinger from RPI, using molecular mechanics calculations (see page 8-116 and Appendix A papers # 241, 242, 284). Figure 18 contains a comparison between results from these two methods. Possible rotation angles are depicted on the left for the different phases (α, β_1, β_2, and the melt). On the right, the molecular dynamics of a single, molecular chain within crystalline polyethylene is displayed as it changes from a regular zig-zag at time 0, through nine picoseconds of motion after instantaneous heating to 316 K (Appendix A, publication #350). Finally, Monte Carlo statistical methods could simulate the much longer times needed to follow crystal and molecular nucleation. These were carried out by Prof. W. Hu after he left UTK/ORNL (see page 9-49 and Figure 47, see also Appendix B, page B-10).

In September 1988, the day-to-day work on the molecular dynamics simulation was helped by the addition of Dr. Bobby G. Sumpter to the ATHAS group. He cooperated with Dr. Noid and handled the link to my polymer research. Bobby had gotten his PhD from Oklahoma University in

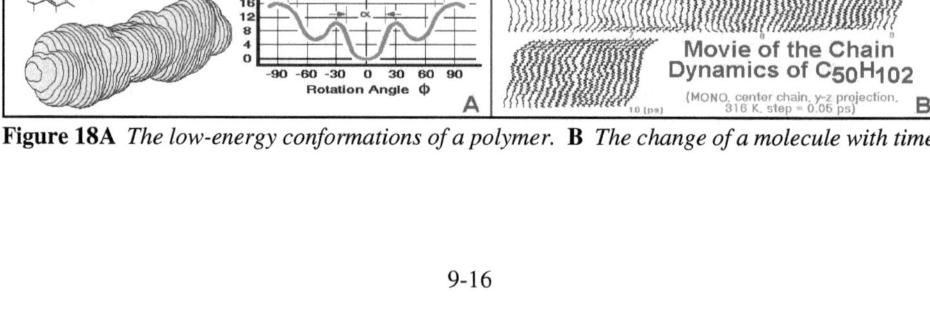

Figure 18A *The low-energy conformations of a polymer.* **B** *The change of a molecule with time.*

1986 and had already spent the summer of 1985 with Dr. Noid at ORNL. In the meantime, he was a postdoctoral fellow at Cornell University (see page 7-2). Being well prepared for his job, he quickly produced exciting results which could be presented already at the 1989 March Meeting of the APS and the Fall Meeting of the ACS (see Appendix C). He kept up this high productivity (see page 8-24). Next, he helped Dr. Noid and me to write a successful proposal to DOE which allowed us to employ Bobby until 1992 (through ORNL) at the increased salary scale of ORNL. After this postdoctoral position, he joined ORNL as a permanent employee. He still contributed to our joint effort until 1996. The work is described in Appendix B, Section E and Figure 15.

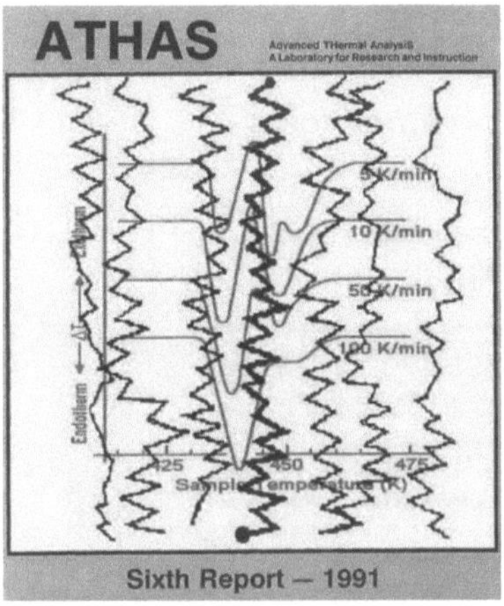

Figure 19 *The changed title page.*

From this work we could 'see' the actual motion of chains within crystals like polyethylene on a picosecond time scale. This was sufficiently important for the understanding of the thermal properties measured by DSC that we changed the title page of the ATHAS reports. Instead of emphasizing the need to know the crystal morphology (see Chapter 8, Figure 69) we now highlighting the additional knowledge about the molecular motion, as is shown in Figure 19.

The *cooperation in the field of X-ray diffraction*, being the area of expertise of Drs. Busing, Habenschuss, and Annis started slowly. Initially, the ORNL contribution consisted of valuable help in the interpretation of our own, semiquantitative diffraction patterns gained at the UTK X-ray facility. Dr. Busing had developed computer software (FIBLS) and set up hardware for full-pattern X-ray analysis using the Rietvelt method. He could not find a clear interpretation of the defects in polyethylene fibers from his prior obtained data.[1] Long discussions convinced me, that the reason had to be a different structure-element, not known before. To make progress, Dr. Busing identified a qualified postdoctoral fellow for us, Dr. Y. Fu, whom I hired in 1991 (see page 9-43). Dr. Busing quickly taught him how to use FIBLS. Even after Dr. Busing retired, they would work together on this project, to be described below. Finally, Dr. Fu's X-ray data, combined with DSC and NMR results, yielded the solution for the structure and defects of polyethylene fibers that were attempted earlier.[1] Almost everyone of the ATHAS Group and ORNL Polymer Group had some contribution by then to this work (see Appendix A, papers #390, 404).

With Dr. Annis, I could jointly purchase a commercial atomic force microscope, AFM, and for a two-week period in 1990 he worked together with a former undergraduate assistant of mine at RPI, John Reffner (BS 1985). John was finishing his PhD with Prof. E. L. Thomas, whom I knew well from visits in his world-famous electron microscopy laboratory specializing on polymer structures (University of Massachusetts, Amherst, MA, and later, MIT, Cambridge, MA). John brought his own copolymer samples, already characterized by electron microscopy, and also started work on our prior discovered extended-chain crystals of polyethylene (see Appendix B, page B-4). With AFM we could find additional information (see Appendix A, papers # 327, 336, 337).

---

[1] W. R. Busing, "X-ray Diffraction Study of Disorder in Allied Spectra-1000 Polyethylene Fibers. *Macromolecules*, **23**, 4608–4610 (1990).

<u>*The work performed by my students and me was recognized*</u> in a special issue of 'Polymer Engineering & Science,' the official journal of the Society of Plastics Engineers with its editor, Professor Roger S. Porter of the University of Massachusetts in Amherst, MA, and the guest editor Professor Stephen Cheng (see the pages 8-99–100 and 8-123–124). It was dedicated to my 60th birthday and contained contributions by my former students, coworkers, and international friends. The cover and back of the issue are reproduced on the right.

Once a tradition was established, it was on my 65th birthday, that a second such special issue was published by the 'Journal of Thermal Analysis' with its editor Professor Judith Simon, from the Technical University, Budapest, Hungary, and as the guest editor Professor Edith Turi, Polytechic University, New York, NY (Vol. 46, Issues 3/4, 1996).

Finally, for my 70th birthday, the symposium held at the NATAS Conference on this occasion was published, this time as a special issue of 'Thermochimica Acta' with Prof. Dr. Christoph Schick of the University of Rostock, Rostock, Germany, as editor and Professor Alan Riga of the Cleveland State University as guest editor. These issues certainly were special birthday presents and also show the successes of my students after they left my direct supervision. It was the best recognition my colleagues and friends could give me.

Another acknowledgment of our work was the 'Award for Research in the Broad Area of Applied Chemical Thermodynamics for 1993.' This award given by the <u>S</u>wiss Society for <u>T</u>hermoanalytic and <u>C</u>alorimetry (STK). It is reproduced with its detailed citation on the next page. The award was presented to me at the meeting of the German Society for Thermal Analysis (GEFTA) and the STK in Munich, Germany (see Appendix C, page C-43).

Polymer Engineering & Science

MID-SEPTEMBER 1992 Vol. 32/No. 17

POLYMER TOPICS

Contributions to This Issue Are Dedicated to the 60th Birthday Celebration of

PROFESSOR BERNHARD WUNDERLICH

PROFESSOR STEPHEN Z. D. CHENG, Guest Editor

CONTENTS

Society of Plastics Engineers

URKUNDE

Die Schweizerische Gesellschaft für Thermoanalytik und Kalorimetrie (STK)

verleiht den Preis für Arbeiten auf dem Gebiet der

"ANGEWANDTEN CHEMISCHEN THERMODYNAMIK"

für das Jahr 1993 an

Herrn Professor Dr. Bernhard Wunderlich
Universität von Tennessee, Knoxville, TN
und Oak Ridge National Laboratorium,
Oak Ridge, TN, USA

Der Preis auf dem Gebiete der angewandten chemischen Thermodynamik wird Herrn Professor Bernhard Wunderlich für seine wegleitenden und umfassenden Forschungsarbeiten zur thermodynamischen Charakterisierung von Festkörpern zuerkannt. Professor Wunderlich hat in mehreren Gebieten der experimentellen und theoretischen Beschreibung des Festkörperzustandes in qualitativer und quantitativer Hinsicht Grosses geleistet. Auch seine wissenschaftlichen Stationen zeugen von der Dynamik des Preisträgers mit Cornell University (1958-63), Rensselaer Polytechnic Institute (1963-88), University of Tennessee, Knoxville und Oak Ridge National Laboratory (seit 1988). Neben Monographien liegen über 350 Publikationen in zahlreichen bedeutenden Zeitschriften vor, ein Volumen, das nur mit dem zielstrebigen Mitarbeiterstab des ATHAS (Advanced Thermal Analysis Laboratory) zu erreichen war. Die instrumentellen Entwicklungen betreffen Thermoanalyse unter extremen Bedingungen: Hohe Drücke bis 3000 bar und Aufheiz- resp. Abkühlraten bis 10^4 K/min. Präzise Messungen der spezifischen Wärme, der Enthalpie und Entropie, insbesondere von linearen Makromolekülen führten zu einer Datenbank mit kritisch beurteilten Eintragungen. Diese Daten bildeten wiederum eine Basis sowohl für Berechnungen und Voraussagen als auch Extrapolationen mittels moderner Rechenverfahren. Wichtige Aspekte der Forschungstätigkeit gehen in Richtungen wie der Beschreibung der Seiten- und Hauptkettenbeweglichkeit als Vergleich makroskopischer Eigenschaften mit molekularen Phänomenen. Computer Simulationen, Untersuchungen mit DSC und Festkörper-NMR sind intensiv zur Berechnung und zur Diskussion von Konformationsänderungen, stressinduzierter molekularer Veränderungen und von Defekt-Strukturen von linearen Makromolekülen eingesetzt worden. Die Charakterisierung unterschiedlicher amorpher Zustände von Mesophasen und speziellen Molekülen wie Fulleren (C_{60}) mit plastisch-kristallinen Zuständen sind ebenso zukunftsweisend.

Das Preiskomitee

E. Marti, H.R. Oswald, [signature], J. Vogt

Dr. E. Marti, Prof. Dr. H.R. Oswald, Prof. Dr. P. Tissot, PD Dr. J. Vogt

Basel, Zürich, Genf, Fribourg, 14. September 1993

An interesting international travel was arranged during the summer of 1988. The last sabbatic in Germany was supported by the Humboldt Award (see page 8-111). It covered the time period from January 1, 1987, but I had already been for some time in Germany, supported by the visiting professor position in Freiburg. This meant, it was possible to shorten the award-supported period in 1987 and spend one month during the summer of 1988 to revisit both, the Universities of Freiburg and Ulm to discuss the achieved progress and arrange further cooperation.

We packaged this opportunity of a supported round trip to Europe around a 10-week stay. It included not only the four weeks supported by the Humboldt Award, but also allowed me to give plenary lectures at conferences in Italy and Israel which fitted at the beginning and end of the stay. In addition, it allowed several other lectures, listed in Appendix C (pages C-33–34). We could also attend the annual award celebration of the Humboldt Awardees for 1988, in Bonn, at that time the "Bundeshauptstadt" (capital) of the FRG. The rest of the time was filled with visits of relatives and friends in Germany and vacation periods, one at our first favorite European location, the Glottertal, staying, as usual, with Mrs. Gabriel, the other in Kössen, the more recently discovered place, staying with the Schlechters (see pages 8-49–50).

During the absence, the Conleys (see page 9-11), looked after our house. We had hoped to miss the hottest months of the season which, we had been told, could be uncomfortable. We found in the following years that one gets used to the heat rather fast because of widespread air conditioning. The peak temperatures we sometimes experienced in Hastings, Chicago, Ithaca, and Troy were just as high, and certainly harder to endure without air conditioning.

To transmit an idea of this ambitious travel, I give on the next two pages *<u>a listing of our itinerary</u>* with brief comments and some snapshots. The figures are collected at the end of the list:

6/18–19 1988 A first-class flight via Atlanta to Frankfurt, Germany paid for by the Humboldt-Stiftung, and on to Milano, Italy, followed by bus to Gargnano on Lake Garda. The bus trip had a small problem. Unbekownst to us conferees, on Sundays, the bus terminated 3 mi short of Gargnano. Usually there were no passengers beyond. After a frantic check we found someone who spoke Italian and could organize taxis.

6/19–24 Plenary lecture and attending the 'European Conference on Advanced Topics in Polymer Science.' The meeting was in the gorgeous Villa Feltrinelli,[1] shown in Figure 20A. Several excursions and a boat ride on Lake Garda gave opportunities to appreciate the beautiful countryside around the lake, meet old, and make new friends among the European scientists, and discuss worldwide polymer science.

6/25–7/2 After the flight back to Frankfurt, we rented a VW Golf and stayed in our 'old' vacation spot, the Glottertal (see Chapter 8, Figure 56). We visited our landlady and landlord from the sabbatic in Schallstadt, near Freiburg of 1986/87, the Burggrafs (see Figure 100A, Chapter 8), and made an excursion to the French Alsace.

7/3–4 A trip back to Quadrath-Ichendorf for a visit of Papa in the retirement home (see page 8-129) and a visit to Irmchen and Henning in Köln, shown in Figure 20B.

---

[1] Built in 1892, painted in gold and terra cotta, reminiscent of a Venetian palazzo, situated on the western shore of Lake Garda. The villa served as refuge for the Italian dictator Mussolini (under German protection). He was ousted in 1943 by King Victor Emanuel III. Close by, Mussolini and his mistress were executed in April 1945 by partisans, ending the saddest chapter of Italian history (see pages 2-18 and 23). Next day, the bodies were hung up in Milano and mutilated.

7/4–7 Attending the 1988 Annual Meeting of the Humboldt Awardees in the Hotel Seminaris, Bad Honnef, near Bonn. In the evening on the Monday of arrival there was a dinner, the Award Ceremony shown in Figure 20C, and a special lecture. On Tuesday, the Opening Ceremony and Plenary Lecture were given at the University of Bonn. The afternoon was reserved for a reception by von Weizsäcker, President of the FRG at his residence, the Villa Hammerschmidt.[1] Wednesday was set aside for a Rhine cruise from Bingen (near Mainz) back to Bonn (snapshot in Figure 20D).

7/8–23 With a stop in Darmstadt, we traveled south again, to Freiburg, to start the short sabbatic visit with discussions and lectures. There, we also met with Hans and Inge from Brandenburg who had gotten a surprising permission to travel to the West. For Hans, it was the first time that he could go to the FRG since the building of the Wall in 1961. Figure 20E shows a dinner outing from Freiburg to the Glottertal.

7/23–29 For the visit in Ulm we could stay with our prior landlady, Mrs. Lusser, in her house in the city. Besides a lecture and the long discussions, we saw also Susi Buhac (see pages 8-83 and 132) who at that time was visiting her mother living not far from Ulm. Figure 20F, finally, is a picture of a get-together with Professor Pechhold's family and Mrs. Lusser (on the right, with dog). Side trips to Rothenburg, the most visited medieval city Germany's, Blaubeuren (see Chapter 8, Figure 102E), and the Cloister Wiblingen with its old, baroque library rounded out our stay.

7/30–8/5 Next was a criss-cross travel through Germany, starting with a return trip to Köln, to, again, visit Papa and relatives after we had seen friends of Heidel's in Pforzheim. The way back called for three lectures. The first during a stop at the Farbwerke Höchst, August 2 (see Chapter 5, Figure 7 and pages 5-4–6). The second at the BASF in Ludwigshafen coupled to a visit with Dr. Heinze 8/3–4 (see Chapter 8, Figure 8). At last, a stop at the Max Plank Institut in Stuttgart, MPI, on August 5–6. There we crossed paths with Prof. Wiedemeier of RPI (see page 8-67) who had connections with the MPI (through Dr. Schneering of Figure 20G), while my contact was Dr. Gmelin (see pages 8-110 and 116). We also met with Heidel's former colleague, Anneliese Kohfink from Northwestern's Deering Library, (see page 6-13, 24 and Figure 25 of Chapter 11, they had gone back to Germany to live in Stuttgart).

8/6–20 We flew from Stuttgart to Munich to rent a VW "Käfer" for a another vacation in Kössen with an excursion to Traunstein, were we picked up Prof. Wiedemeier (see Figure 57 of Chapter 8) who had traveled by train. He also vacationed in Kössen. Together, we made excursions to Kufstein, Kitzbühel, Zell am See, and Kaprun,.

8/21–25 A flight from Munich to Tel Aviv via Frankfurt brought us to the 9th International Congress on Thermal Analysis, ICTA, in Jerusalem, Israel for my plenary lecture. Here we met with many colleagues of mine who regularly attend the ICTA meetings

[1] Located on the Rhine, in the center of the government district. (in 1988 Bonn was still the capital of the FRG). The Villa Hammerschmidt was build 1860 in neo-classical style and owned 1899–1928 by Rudolf Hammerschmidt (1853–1922), a wealthy German merchant. Being vacated by the British occupation forces in 1949, it was owned then by the FRG. In 1994, when the German capital was relocated to Berlin, President von Weizsäcker 'moved' into the "Schloss Bellevue."

(see page 8-23). Tours of the old city (Figure 20 H) and to Bethlehem reminded us of our Christian heritage and illustrated the closeness to Judaism and Islam.

8/26-27 Return flight from Tel Aviv to Frankfurt with an overnight stay near the Airport (in Kelsterbach). There we walked through the small village and could not find "Handkäs' mit Musik" a typical cheese dish that used to be common in every corner-restaurant in the area of Frankfurt. In fact, nothing else we enjoyed to eat during our stay in Frankfurt could be had (see Chapter 5). By then, Greek, Chinese, and other foreign cuisines were in vogue. Then, the first-class return-flight via Atlanta.

On arrival home, the summer in the southern Knoxville was, indeed, over. It was a remarkable trip of high scientific and personal value, as well as showing us many new things, highlighted in Figure 20. Slowly we made up for our inability to travel during our early years in Germany. A problem on return was that the weeds in the garden enjoyed our being away. It took some time to put things in order. For the future, I put a limit of three weeks for out-of-town travel.

Figure 20A *Villa Feltrinelli in Gargnano.*

B *Visiting the Schuers in "Köln" (Cologne).*

Figure 20C *Receiving the Award from Dr. Leussink.*

D *Rhine cruise, Bingen to Bonn.*

E *Visit by Hans and Inge in 1988.*

Figure 20F *Get-together at the Pechholds in Ulm.*

G In Stuttgart, Harry and the Schneerings.

In 1989 the most momentous event was the opening of the "Berliner Mauer" on the 9th of November. After being separated from the West for 28 years, the people of the GDR could visit West Berlin again. Ultimately, the 'Wall' was torn down.

Figure 20H *Street-scene in old Jerusalem.*

How was this possible? Only in January had Erich Honecker (see Footnote on page 3-22), the head of the GDR, officially declared that the "Mauer" (Chapter 7, Figure 5 and Chapter 8, Figure 27) would still be there after 50, even 100 years!

The thawing of the Cold War started with Gorbachev's[1] efforts at reformations in the Soviet Union with "perestroika" (economical and political restructuring, see also the comment on page 7-8) and "glasnost" (a more open government). Next, in 1989, communist Hungary, Czechoslovakia, and Poland increasingly allowed refugees from the GDR to bypass the Wall. Hungary even opened its border with Austria, causing more than 30,000 refugees from the GDR to reach the FRG within the first three months. More and more peaceful demonstrations occurred in Berlin, Leipzig, Dresden, Brandenburg, and other cities of the GDR. The main demands were for increasing freedoms (of expression, to travel, of the press, and of assembly) and for the release of political prisoners. On the celebration of the 40th anniversary of the founding of the GDR (October 7, see pages 3-16 and 28),

[1] Mikhail S. Gorbachev, General Secretary of the Communist Party of the Soviet Union and the last head of state of the USSR from 1985–1991. Announced "perestroika" in 1986 and in 1988 introduced "glasnost" and private ownership of businesses. March 26, 1989 the first elections since 1917 were held to the Congress of People's Deputies as part of his reformed government. He also announced that the Eastern Bloc countries were allowed to determine their own internal affairs.

Gorbachev was invited to East Berlin and urged the SED leadership to institute reforms. But, the GDR tried to follow its own hard-line course, "glasnost" was slow in coming. This is indicated, for example, by his calling the demonstrations in Beijing and other Chinese cities in 1989 'counterrevolutionary.' He supported the use of military force which led to the massacre in Tiananmen Square in June. Finally Honecker was forced resigned, officially citing health reasons.[1]

The reforms were hastened by a live broadcast of a press conference held at 6:00 p.m. on November 9, covering a meeting of the Central Committee of the SED, still in progress then. One of the topics of the meeting was the reading of revised regulations for travel, allowing private trips and emigration without the need to fulfill the usual, strict conditions.

On questioning by a journalist, the topic of the Meeting was leaked by Günter Schabowski, a member of the Central Committee, and on follow-up questioning, he suggested that the easing of travel to the West should be in force immediately. This was explosive news! The following time-line[2] summarizes the excitement: At 8:00 p.m. the ARD (radio and TV network of the FRG) reported in the widely viewed "Tagesschau" The 'GDR opens the borders!' By 9:00 p.m. thousands of Berliners had collected on both sides of the control points and demanded to be permitted to pass, based on Schabowski's statement. But there were no orders to the border guards! By 9:20 the first people were let through the Wall to West Berlin after screening and stamping of their passports. After 10:30 p.m., the control points were overwhelmed and had to open the border for all. There were no orders to the contrary. The only alternative would have been to use force. This would have caused casualties, for which no one wanted to take the responsibility any more. By midnight, all control points to West Berlin were open. The Wall was open. Although order was restored in the following days, the Wall was never closed again. Ultimately, it was torn down. Friends of ours send us shortly thereafter the piece which I photographed as Figure 21, a stark memento of the Wall.

Figure 21 *Remnant of the Berlin Wall.*

Heidel called me by 4:00 p.m. EST with the unbelievable news (10:00 p.m. Berlin time!). We had been in the GDR in 1987 and seen the decay of the country (see pages 8-124–128). We had expected a softening of the oppression in the economically ruined country, but not so abrupt and complete. By evening, we toasted the future of

[1] Erich Honecker resigned on 10/18/89. On 11/8/89 the Chief Prosecutor of the GDR called for his indictment for abuse of authority and corruption, on 12/3/89 he was expelled from the SED (see page 3-9). After German reunification (10/3/90) he was prosecuted for involvement in the deaths of 192 who tried to flee the GDR. After taking refuge in the USSR he was extradited in 1992 (after its dissolution in December 1991). He was released due to ill health after the trial opened in 1993 and emigrated to Chile to live with his daughter and her Chilean husband. Honecker died on May 29, 1994 in Santiago of liver cancer. His widow, Margot (see page 8-124), remained in Chile.

[2] Checked against the data of: www.ddr-wissen.de/wiki/ddr.pl?Mauerfall-Zeitleiste.

our old home country with a bottle of champagne. We could do this with Mutti, who was visiting in Knoxville. Mutti's reaction expressed also our sentiment: "Too bad that Vati could not see the opening of the Wall any more" [and the following reunification of Germany]!

On the next day (November 10), Heidel tried to reach Inge Gottschalk in Brandenburg by phone. As reported in the newscasts, the telephone connection was now quick and seemingly not monitored. Hans answered. Inge had to take some tests in a Berlin Hospital. He had her phone number. Again, fast service. A Nurse answered. Yes, Inge was there, who is calling? Heidel gave her name and, to give her call importance, said I am calling from the USA. Her response, I will get Mrs Gottschalk quickly to the phone, and Heidel could hear her shouting to the other nurses: "I am going crazy, yesterday the Wall fell and today I get a direct phone call from the USA!" Inge came to the phone and was happy that all at once direct, unproblematic connection to us was established. Heidel could even give her greetings from Hans, who had tried several times to reach Inge, but could not get through. At once, foreign calls were not only allowed, but they now had precedence.

Reunification[1] followed in 1990. First, new political opposition parties were established. The "Nationale Front" (see page 3-23) was split into its old parties, with the SED reorganized to the PDS ("Partei des Demokratischen Sozialismus," 1989–2007, now nationwide "Die Linke"). Elections of a new "Volkskammer" of 400 members (see page 3-28) were held with 94% participation.[2] This establishment of a legitimate government was followed by many negotiations between the GDR and FRG and also with the four countries which had different arrangements with the GDR and the FRG (see page 7-7). The latter negotiations produced the 'Two Plus Four Treaty,' signed in Moscow on September 12, 1990, granting full sovereignty to a unified German state.[3] A German-Polish Border Treaty was arranged on November 14, agreeing with the 1945 set borders. On 12/2/1990, the now unified German "Bundestag" (Lower House of Parliament) was elected.[4]

---

[1] The term 'reunification of Germany' is chosen since 'unification' is usually applied to the formation of the second German Reich in 1871 (see page 1-1) and other returns of German territory. Another common German expression used is "die Wende" (the change, or turn-about). Germany was reunified at 00:00 Central European Time on October 3, 1990 when the East Germany and East Berlin joined the FRG. This day is now a German national holiday, "Tag der Deutschen Einheit," 'Day of German Unity,' and replaces the West German holiday of June 17, commemorating the uprising of 1953 (see the Preface and page 4-33) and the East German holiday of October 7, the day of founding of the GDR in 1949 (see page 3-28 and Figure 23 of Chapter 3).

[2] On 3/18/1990. The results were: CDU (41%), SPD (22%), PDS (16%). Compare to the 1946 elections (see page 3-22–23). The LDP (see page 3-8–9) had become part of the BFD ("Bund Freier Democraten") and gained 5% of the vote, but ultimately joined the FDP of the FRG.

[3] The four powers renounced all rights formerly held in Germany as of March 15, 1991 , i.e. the war had ended. Soviet troops were to leave by 1994. Germany agreed to armed forces of no more than 370,000. The 'Nuclear Non-Proliferation Treaty' would apply to the reunified Germany, and no foreign armed forces or nuclear weapons would be stationed in the area of the former GDR.

[4] The 'Upper House of Parliament in Germany' is the "Bundesrat" with representatives appointed by the governments of the states. (Present number of "Bundesrat" members is 69.)

This 12th "Bundestag" represented the first freely elected government of all Germany after WW II. The size of the "Bundestag" was increased in size by 144 members to a total of 662, to accommodate the 'new' states. The party coalition[1] CDU/CSU and FDP of "Bundeskanzler" Helmut Kohl (Chancellor, head of the Government of Germany) was returned into office with a higher majority than before the reunification.

What did reunification bring to the former GDR? The freedoms demanded and listed on page 9-23 were achieved even before political reunification. By March 31, 1990 all Stasi staff members (see page 3-23) were dismissed.[2] The economic, fiscal, and social reunification for the two states, proposed by the FGR in March, was enacted by July 1, 1990. This affected everyone.

Most drastic was the precipitous exchange of one DM-Ost for one DM-West (see page 3-28), applied to the first 4,000 DM-Ost owned by an individual. All savings beyond that and insurance policies were exchanged at a ratio two to one.[3] Currently paid pensions and wages were converted with the ratio one-to-one, up to a given limit. Different rules applied to liquid assets and debts of businesses. This immediately made many of the export goods from the GDR unaffordable to the eastern European and Asian countries. If they had to pay in hard, western currency, they might as well buy western goods.

To cope with the guarantee of jobs which existed in the GDR, most industries, government offices, universities, and institutions were overstaffed. This, together with the financial problems, led to the collapse of the economy and caused major unemployment. Still, the needed qualified work force necessary for the rebuilding was often not available and retraining was not accepted by many unemployed workers because of their rights of refusal. As a consequence, West German firms would often take over the rebuilding of the sadly neglected infrastructure of the GDR. Not only because of their greater financial strength, but also because they had qualified workers.

The privatization of industry and properties was hampered. Often the actual ownership of companies, land, and houses could not easily be established. There were still earlier unresolved ownership claims dating from the 'Third Reich' in addition to the expropriations during the time of the Soviet occupation. Furthermore, there was industrial pollution of the soil of unknown magnitude which the new owner would have to ameliorate when taking over a state-operated facility.

---

[1] Percentages of the governing coalition: CDU (Christian Democratic Union) 36.7%, CSU (Christian Social Union) 7.1%, FDP (Free Democratic Party) 11.0%. Of the opposition: SPD (Social Democratic Party) 33.5%, the Greens (East and West) 5.0 %, PDS 2.4 %, others (5.3%).

[2] The Stasi had 102,000 full time officers when the system collapsed in 1990. In addition, one estimates 500,000 unofficial contributors (informers), about 7% of the population between 18 and 80 years of age. (From a 1991 interview with Joachim Gauck, Commissioner of the FRG in Charge of the Stasi Archives in Berlin as "Sonderbeauftragter der Bundesregierung für personenbezogene Unterlagen des ehemaligen Staatssicherheitsdienstes.")

[3] It is not possible to find a universally 'fair' exchange rate. The black market rate at the time was inflated by the large amount of illegally exported funds to 12:1 (see also page 3-28). Trying to use an exchange rate based on labor productivity yields a ratio closer to 3:1. Purchase-power considerations would have gotten the ratio closer to 1:1, but disregard quality and availability differences und suffer from the fact that prices in the GDR were fixed without relation to cost.

After 40 years of planning by the state of the life of its citizens, most people were unable to cope with the new responsibilities that came with a capitalistic economy. In the first rush to buy the big ticket items, and often on credit, many fell for unethical business practices and purchased things they could not afford or did not urgently need. While in the past, most GDR citizens were saving a good part of their income, the now available goods, previously in short supply or unobtainable, made them to consumers and the investment savings dropped.

In addition, people turned away from the old, established goods of the GDR and chose West German goods, more consumer-oriented, and for a long time in short supply. This added to the demise of the local industry and farming. Finally, the strong unions quickly forced an increasing labor cost, adding to the difficulty of companies to survive, and decreased the attractiveness of locating factories in the eastern provinces. It was often more economical to locate new factories in the eastern European countries with lower labor cost than in the area of the former GDR.

The enormous initial cost of the economic changes had to be born largely by the FRG and caused an excessive debt which even strained the European Economic Community (since 1993, European Union). The FRG, after unification, felt increasing prices and higher cost of services and a growing shortage of jobs. It took much longer than initial estimates suggested to heal the ills caused by the 40 years of central planning in the increasingly corrupt system of the GDR.

Our first *direct information about this exciting period* in the GDR came from friends in form of phone calls, as with Hans and Inge, already mentioned, and from letters, which now came more frequently and were openly dealing with political problems. We were surprised that after a short while, the newly gained freedom was not the main topic of interest in these letters. It seems that the novelty of freedom wears off quickly, despite the fact that the last freedom seeking refugee trying to get across the "Mauer" was shot only on February 6, 1989, and the order to shoot to kill at the German-German borders was suspended by the GDR only on April 3, 1989. Thinking back, this quick taking of freedom for granted was not much different from our own loss of interest in politics after leaving the GDR, described on pages 5-10 and 7-7–8.

Interesting information came particularly from high school friends of Heidel's. This group of 20 girls who considered themselves good friends in 1949, and most of whom stayed in Brandenburg and vicinity, hardly met for most of the 40 years of the GDR. They began to do so regularly only late in 1989. Heidel, although not attending the first "Klassentreffen" (high school reunion), sent a letter, mailed a package to contribute to the festivities, and called them in Brandenburg when they were together for the celebration. From then on, the group met regularly. First every two years, and then every year. From 1991 on, Heidel (and I), made an effort to meet with them every time. In fact, the meetings were often planned such that I could connect it with attending a scientific meeting in Germany, or at least in Europe. Quickly all were good friends again and the political differences were forgotten (or put aside). But not all were happy with reunification and the relationship between East Germans ("Ossies") and West Germans ("Wessies"). The people had grown apart in spirit and character, and it was hard to reunite now, almost a lifetime later.

Heidel kept three letters from one of her more critical girl friends, Annemarie Magritz (see Figure 23B, below), whom I also knew. The first was written on 3/30/1990, shortly after the "Volkskammer" election (see page 9-25). The second, shortly before reunification. The third, after the second "Klassentreffen" (class reunion) in 1991. This was the first reunion Heidel could attend (in annual count, it was the 42-year reunion). Annemarie was teaching school and lives with her husband in their house with a beautifully tended garden, dating from the efforts of her grandparents and parents. Her husband has an engineering degree and worked in a managerial position at the steel mill in Brandenburg. The two sons (with academic training, one in forestry, the other in horticulture) are married, and have children. One son with his family lives upstairs with them.

In the first letter Annemarie commented on the friendships within the group of the 1949 high school graduates. She wrote (the quotations are abbreviated and translated as close to the spirit of her words as I can manage): "In the 40 years since everyone of us has gone her own way, contact was limited by the "unselige" (German for unfortunate, wretched) political system." After detailing the life of the classmates, she commented about the 'dilemma' of all who worked diligently for 40 years in the GDR. There was a proud mention of the 'peaceful revolution,' but a bemoaning of the fact that the elections were financially supported by the FRG,[1] and carrying the condition of no alternatives to reunification. The 'new' parties, including the PDS (see page 9-25), thus, were at a clear disadvantage. She wrote: "Having born the brunt of the revolution, these people have to stay aside and have no chance to participate in a new government. The conditions for a reunification which are presently discussed are based on a common currency for the two states [see page 9-26]. This will make us poor by reducing our savings to a ridiculous level. What good is a world open to us, if we do not have the money to travel? We older people have been cheated a second time in our life and we will not be able to recover—but the youth will, in the future, similarly be cheated. As a result, most of us do not foresee a rosy or positive future, it is a subjugation under other laws and authorities with their own privileges, as with any other government."

The second, 10-page letter Annemarie's was in answer of Heidel's detailed (late) answer of the first. After some general philosophical statements, she commented on her education and problems with the Stasi (see page 9-26 and its Footnote 2). We all knew that her classmate and friend Traudel[2] and two other friends were arrested in 1950 for transporting anti-GDR propaganda. I knew, Heidel had tried unsuccessfully through the "Volkspolizei" in Hohenschönhausen to send her books. Annemarie wrote: "Traudel and her somewhat older friends were often at our house, and I looked upon them as role models, but these friendships were almost my doom. The KgU,[2] located

---

[1] The old parties which split from the "Nationale Front" were supported by their FRG counterparts due to their common roots in the Weimar Republic (1919–1933) (see page 9-25).

[2] Edeltraud Eckert, graduated with us from the same "Goethe Schule." She was arrested in May 1950 by the GDR police as a member of the anticommunist resistance group "Kampfgruppe gegen Unmenschlichkeit, KgU" (Commando Against Inhumanity) and immediately transferred to a Soviet prison. On July 29, Traudel was sentenced by a Soviet Military Tribunal to 25 years in prison. The trial was held without presence of a defense lawyer and without permission to appeal the verdict. Only much later was Traudel permitted to write and preserve her poems and compositions. Due to a work accident and subsequent poor medical attention, she died in 1955. ("Hinter Gittern – ein Mensch," poems, Türmer-Verlag, München, 1969; "Vom Leben trennt mich Schloss und Riegel," biography written by J. Blunck, Langen-Müller, München, 2000; "Jahr ohne Frühling," poems and letters, Büchergilde, Frankfurt, 2005.) In 1993 the Military High Court in Moscow declared that the arrest in 1950 was unfounded, and Traudel was rehabilitated. In 2008, on the initiative of high school students in Brandenburg, a one-hour movie "F wie Freiheit" was produced, documenting her lyrical work and the conditions of prisons in the GDR at that time.

in West Berlin was recruiting members at the "Brandenburg Landeshochschule" in Potsdam (see page 4-6) [where Traudel and Annemarie studied pedagogy]. The goal was to work against the GDR with propaganda and carry out specific espionage orders. I was asked to undertake a secret study to learn how to be a radio operator and hide for a future mission." Further she wrote that she avoided a similar arrest by having a friend at the school to whom she told the proposal by the KgU! He, however, was until 1945 a POW in the Soviet Union and 'converted' to be a member of the 'National Committee for a Free Germany' and had to inform once a week to the Stasi about all his 'friends.' As Annemarie wrote: " When I was questioned by the Stasi myself, they knew about me already. My story was identical to that of the informer and I was sent home as a 'little fish.' For fear, I gave up my study and worked for a while as secretary in a church, but in time, got back to teaching children." She drew the same conclusion as I (on page 3-30): "Hände weg von Politik, dem schmutzigsten Geschäft, das es im menschlichen Dasein gibt" (keep your hands from politics, the dirtiest business of mankind). As a no-party member, she stated: "I nevertheless have served the SED-State for 40 years, trying to give the children I taught, a nature-loving, humane education."

In the description of the advantages of life in the GDR, Annemarie described her view: "With unheard-of hard work and many new ideas, a new republic was built-up from nothing without help of western powers.[1] Our standard of living was frugal, as we can see only now, comparing it with the market economy of the West, but comparing it with the 'brother Eastern Bloc countries,' we had the highest standard of living and were proud of it. The wages were low, but one had to remember, that energy cost, prices of mass transport, and rent were similarly low. In addition, everyone had a right of an apartment and a job."

Then she continued: "The state did everything, even took away the need to think. Everyone could count on his rights, but did not necessarily do his duty. This explains the low work ethics, dirt, and decay at the end. The wages were the same for good or poor effort. [See also the pages 8-124–128]. What seemed at the beginning as laudable idea for which millions died since 1917, was made ridiculous through the dictators who became criminals. In our younger years, we wanted the best for our piece of Germany and one must understand that the older ones are sad because they were so shamelessly defrauded, and 40 years of head-start is too much for us to catch up with. The ideal of socialism has proven itself as utopian, as the time was not ready for it. At present, we have more than one million unemployed workers, and many are on part-time jobs or in an early retirement. Many have fallen victim to slick salesmen and have debts that they cannot pay, particularly if their jobs were lost or curtailed. The result is that the money is gone and the goods are lost also. The GDR population is just too new for the market economy."

The letter closed with a description of the highlights of their extensive travels possible for them earlier, covering eastern Europe, and for her husband, on business, also into the western world, as well as Asia, and Africa. The Magritzes were in this respect certainly exceptionally well off. Already, they also planned a vacation trip to northern Europe, not possible before the "Wende."

The third letter, arriving after the second "Klassentreffen," was not any shorter. It mainly dealt with the extensive renovation and modernizing they were doing on their house and property

---

[1] Here, I think, Annemarie should have written instead 'despite harsh reparations paid to the Soviet Union and trade restrictions.' She probably did not know that the Marshall Plan help which was so beneficial for the FRG, was rejected by both the GDR and USSR (see pages 3-28).

which was finally possible after 40 years of increasing neglect due to shortages of building material for private citizens. A large part of the letter was filled with the description of the efforts to offer her grandchildren learning and playing opportunities, and the travels to West Germany and Austria, as well as planning of travel into the Czech Republic over Christmas (as they had done and enjoyed in the past). Hope can be read from these pages for her husband was allowed to keep his job a few years longer to wind down the business of the steel mill and then start a pleasant retirement. It ends with a comment about the terrible happenings in the disintegrating former Yugoslavia (beginning in 1991 and ending only in 2006, hopefully permanently) and the happenings in Iran, Iraq, and Kuwait (Iran-Iraq war, 1980–1988 and Gulf war, 1990–1991, and still not being resolved). All these countries her husband knew from business trips, and it confirmed for Annemarie how a corrupt leadership can bring untold sorrow upon the population. The letter ends with a question and answer: "Are we doing well? Yes, considering several shortcomings, which are unavoidable because of our still not being fully familiar with the new social system. It is very reassuring when one does not have to turn over every penny and can buy what one has pleasure to own."

How were we personally affected by the reunification? We were not from a country any more to which we could only return for short visits and with special permissions (see the pages 8-46–48, 8-59 and 8-124–128). Now, we were welcome and could go there whenever we wanted to. Our first 'free' visit became opportune after an invitation to give the plenary lecture at the 23rd Europhysics Conference in Stockholm, Sweden in late May 1991 (see Appendix C, page C-38). This was seven months after the "Wende" (see Footnote 1, page 9-25).

On June 1, we flew from Stockholm to Berlin, where we had arranged to meet with four classmates of mine we had contact with. They lived either in Berlin or close by. Two of them lived in the area of the former GDR, Fritz Bockmühl and Fritz Weickert. The other two, Peter Menzel and Horst Rückert, lived in former West Berlin. The graduating high school class in 1949 with all of us is shown in Figure 25. Chapter 3. We met in Peter's house, shown in Figures 22. A long day of discussion showed that all of us were successful in our chosen professions and countries, although it was especially difficult to steer around the pit-falls of politics in the GDR. None had objections about the reunification, although Fritz Weickert, as a school teacher, was complaining about having lost half of the savings, but when we brought him to his car, he drove a fancier, Italian car than all others (since I had to leave the Mercedes in Knoxville, in Berlin I also drove a small rental car).

Figure 22A *Left to right: Peter Menzel, Fritz Bockmühl, Fritz Weickert, Horst Rüchert, and I.* **B** *Group of the school friends with wives. Out of order: Fritz B. (left), Hannelore B. (4th from left).*

After some sightseeing in Berlin, I had three lectures to give (see Appendix C, page C-38–39). Most interesting was the "Institutscolloquium" at the Institute for Physical Chemistry at the Humboldt University, Berlin, on June 6, already mentioned on page 4-35. What a difference from the prior visit in 1968. My old friend, Professor Schmidt, could now freely talk about the increasing problems of the last years of the GDR and the uncertain times to come. In the discussion after the lecture, the professors got together with me, and I noted that all were very unsure of their future. It was clear to them that changes had to occur. The politically engaged colleagues were afraid to lose their positions. The question of funding was unclear, and one could note that the university, doing also industrial research, might have to reduce its size. Everyone had more or less permanent civil service employment, but none of them knew what the rules of the 'new' country would be. Their personal future took precedence over science. Professor Schmidt retired early, shortly after my visit and I had no further connection with the University.

Next, after a brief visit our cultural heritage of Prussia in Potsdam (UNESCO World Heritage List since 1990), *we went to our hometown, Brandenburg*, were Heidel had her "Klassentreffen" (see

Figure 23A,B *Klassentreffen. Excursion on the ferry across the Havel, near Prizerbe, June 1991.*

page 9-27). Figures 23A,B were taken by Heidel of the group of 14 of the 20 ladies who graduated together from the "Goethe Schule" in Brandenburg in 1949 (see Chapter 3, Figure 8), i.e., she is missing in the photos on the ferry across the Havel river. I had delivered Heidel in the late morning to the house of Annemarie (on the left of Figure 23B,) where the group met. The ladies had gone for the afternoon by train to a near-by village, Pritzerbe. For myself, I made a nostalgic inspection of the city of Brandenburg for the rest of the day.

My impression of the city was worse than in 1972, our last prior visit (see page 8-59). Many of the old buildings that were still in good order then, were by now dilapidated from neglect over the years. In many places, one could see a banner promising quick renovation, as shown with Heidel in Figure 24, taken on the next day. Two views I found particularly sad are reproduced in Figures 25A and B. They represent the restaurant next to Vati's old business in the Wilhelmsdorfer Straße 73 (see Chapter 1, Figure 5), and the area

Figure 24 *'Until soon, old house.' (Slated for repair).*

Figure 25A *The restaurant "Gambrinusquelle."* **B** *Franz-Ziegler-Straße from the rear in 1991.*

behind our apartment of 1946–1953 in the Franz-Ziegler-Straße 35 (see Chapter 3, Figure 5). The restaurant was ultimately torn down. Nobody could be found to rebuild it. The block of houses in the Franz-Ziegler Straße was still inhabitable, but in great need of renovation. The backyard, however, shown in the picture, was one of utter neglect. In the background, one can see my elementary school ("Jahn Schule," Figure 13, Chapter 1, since 1995 re-designated von Saldern high school, see page 2-24). I also inspected our first apartment, Wilhelmsdorfer Straße 70. The space of the building was cleared, but used as a trash dump (see Figure 5, Chapter 1). The area of the Werderstraße 28 with Vati's business location until 1953, still in operation on the visit in 1972, was razed and a new, bland apartment block was erected, with an empty lot where the prior Corona Bicycle Factory was located (see pages 2-10–11). Much else was in dire need of improvement, including the whole area of apartment blocks were Heidel lived (see page 4-19) which dated from the 1930s. But also the newly built, nearby housing required overhauling and updating.

After this first visit, we returned annually and could see the continuing progress. It was not as fast as one hoped, and even 20 years after reunification, more work is called for, but Brandenburg is again a nice, clean city with beautiful surroundings, as shown in Figure 26.

Figure 26 *"Brandenburg an der Havel," a photograph from the air at the end of the 20th century.*

Figure 27A *Supper with Hans and Inge and friends.*

Naturally, it was not possible to visit Brandenburg without spending time with Hans and Inge Gottschalk, our most faithful home-town friends. They were much in tune with the changes as they occurred, but also had problems with arranging for the return of the house they had lost by disownment and with the arranging for their pension, which was difficult to adjust to the FRG schedule. But in their mind, these problems were minor compared with the freedom they had gained in their daily life.

Figure 27A is a reminder of this stay on the occasion of a visit to the "dacha" (Russian for cottage) of the Göttels near the Havel river, friends that we also knew from before. The meal was designed to serve all that is difficult to have in the US, starting from the dark bread to the smoked eel and the specialty sausages. The beer supports the points made above. I remember, it was Bavarian, despite the good beers brewed in the area of the former GDR. This was the time to consume the 'better' (advertised and novel) products of the West. The local breweries, as a consequence, had a hard time to survive the first few years after the "Wende."

Earlier, in March, Hans and Inge had made a trip to the USA, impossible before, as was already remarked upon on pages 4-28–29. Besides showing all niceties that could be reached by car, Heidel also arranged a meeting with other school friends from Brandenburg living now in the US, Trautchen Hough (née Thräne) from Atlanta, GA, (see page 9-10) and Dorle Mendheim (née Radikowski) from Chicago, IL, (see page 6-17). Figure 27B shows all four, while signing Heidel's 'guest book' (see page 9-12). In 1979, when a runner was started as a guest book, Hans and Inge had also signed, but at that time we had to take the runner behind the iron curtain, to Prague.

Figure 27B *Inge, Trautchen, Dorle, and Heidel, 1991. The small runner, left, shows the guests in Troy, 1979–1988. For the table cloth, see Figure 16.*

We could now also have other visitors from the former GDR. In 1993, Gabi (see pages 4-16–17, 30 and 5-18) and her niece came for a three week's stay. The following year, two of Heidel's school friends, Lilli Genenz and Irmchen Skodowski (née Gehrke) could also visit us in Knoxville (for a picture of the latter two, see Figure 23A on the left and right, respectively).

Finally, our trip which started with the meeting in Sweden (page 9-30), was concluded by the annual travel to Quadrath-Ichendorf to visit Papa and relatives in Köln, as before. After one more drive along the Rhein and to Freiburg for a lecture (see Appendix C) and a last vacation in the Glottertal (see Chapter 8, Figure 56), we flew back on June 14 with new impressions and the satisfaction to be greeted at the Atlanta airport with a friendly 'welcome home' at the passport control. After quick adjustment to home life, the next lecture away from Knoxville was on June 20.

A result of the "Wende" was also the freedom and growing initiative of the various German scientists to organize meetings between colleagues in the GDR and FRG. A typical example is the *Lähnwitz Seminar*, which expanded quickly to international participation. This biannual seminar on thermodynamics and calorimetry was organized in 1990 when Germany was on the brink of reunification by Professors Schick of the University of Rostock (PhD 1980 from the THLM, see page 8-124) and Höhne, at that time, of the University of Ulm (see page 8-121). The seminar was designed to foster intense exchange of scientific information and discussion of the newest developments, similar to the Gordon conferences (see pages 8-16–17) and was usually held in a remote location, not far from Rostock. In 1994, the third Lähnwitz Seminar, with us in attendance, was held in the "Gasthof zur alten Post" in Serrahn, Mecklenburg, about 35 mi south of Rostock, the harbor city on the Baltic Sea. Figure 28A shows the old country inn (where, supposedly, Napoleon had slept on his flight from Russia). The specific topic of the conference was the 'Determination of Transition Entropies,' and I could introduce the new technique of 'Temperature Modulated Calorimetry,' TMC. The group of the conferees is shown in Figure 28C. At this meeting, we also visited the location of the first "Lähnwitz Seminar," south of Güstrow, (see Figure 28B). Naturally, we also spent time at the Barlach museum in Güstrow(see page 8-47) and went further to "Neubrandenburg," where Heidel had spent some time before leaving the GDR (see pages 4-30–31). From then on, we were invited to every subsequent Lähnwitz Seminar, always to discuss something new. Figure 28D illustrates the attendees of a recent meeting. See Appendix C for details of the Seminars 3–10 (on pages C-44, 48, 52, 55, 59, 60, 63, 64, respectively).

Figure 28A,B *Zur Alten Post in Serrahn, 1994* with *an excursion to the "Forsthaus Lähnwitz."*

Figure 28C,D *The attendees to the Seminar 1994 and in 2008 in Warnemünde, near Rostock.*

Teaching at the University of Tennessee did not have the same importance to me as it had at RPI (see pages 8-13–15, 52–55, 66–75). Unfortunately, the well-developed audio and computer courses could not be continued. As a state university, UTK did not charge a high enough tuition to support the production of the courses and tape equipment for the students, as was possible through the Evening Division at RPI (see page 8-68). By donating a set each of the books and tapes to the libraries at UTK and ORNL, I made sure that everyone could have access to the courses and, naturally, come to my office hours to find answers to any questions. Not many students beyond my research group, however, made use of this learning opportunity.

An involvement in Freshman Chemistry also was not possible, as pointed out on page 8-71. The teaching for the position as Distinguished Scientist was to be one 'live' lecture course of three hours per year on a graduate subject (see page 8-129). In contrast to RPI, the promised teaching 'load' was kept without the need to bring outside support to the department (see the discussion on pages 8-86–88). Since the position was, in addition, a 12-month appointment, there was also no need to find summer support, leaving all outside funding for research and student support.

The courses which I was asked to teach were Physical Chemistry of Polymers (CHEM 595) and Thermal Analysis (CHEM 690). Both dealt specifically with macromolecules and were for graduate students and advanced seniors. Between 3 and 11 (average of ≈8) chemistry and engineering students were registering for the courses. There was no lower limit in enrolment as at RPI (see page 8-67). In my opinion, this meant that such courses could be taught more efficiently as audio or computer courses. The new thermal analysis course I generated, went back to the initial lectures at the University of Mainz (page 8-41) and the later development into the audio courses (Chapter 8, Figures 72A,B). At UTK, recognizing the interest of the engineers in the course, the lectures were expanded in 1998 beyond polymers and called 'Thermal Analysis of Materials.'

The lectures on physical chemistry of polymers, I linked even further back, to my early undergraduate learning at Cornell University (page 7-4–5). To cover the specific field of polymers, I used my research experience, as well as the material for the books on 'Macromolecular Physics,' described on pages 8-41, 8-57–58, and 8-62–63. Next, I took the then two most prominent text books on Physical Chemistry (by W. J. Moore and P. W. Atkins) and went through them chapter by chapter. Assuming that the material was more or less taught in the two undergraduate physical chemistry courses required as prerequisites, I found the chapters which lacked information about the macromolecules. Out of these missing topics needed to understand physical chemistry of polymers, I made my syllabus, each starting with a short general review, and then turning to polymers.

Over the 27 semesters between 1988 and my retirement in 2001, I taught each of these two courses six times. In addition, there was the handling of two semesters of the 'Departmental Seminar' for graduate students (1989/90, CHEM 501, one credit hour each), with lectures given by visiting scientists, and a 'Senior Seminar' (1998, CHEM 401, one credit, 3 students).

Another item discussed in each of my annual reports to the department was *course and curriculum development*. From 1988 to 1990 the Thermal Analysis course was used as base for writing a textbook, initiated during the last RPI sabbatic (see page 8-116, Appendix A, publication #290). To keep the cost of the book low ($35), it was created on my personal computer in ready-to-print form. The text and figures were handled with the new word processor WordPerfect 5.0.

The new generation of personal computers could support teaching even more than described in Chapter 8 (pages 8-66 and 8-69–71). They could store large numbers of illustrations and it was possible to project these in class via a display-board mounted on an overhead projector, wheeled into the lecture room on a cart. In this fashion, the figures could easily be fitted up to the last minute for the current lecture. By 1992, 1,500 illustrations were developed for the two courses filling nine 'floppy disks.' This reservoir of slides was useful also for my research lectures, and with little effort the figures could be adapted to teaching off-campus. For example, in 1992, such lecture series were given in New Jersey, Sweden, and South Africa (see Appendix C).

Another effort in 1991/92 was to try to use my international connections to develop a *study program between European and US universities* which ultimately would result in a joint PhD. It did not succeed. Each partner was interested in a student exchange, but a combined degree program proved impossible unless the student would fulfill both programs, including separate theses in the respective languages. But, my proposal was to efficiently produce qualified PhDs for the research laboratories in the global chemical industry, not to double the effort by earning two degrees.

From 1994 on, the Department of Chemistry had set up a local-area computer network, LAN. As soon as possible, all the information needed to teach my two courses was stored on this LAN. All was entered with the proper presentation software, so that students could preview, review, and copy the information at any time. For the course on Physical Chemistry of Polymers (CHEM 595), which did not have an assigned text, specially prepared lecture notes were written. A process of continued updating and changing was started, not only to keep pace with the development of the science, but also with the quickly increasing computer capabilities. Ultimately, the text was coupled to figures and to hypertext for explanatory and review material. This produced *the first Internet courses* for graduate study. They were also made available to students on CD ROM for study at home (see the stages from 1995–2000 in Appendix A, publications #393, 421, 430, and 464). Finally, the two courses were combined, with the title 'Thermal Analysis of Materials,' and tested in 1998 and 1999. A description was written up and made available to the readers of the 'Journal of Thermal Analysis and Calorimetry' in 2000 (Appendix A, #478).

After the noted decline of interest in science, there was from time to time a call *to improve the standing of the continuously decreasing scores of the US students in science and mathematics* relative to other countries. (See also, the end of the 'Golden Age of America' in the sciences, pages 8-52–55.) This problem is not solved yet. For example, extensive international comparisons were made by the Lynch School of Education of Boston College in 1995, 1999, 2003, and 2007 (TIMSS, Trends in International Mathematics and Science Study, see: *http://timss.bc.edu/*). The results from 20 countries for the level of education in the 12th grade, i.e., the performance just before the student enters the university are given in Figure 29. This result is truly devastating for the US. With this poorest performance of all developed counties listed, the efforts of the universities are seriously hampered to educate the quality and quantity of American scientists and engineers required for the future.

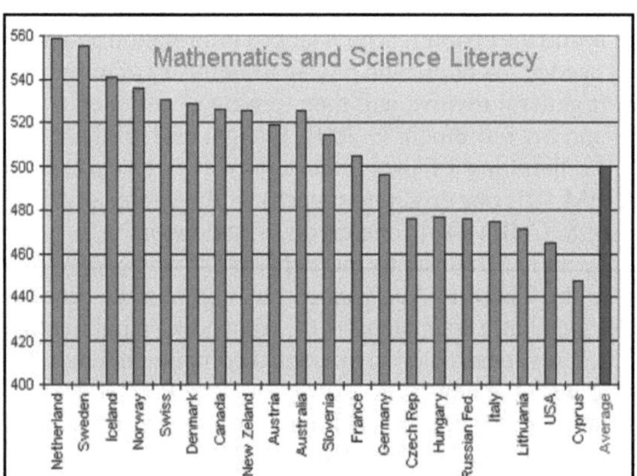

Figure 29 *Redrawn data from TIMSS for 2003.*

A direct experience of mine with the politics of improving education dates back to 1990. At that time, the goal of President G. H. W. Bush was to be the

'Education President.' In his 'State of the Union' address of 1990 he said: "By the year 2000, every child must start school ready to learn. The United States must increase the high school graduation rate to not less than 90%. In critical subjects, at the fourth, eighth, and 12th grades, we must assess our students' performance. U.S. students must be the first in the world in math and science achievement. Every American adult must be a skilled, literate worker and citizen. Every school must offer the kind of disciplined environment that makes it possible for our kids to learn."

Figure 30 *President G. W. Bush, N. R. McWherter, Governor of Tennessee, Lamar Alexander, President of UT[2], and Distinguished Scientists. (I am placed as #3, and barely visible is Dr. Guichon, of page 9-6).*

On February 5, just a few days later, a private meeting was arranged for the nine Distinguished Scientists at UTK with President Bush[1] and the President of UT[2] to discuss the advantages of the Distinguished Scientist Program.

Usually I stay away from chances to see high officials, unless they have something of importance to say, or I can talk to them. Here, I hoped to be able to express my agreement, and offer thoughts on how to solve problems of teaching . But, only 20 min were allowed for our discussion. The rest of the two-hour visit of the President was used for inspection of laboratories. Of the 20 min, 13 were needed for introduction of our program. There was no discussion outside the handshake before and after the seating around the table, shown in Figure 30! For me, this was upsetting; I had hoped to show the impossibility of a quick turnaround as proposed. During the goodbye handshake, I mentioned that there was no time to state my opinion, and asked if it was possible for the President to join me and Heidel for dinner and discussion. As a good politician, he declined gracefully by saying: "I am expected by Barbara, but should you ever be in Washington, try to look me up." Fortunately, I knew that chances of a meeting in Washington were sufficiently slim that I did not attempt a separate trip.

My estimate was, it would take at least 20 years for the proposed improvement. First, new teachers needed to be educated, then a new generation of students must be taught. Finally, and most importantly, it would take changes in societal values. These must reverse the trends enumerated on pages 8-52–55. There must not only be special support for challenged students, but also for the above average students. Last, and not least, secondary schools must not spend excessive time to develop 'self-esteem,' but to pay attention to solid academic training. This involves the discipline to *study* the properly selected material *and to retain* the knowledge. From my experience, I felt that our students may be well taught by being presented with much of the required material, but Figure 29 shows that they certainly do not retain the knowledge (see also page 8-73).

---

[1] President from 1989–1993. Also, see in this connection the discussion of the Distinguished Scientist Program at UTK, page 9-6, and the CAREER program of NSF, started in 1992, page 9-13.

[2] Lamar Alexander, former Governor of Tennessee (1979–1987), President of UT (1988–1991), later US Secretary of Education for President Bush (1991–1993), US Senator (2003–).

Figure 31 *Stefan and Alyssa, 1989.*

These first few years in Knoxville saw also **a number of changes in our personal life**, which need to be recorded. On top of the list is that our second grandchild, *Alyssa was born on May 1, 1989*, in Syracuse, NY. Figure 31 shows the two proud grandmothers and Anna on the occasion of the christening in front of the Volodymyr Ukrainian Church in Utica, NY (see 8-82–84). This enlargement of Brent's family called for a number of additional visits between Syracuse and Knoxville. The Figures 32 show the quick development of Alyssa. These pictures were taken on a visit to the Burnet Zoo in Syracuse in April 1990, about one year later. On the way home, we stopped off in Ithaca, NY, to attend the wedding of John Buhac at Cornell University (Son of Susi and Iwo Buhac, see page 8-83 and 112).

Our complaint about the long distance driving from Knoxville to Syracuse was heard. In 1993 Brent took a new job in Columbia, SC, only 300 mi from Knoxville, with Policy Management Systems Corporation, PMSC.[1] One of

Figure 32A *Brent's enlarged family and Heidel* **B** *Alyssa with Grandmother Heidel, Burnet Zoo.*

his first jobs was to work for half a year with the "Gothaer Lebensversicherung AG" in Göttingen, Germany, to develop a new computer software program. His German, learned at home (see page 6-25) and in Germany during the sabbatic 1967–1968, and also in secondary school and college, was sufficient to succeed with the job with only few emergency phone calls to me for help.

---

[1] Now renamed to Computer Sciences Corporation, CSC. It provides information technology, personnel staffing services in systems integration, enterprise-application development and management, application-software for the financial services industry, business-process outsourcing, managed hosting services and applications. The CSC serves Fortune Global 1000 companies in fifteen industries and national and local governments. The CSC employs close to 98,000 people in 93 countries. It is presently headquartered in Falls Church, VA (since 2008).

Families do not only increase. It should be the natural development that *one generation dies after the other*. It was mentioned before that all my four grandparents, had died before we left Germany in 1953 (see Chapter 2, pages 2-3–4 and Figures 2 and 3, and Chapter 11, Figures 3 and 5, and page 4-31). Heidel's grandparents died earlier, so that I never met them (for information, see page 11-2 and Chapter 11, Figures 5, and 6).

Figure 33 *Mama's Grave in Bergheim.*

The next generation was represented by our parents. Mama passed away in 1979, as mentioned on page 8-110. She was buried in Bergheim and her grave is shown in Figure 33. Vati died in 1985 as described on page 8-108 and documented with Figures 97 and 98 of Chapter 8. By 1985, thus, only half of our parents were still with us (Figure 99 of Chapter 8). Papa lived with us in Knoxville, until he entered a senior citizen home near Bergheim at age 88 when he did not like to travel any more (Figure 108 of Chapter 8 and page 8-129). Mutti stayed in her home in Hastings, NE with Joachim looking after her after Vati had died. On visiting Mutti in 1992, Figure 34, below, was taken in the house in Hastings, NE (see Figure 25 of Chapter 6). Tante Lieschen was at that time 90 years old and was living in a home for senior citizens in Kenesaw, a few miles west of Hastings. A few years later, when Mutti could not handle her own house anymore, she joined Tante Lieschen in the same home.

When in 1992 we were on *a two-week trip to Sweden* for a series of lectures in Stockholm (see Appendix C), we were given accommodations in the former home of Gunnar Wallquist who had donated it with all its furnishings and valuables to the Royal Institute of Technology, his alma mater. Figure 35 shows the living room. Besides the exquisite furniture, carpets, and pictures, there was a cabinet full of netsukes, one of my collecting hobbies. I would have liked to take home a few of them. Since this was not possible, I increased my collection after getting home (see page 9-3).

We liked the stay in Stockholm because here was a big European capital that was not devastated during WW II. It still had its original character. At the end of the stay we had scheduled the usual annual visit to Papa in Quadrath-Ichendorf. During one of my last lectures, however, I

Figure 34 *Heidel, Joachim, Mutti, and Tante Lieschen for coffee in the kitchen, visit in 1992.*

Figure 35 *Heidel in the beautiful antique apartment of Gunnar Wallquist, Stockholm, 1992.*

Figure 36A *The funeral arrangements in the church.*

received notice that Papa had died suddenly on June 7. The phone call was relayed from the home in Quadrath-Ichendorf to Brent in Columbia, SC, since we could not be reached at home. With Caryn's help, the telephone number of the Royal Institute in Stockholm was found and the message relayed. After calls of Heidel to the relatives in Cologne to make arrangements, it turned out that the funeral could easily be set for the time when we wanted to be there anyway. The Figures 36A to C show the funeral, the get-together of relatives and friends in Bergheim, and the ultimate resting place of Papa.

B *The funeral guests at the "Jäger Eck."*

C *The grave and stone in the cemetery in Bergheim.*

Figure 37 *Frank's funeral, on January 14, 1992.*

This was not the only sad occurrence. Early in January, when we were visiting Cancún, Mexico, <u>*Frank, my older brother, died suddenly*</u> in his sleep. We managed to return hurriedly to give him a proper send-off (see page 9-66–67). Figure 37 shows the three remaining brothers and other close family (front row left to right: Shahla, Eberhard, Dagmar, Heidel; back row: Brent, Joachim, Manfred, Caryn, and I). Figure 38 is of his grave, photographed later.

The year 1992 ended with additional sad news. When we returned from the Meeting of the Society of Polymer Science of Japan, held in Yokohama, Japan (see Appendix C), we were informed that Al Narten, our Group Leader at ORNL (see pages 8-112 and 9-14, 15, 16), had died from complications during surgery. He was missed by the Polymer Group. Much of the group spirit generated through his effort by the weekly meetings was lost.

Having started these unhappy paragraphs, one might as well continue with other bad news. In 1994, Hans Gottschalk (last picture with us is Figure 27) died of a heart infarct when traveling in Quedlinburg (see page 8-128).

Figure 38 *Frank's grave in Columbia, SC.*

Mutti survived Vati by 14 years and in the last years we visited her once a year in her retirement home in Kenesaw, NE, near Hastings. Joachim drove there almost daily and made sure that she had good care. The two sisters, Mutti and Tante Lieschen, had rooms on the same corridor. Mutti died 93 years of age on May 30, 1999. Her and Vati's resting place in Hastings is shown in Figure 39. The two rocks were found by Joachim in Colorado near Mutti and Vati's favorite vacation spot (see page 8-18). Tante Lieschen survived Mutti, reaching the age of 104.

Figure 39 *Vati and Mutti's grave in Hastings, NE.*

About 50 years after our grandparents had died, all our parents had left us. With the premature death of Frank (see Figures 37 and 38) this left me at the turn of the century as <u>*the oldest surviving male member of the Wunderlich Family*</u>. Despite the dangers of the 20th century, all male members of the family had lived a normal life. Only Frank was a soldier for a short time (see page 2-25).

The conclusion of Vati and Mutti's life, as marked by Figure 39, may be a proper moment to summarize the accomplishments of the family they started. All their four sons married and gave them six grand children, which, in turn, also all married and, by now, added six great grand children of which one already is married. Vati was the last member of the family who only had the usual 8-year primary schooling in Germany ("Volkschule"). This he followed by a thorough trade education, ending with a certification as master roofer ("Dachdeckermeister," see Figure 26, Chapter 1) and a sideline, as tinsmith. Frank's and my schooling were the first that advanced from "Volksschule" to "Mittelschule" (junior high school) and beyond. Looking at the 20 direct members of the Wunderlich family and their spouses who were older than 25 years by 2010, all have earned a bachelor's degree or equivalent. Nine ended their education with a doctorate, ranging from science to engineering, literature, and medicine. All are US citizens with a strong international background, reaching from German to Ukrainian, Iranian, Irish, and Chinese, as mentioned on page 8-84.

Research at UTK and ORNL in the last decade of the 20th century continued at the pace set in the first few years of the transition from RPI to UTK/ORNL described on pages 9-14–19. The scientific details are summarized in Part E of Appendix B, and extended in Part F to cover the time to 2001. The full group of researchers as it appeared in 1990 is shown on Figure 40, including the ORNL group. The details of the cooperation with ORNL and the new directions the research had taken, have already been outlined on pages 9-16 and 17.

The effort in _molecular dynamics_ simulations which was initiated under the leadership of Dr. Don Noid is summarized with Figure 18B, above, and Figure 15 of Appendix B. It allowed us to see the actual motion inside polymer crystals by creating movies in super-slow motion. Time was stretched by a factor of 10^{12} which makes fast processes of 10^{-12} s visible by taking 1.0 s, or producing a series of snapshots, as shown in Figure 18B.

Figure 40 *The Polymer Research Group from UTK and ORNL. Front row, L to R: Dr. M. Varma-Nair, Dr. K. Roles, Dr. B. Wunderlich, Ms. J. Hickson (Secretary) Dr. A. H. Narten (Leader, ORNL Polymer Group), Mrs. C. Sumpter-Getino. Middle, R to L: Dr. B. G. Sumpter, Dr. A. Xenopoulos, Dr. J. Cheng, Dr. A. Habenschuss (ORNL), Dr. L. Liang, Dr. W. R. Busing (ORNL). Top three, L to R: Dr. D. W. Noid (ORNL), Dr. B. K. Annis (ORNL), and Dr. Y. Jin. Photo of 1990.*

The day-to-day work was done by Dr. B. Sumpter (see pages 9-16–17) and then by 'Luke' Liang, a new graduate student (PhD 1990–93), who followed his PhD with an additional year as postdoctoral fellow. In a total of 15 papers the unique observations of his thesis were documented. They included the atomistic dynamics of crystals of 9,600 flexible chain atoms and their various conformational defects, dislocations, and slip planes. Simulations were made mainly on the skeletal motion, using the united-atom model with the CH_2-groups replaced by a similarly sized atom of mass 14. Simulations were done below and above the melting temperature with and without external deformation (see Appendix A). Finally 'cooling curves,' as shown in Figure 15 of Appendix B and 'Thermal Analysis via Molecular Dynamics Simulation' were accomplished (see Appendix A, #406). The link of molecular dynamics to the ATHAS effort was expressed by Figure 19. In addition, some effort was made to explore the full molecules, including the group vibrations, increasing the number of atoms in the crystal to almost 30,000.

A complete defect theory of linear, macromolecular crystals was authored by the molecular dynamics group in 1994 (Appendix A, #350). This cooperation continued until the ORNL mainframe computer system was changed to parallel computing. This change would have called for major reprogramming, a task not of interest to Dr. Noid, who had turned to other computational enterprises and felt bypassed by the polymer applications based on his work. The later parts of the thesis of Dr. Liang had already to be carried out on the UTK computer, which, however, was less accessible to us because of the larger number of users and our need for long computation times. An effort to continue some calculation involving Dr. Kreitmeier (see pages 9-48), a visiting professor from the University of Regensburg (part of a German Supercomputer Conglomerate), did not come about since Dr. Noid did not permit his proprietary computer program to be disseminated at other universities. In 1996, the last paper about molecular dynamics of polymers was published.

Other computational efforts are listed in Appendix A. They involved computer simulations of the atomic force microscopy and the use of neural network programs for the inversion of the Tarasov Function used for the computation of polymer heat capacities and extrapolations of heat capacities to very low temperatures, difficult to reach experimentally (Appendix A, #348, 328, 319). These applications of modern computation methods showed promise, but were not further pursued although they indicated the possibility of enhancing thermal analysis. Besides cooperation in the ATHAS effort, Drs. Noid and Sumpter, independently, developed in several papers the use of the computational neural network paradigm for predictions in chemistry and materials science from limited data, including thermal properties as well as structure-activity-property relationships.[1]

The major cooperation based on *full-pattern X-ray diffraction* with Dr. Busing (see Figure 40) started with Dr. Yigang Fu as a postdoctoral researcher (1991–1995, see Figure 46, below). Dr. Fu obtained his basic training in China (Nanjing Chemical Engineering Institute and Wuhan University of Technology) and then was admitted to the University of Kentucky for his PhD work (1986–1990) in the laboratory of Professor C. P. Brock, well known to Dr. Busing. For his thesis, he worked in the field of X-ray crystallography of the thermal motion, disorder, and phase transitions of small molecular crystals. This made an ideal fit to the advanced Rietveld technique developed by Dr. Busing (see page 9-17).

Figures 17A–D of Appendix B illustrate the first success of the Rietveld method applied to oriented fibers of poly(ethylene terephthalate), PET. The X-ray diffraction patterns were separated into their three components, originating from the amorphous, crystalline, and an intermediate phase, a mesophase. This structure information could then be connected to the mobility derived from solid-state NMR, heat capacity, as well as mechanical properties. To everyone's surprise, the mechanical properties were not determined by crystallinity, as earlier analyzed based on two phases only, but by the amount and orientation of the mesophase, as is shown in Figure 41. This could be understood from the overall picture of the fiber structure. The crystals and the remaining amorphous PET are separate phases, molecularly connected by the continuous, load-bearing, oriented mesophase. The oriented mesophase is related to the rigid-amorphous phase (Appendix B, page B-15 and Section F). Below its glass transition temperature, it becomes a strong, solid fiber.

This job on PET, the common 'polyester,' was followed with similar results by an effort on gel-spun ultra-high-molar-mass polyethylene fibers (see Figure 44D, below, and Appendix B, page B-20). Such fibers are used for high-strength applications, such as bullet-proof vests. The five-year

[1] B. G. Sumpter, C. Getino, and D. W. Noid, *Ann. Rev. Phys. Chem.*, **45**, 439–481 (1994); B. G. Sumpter and D. W. Noid, *J. Thermal Anal.*, **46**, 833–851 (1996); B. G. Sumpter and D. W. Noid, *Ann. Rev. Mater. Science*, **26**, 223–277 (1996).

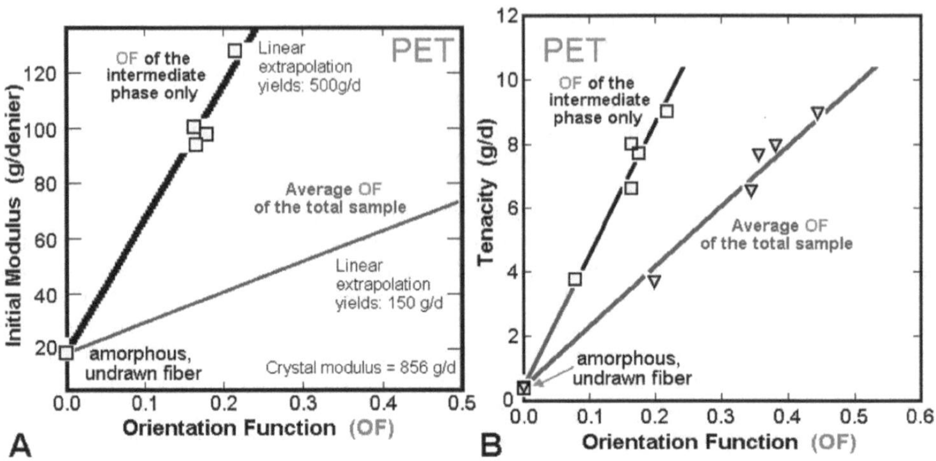

Figure 41 *Mechanical properties of PET fibers.* **A:** *The initial modulus (stress divided by strain).* **B:** *Tenacity (strength at break). The denier is a measure of the fineness of the fiber (in grams per 9,000 meters). Note that extrapolation of the orientation function of the total sample to 1.0 should correspond to a single-crystal. The prior used model of two phases only (crystal and amorphous phase) leads in Figure A to an extrapolated modulus which does not agree with a single-crystal (from X-ray analysis). Without knowledge of the mesophase, one assumed that PET fibers could never reach a modulus beyond 150 g/d. Similar reasoning applies to the tenacity. The old model stifled advances to better fibers, the new one suggests that considerable improvement is possible.*

work of Dr. Fu was a big success. At that time he was the only polymer crystallographer who could handle the full-pattern X-ray diffraction analysis. Still, he had difficulty interesting industrial laboratories to employ him. After many polymer crystal structures had been determined with X-ray diffraction in the 1980s, the interest on industrial samples had waned. It was too often that the crystallinity or average orientation function of the samples did not yield the expected link to sample performance, as indicated in the legend to Figure 41. Ultimately, Dr. Fu found a better opportunity as software project engineer in the Silicon Valley in California.

Initially, a *direct view of the crystals of polymers* in low magnification was possible in our laboratory with interference microscopy (see page B-4 of Appendix B, Figure 30 in Chapter 7, and in Chapter 8 Figure 69 and page 8-2). Higher resolution became possible with electron microscopy. My own knowledge of electron microscopy went back to my time in Berlin (see page 4-25, and also 5-12). In our own laboratory at RPI we kept a simple electron microscope (see pages 8-24–25). The higher resolution pictures were then taken at the institute facilities at RPI, UTK, and ORNL Part of our work was also done by Professor Bu at Fudan University (see Figure 43 on the next page).

The low-resolution facility was replaced by buying jointly with ORNL an atomic force microscope, AFM (also called a scanning force microscopy, SFM). Dr. Annis of ORNL (see Figure 40) set up the AFM, laboratory with Dr. Reffner (see page 9-17). The analyses included the microphase structure of block copolymers, the morphology of folded-chain and extended-chain crystals of polyethylenes, the surface structure of the small-molecule model-compounds of tetraalkylammonium salts, and poly(lactic acid) (Appendix A, # 327, 336, 337, 388, 547). In 1998 the AFM could be coupled with a microcalorimeter tip, as shown in Figure 42A. This μTA allows a first step toward the thermal analysis of specific areas of an observed microphase. The tip consists of an exposed Pt-wire core with an area of ≈0.04 μm^2. By finding the DC power needed to keep the

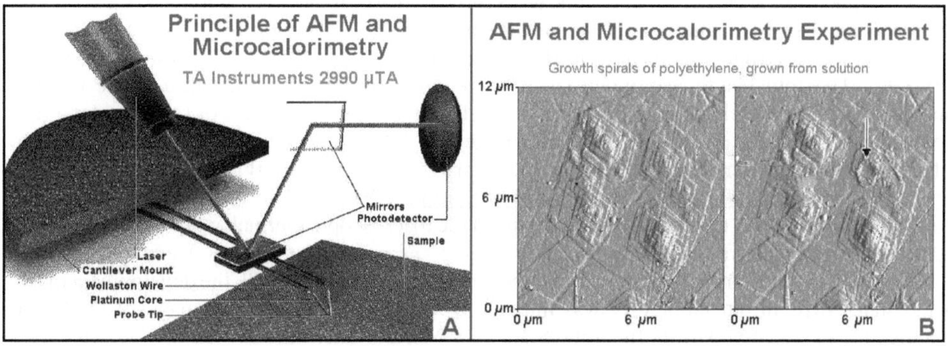

Figure 42A,B *Microcalorimetry and AFM experiment on polyethylene (↓ indicates heated area).*

temperature of the Pt wire constant while scanning the sample with a constant contact force, a measure of the thermal conductivity can be obtained. In addition, the usual image of the topography is gained from the changing vertical positions of the probe tip during scanning. Figure 42B illustrates the effect of local melting of a part in a growth spiral of folded-chain crystals (compare to Figure 30 of Chapter 7).

Single-molecules single-crystals were first analyzed with electron microscopy at RPI by Professor Bu (see Chapter 8, Figure 101 and page 8-123). He continued this work in China (Appendix B, page B-22). Figure 43 shows two of Professor Bu's electron micrographs of single-molecule particles of polystyrene. On the left, a broad distribution of atactic polystyrene was dissolved in benzene to 2×10^{-4} wt.-%. This concentration so low that the dissolved molecules are fully separated from each other. The solution was then spread drop by drop on a water-filled shallow trough where the insoluble benzene evaporated. The floating, glassy polystyrene spheres

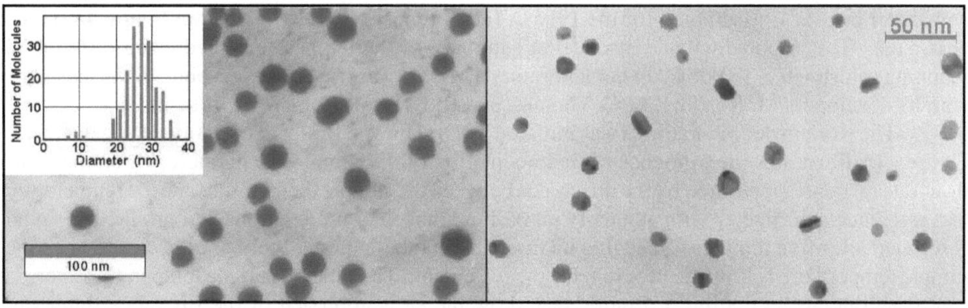

Figure 43 *Single-molecule particles of polystyrene. Left figure, atactic, amorphous, glassy spheres. Right, isotactic single-molecule single-crystals (Appendix A, publications #300 and 374).*

could then be transferred to an electron microscope grid for imaging. The distribution of the diameters of the spheres shown in the left figure matched the distribution measured by size-exclusion chromatography. In another preparation, glassy, isotactic polystyrene spheres were produced in the same fashion, but in this case, the isotactic particles could be crystallized on the carbon support of the electron microscope grids by heating to 448 K. The right figure illustrates the obtained lamellar, 14 nm thick, single-molecule single-crystals. Their morphology, fold length, and crystal structure (from microbeam electron diffraction) matched the results known from bulk crystals. The fold length was much less than required by equilibrium, i.e., a single polymer molecule can also not reach equilibrium on crystallization. Later, particles of single and multiple molecules were produced by electro-spraying methods. At this time, the particles were analyzed by electron microscopy and also by AFM, as in summarized in Appendix B on pages B-36–37.

Detection of the motion in mesophases by solid state NMR was the fourth major microscopic tool added to our effort to understand the details of the macroscopic thermal analysis of the ATHAS data. Its importance was recognized already when planning my last sabbatic leave (pages 8-109, 110). After the return from the sabbatic and the move to Knoxville in 1988, the development of ^{13}C, solid-state NMR in our laboratory was initiated by Dr. Jinlong Cheng (see pages 9-15–16 and Figures 17, 40, and 46). His thesis 'Solid State ^{13}C NMR and Thermal Analysis of Conformational Motion and Disorder in Small and Large Molecules' and later his postdoctoral work, as well as further cooperation during his first employment with RayChem Co. in Menlo Park, CA, contributed to the analysis of many of the molecules looked at during this time period (see page 8-24, Appendix B, page B-20, and Appendix A). His work was continued by Dr. Wei Chen (PhD thesis 1992–1996: 'Characterization of the Thermotropic Mesophases by Thermal Analysis and Solid State ^{13}C NMR') and subsequent work as postdoctoral fellow (until 1999, see also pages 8-24 and Figures 45 and 46, below). During the analyses in his thesis work and the extensive review of the earlier literature, we came up in 1999 with *a first definition of nanophases*, published in a review paper, listed as #440 in Appendix A: "In nanophases, the opposing surfaces of a phase area are sufficiently close to interact." This allowed later to complete the overall description of materials, starting from molecules, going to phase types, and ending with phase sizes (see the pages 8-66–67 and 10–4).

These four modern experimental methods highlighted in the previous paragraphs [Molecular dynamics simulations (page 9-42–43), full-pattern X-ray diffraction (page 9-43–44), direct view of the crystals (page 9-44–46), and solid-state NMR] gave *new dimensions to the ATHAS effort*. It led to a better interpretation of the quantitative thermal analysis. To advertise the unique progress, a new cover page was assembled for the 1995 ATHAS report. It is reproduced in Figure 44 on the next page. This report included also a first summary of the temperature-modulated differential scanning calorimetry, TMDSC. In our laboratory TMDSC was used in 1992, soon after discussions with its developers M. Reading, L. C. Thomas, and R. L. Danley of TA Instruments, Inc.

The work in our laboratories was initiated by Drs. Jin and Boller, as mentioned on page 9-15. Figure 44A illustrates the influence of the modulation of the sample temperature, T_s, on the heat-flow-rate response (measured by the difference between the sample temperature and the temperature of a reference calorimeter with an empty pan). For sinusoidal modulations, a reproducible signal is reached after five minutes. Speeding up modulation takes a larger number of cycles to have the sample temperature follow the program for measurement. The plots of sample temperatures versus heat-flow rate at the bottom of Figure 44A show the slow approach of the Lissajous figures to reproducibility. Thereafter, the data are usable for quantitative TMDSC (see also page 9-50, below).

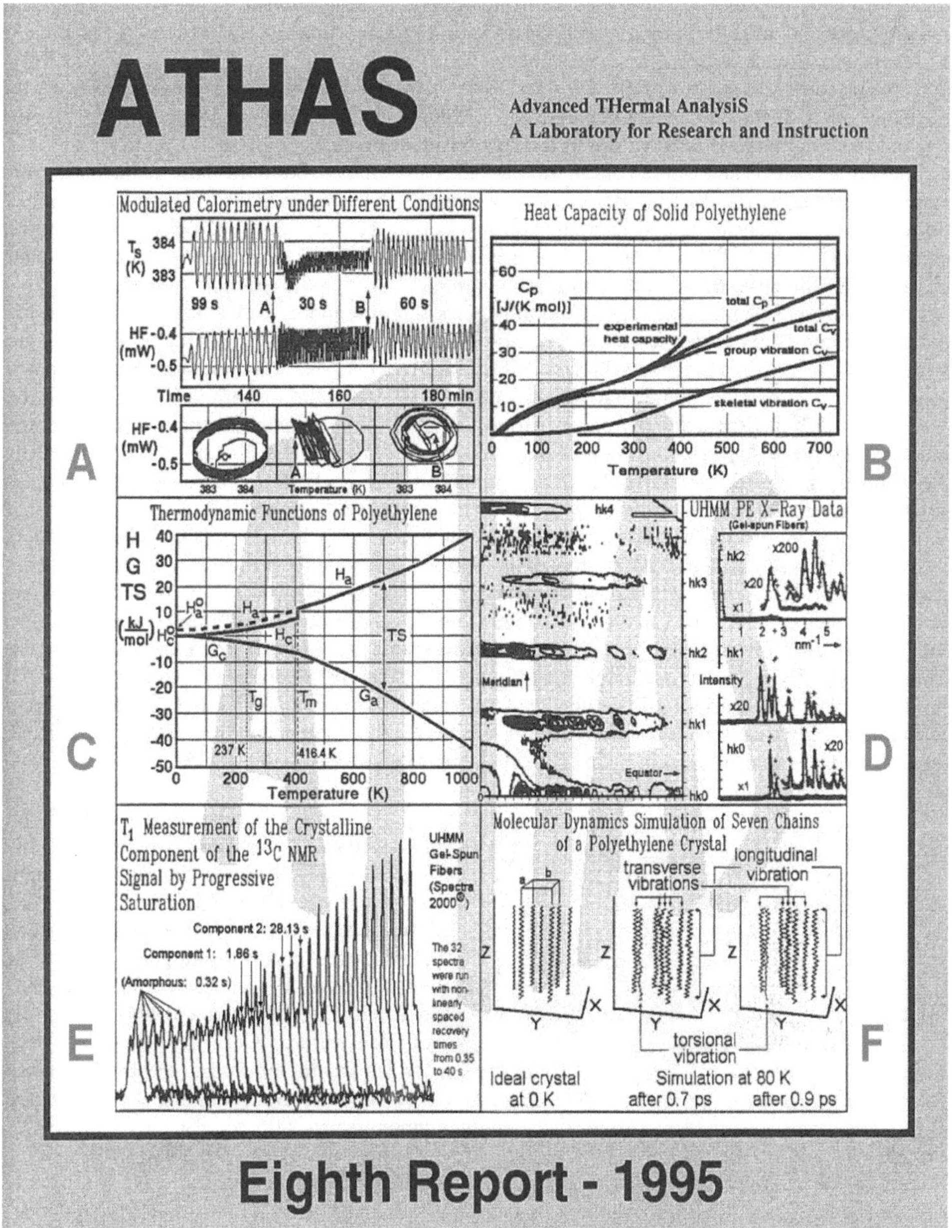

Figure 44 *Title page of the last printed ATHAS report. Later reports were posted on the internet. Earlier title pages of the ATHAS reports are reproduced in Figure 19 and Chapter 8, Figure 69.*

Figures 44B and C are copies of standard DSC results of heat capacity of crystalline polyethylene and the three integral functions derived from heat capacity, as detailed in Figure 4 and 5 of Appendix B with the appropriate references. The modulation in TMDSC, shown in Figure 44A, adds the capability to standard DSC to establish the degree of reversibility in the time scale set by the modulation. If standard and modulated data are identical, one can conclude reversibility. If not, one has to study the time dependence of the absorption or evolution of heat.

Figure 44D is an example of the full pattern X-ray analysis for ultra-high-molar-mass, gel-spun polyethylene fibers. The main deviations between the calculated diffraction pattern for orthorhombic crystals and the measured pattern is seen in the meridional 002 peak (and not in the equatorial *hk*0 peaks). This accounts for the presence of an intermediate phase of parallel molecular chains with little or no lateral order, a structure similar as seen in the structure of mesophases.

Figure 44E displays the varying mobility of the three components discovered from the X-ray pattern in Figure 44D as measured by ^{13}C NMR. There is a small fraction of an amorphous phase with the fastest recovery time (0.32 s). The intermediate, partially ordered phase, named 'component 1,' recovers somewhat slower (1.86 s). Finally, the orthorhombic crystal phase, 'component 2,' has a much longer recovery time (28.13 s). According to low-angle X-ray data, the intermediate phase has a longest extension in the chain direction and is dominant in setting the mechanical properties, as was also shown in Figures 41 for PET. (Links to the publications on these fibers are given in Appendix B, page B-20.)

Figure 44 F, finally, illustrates the molecular dynamics of the skeletal vibrations of a polyethylene crystal, as seen from computer simulation. The drawing of the chains after 0.7 and 0.9 ps are similar to those of the title page of the 1991 ATHAS report in Figure 19. More details are given on pages 9-42–43 and can also be gained from Appendix B, Figure 17 and pages B-18–19.

A picture of *the coworkers at UTK* with their spouses at our 1996 annual dinner at the Mandarin House in Knoxville is shown as Figure 45. Since the majority of the members of the group were of Chinese origin, and the choice of the restaurant was by majority decision, this Chinese restaurant was frequently our common meeting place. Earlier experimentation with a typical steak house found less approval. Chinese are not as eager to experiment when it comes to foreign cuisine as European and (some) American visitors.

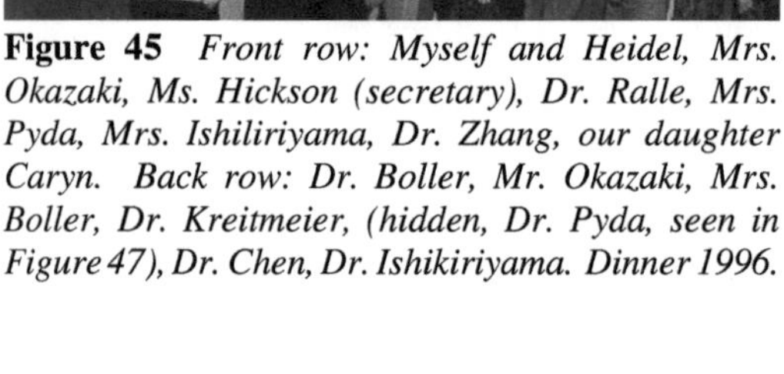

Figure 45 *Front row: Myself and Heidel, Mrs. Okazaki, Ms. Hickson (secretary), Dr. Ralle, Mrs. Pyda, Mrs. Ishiliriyama, Dr. Zhang, our daughter Caryn. Back row: Dr. Boller, Mr. Okazaki, Mrs. Boller, Dr. Kreitmeier, (hidden, Dr. Pyda, seen in Figure 47), Dr. Chen, Dr. Ishikiriyama. Dinner 1996.*

The new coworkers in Figure 45 are: Dr. Andreas Boller from Switzerland (PhD student 1992–1996, postdoctoral fellow until 1997, see also pages 8-24, 9-15, and 9-46), Mr. Iwao Okazaki, (1994–1996 student on leave with support from Toray Industries, Otsu, Japan; the work later earned him a PhD in Japan), Dr. Kreitmeier (1994–1996 Univ. of Regensburg, Germany, see also page 9-43), Dr. Marek Pyda, (1994, educated in Poland, he stayed until the closing of the laboratory in 2006. see also pages 8-24, 9-15, 49 and 10-5–6), Dr. Kazuhiko Ishikiriyama, (1995–1996 on a one-year leave from Toray Research Co., Otsu, Japan), and two Chinese graduate students, Mr. Ge Zhang who continued our work on proteins (PhD 1994–1996 and then as postdoctoral fellow 1996–1997, see also page 9-15), and Mr. Wei Chen (see pages 8-24 and 9-46). Besides the annual restaurant dinners, all coworkers were occasionally invited to our house. By now, the group was so large that we scheduled an open house, partitioning the group. Figure 46 is a photo on such an occasion (compare to Chapter 8, Figure 16). The group shown includes Ms. Hong (a prospective graduate student, left of Heidel), Dr. Maria Mucha (on the other side of Heidel, with son Jarek in back. Dr. Mucha was from 1973 to 1974 a postdoctoral coworker and, in 1992, a visiting professor from the Technical University in Łódź Poland, (see also page 8-60). Also shown in front is Professor Nobuyuki Tanaka (1991–1992, visiting from Gunma University, Japan). The others are graduate students (from left to right) Dr. Fu (elevated, see pages 9-43–44), and Drs. Cheng (center middle), Jin, and Liang (above), and Xenopoulos (far right, with friend on his right, Dr. and Mrs. Chen and Mrs. Liang, in front). The students were introduced with Figures 40 or 45.

Figure 46 *One of two groups of UTK and ORNL coworkers at an open house in 1992 with some of the missing students in Figures 40 and 45, shown above.*

A final group photo, also taken at the Mandarin House, is reproduced in Figure 47. It shows in front (center) Dr. Jeongihm Pak, from Korea (MS and PhD 1996–2001, followed by a postdoctoral period until 2004) and the Postdoctoral Research Associates: Dr. Wenbing Hu (2000–2001, now at the University of Nanjing, China), Dr. Alexander Buzin (2000–2002, Russian Academy of Sciences, Moscow), Dr. Pawel Kamasa (2000–2001, Research Institute for Solid State Physics and Optics Budapest, Hungary), Dr. Marek Pyda (see also page 9-48). In the intermediate years, four additional international postdoctoral fellows worked in our laboratory: Dr. Yon Ku Kwon, (1996–1999, Department of Polymer Science and Engineering, Inha University, Korea). Dr. Il-Kwon Moon (1997–1999, from the NanoSystem Laboratory, Suwon, South Korea), Dr. Renè Androsch (1997–1999, from the University of Halle, Germany), and Dr. Maria Laura Di Lorenzo (1998, from the Istituto di Ricerca e Tecnologia delle Materie Plastiche, Naples, Italy). Overall, the group of 57 postdoctoral coworkers listed in Appendix A consisted of a most international group. From the first postdoctoral fellow (1961, Dr. E. A. James, see page 7-6), to the last (2006, Dr. Wulin Qiu, see page 10-5), this group represented 15 different countries and supported my work with ≈112 person-years. During the same time period, my graduate students spent close to 200 person-years on their MS and PhD theses (see Appendix A, Part E).

Figure 47 *The UTK group at the Mandarin House in 2000. In the center Mrs. Pak, my last graduate student. In back, left to right, Drs. Hu, Buzin, Kamasa, Pyda, and myself.*

This changing group of outstanding coworkers was involved in *the main progress in calorimetry in the 1990s, the temperature modulation* of DSC, the TMDSC. This is the main topic in Section F of Appendix B. Many firsts in the field are reported and explained with 35 graphs.

Section F has five parts. From page B-24 on, the new instrumentation and the mastery of temperature modulation and data deconvolution is dealt with (Part a). The increased precision of heat capacity measurements and information on reversibility were the main findings.

Parts b and c, (starting on pages B-25 and B-29) bring a summary of the experience gained from reversible heat capacity measurements and their separation from irreversible latent-heat effects. They led first to a better understanding of the glass transitions and then, the order-disorder transitions, involving crystallizations and mesophase formation, as well as isotropization. In many cases, an intermediate phase was discovered and could be studied with the auxiliary measuring techniques added to our tools, as described on pages 9-42–46.

One of the intermediate phases is the rigid-amorphous phase (or fraction), RAF (Part d, from page B-32). It is a nanophase with a separate glass transition. In case it is partially oriented, it shows a latent heat of disordering on isotropization, as found in the fibers of PET (pages 9-43–44) and polyethylene (page 9-48). It was observed first by heat capacity measurements in the early 1980s (Appendix A, publications #170, 219).

In the final Part e (from page B-34), the expansion of the information gathering for the data bank is described as it applies to the newly analyzed biological macromolecules and precisely structured copolymers. The latter lead to Section G, describing the understanding of the important concept of decoupling of molecular chain segments, also mentioned in Chapter 10 (see page 10-4).

Finally, research into single-molecule single-crystals, described on pages B-22–23 and the use of microcalorimetry on very small samples are mentioned and summarized on pages B-36–37 (see also page 9-45–46 and Figure 43). This direct calorimetric information on small and metastable systems is helped by using fast measurements. This progress was predicted for the 21st century (Appendix A, publication #450), based on our early results (Appendix A, # 34, 81, and 259). A description of super-fast calorimetry is also given in Section G of Appendix B.

Looking at *our multi-directional broadening of the approach to thermal properties*, one can see that this 'systems approach' of working with competent, young associates who have good basic education in the specific fields was very efficient. It contrasts the often chosen path of attempting to create new interdisciplinary investigation directions by establishing a 'research center' with a leader to direct senior researchers already established in the basic fields.[1]

In our approach, the younger associates who did the experimental work were under my direction. Only a second step of cooperation involved advice from the established experts of ORNL, covering the new experimental techniques and the interpretation of the results. The 'center' approach reverses this chain of information flow. The leader of the center coordinates a new direction and imposes it on independent senior personnel. The latter will then direct the experimental work. By having my postdoctoral coworkers and students report directly to me, it was possible to be more focused and efficient. When warranted, improvements in research direction and personnel could be made quickly. In this mode, almost 250 publications were generated in the last decade of the 20th century. This broadening of the research from single-direction instrumentation and interpretation can also be followed in the sections of Appendix B.

---

[1] As discussed in relation to the polymer science development at RPI (pages 8-85–86).

Into the middle of the just described period of research fell my 65th birthday. This year of 1996 was a specially eventful year. It started with a rare, big snowfall (see also page 9-3 and Figure 4). Snow in this amount is so rare in Knoxville, that I had to take a number of pictures again. Figure 48 is an example of the beautiful view from our patio.

After this snowy beginning, we had to plan for my birthday celebration in May. The travel plans, as so often, did not place me in Knoxville on the given day (see Appendix C). During our visit in China in 1984 we had overcome the superstition of celebrating birthdays before the actual date (see page 8-99–107). At that time, Stephen Cheng had arranged the celebration a few days before by inviting the polymer faculty of the East China Institute. On the actual day, we were traveling to Hefei (Appendix C, page C-28). Since this went well in 1984, we tried the same now. The birthday was celebrated some 10 days earlier at home. This was followed by a number of later parties, small and large, the last being in August!

Figure 48 *Snow in Knoxville in 1996.*

During the birthday we were traveling in Europe and stopped in Berlin. There we went after 43 years to my old student quarters, Am Friedrichshain 10 (see page 4-9). The building looked still as before, just, that sometime after I left, the facade had been newly plastered. There were no bullet holes in the walls anymore, as seen from Figure 49A, but by now it could stand another face lift. Entering the building, we found out that my landlady, Fräulein Schmidt, had married and was now widowed, but still lived in the same apartment. We could take the picture with her from 'my' balcony as Figure 49B. The big difference was not in

Figure 49A *Am Friedrichshain 10* **B** *My 'old' landlady (now Mrs. Prang) with Heidel, 1996.*

the house or the ladies (compare to Chapter 4, Figures 4 and 10). Even the balcony was still the same (top floor, the left half of the center balcony, separated by the glass partition in Figure 49A). It was the green park in the background of Figure 49B, the "Friedrichshain," which in 1953 was still called the 'rubble' mountain (see page 4-15).

The next stop was the Fourth Lähnwitz Seminar, with the topic of 'Temperature-modulated Calorimetry.' It was held on June 4–7 in the "Schloßpark Hotel," in Krugsdorf, NNO of Berlin, near Pasewalk (see also page 9-34). The newly refurbished hotel is depicted in Figure 50. There, the birthday was celebrated once more with many of my colleagues. Since the Lähnwitz Seminar is usually held at the end of May or early in June, it often includes a recognition of my birthday.

Figure 50 *Location of the Lähnwitz Seminar in 1996.*

The biggest party, however, was held at the 11th ICTAC Meeting in August (see Appendix C, page C-47, for other ICTAC meetings see also the pages 8-23, 57, 65, 79, and 9-21). During the opening session, I received the TA Instruments/ICTAC Award for the work described above and in Section F of Appendix B. Figure 51 is the remembrance of the ceremony on August 12.

In addition, Dr. Turi invited many of my past and present coworkers and friends to contribute to the meeting. Before, she had collected a large number of papers from my students and colleagues for a 65th birthday issue, published in the Journal of Thermal Analysis (see page 9-18).

Figure 51 *Opening of the 11th ICTAC with Dr. David Chalmers presenting the TAI-ICTAC Award.*

After our papers were presented, all my students who could attend, came together with the Turis (see Figure 52) for a memorable 65th Birthday Dinner Party in the famous White Dog Café. Figure 53 is a montage of the birthday card, a snapshot of our hotel room with flowers from Dr. and Mrs. Cao who could not attend, and a gift brought from China by Dr. Stephen Cheng. We appreciated the large vase even more after we could imagine his problem on handling it on the long flight from China he had just returned from. We only needed to manage it on the flight home. (The flight was cancelled after boarding because of a thunderstorm in the area. This forced us to stay for the night at the airport hotel—with the vase). The signatures in Figure 53A are: *(Left, top to bottom:)* Larry Judovits (PhD 1985), Joseph Menczel (PD 1979–80, and 1985), Edith Turi, Stephen Cheng (PhD 1985), Ted Davidson (PhD 1967), Janusz Grebowicz (PD 1980–85), Manika Varma-Nair (PD 1988–93), Suk-fai Lau (PhD 1982), Kazuhiko Ishikiriyama (PD 1995–96), A. Boller (PhD 1996), Marek Pyda (PD 1994–06). *(Right, top to bottom:)* Richard Bopp (PhD 1993), Paul Turi, Mike Jaffe (PhD 1967), Mark Altman (PD 1974–76), Bruce Prime (PhD 1968), Ravinder Nair, Wei Chen (PhD 1996), Ge Zhang (PhD 1996), Iwao Okazaki (PD 1994–96).

Figure 52 *'Drs. Edith and Paul Turi.*

Figure 53A *Signatures of the guests.* **53B** *Memorabilia from the 'White Dog' Birthday Party.*

Pictures of some of the attending former and present students at the Café are reproduced in Figures 54A–F. It was an interesting meeting and nice to see so many successful coworkers in one location. These and Figures 17, 40 and 45–47 of Chapter 9 and Figures 1, 10, 16, 29–31, 33B, 93B, 95A, D, E, and 101 of Chapter 8 cover a majority of my students. (For a full list see Appendix A.)

Figure 54A,B *In the White Dog Café: Myself (left)and Heidel (right), with the coworkers Drs. Prime, Davidson, and Judovits, at the 11th ICTAC meeting in 1996 (for signatures see Figure 53A).*

Figure 54C,D *Former coworkers, Drs. Menczel, Bopp, Ishikiriyama, and Lau.*

Figure 54E,F *Drs. Grebowicz, Boller, and Varma-Nair with husband (and Dr. Zhang in back).*

The increase in research activity during the 1990s certainly did not bring a decrease in the travel activity, although the frequent consulting trips to various locations had stopped because of the decrease in polymer research at the DuPont(see Appendix C, page C-4 and also page 8-129). The travel listings in Appendix C show my appearance at many further national and international meetings. One slow change, however, can be noted. The **travel on business was increasingly augmented in the 1990s with travel for the 'the need to see.'** Heidel, whenever possible, would not only make all the often rather complicated travel arrangements, but also accompany me on the longer trips. The time for sightseeing, even when carefully incorporated in business travel, was usually limited. In Europe, extra time was kept to visit relatives and to stay in our favorite vacation spots (see on pages 8-49–50). There were, however so many locations in the rest of the world we liked to see which are not near universities or research laboratories. In this last decade of the 20th century, we increasingly attempted to remedy this. More and more we tried to add side trips, and finally, we also undertook some fully private sightseeing travel, to be briefly summarized below.

Figure 55 *Golf-course meeting at Shuzenji in 1990.*

During this time, <u>*the most common business travel across oceans went to Japan*</u> and its most important cities, Tokyo, Kyoto, Osaka, and Fukuoka became quite familiar to us (see Appendix C for lectures presented). Overall, eight trips were made to Japan. Most impressive were those in 1966 and 1983 (see the pages 8-25–30 and 8-97–99).

Trip three was in November 1990, to give the plenary lecture at the Conference of the 'Association for the Progress of New Chemistry' at the Laforet Shuzenji Golf Resort in the center of the mountainous Izu Hantu peninsula, about 75 mi SW of Tokyo. This was an interesting location. The top of a mountain had been flattened to create the golf course shown in Figure 55. With the beautiful landscaping and views, it was quite impressive and probably expensive. Since I did not have to foot the bill, I cannot tell, but the travel expenses permitted for the extra days were generous. Outside the three days of the Gordon-conference-style meeting (see page 8-16–17) of about 50 attendees, most from industry, I had two days at the Tokyo Palace Hotel and met with Dr. Hidematsu Suzuki (PD 1982–84) for discussion and sightseeing. After one week, I was home again.

For the trips 4–8 in 1992, 1993, 1994, 1996, and 2002 Heidel could join me. Again, details of the 'business parts' of these trips can be found in Appendix C, pages C-41, 42, 45, 49, and 58. The 1992 meeting was held at the Hotel International in Yokohama. This remarkable skyscraper has a sail-like structure, erected over a six-storey base, all overlooking the bay. Many of our international colleagues were invited, so that a good exchange of ideas could take place. Thereafter, Heidel and I went for a lecture to Gunma University and a few days of relaxation with Professor and Mrs. Tanaka (PD 1991–1992). A new experience was our stay at the Hotel Tenbo, a Japanese-style "ryokan" (inn). The management of the hotel took pity on us and arranged for a half-hour private use of the communal bath, otherwise available for the whole floor of the hotel. Trips to the Japan Mushroom Park, Okawa Art Museum, and to Mount Akagi and Haruna closed our visit.

Next was the 1993 meeting to celebrate the founding of the polymer department of Osaka University, 30-years ago. Several of my former coworkers made sure that we could complete the inspection of Kyoto. Figure 56 was taken in the Katsura Imperial detached palace (villa) where access needed permission from the Imperial Household with months-long waiting time (and years for Japanese visitors). We had visited there in 1966, but could not fully enjoy the beauty because of renovations (no water in the ponds!). This time, all looked perfect (too perfect?). Another highlight was the visit of the Midera Temple on the slope of Mount Hiei. Figure 57 shows the view as photographed by Dr. Minori Todoki of the Toray Research Company in Otsu. In the following years he sent two of his coworkers to Knoxville for study (Okazaki and Ishikiriyama, see Figure 45 and Appendix A).

Figure 56 *Tea-house of the Imperial Villa of Katsura, revisited in 1993.*

Our sixth trip, in 1994, was centered around an international conference in Okayama. We stopped in Tokyo, and Hiroshima. Naturally, as before, we had to visit the Peace Memorial Museum (see page 8-99) and were rewarded, as before, with a drive to the Inland Sea (Seto) and a boat ride to the Itsukushima Shrine on "Miyajima," the Shrine island. Since 1996 it is a UNESCO World Heritage Site. Figure 58 is my photo of the shrine's "Torii."

The seventh trip was made shortly after the birthday party in 1996 (see the pages 9-52–54). It centered, as so often before, around a meeting supported by the IUPAC (see page 7-22). In the present case, it was the 14th IUPAC Conference on Chemical Thermodynamics (ICCT-96) in Osaka. We traveled directly to Osaka. After the meeting, we visited with my former coworkers and Dr. Todoki of Toray Research Co. Figure 59 shows an overview of the large garden (133 acres) of Shūgaku-in Imperical Villa. The bridge, Chitose-bashi, on the left of the figure, I had admired

Figure 57 *On a sightseeing trip during the 1993 travel to Japan. On the slope of Mount Hiei, overlooking Otsu and Lake Biwa.*

Figure 58 *View from the Shinto Itsukushima Shrine, Miyajima.*

Figure 59 *The upper gardens of Shūgaku-in Imperical Villa in the eastern suburbs of Kyoto, 1996.*

already 30 years before as one of as the most masterfully designed structures (see Figure 33D in Chapter 8 and page 8-28). This assessment still held in 1996. From the Osaka area we went back to Tokyo on behalf of TA Instruments, Japan, Inc., to participate in a thermal analysis exhibition.

Because of the dominance of adiabatic calorimetry based at the University of Osaka and their successive world-renowned department leaders Professors S. Seki, H. Suga, and T. Matsuo, and the criticism of other Japanese experts, differential scanning calorimetry, DSC, was slow in catching on. Knowing the Japanese work well, and having repeatedly visited the University of Osaka, I was able to show the increased precision in quantitative DSC, and now also point to TMDSC as a means to check on thermodynamic reversibility (see page 9-50). This helped promoting TAI's DSC. The flight back went from Narita, Tokyo's airport, a good distance outside the city.

A final, eighth, visit to Japan was possible in 2002. It went to the third International Meeting on Polymer Crystallization. The meeting was mentioned on pages 8-63 and 10-8 and details are given in Appendix C, on pages C-22, 41, 58 and 64. Most impressive was the nearby Mount Fuji.

Three other transcontinental trips outside of Europe and on 'business' went to South Africa (1992), Australia (1995), and China (2000). All three had a certain ' need to see' component. As always, the 'business component' is detailed in Appendix C (pages C-41, 47, and 56). Naturally, not having set foot on the African continent, we gladly took the opportunity to visit South Africa when it was offered in 1992. *The inviter was the South African Thermal Analysis Society*, SATAS.

We left on October 21. All airlines offered only flights with a layover of about 10 h. While airlines arrange a daily round-trip to Europe with one plane, Europe to South Africa did not seem to leave enough time for a daily turnaround. Perhaps, we had forgotten that the African Continent stretched some 4,900 mi to the south, more than the distance from Atlanta to Munich.

Sitting half a day after a whole night's flight at the airport in London, Frankfurt, or Zürich was not our idea of a pleasant stopover. Just think, how one would feel the following morning in Johannesburg, RSA (Republic of South Africa). Fortunately, we found out that Swissair, one of our favorite airlines (which since 1997 has been reorganized several times), flew only every second day to Johannesburg. This meant, we could stay for one day in Zürich without being charged for two separate flights. They did not fly on the 22nd, so the continuation flight was on the 23rd. We arranged to have a hotel room reserved by our good friend and my colleague in thermal analysis Dr. Hans-Georg Wiedemann (see Appendix A #208, 236, 237, 243, 333), who lived near Zürich. We arrived before noon, checked into the Mövenpick Hotel, and slept well into the late afternoon. Then we met with the Wiedemanns for dinner, and went back to the hotel to sleep again and fully adjust to the six-hour time difference (and all for only a one-day hotel bill!). The next morning and afternoon, we made a boat tour of the Zürichsee, visited the "Kunsthaus Zürich" which had an exhibition of works by Gustav Klimt (1862–1918, Vienna Art Nouveau painter), and inspected the "Zürcher Bahnhofstrasse" with its world-renowned shopping opportunities. This brought us to the evening-flight to Johannesburg, arriving there in the morning of October 24, a Saturday.

In Johannesburg we had time for a city tour, including a view of the huge mountains of the dumps of the "Witwatersrand" gold mines. Three observations gave us lasting memories. First, the streets near our hotel, the Rosebank, were fully lined with blooming Jacaranda trees. Second, most houses were strongly protected with barbed wire fences. Our visit fell into the final years of "Apartheid," and streets were not safe. Third, it gave us a special feeling when going through the crowded streets and finding ourselves to be the only white people.

Figure 60 *The Mzumba Dancers at the Safari Ranch.*

On Sunday, a trip was arranged to tour the capital Pretoria, and the Heida Safari Ranch. With its many monuments, Pretoria gave us an insight into the time of the founding of the Boer Republics in the 19th century and the Second Boer War at the beginning of the 20th century (see page 3-38). The ranch gave an impression of farming in the desert. Figure 60 depicts a lively show of the dances of the native Zulus, produced for the tourists.

The following week was filled with discussions at the Plastic Federation of South Africa and the Atomic Energy Corporation. The AEC was just reorganized from a secret research organization funded by the state, just like Oak Ridge in the 1940s. As such it had the (failed) goal to make the RSA an atomic power. Now it was to become a private company. The aim was to start self-sustaining productions within a few years, making use of the old facilities. The enthusiastic and well-educated research workers were to create novel materials, including polymer products. One of their noteworthy ideas was to make a cheap gasoline-tight polyethylene tank by perfluorinating the inner surface using vapor-deposition of fluorine. A second product was an appropriately cross-linked poly(ethylene oxide), capable of swelling in water and serving as a reservoir to reduce the frequent need to water plants. A third example I remember, was the adding of a hard, metastable diamond layer on the surface of glass objects, also by vapor deposition. All projects seemed feasible and were fun to discuss.

The remaining three days of the week were reserved for a tour of East Transvaal and the Kruger National Park, established in 1898. Besides the scientific content of the visit to the RSA, this was the highlight of the trip. Figure 61 is a copy of a large, framed collage of some of the animals I had photographed. It was displayed in my office at UTK for many years. A 400% magnification helps to better see the originally 8×10 photos in the picture frame. Overall, we tracked the sighting of birds and animals on a check list provided by the Wildlife Safari operator. The trip was led by a well-educated biologist who pointed out each of the species. We marked 28 species of birds, most unknown to us, and 25 animal species, only two of them we did not know. The trip started from Johannesburg with a minibus of seven guests and reached the park at about noon and continued with a four-hour safari into the park with stops for the observation of the wild life. In the evening we reached the Lower Sabie Camp which was surrounded by miles of wilderness for the animals. The elephants, lions and cape buffalos were kept out by a high and sturdy fence, perhaps 1,500' in diameter. In the evening, the compound was closed. Wandering outside the fence on foot was forbidden. When we toured inside the fence at night, it felt as if we were in a cage, being observed by the freely roaming animals on the outside. Their eyes shone in the flashlight of the guide. The next day was filled with a morning and an afternoon safari to different areas of the park, and on the third day the visit was concluded with a fourth ½ day safari and before we went back through East Transvaal to Johannesburg.

Figure 61 *Photos from Kruger National Park, RSA, 1992. The hippopotamus (left center) needs some magnification and the zebra (left bottom) is hard to discern in the chase.*

After the safari we moved to Pretoria, the location for the SATAS Meeting (detailed in Appendix C, page C-41). This was followed by visits and lectures with industrial researchers. Ultimately, we ended up in Cape Town where the organizers had arranged some additional sightseeing, including the city of Cape Town, Table Mountain, Clifton Beach, Stellenbosch Wine Farms, Kirstenbosch Botanical Park, and the waterfront close to the Cape of Good Hope, (almost) the southernmost tip of Africa[1]. On November 7 we started our long flight back from Cape Town via Johannesburg, Zürich, and Atlanta to Knoxville. All together, it was a trip of an exhausting 37 h (this time without a recovery stay at a hotel in Zürich). A total of 13 h of the time was spent on waiting and ground transportation at the various stopovers, mostly in Zürich and Atlanta.

The next 'business trip' went to Australia, a just as long travel of about 24 h in the air and over 30 h total time if one includes the ground travel and waiting at the airports. The major event on this trip was an invitation to give the plenary lecture of the Polymer Division at the 10th Royal Australian Chemical Institute Meeting (RACI), in Adelaide, South Australia. In addition, there were two lectures and three workshop contributions. This made for a quite busy work week, starting on Wednesday evening, September 27, 1995. The meeting continued through the weekend, as is common for this meeting. It ended on Monday afternoon (see Appendix C, page C-47).

---

[1] The Cape of Good Hope is south of Cape Town, forming the tip of the Cape Peninsula which is bordered on the east by the Table Bay. The true southern tip of Africa is Cape Aghulas, about 100 mi southeast of the Cape of Good Hope, too far for us to reach on this afternoon in 1992.

Figure 62 *Honor Garde in Adelaide, Australia, 1995.*

On our arrival we were greeted in front of the Hotel Inter-Continental by a parade of the honor garde of the government of South Australia. The leading horses are shown in Figure 62. They were followed by an impressive band and more than 100 foot soldiers. Just when we got ready to acknowledge the reception from the balcony of our room, we noticed that someone else was honored by the Lady Governor. The Parliament House was just across the street and it was not our reception. Still, we also were well received by colleagues from the University of Adelaide, though not by the hundreds. Professor Jani Matisons and Dr. Peter Folland were our hosts. The staff of TA Instruments Inc. (c/o Waters Australia Pty Ltd), re-enforced from the US, were organizing the Thermal Analysis Workshop.

Figure 63

The rest of the second week was considerably more relaxed. First, we explored the beautiful city of Adelaide, a city with a population of over one million. It was established in 1836 as the oldest municipality in Australia. We met HM Queen Adelaide,[1] the patron of the city, in the Town Hall near our hotel. This special occasion is documented in Figure 63, shown with a special, computer-enhanced sign (the original text was: Smoke Free Area).

After several interesting discussions at the University of Adelaide, the outside funding available for the meeting was spent on an outstanding dinner with our hosts. It was held in one of the finest restaurants NE of Adelaide, the Drumminor in Ridgehaven. Their dinner included Australian wines from their superb cellar. The party is shown in Figure 64 under the gazebo of the restaurant.

Figure 64 *Private conference dinner. The Matisonses on the right, the Follands next to Heidel.*

[1] (1792–1849). Born as Princess "Adelheid von Saxe-Meiningen." Married in 1818 to the Royal Duke Prince William of Clarence. On the death of King Georg IV, William's elder brother, they were crowned in Westminster Abbey in 1831, as King William IV and his Queen Consort.

Figure 65A *Koala at work on his daily eucalypt diet.*

We found it interesting that the population of Adelaide found it important to point out to us that they are not descendants of the criminals deported from England to the prison colonies. Their customs we found typical British, but on discussing these, everyone was eager to prove that they were Australians and not British.

Naturally, we also met the native animals of Australia in zoos, sanctuaries and in the wild. Figure 65A is an example of a 'wild' marsupial, the bear-like Koala seen in the Urimbirra Wildlife Park in Victor Harbor, some 50 mi south of Adelaide. On this trip we also went to the off-shore islands on the Southern Ocean, open to Antarctica (≈3,000 mi south, see Footnote, page 9-85). The distant ice could be felt by a very cold breeze from the south. The island was a breeding ground for penguins and we could see some of the young, waiting in their caves for the feeding.

Another trip with the Matisonses went to the nearby, quaint village of Hahndorf, settled in 1839 by German immigrants, noted by the village architecture and their cuisine. When they had to change the name of the village during WW I to 'Ambleside', they worked on changing it back to Hahndorf as soon as politics allowed it (in 1935). Hahn was the name of the captain of their ship who made sure that they would settle safely in a fertile area (after a harrowing sea voyage).

The last days we were on our own. First, we made a day-trip to one of the most famous wine areas, the Barossa Valley, NE of Adelaide, also a German settlement. In the 1840s, it was called "Neu-Schlesien." Figure 65B is a view of the valley. The tour included visits to about 40 wineries. By the end of the tour we were convinced of the quality of the wines. My particular favorite was the 'Grandfather Port,' by Penfold. We were happy that we did not have to drive home by ourselves.

The last excursion was by plane to Cairns, a tourist town on the Coral Sea. It is a town of 50,000, located 1,500 mi north of Adelaide. The surroundings are defined by the Tableland Mountains, the Great Barrier Reef, a UNESCO World Heritage Site since 1981, a rainforest (see also page 10-17), and extensive sugar plantations. We made it a 2½ day visit. Figure 65C is a view into the Coral Sea, showing a distorted image of the fishes on their way down to the coral reef.

Figure 65B *The view of Barossa valley from Mengler's Hill, near the city of Tanunda, 1995.*

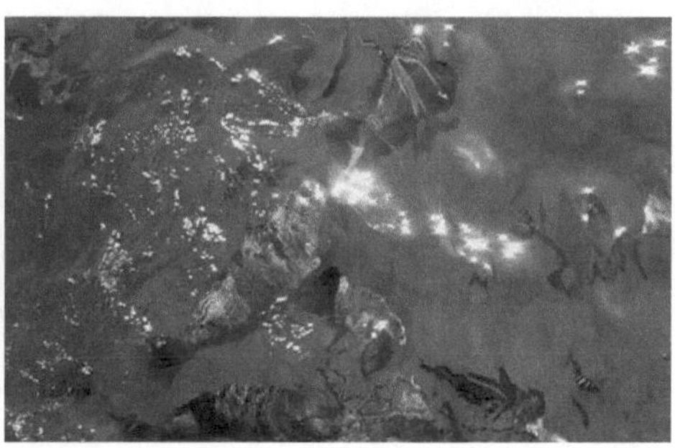

65C *Photograph directed toward the coral reef.*

Our trip home almost ended before it started. The airport taxi was late in getting us to our flight. When we told the driver we would go to Sydney, he hurried us to the national terminal of Cairns. At the counter, we were told, you are late *and* in the wrong place. Your flight goes from the international terminal, about two miles away, on the other side of the airfield.

Since our taxi had left, Heidel tried to convince another incoming taxi that we would be glad to give a generous tip if he could get us out of this emergency. He understood, and went lightning-fast to the other side. When we got to the desk, they said the plane is closed. After explaining that this would mean missing our international flight, the supervisor managed to hold the plane, grabbed part of our luggage, and ran with us to the plane. Before we even sat down, the plane rolled onto the runway. The explanation was simple. When we arranged our flight to Cairns from Adelaide, there was no available seat. The inventive travel agent, however, got permission to put us on an international flight from Japan to Sydney with a brief stop in Cairns to drop off tourists. According to common practice, this plane should not have taken on additional passengers. When we got into the plane, we found mostly sleeping Japanese businessmen on their way from Tokyo to Sydney. On the way, we could admire the coast-line and the Barrier Reef from above. We arrived in Sydney in good time to catch our flight to Los Angeles. It was even possible to buy tax-free 'Grandfather Port.' Dead-tired we arrived in Knoxville, about 30 h later (minus a day, see page 10-16–17).

The final trans-oceanic business trip of the 20th century went to the International Symposium on Polymer Physics, PP'2000, in the city of Huangshan in China. A selection of pictures is given with Figure 66. This was *our second visit to China, 16 years after the first memorable visit* (described on pages 8-99–107). What a difference awaited us! The old Hongqiao airport was overshadowed by the new Pudong airport, built in 1999 (Figure 66A). We arrived in the morning and were greeted by Professor Bu (see Chapter 8, Figure 101, page 8-123 and also 9-44–45). He brought us to the new Radisson Hotel.

This was not the 'old' Shanghai anymore! The city was now a bustling metropolis of a population of 13 million. It had almost doubled in size since we saw it last. Figure 66B shows the growing city from the hotel window. There were no skyscrapers in 1984! Now, the inner city, seen in the distance, looked like Manhattan. Before, many roads were tree-lined and without lights at night; private cars were prohibited. The change was miraculous. One had to search for the 'old Shanghai,' it may be gone tomorrow. Our first inspection was on the afternoon of the arrival day. With help of the concierge we found a taxi with a lady guide who took us on an impressive private tour to compare the old Shanghai we still remembered with new, modern Shanghai.

Figure 66A *Arrivals at the large and super-modern, new Pudong Airport, as it greeted us in 2000.*

Figure 66B *View from the Shanghai Radisson.* **C** *Traditional houses on a waterway of Suzhou.*

D *Tourists traveling on Hangzhou's West Lake.* **E** *Near Mt. Huangshan, the Tangyue Archways.*

F,G *Visit of the 450 sq. miles of the Huangshan Mountain Range (First Mountain under Heaven).*

The first 'business' to be completed was a lecture and discussion at Fudan University which Professor Bu had arranged (see Appendix C, page C-56). Here the progress was also evident, but more important than the improvement on buildings was the open and animated behavior of students and professors. Professor Bu took good care of us for these first two days.

At the weekend, we were to assemble with a few colleagues for several days of sightseeing by small bus on invitation of Dr. Stephen Cheng who also had arranged our previous trip to China and then accompanied us (see also the pages 8-123–124). He now was co-chairman of the international meeting in Huangshan. Professor and Mrs. Bu delivered us by taxi to the fancy Rainbow Hotel, where we checked in.

We did not understand a discussion on check-in, but noted that we seemed to have had no reservation. When we asked Professor Bu to inquire if any of the others to accompany us had checked in. We were told, Dr. Cheng had, but he was not in his room, and the others had not arrived. This did not look right. Heidel insisted we should not unpack. Close to dinner time, Heidel was asking for the manager. He spoke English and we found out that still none of the others had checked in, but also, none had reservations! Also, Dr. Cheng was still not in his room, but he was from Japan (wrong Cheng!). We were not in the right hotel!! What to do? We had no telephone number to call. Finally, Heidel had a grandiose idea. Recalling some difficulty of Professor Bu's with the taxi driver in telling him where to go, she asked: "Is there another Hotel with a name in Chinese sounding similar to 'Rainbow Hotel'?" "Yes, there is, the Hong Qiao State Guest House."

The manager called the Guest House, and yes, we had reservations there, and all the others were already checked in. After paying the hotel bill for ½ day, a taxi was arranged and in 20 min we were in the right place, a beautiful guest house within a large garden area. Before we called, the group had already gone to a restaurant for dinner, but had left no address. Since the Guest House had no restaurant, we had to walk a considerable distance before we found a suitable place to eat and where we could order an acceptable meal (with difficulty). Back by taxi, and after a good night sleep, we were ready to go on our trip by the next morning.

Soon we noted that some of the old, narrow roads were now superhighways. A good number of the peasants had also gotten sufficiently prosperous to built affluent new houses in the countryside. The first stop was the ancient city of Wuxi south of the Yangtze River, with many lakes and canals. This was the city where Stephen's ancestors had lived, while he was born in Shanghai. We walked through several of the beautiful gardens bordering lake Taihu. Most impressive was Jichang garden. At the 28-storey Grand Hotel, we could overlook the quickly growing industrial city.

Next day, we went to nearby Suzhou (Chapter 8, Figure 95E and page 8-105). Figure 66C shows the traditional houses along one of its many canals. Naturally, we also saw several of the over 200 gardens in Suzhou, some dating back more than 2,000 years, and the over 1,000-year-old, seven-storey leaning Yunyan Pagoda and the 9-storey Beisis Pagoda (built in 1162). In the next morning, in front of the Hotel Lidu, we could observe the rush hour traffic of hundreds of bicycles and motor scooters. The industry of Suzhou also was developing rapidly, as did that of Wuxi.

The third city we visited was Hangzhou (see also page 8-107). Figure 66D shows a traditional tourist boat we passed on a tour of West Lake. The skyline of Hangzhou, when observed from the lake, had become more prominent (and hazy) than in 1984 because of the many new tall buildings and the industrial smog. After visiting the 1,600 year-old Lingyin Temple complex with its highly prized cultural relics, we visited my favorite industry, the making of green tea. We were shown the stages from growing the tea plants to the final packaged product. Naturally, we were also offered a taste of their best.

After another night at the State Guest House in Shanghai, we were ready for the main attractions of the trip, the international meeting in Huangshan City (see Appendix C, page C-56), located in southern Anhui province at the foot of the Yellow Mountains (Huangshan Mountain Range, a UNESCO World Heritage Site since 1990). We flew to Huangshan, and before the conference, there was a ½-day trip through Shexian County with its many historic houses and the seven archways dating back to the Ming and Qing Dynasties, illustrated in Figure 66E (1368–1912). After the conference, an overnight's stay was arranged in the Shilin Hotel, in the center of the uniquely shaped mountain range. Figure 66F is one of the many photographs I took. In the morning, I was enticed to get up early while it was still dark and to join the big crowd clambering up the mountain, to see the glorious sunrise. The sunrise was not very spectacular, but the crowd on the narrow path, hit by the first rays of the sun was. Figure 66G gives an impression of the 'view.'

Figure 67 is a final remembrance of the Yellow Mountains. We fell for a quaint tourist offering. We bought a special lock with two keys and closed it onto a steel chain where already hundreds were hanging. In this way, both of our hearts were supposedly locked together unless we both go back up to the mountain, found the lock, opened it with our separate keys, and removed it. Both keys can be seen stored in our album on top of the picture of Heidel turning the second key. This is the second Chinese myth that should help us in our life. According to the proof in Chapter 8, Figure 96B, we should be blessed with long life (see page 8-105), and now, we should also be able to stay together forever. Indeed, we had our 50th wedding-anniversary in 2003 and are looking forward to our 60th in 2013. Since, however, we are not superstitious, we also had special celebrations of the 40th and 55th, just in case.

Figure 67 *Locking our hearts on Huangshan Mt.*

In the evening of the 18th of September we flew back to Shanghai for a day's rest before our return flight to the US. Dr. Bu was awaiting us, and insisted on reimbursing the extra expenses we had due to the wrong assignment of the hotel and also paying for the last night's stay. The next morning, the weather was good and the flights were supposed to be on time. What could go wrong? At the airport there was an enormous crowd. For the last two days the planes were cancelled due to a typhoon over the East China Sea, and now, the backlog of passengers had to be taken care of together with the current passengers to be checked in.

At this time we had begun to purchase three tickets for the two of us for the longer trips, to make travel more comfortable. Heidel could then stretch out over the whole row and sleep, and I could sit sideways, so that I also could fully stretch my legs. Sitting in the closely spaced tourist class seats, frequently, I could not get my legs under the seat in front. With these three tickets, we went to the business class counter which was not as crowded, and offered the three tourist tickets in exchange for two business class tickets. Since tourist class was more in need of extra seats, they made the swap, and we flew without delay and with much more comfort. To our surprise, on arrival in Atlanta, we had a similar problem. The plane was overbooked. Anyone waiting was asked to consider giving up their seat for a generous reimbursement. We gladly 'sold' our third seat reservation, in the two-seat rows of the small planes three seats were of no advantage anyhow.

<u>The question which arose out of what we had seen on this trip to China</u> was, what does the future bring? It became clear that with minor adjustments in the dictatorial rules, China had created a burgeoning economy, relying to a good degree on cooperation of the free world as a customer and supplier of technology. We had seen this system in its infancy, as mentioned on pages 8-103–106.

I remember remarking in 1984: "Why not strengthen the free market initiative by allowing entrepreneurs to keep some of their gains?" The quick answer by our Chinese friends was then: "This is not in accord with the communist system!" To this, I suggested that it should not be difficult for a dictatorship that 'convinced' a country of the benefits of the disastrous 'Great Leap

Forward' in 1957 and the terrible 'Great Proletarian Cultural Revolution' in 1966 (see Footnote 2, page 8-99), to call now, in 1984, for a new socialist development which would make use of the two most important, opposing human traits, laziness and greediness. It should allow enterprising members of the proletariat who produced above the 'norm,' to keep some of the 'profits to society' as a reward for overcoming laziness with greediness. These human traits were discussed earlier when analyzing the 'refugee mentality' on page 5-9. Perhaps, the authorities either were told of my remarks (unlikely) or had similar ideas when assessing the demise of one communist country after the other because of their economic collapse in the late 1980s (more likely).

Such juggling of the economy must be balanced by limiting corruption which, left unchecked, seems to grow more quickly out of dictatorships. Looking at the continuing growth of the Chinese GNP beyond 2000, it should soon exceed the GNP of Europe and the US. In case their leadership can also make progress toward developing more personal freedom and keep the corruption within limits, the rest of the world may see a new leader and nuclear superpower. This simplistic assessment must be tempered by the observation that an economic upturn benefitting a large part of the population weakens the drive toward change. On the other hand, an economic downturn and increasing corruption strengthen the opposition to any political system.

In addition to the travel connected to the 'normal' life of a professor of chemistry with many international connections, we now also undertook **the first vacations for private explorations**. They led us to Cancún (1992, 1994), Alaska (1996), Belize (1997), Norway (1998), and Hawaii (2000). Before, for us vacations were one-week's rests whenever we got close to our preferred vacation spots, described on pages 8-49–50. Now we could afford, and felt it was time, to occasionally arrange for independent travel and to go to places for which we felt a 'need to see' to round out our picture of the world (see also the pages 8-50, 9-55, and 57, as well as Appendix C).

The *trips to Cancún and Belize*, both on the Caribbean coast, were to refresh and extend my knowledge of the Mayas (see pages 8-94–95). On the first trip to Cancún Mexico, state of Quintana Roo, we made a bus tour to revisit the 110 mi distant Chichén Itzá, a UNESCO World Heritage site since 1988. Figure 68 shows the less excavated and restored part of the large site which is the showplace of late Maya culture, centering around the mouth of the "chichén itzá" (sacred well). Cancún itself has several Maya sites strewn in-between the row of tourist hotels. Figures 69A and B are pictures of the Ruinas del Rey, now mainly populated by iguanas and few tourists.

Further exploration on the first trip to Cancún was interrupted by a phone call by our nephew Eberhard, informing us that my elder

Figure 68 *Chichén Itzá.*

Figure 69A,B *The Ruinas del Rey and their iguanas, Cancún 1992.*

brother Frank had suddenly died. As if we had anticipated an interruption of our first vacation, we had taken out a trip-insurance and taken the car to Atlanta for the non-stop charter flight. Because of this insurance, we had no problem arranging for a return flight (in Spanish, of which we do not even have rudimentary knowledge). All was arranged quickly. In the next morning, we boarded a regular commercial flight to Atlanta via Miami, provided by the insurance. There, we picked up our car at the motel, drove to Columbia, and were in time for the funeral described on page 9-40.

After two years, we thought it was time to continue our Cancún vacation, but this time without insurance. In addition to, again, visiting the Ruinas del Rey, we discovered the archaeological area of San Miguelito in the middle of the chain of hotels illustrated by Figure 70.

Figure 70 *Maya site, next to the modern hotels in Cancún, facing the beach on the left.*

Of great interest was then a 30 mi bus trip to the Maya ruins of the walled city of Tulúm, built from 1200 to 1450. Tulúm is the third most visited archeological site in Mexico. It represents a majestic sight along the coastline of the Yucatán peninsula, as shown in Figure 71A. A specially beautiful head of Maya art caught my eye and is captured on Figure 71B. In 2004 we traveled a third time to Cancún, this time to contribute to a conference on materials research (see Appendix C, page C-61).

These and earlier visits, however, had only covered the important archeological Maya sites in Mexico. The area of Mesoamerica occupied by the Mayas, however, was much larger. It extended all the way to the south of the Yucatán peninsula into what is now part of Guatemala, Belize, El Salvador, and also Honduras. While the Mexican sites are mostly post-classic, the southern sites belong to the classic period of the Maya civilization, lasting until about 900 AD.

Figure 71A *Heidel in Tulúm, Quintana Roo.*

71B *Maya God, Tulúm, 1994.*

When we traveled to Belize in 1997, we got to know an interesting country. In 1981 it gained independence (from being the British Honduras Crown Colony), but it remained part of the Commonwealth. The size of Belize is about that of Massachusetts. In front of its Caribbean coast lies the Barrier Reef, 200 mi long, with 450 islets and islands, the second longest reef in the world, and also a UNESCO World Heritage Site since 1996.

Figure 72 *Coastal mangrove swamp in Belize, 1997.*

What does one do in Belize City for somewhat more than a week? It is the largest city in Belize with a population of ≈60,000. First, we explored the city. It is a typical Latin American city with one difference, English is the official language. This makes it easier to get around. Then, there are the Belize river and the coastal swamps. We visited them by boat, as shown in Figure 72. This was followed by an afternoon in the delightful 'New Belize Zoo,' a small private enterprise started in 1991 and featuring over 100 indigenous animals of Belize, with a particular effort to teach awareness of the endangered species. The zoo is located ≈30 mi inland, not far from the capital, Belmopan. Next, we made an 80-mile trip by minibus into the jungle to inspect the Maya site of Xunantunich, remembered with Figure 73A. From the top of the pyramid, one could see the jungle of Guatemala with its disputed border to Belize.

Figure 73A *The 1,400 year old stucco frieze on the Maya pyramid of Xunantunich, 1997.*

73B *View of the Maya city of Tikal in Guatemala with its famous pyramid placed in the background.*

The following day, we flew with a single-engine plane to Santa Elena, Guatemala, to tour the Tikal National Park, a UNESCO World Heritage Site since 1979. The overnight stay was in the Petén Esplendido Hotel on Lago Petén Itzá. Figure 73B does not have the commonly photographed Pyramid of the Grand Jaguar in the foreground, but the interesting ruins of the ancient Maya city.

Two more tours into the jungle were undertaken in Belize. One by boat, to "Lamanai" (submerged crocodile), shown in Figure 73C. Indeed, we saw crocodiles. The other, by minibus to "Altun Ha," is described in Figure 74. It ended our Belize vacation.

Figure 73C *Ruins at Lamanai.*

Figure 74A,B *Pyramids at Altun Ha. They are reproduced on the label of the best local beer.*

The next two trips went *as far north as one could go in North America and Europe*. The philosophy behind planning these trips was to get as close as possible to the cold Arctic Ocean without extraordinary efforts. For us, the shores of the Beaufort Sea at Prudhoe Bay, AK, and the Barents Sea at the "Nordkapp" (North Cape) in Norway served this purpose.

On July 13, 1996, we boarded the m/v Crown Princess in Vancouver, Canada, for a seven-day trip along the "Inside Passage" to Seward, AK, located a short bus-ride from Anchorage, AK. *It had taken over 40 years until we again boarded an oceangoing ship*. The first experience on the "Gripsholm" was too rough a crossing of the Atlantic to make us fans of ocean-going ships (see Chapter 5, Figures 36A,B and page 5-21). The Gripsholm had only 18,600 GRT, compared to the huge Crown Princess of 70,000 GRT which was to bring us to Alaska. Figure 75 shows our ship traveling through Glacier Bay. She was rather new, built 1990 in Montfalcone, Italy, and was later renamed for different cruise lines. In 2002 to m/v Arosa Blu, 2004 to m/v AIDAblu, and 2007 to m/v Ocean Village Two. Her speed was comparable to the Gripsholm since the era of the fast Atlantic crossings had past. Most of the big passenger ships were now built for cruising and, accordingly, are relatively slow, but stabilized, and also amazingly maneuverable with side thrusters. They are still powered by diesel engines, but many of the more modern motor vessels, such as the m/v Crown Princess, use electric propulsion to drive the ship (just as in hybrid cars).

The first port of call in Alaska was Ketchikan, a town of little more than 7,000 population. The monstrous ten-storey cruise ship towered over the small houses, and the passengers, ready to investigate the city, flooded the stores. More interesting was an excursion to an Indian village. It

Figure 75 *The cruise ship m/v Crown Princess entering Glacier Bay with us on board in July 1996.*

Figure 76 *Clanhouse replicated in the Totem Bight State Park, Ketchikan, as used by the Tlingit Indians.*

gave an impression of the changes of the ways of life of the native population when going from South America (see pages 8-33–35) to Central America (see the pages 8-94–95 and 9-66–69), to North America (see page 8-95) and now to the homeland of the Inuits. A typical replica of an assembly house is shown in Figure 76. The area is also known for its bald eagles, easily observed, nesting in the tall trees along the coast line.

The next stop was Juneau, the capital of Alaska, a city which can only be reached by sea (and air) with a population of 30,000. Here we could see the Juneau ice field with an area of 1,500 square miles. It feeds, the nearby Mendenhall Glacier. We could have surveyed it by plane, but felt the plane was too small. This was the first big glacier we saw. An interesting attraction was also the widespread salmon fishery.

The third stop was Skagway, AK. It is known from being the gateway to the Klondike of the Alaska Gold Rush fame. The city had ballooned to a population of 20,000 in 1897, 30 years after the purchase of Alaska. Today, it is a lonely town of 700. From here the terrible climb across the 3,202' White Pass had to be undertaken to reach the head-waters of the Yukon River. We repeated this beautifully scenic trek—but by bus.

Next followed a two-day *cruising through Glacier Bay National Park* (see Figure 75), and the even more spectacular Prince William Sound and College Fjord. The wind-direction was favorable, so that the big ship could approach the glaciers and we could observe, photograph, and hear the explosive calving of the glaciers. The final day on board brought us to Seward.

We were among the ≈20 passengers of the total of 1,600 who opted for a flight to Prudhoe Bay from the Anchorage Airport, which is a pleasant three-hour ride by bus from Seward. After the 1,100 mi flight north, we arrived at Deadhorse Airport, the beginning of the Alaska Pipeline. The Princess Line had erected a hotel, the Arctic Caribou Inn, consisting of 75 rooms based on interconnected, mobile-home-like modules. An afternoon excursion was made to see the oil field installation and the countryside. Three interesting observations characterized the stay. First there was a walk to the shore of the Arctic Ocean, as seen in Figure 77. Second, the sun had not set at midnight. Third, the cafeteria-style dinner looked limited and much like home-style, but consisted of a delectable caribou stew.

Figure 77 *Heidel at the Beaufort Sea at Prudhoe Bay, the most northern spot we reached on 07/20/96.*

Figure 78 *Coldfoot 'newspaper,' accommodations in the Arctic Acres Inn, and a visit to the general store.*

In the following morning a two-day bus ride began. We went south, across the Brooks Range, to the center of Alaska and its biggest city, Fairbanks. The trip followed the 414 mi long Dalton Highway, a gravel road, as is common in Alaska. Dalton Highway goes largely parallel to the pipe line. The single bus driver and guide had to stop every hour to clean the windows from the mud and dust of the road. At each stop we could have a quick walk around the bus and inspect the huge number of animals within sight—a dense cloud of very hungry mosquitoes. Otherwise, the fauna made a good job of hiding. We saw no member of the 'biggest caribou herd in the Arctic tundra.' After crossing the Arctic Divide (4,800'), we arrived in Coldfoot with a population of 13. It was described in the itinerary as 'a basic truck stop with some accommodations of modular construction of basic nature, though clean and warm' (as in Deadhorse). Figure 78 represents the page of Heidel's Alaska travel album dealing with Coldfoot, AK. The general store was the biggest building in town, open for business by demand only.

Continuing the trip, we crossed the Arctic Circle with a proper Champagne ceremony and stopped at the banks of the wide Yukon River. After another 9 h ride we made it back to civilization at the Fairbanks Princess Hotel for a two-night stay enlivened by sightseeing during the day. Just outside Fairbanks, the gravel road had changed to blacktop and the pipeline disappeared, bypassing the city.

In Fairbanks, we took a sternwheeler trip along the Tanana (a tributary to the Yukon) and the Chena Rivers with many historic native and early settler's homes and fishing installations. Naturally, we also visited one of the few still operating gold mines to pan for gold. Befittingly, the mine was called 'El Dorado.' I am sure the sand we were given to work on was spiked with gold at the value of about half the cost of admission. Everyone found some gold, and special golden chains and capsules were for sale to display the ½-hour effort.

Two days were then spent on traveling with the Alaska Railroad using the Midnight Sun Express with two-storey 'Ultra Dome' observation cars, built especially for the Princess Line. Via Denali National Park, we made our rail-journey to Anchorage. Again, the advertised animals were scarce, even in the National Park. The countryside was impressive, although Mount McKinley hid from view. We stayed another two days in Anchorage before flying back to Knoxville on July 27.

This Alaska trip was certainly a great experience. It was well planned and executed. For us, it also removed our fear of ocean liners, so that we could follow this trip with other enjoyable ocean journeys. These first travels of ours to a very sparsely settled tundra, snow-covered mountains, and wooded wilderness were impressive, but we also saw that some 130 years earlier, when Alaska was purchased from Russia, one might complain that "it contained nothing of value" and "is a frozen wilderness" (quotation from the New York Tribune of the period). Comments like this gave rise to call the purchase 'Seward's Folly' or'Seward's Icebox.' We could observe ourselves, there are still signs of a frontier in Alaska. A nice characterization came from our lady guide on the drive from Prudhoe Bay. When asked to comment on the surplus of the male population, she commented: "The odds to find a husband are good, but the goods are odd." A frontiersman seemingly is not a civilized family man.

The value of Alaska has changed with the 19th-century gold rush and the 20th-century oil finds. The purchase was a good investment, although for the small US government budget, the price was steep. Recently, we heard that in Soviet school books it was stated that the 'US-enemy has stolen the Russian Alaska.' It is, however, easy to verify the check of Figure 79 and the other documents of sale.[1] Surprisingly by many, the opinion that it was a 99-year lease and no sale is still held today. For example, stated by Prof. Igor Panarin in an interview with Izvestia (11/24/2008). He is of the Russian Foreign Ministry's academy for future diplomats (formerly a KGB analyst).

Figure 79 *The document the United States government used to purchase Alaska from Russia in the 19th century. The source of the public domain internet document is: http://www.ourdocuments.gov*

[1] The purchase was initiated by Tsar Alexander II of Russia and negotiated by Edouard de Stoeckl, Envoy to the US, and William Henry Seward, Sr. (1801–1872),Secretary of State of the US (1861–1869 under the presidents Lincoln and Johnson). The negotiations ended March 30, 1867. The treaty was proclaimed by the US on June 20, 1867. The purchase price was set at US$ 7,200,000, and the check, dated August 1, 1868, is reproduced in Figure 79. No reservations to the cessation of the territory and dominion therein are listed in the Treaty of 03/30/1897 (Article VI).

Having seen the midnight sun in the US, looking at the map in my home office (see Figure 3A), I could see that *Norway reaches just as far north* as Alaska (71°). Thus, in May 1998 we embarked on a trip to the "Nordkapp" (North Cape). A plane brought us to Bergen, Norway, an old city founded in the 10th century. In the 13th century, it became one of the Hanseatic League's[1] trading posts for dried cod from the north. We could still inspect parts of the buildings at the old quayside which reach back to this time (Bryggen, a UNESCO World Heritage Site since 1979). On these walks through the town, Heidel felt a first pain in her right leg, which slowly got worse and made it difficult to walk longer distances. Still, on the next day we boarded our ship for a seven-day voyage to Kirkenes, close to the Russian border. The ship, depicted in Figure 80 was appropriately called the M/S Nordkapp. It was one of the well known "Hurtigruten Coastal Steamers," but not an actual steamer anymore, rather a modern, 404' long motor ship (M/S) of 11,386 GRT, built in 1996. It could carry cars (45), people (690 passengers and tourists, 480 beds) and goods at a speed of 18.5 kn. The ship was decorated with beautiful paintings of coastal scenery by the renowned North Norwegian artist Karl Erik Harr. The multiple duties of the ship called for many stops to load and unload, as shown in Figure 81. Ports with longer stays were Ålesund, Trondheim, Bodø, Svolær, Harstad, Tromsø, Hammerfest, and Honnigsvåg. The days were longer than at home, and above the Arctic Circle, we experienced the midnight sun (although most nights were cloudy). Soon, we were enveloped in the beautifully structured fjords. Further north, they became snow covered, and finally, in the inside passages, glaciers extended down their sides.

Figure 80 *The M/S Nordkapp on its stop in Tromsø.*

Figure 81 *Loading and unloading at one of the stops.*

After seeing the ship's physician, Heidel was asked to go to the hospital to have her leg checked when we arrived in Trondheim. All arrangements were made, so that we could take a taxi and were admitted immediately. At the hospital there was not enough time for detailed testing, but the physician diagnosed a phlebitis in the leg,

---

[1] The Hanseatic League or Hansa was the 13th to 17th century alliance of international trading cities in northern Europe. The base was the town Lübeck, Germany, at the Baltic Sea. The network of alliances of cities within the Hansa fluctuated between 70 and 170 cities. One of these was our hometown Brandenburg (the Old and New Towns joined the League 1314 and 1315, respectively).

caused by the long overseas flight. She remarked that this is a common ailment of tourists coming to the hospital. We should have had more exercise on the plane and been able to elevate our legs when sitting for the 10 hours of the flight. After a careful inspection of the leg, she offered an opinion that the phlebitis was caused by the outer veins and it might not need an overnight hospital stay to check for problems with the deep veins. She commented that this would have meant to take a plane on the following day to catch up with the ship at the next port. She prescribed a surface treatment and rest with an elevated leg, as documented in Figure 82. In addition, in case there was no improvement, we were to have a more detailed check at the next longer stop on our trip, which happened to be Berlin, about one week later.

Figure 82 *Heidel with elevated legs skipping the walk through the Museum of the Tromsø University.*

After having crossed the Arctic Circle with the customary celebration and certification on May 28, my birthday, we approached Bodø, a trade center for the fish caught at the Norwegian coast. Figure 83 illustrates the big drying racks for cod to produce stockfish, still a main trade item of the area, shipped through Bergen (see also page 9-73).

Figure 83 *Drying cod at Svolvær harbor.*

The captain, proud of his ship with the most modern GPS navigation, up-to date diesel engines, and side thrusters, showed off the bridge and engine room to the passengers. After passing the breathtaking Lofoten Wall mountain range, the captain illustrated the ship's maneuverability. He entered an impressive, narrow, dead-end fjord. Just before reaching the end of the fjord, he made a U-turn, with no more than three feet to spare on bow and stern. Figure 84 pictures the critical moment of the turn. We were impressed and felt safe on the ship.

Figure 84 *Maneuvering the U-turn.*

Much later I found an Associated Press news item from Oslo, Norway, dated Wed., Jan. 31, 2007: "A Norwegian cruise ship carrying nearly 300 passengers, including 119 Americans, ran aground on a remote Antarctic island and damaged its hull before getting free of rocks, officials said Wednesday. No one was injured. The M/S Nordkapp got off the rocks under its own steam and sought shelter in a nearby harbor, where the 294 passengers were being transferred to a sister ship as a precaution, said Hanne K. Kristiansen, a spokeswoman for Norwegian Coastal Voyage. She said there was no danger to those aboard or to the Nordkapp after the incident in the Southern Ocean. Another Norwegian Coastal Vessel, the M/S Nordnorge, sailed into Walker Bay and the passengers were being transferred onto it in small boats usually used for sightseeing. The Nordnorge will then take them to Ushuaia, Argentina, a roughly 40-hour trip." "We are having a fine time. In fact, it is very nice," Norwegian passenger Terje Johansen told the Associated Press from the Nordkapp earlier in the day. He said passengers initially were nervous when the ship ran aground, but quickly understood that there was no danger. "There is a little bit of waves, so they are waiting with the transfer. Right now, I'm in my cabin trying to get a little sleep," he said by ship's telephone.

Figure 85 *Statue of Roald Amundsen.*

My thoughts on reading the story were that, maybe, the navigation system was not so good in the Southern Ocean? The reaction of the Norwegian was also typical. Comments from American passengers might have been different from those by Johansen.

The next, longer layover was in Tromsø, the 'Arctic Ocean City' (see Figure 80). Many polar expeditions set off from Tromsø. Figure 85 is my photograph of the statue of Roald Amundsen (1872–1928), the first person who reached both, the South (1911) and North Pole (1926).[1] He started his career by succeeding to traverse the North-west Passage in 1903.

Polar expeditions were Vati's interest (see page 1-25), so that I also was familiar with them. Before we left Norway from Oslo, I could learn more about polar investigations from a visit of the "Fram" Museum, displaying Amundsen's ship used for the Antarctic exploration. The ship's name translates to 'Forward.' The Fram was built 1893 for Nansen[2] to withstand being frozen into the ice and drift across the North Pole (unsuccessfully). As such, it is the strongest wooden ship ever built.

---

[1] Robert E. Peary claimed to have reached the North Pole (1909) and Richard E. Byrd, by plane (1926). Both claims are questionable. Shortly after Byrd, the pole was definitely reached by Roald Amundsen and, Oscar Wisting (Norwegian explorers), Lincoln Ellsworth (American sponsor), and Umberto Nobile (Italian airship designer and pilot) with 12 other crew members. They crossed the pole in the Italian, 300' long, hydrogen-filled, semi-rigid airship "Norge" starting from Wadsø. The first ships at the pole were the Nautilus (1958) and the Skate (1959), U.S. nuclear submarines.

[2] Fridtjof Nansen (1861–1930) was a well-respected Norwegian Arctic explorer, scientist, and politician (1922 Nobel Peace Prize as High Commissioner of the League of Nations).

During the stop in Tromsø we also visited the University Museum (see Figure 82). It keeps many artifacts of North Norway and the native Sámi people[1] (also called somewhat pejoratively: Lapps). There, I also saw my first example of a Narwhal tusk. A look at the modern Arctic Ocean Cathedral, completed in 1965, and other points of interest were also squeezed into the round trip through the island city with a population of 50,000.

The "Nordkapp" is found at the tip of the island of Magerøya. It is (almost) the northernmost point in Europe. Knivskjellodden, 5 mi further east on the same island reaches about one mile further north, but it is less photogenic than the view from the Nordkapp, shown in Figure 86. From the small fishing port of Honnigsvåg we took a 4 h round trip to the North Cape Plateau with (touristic) exhibitions of reindeer and Sámi[1] camps. North Cape Center is a tourist center, largely built underground into the 1,000' high cliffs. The view was not the greatest, but we were told to have a sunny, clear day at the Nordkapp is equal to winning the lottery.

Figure 86 *View to ENE from the Nordkapp, Knivskjellodden in back.*

The last stop was in Kirkenes from where we took the plane to Oslo, to complete our Norway trip. Heidel and I visited the "Munch Museet," and, as usual, had no problems finding pictures we would have liked to take along (see pages 8-88–89). In the afternoon, I visited by myself the Viking Ship Museum, the Fram Museum (see page 9-75), and the Kon-Tiki Museum (showing the raft Thor Heyerdahl used in his mistaken 1947 'proof' that the Polynesian Islands might have been settled from South America, see also page 10-15). Heidel could during this time rest her leg for the flight to Berlin.

Figure 87 *Penguins in Europe, 53.5° N. Berlin Zoo.*

One more picture in my collection, Figure 87, shows penguins. Are penguins supposedly not to be living in the northern hemisphere? This obviously is not true if one includes the rambunctious penguins in Figure 87. They were photographed in the "Berliner Zoo." We always like to come back to Berlin for sightseeing, to remember our student days, and see old school friends.

This time, however, we made an appointment with Uwe Gottschalk, MD, our friend Inge's son who was the specialist to carefully test Heidel's leg and prescribe a more aggressive cure, so that by the time we got home, things were all right again.

---

[1] The Sámi are the indigenous people of northern Europe. Recent genetic studies indicate that the Sámis are among the first population of Europe, related to Basques and the Catalonians.

Berlin was not the last stop on this trip. While earlier. we often added some short sightseeing days to business trips, this time we had added a week's business to the vacation trip. After the three days in Berlin, we drove to the 5th Lähnwitz Seminar in Kühlungsborn at the Baltic coast, dealing with temperature-modulated calorimetry (see the pages 9-34, 52, and Appendix C, page C-52). In Kühlungsborn, Heidel could also meet with Inge and thank her for her son Uwe's help at the "Krankenhaus im Friedrichshain," one of the premier hospitals in Berlin (which also sent us no bill). In addition, we could on this occasion initiate a smoothing of an old family problem between Inge and her son Uwe. The travel was completed by the flight back from Berlin to Knoxville.

For our last vacation trip in the 20th century we went in the year 2000 to *the Hawaiian Islands*, or better on a trip around the Hawaiian Islands. The islands were discovered by James Cook in 1778 and named the Sandwich Islands (in honor of the 4th Earl of Sandwich, the one being fond of sandwiches). We docked at four of the seven inhabited islands of Hawaii,[1] the southeastern, windward islands of Kaua'i, O'ahu, Māui, and Hawai'i). Hawaii is so far from the centers of human development that for long it was a mystery how some 1,500 years ago it was possible to settle these islands until the feat was duplicated in modern times (see the previous page). The Polynesian settlers came from the Marquesas and Tahiti, some 2,500 mi away. Their early emigration was accomplished in double-hulled canoes.

This trip to Hawaii caused us to undertake future travels to other interesting islands in the South Pacific, the Galapagos Islands, Easter Islands, Samoa, and New Guinea (pages 10-13 and 15–19).

The boat trip started on February 26 with the flight to Honolulu on O'ahu and the boarding of the SS Independence,[2] as shown in Figure 88. As is traditional, we were greeted with beautiful garlands of flowers,

Figure 88 *Being greeted on ship board in Honolulu.*

---

[1] The U.S. State of Hawaii in the center of the North Pacific is an archipelago of 19 islands and atolls, numerous smaller islets and seamounts, extending some 1,500 mi from the Island of Hawai'i, which gave the chain its name, to the Kure Atoll (excluding Midway, which is a US territory). The archipelago was formed consecutively by volcanic activity during the 5 mm/year east-to-west movement of the Pacific plate over a hotspot in the mantle (presently located and active under Hawai'i). With almost 2,000 mi to the next continent, it is the most isolated group of islands.

[2] The ocean liners SS Independence and her sister ship SS Constitution were built in 1951 for the New York–Mediterranean route and cruised at speeds of 26 knots, powered by two steam turbines. They were the first fully air-conditioned liners. The ships sailed fully booked until the 1960s and were featured in movies (An Affair to Remember), and TV shows (I Love Lucy, Magnum P.I.) and moved Princess Grace Kelly and her bridal party to the fairytale wedding in Monaco in 1956. The SS Independence had 20,220 GRT after her refit in 1980, was 682' long, could carry 750 passengers, and was used for 7-day cruises around Hawaii until 2001 (shown in Figures 89 and 91).

Figure 89 *The SS Independence in 2000, her last year of service cruising around the Hawaiian Islands.*

the leis. In Kahului Harbor on Māui, I could photograph the graceful, classical steamer (Figure 89). The ship was not going at trans-Atlantic speeds anymore so that we had a leisurely day at sea (Sunday). The next day was spent on Kaua'i with a brief look far across the waters to Ni'ihau, the small, 'Forbidden Isle' (off-limits except for the owners and invited guests). Kaua'i certainly earned its nickname, the 'Garden Isle,' with its rocky coastline, Fern Grotto, Waimea Canyon, and abundant tropical vegetation. Late in the afternoon, we left for Māui, and arrived there next morning.

Māui, the 'Valley Isle,' was a two-day stop. We docked in Kahului. The days were filled with more activities than one could participate in. We took a sightseeing bus trips with lectures to the Iao Valley Plantation, to the Haleakalā Crater National Park (10,000' high and a UNESCO Biosphere Reserve), to the Māui Ocean Center with its big aquarium and other native marine animal exhibits. In the evening we enjoyed the Old Lahaina Lu'au (except for the poi made from the taro root, brought by the early settlers from Polynesia). We skipped the excursion to the undeveloped tropical paradise of Hana (40 mi along the coastal mountains and beaches), rides to Moloka'i by helicopter, known as the residence of Father Damien, the Belgian priest (1840–1889, sainted in 2009). caring for sufferers of Hansen's disease (leprosy), and enterprises like bicycling down Haleakalā Mountain, kayaking, whale watching, diving, and snorkeling. In all, this was a perfectly planned stop which should have lasted longer. We could easily have spent a week in Māui and continued with the next ship.

The following two days were spent on Hawai'i, the 'Big Island.' The first day we anchored in Hilo and spent the morning at the Tropical Botanical Gardens and the Mauna Loa Macadamia Nut Plantation. The afternoon was used for the Volcanoes National Park, a UNESCO World Heritage Site since 1987. Here we could see two of the most active volcanoes in the world, Mauna Loa (13,680') and Kīlauea (4,100'). When measured from the nearby ocean floor, Mauna Loa is higher (33,360') than Mount Everest (29,030'). We explored the 800' high volcanic cinder hill in the eastern rift zone of Kīlauea, grown after 1983. It has the interesting Hawaiian name "Pu'u'Ō'ō." More recently, a large amount of Kīlauea's magma drains through lava tubes into the ocean. During the night, the ship moved to the east coast, and we had the spectacular sight of the molten lava plunging into the sea. In the morning, we anchored at Kailua-Kona. From the large offering of things to do, we chose to delve into the history of the settling of Hawaii. Figure 90

Figure 90 *Kona coast at Pu'uhonua o Hōnaunau.*

Figure 91 *The SS Independence at the Kona Coast.*

illustrates one of the many sites maintained in the Pu'uhonua o Hōnaunau National Historical Park, the walled 'Place of Refuge.' Until the early 19th century, Hawaiians who broke a "kapu" (one of the ancient laws of the gods) could avoid certain death here. The offender might be absolved by a priest, and freed. Along this coast is also the Kealakekua Bay where Captain Cook was tragically killed on his second landfall in Hawaii in 1779 (having broken a "kapu"?).

Figure 91 shows the coastline with the anchored MS Independence, ready for the ship-to-shore traffic by launch. In the afternoon, I undertook a bridge and engine tour. It was the last year when safety regulations allowed passengers to inspect the ship's bridge and engines. In the engine room, it was apparent that she once was a fast ocean liner, but now was devalued to a leisurely cruise ship. Three of the five boilers were out of commission. Two sufficed to cruise between the islands, although we were told, it occasionally would be nice to outrun an oncoming hurricane. Also, the end of the life of her became obvious. In the bottom of the ship a considerable depth of water was collecting and the engineer commented that without a major overhaul the ship could not pass next year's coastguard inspection. Indeed, it was her last year of service.

The next morning (Saturday) we arrived back in Honolulu, O'ahu (nicknamed the 'Gathering Place'). We began the day with a tour by bus to Pearl Harbor and the Arizona Memorial, a sobering reminder, that this island paradise was not safe from the horrors of World War II.

By noon, our baggage was processed and we could catch our plane home. To the selected Figures 88–91, one must add the ≈180 photographs in Heidel's travel album which helped me in recollecting the extraordinary place names to be learned during this travel to a beautiful island.

These 'vacation trips' laid the seed for my changing from traveling as an internationally active scientist to a world traveler who had slowly developed an agenda of 'need to see' (see page 9-55 and 66). As mentioned above, other Pacific islands were of interest. But also, we had not seen the southern part of South America with its Cordillera of the Andes and Patagonia, Tierra del Fuego, and "Kaap Hoorn" (Cape Horn, see Figure 96, below) and its special animals such as the albatrosses and penguins (not only the ones exported to the north, seen in Figure 87). Also, we had not taken a trip around the world (in 80 days or less?). Finally, in the Mediterranean, the source of Western Civilization, are still many sites for which we have a 'need to see.' For all this, one requires more time than the life of a busy university professor can spare. Despite the professional activities, Chapters 8 and 9 reveal that we did at least a step toward satisfying this 'need to see.'

From all this and other points to be addressed, **the question arose when should one retire**? Already during my 65th birthday party (page 9-52–54) some thought, it was to be a retirement party, others asked about my plans for the retirement. But now, the 70th birthday drew near!

In the 1960s to 70s, the general question of retirement came up occasionally during lunch and dinner conversations while on consulting visits to the DuPont. (Appendix C, pages C-4–5). In the earlier years of these discussions, most research laboratories and universities still had mandatory retirement at age 65. Even earlier retirement has always been in effect for the armed forces, police, firemen, athletes, and pilots (for reasons of decreasing physical and/or mental fitness). It was well known that in these latter cases a first career was to be followed by a second.

The setting of a retirement age, however, is somewhat arbitrary. Everyone seems to have a 'natural retirement age.' Some students retire soon after they finished their degree and do not perform well in their chosen profession. Others retire later, but long before mandatory retirement age. Out of tradition, some were ready to retire at 65, but many also desire to work longer. Three examples of the latter I am familiar with are: Professor Peter Debye (1884–1966, see page 6-36), who gave his second prestigious Baker Lecture at Cornell University at age 77 (see page 7-3; the first was given in 1939 on 'Molecular Structure by [X-ray] Interference Methods'); Professor Paul Harteck (1902–1985, see page 8-12) who retired in 1982 shortly after successfully finishing research and being granted patents on commercial nitrogen fixation using nuclear and solar energy; Professor Hermann F. Mark (1885–1991, see page 7-27) who still gave, what all of us thought, was the best invited lecture at a symposium at Rensselaer in October 1982. At that time he had reached age 97.

The 'Age Discrimination in Employment Act' of 1967, ADEA, opened the discussion of the legality of the forced retirement in the US. By 1978 the ADEA prohibited mandatory retirement in most business sectors, and by 1993 it had also phased out mandatory retirement for tenured university faculties. This left the question of my retirement to me. My Distinguished Scientist position (see page 9-6) was a tenured professorship and had, as such, no age limit. A number of countries, as for example Great Britain[1] and Japan, still had not accepted such laws in 2008. To me, this became clear when I made a longer trip through Japan in 1983 (see the pages 8-97–99 and Appendix C, page C-26). Let me quote the pertinent paragraph about the retirement in Japan which I reported to the Japan Society for the Promotion of Science, the sponsor of my travel: "A surprising observation was the rather strict retirement policy at universities. While at present the generation of polymer scientists which made Japan's greatness in this field is forced to retire rather abruptly at 62 or 63 years of age with little input in further research and education, I am used to see the elite senior scientists of other countries to contribute into their 80s. The losses to science and education are obvious, the productive period of top scientists may be shortened by 1/3 or more. It should be possible to find ways to extend the active life of scientists without hindering the development of younger researchers." I received no comments in return (may be, nobody read the report).

To change the mandatory retirement to a different age, abolishing it completely, or finding physical and mental tests to find an employment limit, naturally, has its own problem which I do not want to discuss. The thoughts about my own retirement age are expounded next.

A number of *key questions had to be answered to find a solution* for a retirement age which is fair to the scientist himself, students, institutions, state, and country. One may ask: (A) What should I do after retirement? (B) Is my financial security maintained? (C) Am I still able to work creatively and teach effectively? (D) Am I causing harm to the students, institution, and profession by requiring financial resources and blocking young, talented scientists from an earlier start? These four questions, and possibly several more, needed to be addressed to fix my retirement age.

Question (A) was the one I usually asked in the early discussions at the DuPont, mentioned above. For colleagues who had a clear picture of wanting to start a new life after retirement from

---

[1] See for example: P. Lawrence, 'Retiring retirement,' in *Nature*, **453**, 588–590 (2008), where a case is made to end mandatory retirement worldwide for scientists, with many example cases and a comparison to the US and Australia, where this has been accomplished long before.

science and follow interests or hobbies they had already, the question could be answered as soon as point (B) was resolved. When financial security has been reached, a new direction can be taken easily. The new career may be the start of a business, changing into administration, run for political office, taking a chance in writing, music, painting, or even traveling. But often, the answer I heard was just to take it easy, go fishing, play golf, etc. This would in the long run not be satisfying.

For me, the key interest had always been science and teaching. On the other hand, I also developed, together with Heidel, a 'need to see' and understand the world. Still, I do not want to underestimate the fun I had earlier with collecting coins (see on the pages 1–2, 8-35 and 49). After moving to Knoxville, my collection has never been brought home from the safe-deposit box in the bank. This means, I am still waiting to have enough time to update and improve the coin collection with new acquisitions, computer listings, digital pictures, etc. My first retirement, at age 57, saw only a shift in direction of research and teaching. It resulted in a broadening of my research direction, and a reduction of the teaching to graduate subjects, as described in this Chapter 9.

The pension I collected from RPI supported this second, slightly different career at UTK and ORNL, despite that the new position was connected to a better salary. The pension allowed more joint traveling with Heidel and indulging in my hobby of fancier cars (see page 8-38–40, 9-10–11 and 83). The financial security was also improved by our lifestyle. After the initial needs were satisfied, we always limited our standard of living to less than permitted by the salary level attained. The only credits we ever needed were for our emigration, the first new car, and our mortgages. The first two questions about interests after retirement and financial security, thus, were of no concern.

To me, question three (C) had a surprising, three-tiered answer. After 1994, no new graduate student knocked at my door to possibly choose me as their research director. After some inquiry, I found that the students were afraid I would retire and leave them with an incomplete thesis (despite the fact that I always had sufficient funding and converted all teaching assistants to research assistants as soon as possible). Mrs. Pak was an exception. She was not qualified to teach because she failed the English language speaking test. When she came to me, I took a gamble and offered her a research assistantship with the goal of a master's degree. (I understood her English and did not want to send her back to Korea.) Five years and two babies later, after an exceptional performance, she had graduated with a PhD degree, and then continued as a postdoctoral associate.

The second tier concerned research support. This was no problem since in addition to the funding connected to my position (see page 9-6), I had consistently received support from NSF.

The third tier was more difficult to resolve. It deals with the changing ability to solve problems and recall facts and names for effective lecturing and teaching. I noted no difficulty with solving research problems. A definite asset was the increasing experience. For lecturing, I noted that the speed to recall names and events was slowing. I never had a good memory for names. This possibly provided one of the reasons why I chose science as a career. Opposite to the belief of many, one does not have to have an excellent memory. Most facts, once understood, can be quickly derived from a small base of information which is best learned by rote early in life. Now, I observed that occasionally the recall of names was slower and when, by chance, the German word in a sentence came to mind first, the corresponding English one was not retrieved instantaneously, or vice versa. This was limiting for discussions and lectures. Not that I couldn't hide or avoid most of these problems, but I knew them. Besides, it required more time and care to prepare for lectures and discussions. My own assessment for question (C) was that the lack of graduate students and what might be called colloquially 'senior moments' were negatives. That I still commanded research support from outside, and had improved in research ideas and solutions were positives.

The question (D) was hardest to answer. Certainly I wanted to make sure to retire early enough so that my productivity was still close to the level it had reached over the years. Nobody was to be able to say: "This is one of those professors who reached a top position and pay, but he is long past his prime. Now he only collects the money and keeps younger scientists from moving up and increasing their productivity as he once could!" Checking my detailed, up-to-date records, the progression of publications written, in steps of 50, followed the approximate time line: 1957, 1967, 1973, 1981, 1983, 1986, 1990, 1993, 1995, 1998 (Appendix A from #1 to #451).[1] They seemed to show my initial professional growth, but no noticeable slowing by the year 2000. Similarly, according to Appendix C, the annual outside lectures changed from a total of 47 in 1995, to 37 in 1996, 32 in 1997, 35 in 1998, 30 in 1999, and 28 in 2000.[2] Perhaps there is a small decrease over the years, but the travels went to further removed locations with fewer centers of research, and our separate vacation goals were extended.

Overall, it looked that combined with the other three points, perhaps, a partial retirement might be called for. This was also suggested by needing longer for doing things and an increasing aversion to do extraneous duties that could not be avoided. Finally, there also was the wish to do more extensive travel unrelated to 'business,' and to have some leisure time for my hobbies.

Unfortunately, as a Distinguished Scientist, I held an appointment that could not be gradually faded out of. It was only possible to retire fully, and then also lose the research support connected to the position. Under more flexible conditions, a solution would have been to slowly decrease the salary and let an increasing use of the pension and Social Security make up the difference.

Such solution I had proposed often at the early discussions at the DuPont laboratories to my elder colleagues, although I did not get much support for this suggestion. Could I now do much else but to show them wrong? An even more 'ideal solution' would have been to request that the funds freed by the decrease in my salary to be transferred into the portion supporting my research. This would have allowed to have additional coworkers to follow up the many ideas I had in the last years. When I posed this solution, I was told that this was not possible. Federal and state governments were involved in financing my position, and as one knows, they seldom permit flexibility.

Combining these analyses of these questions (A) to (D) led to my decision to retire from the Distinguished Scientist position soon after the end of the 20th century. After discussing this change with Heidel, I informed the Department Chairman in the year 2000 that I would retire on my 70th birthday, which also was the end of the academic year 2000/01.

With this decision made, *planning for the future* was the first order of business. Figure 47 showed the five major coworkers active during 2000 and 2001. The NSF Grant I held in 2001 was extended to April 30, 2003. By this time, the last student, Mrs. Pak, could easily finish her degree and the appointments of the research associates Drs. Buzin, Kamasa, and Hu had run out. The Deparment of Chemistry of UTK and the Chemistry Division at ORNL promised to continue their support of our research by maintaining the research labs. In addition, the cooperation with Dr. Habenschuss, by then my ORNL Group Leader, also lasted until his own retirement in 2007.

---

[1] This progression continued, despite the retirement decision, from publication #483 in the year 2001 to #533 in 2003, and #583 in 2008. Optimistically, I gave this book the number 600.

[2] The outside lectures changed after retirement to 20 in 2001, 22 in 2002, 12 in 2003, 19 in 2004, 18 in 2005 and 14 in 2006. In 2007 and 2008, five annual lectures were scheduled.

My secretary of many years, Ms. Joann Hickson (see Figures 17, 40, and 46), had left before, taking a position outside the Department of Chemistry. In time, the WordPerfect word processor she worked so ably in 1988 had developed sufficiently that everyone (including I) did his or her own typing. This left, at best, half-time work for her. Trying to find another half-time job for her in the Department was unsuccessful (even after paying her out of my grant for a one-year trial period). The secretarial levels available just paid too little, or she was not qualified, or she had too much seniority. After she left, we could manage reasonably well by having Dr. Pyda, who was by that time functioning as the Laboratory Director of ATHAS, placed at her desk with a newly installed telephone answering system. Dr. Pyda was also accepted by the Department as a Research Professor (without additional pay). This meant, research was ready to continue at the reduced level of support which NSF kept up until 2006 (see Appendix B, Part G and Footnotes of page 9-82).

Planning our personal future was a bit more uncertain. One of my hobbies, the photography (mentioned on pages 3-30, 5-7–8, and 6-14) got a big boost by the recent availability of increasingly more sophisticated digital cameras of high resolution, and by the processing of the pictures moving to personal computers. In the distant past, the excitements that led me into the study of chemistry had come in part from photography. It was created by the understanding of the capture of the photons by defects in $AgNO_3$ crystals, followed by reduction to metallic silver in the developing step of the negative, and a similar process in enlargement and printing of the positives. In time, my coveted single-lens Exakta Varex (see page 4-27 and 5-8), and a subsequent Olympus SLR, were retired and replaced by a sequence of digital Olympus and Nikon cameras which could use or duplicate the old wide-angle and telescopic lenses. Ultimately, the digital cameras made the old 35-mm film obsolete. Since 2001, my annual directories of digital images have reached 1,000 pictures or more. The thousands of old, chemistry-based negatives and paper copies await scanning and conversion to digital images, a process to be tackled sometimes in the future.

The hobby of coin collecting, had not been activated by the time I wrote this paragraph in 2009. Earlier, I mentioned my interest in engines (see page 3-12) and automobiles, the latter as long as they carried the name Mercedes (see the pages 1-32, 8-38, 40, as well as 9-10–11). This 'hobby' was continued with a wish for a special 70th birthday present. It took on the form of my 'dream car,' modern, powerful, mechanically, and electronically the state of the art. If I wanted this to come about, I had to act more than one year in advance. Early in 2000, I went to our Mercedes dealer and offered to trade my two classic models (see Figure 14A and B) for the best there was, a CL500. The deal was made, I turned in the 300SD diesel, and waited for my birthday. After one year, I was told that there was no 2001 model anymore, I had to opt for the new 2002 with slight improvements (as if this were possible). A few months after my birthday, I traded the 560 SEC for the car in Figure 92, a true marvel of technology and electronics with a thick book of instructions I have not fully worked through yet.

Figure 92 *My Mercedes CL500 was delivered in Knoxville on August 13, 2001 after a 15-months wait.*

Travel plans outside of 'business,' we experimented with during the last few years (pages 9-66–79), were now worked on in earnest. In August 2001, two months after the start of retirement, we went to the Normandie in France, with the Northwestern Alumni Association, by now, our favorite travel-group. We spent eight days at the 'Alumni College' in Lisieux, the 'heart of Normandy' with lectures in the morning followed during the day by inspecting what was discussed (see Appendix C, page C-66). This is an ideal way to learn about a country and its history. In the Normandy many interesting points were linked to foreign interventions. They included the immigrations of the Vikings from the 9th to the early 11th century which shaped the character of population, the battle of Hastings in 1066, as highlighted in the Bayeux Tapestry of 1077 shown in Figure 93, and most recently, the battle fields and monumental cemeteries of WW II, dating from 1944. In Figure 94, the German cemetery is depicted, a more somber view than the well-known American Military Cemetery and Memorial in Colleville-sur-Mer, overlooking Omaha Beach.

Figure 93 *A short excerpt of the impressive, 230' long tapestry which is displayed in Bayeux.*

Figure 94 *"Cimetiére Militaire Allemande, La Cambe.' The bodies of 21,500 soldiers who fell during the fighting in 1944 are buried here.*

This first travel was to be followed in 2002 with a trip through the Panama Canal, the largest engineering project of the early 20th century which, when looking at Figure 95, seems to be in dire need of enlargement. Our ship, the MS Norwegian Sky had certainly reached the size 'panamax.' She was built in 1999 in Bremerhaven, Germany, with 77,100 GRT and cruised at 21 kn. This trip left from Miami, FL, and included stops in Jamaica and Aruba. Beyond the Panama Canal, we docked in Costa Rica and the Pacific Coast of Mexico to end the trip of 16 days in San Diego, CA (see Appendix C, page C-66).

Figure 95 *The Norwegian Sky squeezing through the Gatun lock of the Panama Canal on April 12, 2002.*

The third trip was to satisfy a 'need to see' mentioned on page 9-79. We planned for 2003 a trip from Santiago, Chile, to Buenos Aires, Argentina, around Cape Horn and via the Falkland Islands and various stops along the coasts as listed in Appendix

C, page C-66. The furthest point south at 55° 55' is illustrated with Figure 96 (compare to the North Cape at 77° 10' N and the Cape of Good Hope, 34° 21' S). When we got there, it was surprising that Cape Horn is an island, just as the North Cape, shown on Figure 86. Having heard about the difficulties to sail around the cape before the opening of the Panama Canal, we could not count on the beautiful day we had, sunshine and a quiet Pacific, Atlantic, and Southern Ocean[1]. The upper left photo of mine is the southernmost tip of the cape with an enlargement of its Albatross Memorial on the right. The ship was the MS Norwegian Dream (Built 1992 in St. Nazaire, France, refitted in Bremerhaven, Germany, 1998, 50,750 GRT, 21 kn).

In these three years there were still additional, long oversea trips connected to scientific lectures in Japan (2002) and Thailand (2003), mentioned on page 10-9, in addition to several European travels, as are listed in Appendix C, pages C-57–60.

On December 31, 2000, the 20th century came to an end. Surprisingly, most people thought the 20th century ended with the year 1999, so that the 'big' millennial celebrations were held one year too early. The dire 'end-of-the-world' forebodings common for such major anniversaries, thus, were not taken seriously by me. How is such an error by one year possible?

Figure 96 *"Cabo de Hornos" ("Kaap Hoorn," Cape Horn). Upper left, view on 1/27/2003. Upper right, Albatross Memorial barely visible on the left. Bottom, map showing the cape with sketches of the ships sunk.*

It results from the poor teaching! Not caused by the recent decline in teaching efficiency and quality, as described on pages 8-52–53 but dating back much longer. It is linked to the limited understanding of 'zero.'

[1] The Southern Ocean is also called the Great Southern Ocean, Antarctic Ocean, or South Polar Ocean. The International Hydrographic Organization (IOH) has designated the Southern Ocean as an oceanic division encircling Antarctica at a latitude of 60° S. The British Hydrographic Office uses 55° S, making the Southern Ocean butt against Cape Horn. The Australian Cartographic Authority makes the South Australian coast its northern border. The 'Seven Oceans' (North and South Pacific, North and South Atlantic, Indian, Arctic, and Southern) demonstrate the expansion of maritime travel since the 'Seven Seas' of Antiquity [Black Sea, Caspian Sea, Persian Gulf (today not considered a 'Sea'), Red Sea, Mediterranean Sea, Adriatic Sea (part of the Mediterranean), and Arabian Sea (part of the Indian Ocean)]. The IOH lists more than 100 bodies of water as 'seas.'

Should we not have learned many years ago the difference between counting numbers, and intervals? The sketch below represents the functioning of numbers:

1 2 3 4 5 6 7 8 9 10

(0) 1 2 3 4 5 6 7 8 9

The ten numbers at the top are linked to only nine intervals, listed at the bottom. Historically, when measuring time, unfortunately, one uses two methods. Seconds, minutes, and hours are counted after the end of the time interval. Six hours past midnight, it is six o'clock, i.e., one uses the bottom numerals. For days, months, and years, however, one uses the top numerals, i.e., a year, month, or day starts at midnight after completion of the *previous* day, month, or year.

When our calendar numbering system, the era, was set, there was no zero in the system of numbers in the Western World.[1] The Christian era (AD, "Anno Domini," or CE, Common Era) starts with '1' as the year when Christ was (supposedly) born and not with zero, then, an unknown numeral. The era Before Christ (BC) is counted backwards, i.e., 12/31/01 BC is followed by 1/1/01 AD, there is no year 00. This outdated, system should have been updated. If fixed centuries ago, it would be taught in kindergarten by now and there would have been no further problem.

This difficulty in counting became obvious to me when I was in the hospital (pages 2-15–16). The nurse explained that to take the pulse, one counts the heart beats for 15 s and then multiplies with 4. But she multiplied beats, not intervals, and accordingly, she got a value too high by four. Later, medical professionals who understood the problem told me, this makes no difference, all our guidelines take that 7% mistake into account. (Not very scientific!)

Note two more serious examples: The lack of teaching the international system of units (SI) cost NASA in 1999 a $125 million Mars orbiter. It crashed because one engineering team used SI units for spacecraft operation, while another, the old English units. Much more serious was the 'Challenger' space shuttle disaster in 1986. In this case, the launching crew did not appreciate that on a cold day a rubbery O-ring can become a solid glass and lose its ability to seal. Even today, one can find fellow scientists and engineers who cannot properly assess these differences between solids and liquids, a subject I have frequently dealt with (pages: 7-20, 8-67, 92–93, 109, 9-16, 10-4).

My own experience was that in daily life, and also in my profession, I had to unlearn SI units taught in Germany, and deal with, for example: °F, instead of °C; inches instead of m; mph, instead of km/h; mpg instead of L/100 km. This involves not only a cumbersome conversion factor, but one also loses the intuitive feelings one had acquired. Does −15°C not seem much colder than +5°F? Is the fuel economy of my Mercedes on the highway with 8.5 L/100 km the same as 27 mpg?

Similarly a colleague of mine at UTK, Dr. Al Hazari, complains in a book "Misconceptions in Chemistry, Addressing Perceptions in Chemical Education" (Springer, 2009) that in high school not only does one often transmits outdated knowledge, but one also bases teaching on outright misconceptions. These all are examples that *back-integration of new knowledge into daily life is necessary*. In fact, the back-integration of knowledge into teaching is as important as its discovery.

[1] Zero is the least non-negative number *and* at the same time a digit (place-holder) in a positional numeral system. The oldest known (Indian) text using a decimal place-value system involving 0 is dated 458 AD. This knowledge moved then into the Islamic civilization and from there to the Western World during the 12th century. (A zero was used long before by the Mayas.)

CHAPTER-10

Epilogue

On January 1, 2001, the 21st century began. As discussed on page 9-85, this was one year later than public opinion indicated. This date closes the main narrative of this book detailing my experiences in the 20th century while living in Germany and the USA. Reaching the new century was a day of celebration for us. Against all odds Heidel and I had survived seventy years of this terrible century and achieved more success than anyone would have thought possible during our early years. As everyone, we are hoping for something better in the new century, although its beginning did not fulfill this expectation, as mentioned already on page 3-38.

On the first two pages, this Epilogue starts with *a short reexamination of the stages of our life*. This is followed on pages 10-3–13 by a summary of the first years of the new century. In this period fell also our third 'trip of a lifetime' (for the others, see the pages 8-25–30 and 99-107). It was a world tour in 21 days, described on pages 10-13–24. A final statement closes the Chapter.

For us, the first third of the 20th century was brought to life through direct information from our parents and grandparents (see Chapter 1). In Europe, this time was dominated by WW I and its unfortunate conclusion with the treaty of Versailles. This led in the next 12 years to the dictatorship of Hitler, as outlined in Chapter 1. Although these initial years were difficult times, they began with, what we still considered a 'normal childhood.' My brother and I grew up in a well-functioning, disciplined family. In a strict school environment we were taught how to read, write, and calculate. By 1939 we gradually had learned our duties and how to function within the family and had clear ideas about our later life. But, we also were instructed by example of how to survive state politics.

Everything changed with WW II. The 'normal childhood' came to a rapid end. In the beginning, we did not recognize the terror the dictatorship and modern warfare was capable of, as described in Chapter 2. But by the acquired skills and an uncommon amount of luck, we all survived this dark period. By 1945, the draft-age was closing in with early pre-military service. Comparing this time with the life of today's children of similar age in Germany and in the US, I still would not want to exchange my past for their present. I learned much at an age when learning was easy, and through the almost heroic efforts of Vati and Mutti, my brothers and I did not suffer harm through hunger despite the widespread scarcity of food. Even more importantly, we were cautiously steered away from the official effort to instill us with the "Nationalsozialistische" (Nazi) ideology. At the end of WW II we were prematurely grown up, and today's 'terrible teenage period' was missed. We knew and gladly accepted the duties and responsibilities of adult life (see page 2-31).

The most formative period in my life, then, was the time to 1949. As expressed in Chapter 3, life in Germany started at level *zero* in 1945. The country and economy were destroyed, and even bigger food shortages existed than during the war. Most of the harsh restrictions initially imposed on Germany by the victorious Allies seemed worse than the conditions imposed after WW I.

Fortunately, the dire limitations on civilian industry were soon lifted due to the immediate start of the Cold War (see page 2-23). The USSR was an ally of Germany until 1941, gaining territory from Poland, Finland, and Romania, as well as annexing the Baltic republics. Then it became a partner of Great Britain and the US, taking most of the spoils of WW II in territory, manpower, and valuables. During the Cold War, it now was the key adversary of the old allies.

For me, the goal to become a farmer was blocked by the land reform (page 3-9). Then, the limits on German industry, proclaimed at the Potsdam Conference in 1945, strongly affected my future. The gradual disregard of these restrictions came too late to influence my choice of a profession (pages 3-29–31). The newly created democracy in West Germany, the FRG, prospered as a partner of NATO, while East Germany, the GDR, joined the Warsaw Pact (see page 5-10). The early democracy in East Germany was exhilarating to us, but was quickly suppressed by a new dictatorship. Surprisingly, many valuable lessons can be learned out of such a terrible past. Today's children seemingly have too many diversions as a detriment to their later adult life.

The difficulties to succeed in a communist system are displayed in Chapter 4. As a student one was classified as an enemy of the people when one did not actively support the official, communist party-line. Ultimately, even apolitical students were not tolerated. Again, with help from home, I survived and gained sufficient knowledge in chemistry, so that after fleeing East Germany in 1953 (see the Preface), I could quickly be integrated into the West German student body and society (see Chapter 5). By the end of our time at the Goethe University in Frankfurt, my professional goals were well established. I was married to Heidel, and both of us together, had a clear view of what we wanted for our future in the USA or if need be, after a return back to the FRG. Leaving Germany for the US in 1954 was an experiment which could have been reversed, but a return was never considered (see page 5-13).

Over the years, we did not seriously contemplate the otherwise interesting questions '*what would have been if* ?' These are idle exercises, just as asking: What would be if Nazi Germany would have won the war? If Hitler would have stayed allied with the Soviet Union? If the Soviet Union would have discovered the road to economic success while remaining a totalitarian state, as China has succeeded in doing? These are examples which, I am sure, would have caused even greater problems. The more positive alternatives would have been if the family-related monarchies of Europe had prevented WW I instead of destroying themselves by letting it take its course; if the League of Nations, formed after WW I succeeded to forestall the scheming of Hitler and Stalin to gain world domination; finally, for the present and future, what would be if the UN, formed in 1945, as mentioned on page 2-23, could be a true super-government with a world court to settle the problems for all nations, and an independent military to police the properly enacted laws?

The periods of time in the US were centered at Hastings NE and Chicago IL (Chapter 6), Ithaca, NY (Chapter 7), Troy, NY, (Chapter 8), and Knoxville, TN (Chapter 9). Two sabbatic leaves were in Germany, one in Mainz (summarized on pages 8-37–49), and one in Freiburg and Ulm (see the pages 8-112–128). This development of our life saw a close interplay between a satisfying family life and my professional growth, the latter dominated by research, teaching, and travel. In the pertinent chapters it is shown that with careful planning and a combined effort of the whole family, a unique common experience could be established. The politics that strongly influenced our early life was traced to the end of the 20th century. Intertwined are comments on the experiments and thoughts on improvements of teaching and the development and accomplishments of my scientific researches. In Appendix B, the latter are further detailed for scientists among the readers.

Before closing the book, I like to give **a short summary of what the first tenth of the 21st century has brought**. The decision to retire with my 70th birthday was made earlier, and its discussion is given on pages 9-79–82. During the general 'Recognition Dinner' for the faculty of the University of Tennessee, the first acknowledgment of my retirement, was made in March 2001. The retirement gift consisted of a special UT ball pen. It did not make a big impression, but, the pen still writes well. Next, came my 70th birthday. It was celebrated in Kössen, our favorite vacation spot (see page 8-50). For this birthday, I had wished for, and ultimately received, my best new car ever, shown on page 9-83. Since my 60th birthday, these 'round' birthdays were celebrated more than once and ended with a meeting with students and colleagues. In 2001, this celebration was held at the 29th NATAS Meeting in St. Louis, Mo, September 24–26. The celebration was at a dinner party, such as described in more detail for the 65th birthday on pages 9-52–54. Many of the contributed papers to the special seminars in my honor were published. The bound-copies from birthday 60, 65, and 70 take a central place on my book shelf (see also page 9-18).

Just before leaving for the meeting in St. Louis, my fellow Distinguished Scientist, Professor Georges Guiochon (see page 9-6) had arranged a retirement party. He had invited the colleagues from the Department of Chemistry and also Oak Ridge National Laboratory and my remaining students to the UTK Faculty Club. Figure 1 is a reminder of this party. It represents only one of the many photographs that were taken. All guests had contributed to a very special present, a copy of the modern English translation of the original "Four Books on Architecture" by Palladio (1508–80). The book describes, besides the architecture, the materials and techniques used to erect the ancient Roman buildings. The book is based on the manuscript "De Architectura" by Vitruvius, an architect who was active in the first century BC. The 1997 translation contains facsimiles of the over 200 wood cuts of the original Palladio book.

Figure 1 *Heidel and I at the retirement party. In the background Dr. Annis, ORNL, the Dadmuns, UTK.*

Shortly after the delivery of my birthday present Mercedes (Figure 92 of Chapter 9), my daughter Caryn and I were invited to a special reception by the dealer, where Figure 2 was taken. This can be taken as a fitting ending of *the birthday and retirement festivities of 2001*.

Figure 2 *Testing an antique Mercedes with Caryn.*

__Research after retirement__ continued to the end of the NSF research grant in 2003, as was projected on pages 9-82–83. This was followed by a new successful NSF grant proposal based on the title: *Coupling Within and Between Nanophases of the Global, Metastable Structure of Polymers*. This grant allowed research to continue until May 2006 (see Section G in Appendix B).

The main research after retirement dealt with three topics: (1) The fit of nanophases, which now had an operational definition (see page 9-46), into the general scheme of simplifying the basic knowledge about the molecules and phases. (2) The thermodynamic reasons for the different crystal structures of nanophases and their changes of properties within their volume. (3) How do macromolecules, being longer than nanophases couple across the phase boundaries?

An early development of the topic (1) was indicated on pages 8-66–67. Topic (2) is reviewed in publication #584 (see Appendix A), based on a lecture given in 2008 at the 10th Lähnwitz Seminar in Warnemünde (see Chapter 8, Figure 28D). Cases of coupling, the issue of the topic (3), range from little to strong coupling. It was shown that *intermediate coupling* could lead to various degrees of stress transfer across phase boundaries, resulting in unique, new properties.

The thermodynamic description of successive physical and/or chemical processes in such multiple-phase systems can, in addition, be governed by the *kinetics* of the coupling. Processes may vary from being practically simultaneous to being fully independent. Such changes have consequences on, for example, the transition temperatures. This topic was discussed first in 2005 in a Keynote Lecture in a symposium on "Morphology and Structure Development" at the 21st Polymer Processing Society Meeting in Leipzig, Germany. The manuscript of the lecture was subsequently published, as is shown in Figure 3. More scientific details about this work are described in Appendix B, Part G (for papers and lectures, see also the Appendices A and C). As mentioned on page 9-82, the laboratories at ORNL and UTK continued with maintenance support.

In 2002, __an expansion of the laboratory facility at UTK__ was proposed on the suggestion of Profs. Jimmy Mays and Mark Dadmun (Department of Chemistry). The plan was based on the growth of interest in synthetic and biological macromolecules in the Chemistry Department as well as other divisions on campus. In addition, there was an increase in requests from outside research laboratories of industry and from academic institutions for support with measurements on polymer systems as well as for advice and discussion of special polymer problems. Such cooperation was usually combined with offers of financial support.

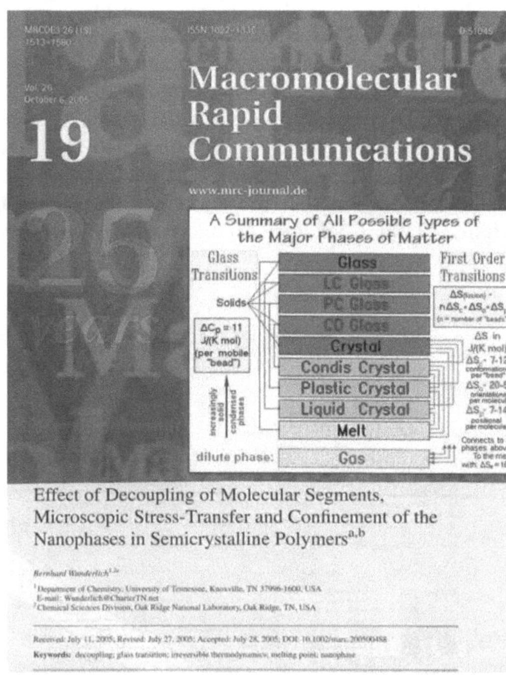

Figure 3 *Magazine cover and details about the publication #556. (Details are listed in Appendix A.)*

With the support of the ATHAS Director, Dr. M. Pyda, and myself, it was agreed, to develop such comprehensive *Polymer Characterization Laboratory, PCL*. This laboratory was to be based on the existing ATHAS Laboratory, founded in 1980 at RPI (see page 8-63). By being part of the larger characterization facility, it was hoped that our work could be continued beyond my ultimate retirement. To start the effort, the major portion of Dr. Pyda's salary, now the Director of PCL, was paid using NSF and our industrial support. In addition, all ATHAS laboratory space was made available to the joint enterprise. Finally, the pertinent equipment of all parties became assets of the PCL. In the future, external support was to be generated, to make the new laboratory self-sustaining. The oversight of the PCL was to be done jointly. A proposal to the Tennessee Advanced Materials Laboratory at UTK, TAML, was submitted in May 2002. It provided partial support for two years, including the hire of a retired high school teacher as technician. The TAML funding started on April 1, 2003. The technician, Mr. T. Malmgren, began his job in August 2003.

Figure 4 *The last group of researchers. On my right, Drs. Pyda (until 3/2006) and Qiu (until 4/2006), on my left, Dr. Ma (until 9/2005), during the last group dinner at the Mandarin House, 9/16/2005.*

In the meantime, work had increased to such a degree that we hired Ms. E. Nowak-Pyda (wife of Dr. Pyda, see Figure 45 of Chapter 9) as a technician in March 2003, and Dr. Jeong-ihm Pak (see Chapter 9, Figure 47, and page 9-81), now as a postdoctoral fellow. Two newly selected Chinese postdoctoral associates had visa difficulties due to the tightened immigration laws by 2002. It took until April 2004 for Dr. Wulin Qiu, who worked as a postdoctoral fellow in Japan, and until June 2004 for Dr. Jisheng Ma, who before had a postdoctoral appointment in Australia, to join us. Figure 4 shows the final senior researchers. Their work is listed in Appendix A (2004–2008) and can be seen in Appendix B, Parts G and H.

The first two years of the PCL, the Polymer Characterization Laboratory, was successful. New instrumentation was acquired to complete the characterization ability. The thermal analysis was expanded to thermogravimetry, dynamic mechanical analysis, and modernized TMDSC. The molecular characterization included means to determine molar mass and its distribution, multi-angle light scattering, and differential refractometry. About 10 external and 11 internal group users had utilized the PCL facilities. Approximately 1650 hours of payable usage had been recorded by 9/30/2004. This was close to self-sufficiency (if all internal users would have paid their assessment).

On Tuesday, Nov. 2, 2004, however, I was informed by Profs. Mays and Dadmun (without prior warning or being invited when prior discussions were held) that they had decided that Dr. Pyda was to be terminated as of January 1, 2005. This would have caused extreme hardship for my coworker of more than 12 years. Also, it would have jeopardized our research to be carried out under the NSF grant, which relied upon Dr. Pyda's running the laboratory, his teaching of the new postdoctoral fellows, and his cooperation in the Data Bank calculations and administration. My opposition had little effect. This was *the demise of the PCL*. The following letter was given to me at the meeting and was delivered with minor updates just three hours later to Dr. Pyda.

Dear Dr. Pyda:

We regret to inform you that unfortunately we can no longer continue to support your appointment as Director of the Polymer Characterization Laboratory (PCL). The original seed funding for combining the existing Advanced THermal AnalysiS (ATHAS) laboratory with new personnel and instrumentation to form the PCL has been terminated. We have therefore had to restructure and reorganize the PCL, such that it will no longer be possible to maintain the connection between ATHAS and the PCL. The termination of our support for you as director is based on this lack of funding.

We are happy to see that you will be able to continue as the director of the ATHAS laboratories and personnel. We would like to thank you for the time and effort you have spent in setting up and operating the new equipment that was acquired in the two years since the inception of the PCL. Especially you did a superb job in the training Mr. Tom Malmgren who was new to the field of polymer characterization and is now able to work independently on both, gel permeation chromatography and the thermal analysis equipment. The PCL laboratory was set up and became a functioning entity largely through your efforts.

We are sorry that the financial support did not develop as planned and wish you the best for the future.

Sincerely,

Please make sure that the letter is printed on a Departmental Letterhead and signed with name and title of the two present directors of the PCL [Professors Mays and Dadmun].

This surprise was followed on November 3 by an e-mail of Prof. Mays, revoking Dr. Pyda's right to schedule equipment of the Polymer Characterization Laboratory (PCL) which was bought whole or in part with Dr. Mays' funds. After several meetings, involving also the Department Chairman and the Office of Human Resources of the university, we found an amicable (?) solution. Dr. Pyda could not be dismissed that abruptly, since he then could sue the University. Ultimately, in order to keep university benefits for his family, he was permitted to take early retirement from the university. In the meantime, my NSF grant could stretch his employment until March 2006, together with industrial funding and a recovery of the collected moneys from service work carried out during the existence of the joint PCL. The ATHAS equipment and research space was separated again, and the trade-in value of the ATHAS equipment for newer devices was evaluated and converted into time of our use on the new equipment of PCL. It was still not exhausted by 2010.

Thus, *the future of ATHAS at UTK found its end in 2006*. I did not apply for further extension of our NSF support, although it would have been considered by the agency. By 2006, I would not have been able to collect a new group of coworkers and supervise them daily in the learning phase for advanced thermal analysis. Many of the specialized computer programs used in our research and for the handling of instrumentation were developed over the years when all laboratory research was done by my coworkers. For me, to be up to date again, it would have taken to learn these details anew from the bottom up. By then, however, I was only twice a week in my office at UTK. Dr. Pyda could have solved this problem easily. He handled all these programs during the last 10 years and wrote several of them himself. But, he was not eligible for external funding since he was not employed as a full-time faculty member at UTK and had advanced in rank beyond postdoctoral fellow. The solution that was finally arranged for the future of ATHAS and the Data Bank was to move all equipment, programs, and the collection of pertinent literature to the Department of Chemistry at the University of Technology in Rzeszow, Poland, where Dr. Pyda had been offered a faculty position. For the move, Dr. Pyda was granted support from European NATO funds, and UTK allowed the transfer without cost recovery. Naturally, I keep contact, but the UTK and the US has lost a valuable resource for macromolecular physics, chemistry, and engineering.

The laboratory and research connection with ORNL was made use of by Dr. Ma until 2005. After his departure, there still was some cooperation left with Dr. A. Habenschuss which ended with his retirement in 2007.

During this time period from 2001–2006 about 70 papers and conference reports were contributed to the literature which took to 2008 for final publication, as listed in Appendix A and summarized on page 10-4.

The teaching effort continued after the retirement in 2001. As before, my interest was to address the problems of university teaching and learning as it arises out of the growth of scientific knowledge. These problems were disseminated earlier on pages 8-71–75 and 9-36–37. Possible solutions were suggested based, first, on experiments with audio courses, as described on pages 8-67–69 and later, on computer courses, as presented on pages 8-69–71 and 9-35–36. This work had the goal to broaden academic teaching which has grown in many fields in volume beyond the limits of most universities. In both cases, these courses were constructed as learning tools for students already interested in acquiring the knowledge they urgently needed, rather than to create the excitement necessary to select this topic as a study area. Naturally, such courses offer also a possibility to contribute to the teaching effort after retirement.

As a consequence of my earlier thoughts about learning without live lectures, the first job undertaken was an update of the computer course "Thermal Analysis of Materials" (Appendix A #464) which was developed from 1998 to 2000. This teaching with computers was initiated in 1983–84 at RPI (Appendix A #201–204 and pages 8-69–71) and then further improved, expanded, and published on the internet by 1995–1996 (Appendix A #421, 393, 430). It was used as a direct replacement for low-attendance graduate lectures. The new round of updates was completed in 2005 with continual further improvements until 2007. At last, there was a change to color for all computer graphs to increase the clarity, and seen in Chapter 9, Figures 18, 41–43. The course is listed in Appendix A as #560 with its various internet addresses for downloading. Its title page is displayed as Figure 74 of Chapter 8. Parallel to this work, all the course material was combined into a ≈1,000-page, 'old-fashion' book with a much broader link to the related literature, published in 2005 under the title "Thermal Analysis of Polymeric Materials" (Appendix A, #500).

<u>A major event after the closing of the laboratories in 2006</u> was a Symposium on "Semicrystalline Polymers, Two Phases or Three?" It was given in honor of my 75th birthday and coupled with an award for "Lifelong Accomplishments in Thermal Analysis and Calorimetry," at the 34th Annual Conference of the North American Thermal Analysis Society in Bowling Green (see

Figure 5a *Official picture of the award presentation.*

Figure 5b *Visit at the Corvette Museum.*

Appendix C). Figure 5a is a picture of the ceremony with Dr. Joseph Menczel, my former coworker and Dr. Rick Seyler. Both were Cochairmen of the Awards Symposium. Figure 5b is a snapshot in front of the Corvette Museum, close to the Conference venue, where I could compare my car, in the foreground, with the production examples of the best known US sports car. As before, this again was part of a 'round birthday celebration.' These 'special birthdays' seemed to come faster and faster. (Compare to pages 10-3 and 9-18, 52–54 for the earlier 'round' birthdays).

<u>Closing the laboratory in 2006, did not mean that all my scientific work stopped</u>. The research for over 50 years had clarified much, but there was still a need to fit the results together and to connect them to the existing knowledge on reversible and irreversible thermodynamics. A number of topics were due for extensive reviews. Part H of Appendix B contains a listing of these reviews together with the appropriate references. One special lecture was at the 234th ACS Meeting in Boston, MA, 2007 on the "Fifty-year Development of the Understanding of Motion and Defects in Macromolecular Crystals Based on Thermal Analysis, Structure Analysis, and Computer Simulation." This presentation was given as is shown in Figure 6. The meeting was the 4th international discussion on polymer crystallization mentioned on page 8-63. All four are listed in Appendix C (on pages C-22, C-41, C-58, and C-64). The lecture

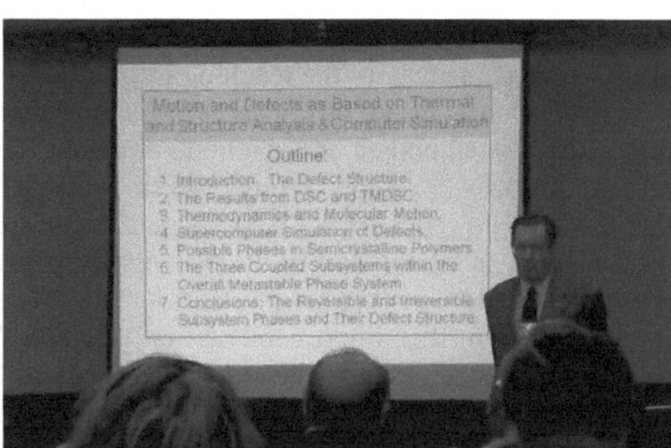

Figure 6 *Beginning of my lecture at the ACS meeting in Boston, MA, in 2007 (taken by a student of mine).*

was digitally reproduced as part of the CD with the proceedings of the meeting (see Appendix A #577) and later published in a more extended form (see Appendix A #581).

The continuation of the lecturing outside the university after retirement in 2001 and after closing of the laboratory in 2006 is summarized in Appendix C (pages C-57–63). My 8th trip to Japan in 2002 was to the 3rd International Discussion Meeting on Polymer Crystallization in Mishima, Shizuoka.

Besides the frequent trips to Europe, one more overseas-travel was to the Pacific Polymer Conference in Thailand in 2003 (see Appendix C, page C-60). Figure 7 is a group photo of the hosts

Figure 7 *Photo of all invited lecturers at the 8th Pacific Polymer Conference in 2003 in Thailand.*

and invited lecturers, and Figure 8 is a snapshot of the cultural presentation during the meeting. A number of excursions before and after the meeting gave us a good picture of this never colonialized country and its past under the Siamese Ayuttaya Kings. My slide show of 329 pictures is a record of an unforgettable trip to a country with a great past, struggling to reach a bright future by working on industrial development and creating a science base. This trip rivaled in travel time and excitement the earlier ones to South Africa (1992), Australia (1995), and China (2000), detailed on pages 9-57–66.

Figure 8 *Traditional (all male) representation of one of the many Thai fables, beautifully presented to the conferees at the Polymer Conference in Bangkok.*

Going through our annual personal picture albums, a number of events must be mentioned. First, there was our Golden Wedding Anniversary in 2003, about thirty years later than the analogous anniversaries of our parents, pictured on page 8-84. To be sure that we could have a similar gathering to celebrate our many years of marriage, we had arranged a trial festivity five years earlier with many of our relatives. Thus prepared, the fiftieth proceeded even better.

The family picture taken on this day is shown in Chapter 11 as Figure 28. Figure 9 depicts the head of the table with Stefan and Alyssa, our grandchildren. The 23 additional guests were placed around a horse-shoe arrangement in the side-room of the 'Copper Cellar West' which was voted for 2003 'Knoxville's Best Restaurant.' After the greeting with Champagne and exchanging of information about the families and friends, we sat down as is shown in Figure 9, each to order the special dinner to their liking. During this interlude, I had time to present a slide show, reminiscing about our life in the past, using digital projection. Many of the pictures are in the collection of this book. Figure 10 is a reproduction of the motto for the evening.

Figure 9 *The Golden Wedding Anniversary showing Stefan and Alyssa, and the white lilac centerpiece.*

Figure 10 *Time to celebrate. Left, a picture of our wedding bouquet. Right, lilac of the silver anniversary.*

It can be seen that the unusual white lilac was repeated on the three occasions which marked our married life. The choice of the spring flowers for the bridal bouquet of the winter wedding was out of necessity (see the pages 5-14–16). So short a time after Christmas, there were few other flowers available. Frankfurt seems to have had a big lilac bush blooming at the year's end in a hot house. When I wanted to duplicate the lilac for our silver anniversary, it took quite some effort to procure what is shown on the right in Figure 10. The florist told me that they had come from Frankfurt! (From the same bush?) By 2003 the international flower business was better organized. All types of flowers were available in winter from the southern half of the Americas, just that lilac is not very much in favor since it does not ship well.

After the dinner, gifts were presented. I have to mention the most appreciated one. Brent had, with Caryn's help, gotten hold of Heidel's address books reaching back into the 1950s and written close to 300 letters asking to contribute a greeting card, picture, or anecdote out of our past. Many answered him from all over the world. The overwhelming response filled two large-size photo albums that Brent and Anna prepared beautifully. The long list of addresses is easy to understand. Heidel always has a list of about 150 of relatives, former students, and friends with whom at least Christmas greetings are regularly exchanged.

The grandchildren could also contribute to a special gift. Brent's whole family sat together on many an evening and folded 1,000 colorful origami cranes, a symbol for good luck. They now stand on our fireplace mantle in two large glass urns.

Heidel wished for something special: A nostalgic trip to Germany with the family. We went in 2004 and spent a memory-rich two weeks in our "Heimat." Lufthansa had just begun trial service from Charlotte, NC, to Germany at an exceedingly low price. The grandchildren were by then old enough that they wanted to know all the tales we could tell about the cities, the destruction, and our study in Berlin. Stefan was even old enough to order his first 'legal' beer in the restaurant (in Germany and in the presence of parents). Overall, this anniversary served to reunite the family and friends. As mentioned on page 10-9, we had a trial for the 50th anniversary in 1998. To continue this planning, in 2008 we managed to celebrate a preview of what we hope to be our 60th in 2013.

<u>Other family festivities</u> included the graduations from high school of Stefan and Alyssa in 2005 and 2007, respectively. They were (now) special events for the family. The Figures 30 and 31 of Chapter 11 remind us of these. My 75th birthday in 2006 needed some additional attention beyond the festivities of page 10-8. As stated, we were in Europe at the time of the actual birthday. In addition, we also wanted to recognize Heidel's 75th birthday which was a year earlier, but could not be celebrated properly since one year before, our sister-in-law, Dagmar, had her 75th birthday celebration with the whole family. To coordinate such an event needing travel from the central and north-east US each year, however, is hard to accomplish. So, we took a page out of the birthday arrangements of the Queen of England. Figure 11 shows the invitation, not to Heidel's or my birthday, but to our combined 151st birthday, and not on either's birthday, but on a suitable day for travel, namely July 1, so that most could make use of the extra holiday on July 4 to travel home.

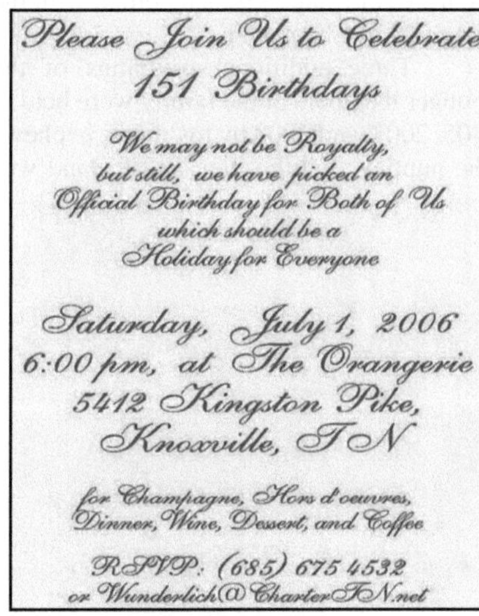
Please Join Us to Celebrate
151 Birthdays

We may not be Royalty,
but still, we have picked an
Official Birthday for Both of Us
which should be a
Holiday for Everyone

Saturday, July 1, 2006
6:00 pm, at The Orangerie
5412 Kingston Pike,
Knoxville, TN

for Champagne, Hors d'oeuvres,
Dinner, Wine, Desert, and Coffee

RSVP: (685) 675 4532
or Wunderlich@CharterTN.net

Figure 11 *Invitation*

For the early arrivals, traveling by automobile from upstate New York, there was a brunch at home, shown in Figure 12, before the dinner at our favorite restaurant in Knoxville which is documented with Figure 13 with most guests (from left to right: Anna, Heidel and I, Caryn and Brian, Connie (a friend of Heidel's), Galya Ritter (neighbor of ours in Knoxville, her husband was out of town), Jim and Lotte Moore (barely visible, better seen in Figure 12,), Carol and Manfred (on their first visit to Knoxville), Shahla and Eberhard (the empty chair, indicates that Eberhard was the photographer), Susi and Ivo Buhac (see Figure 12, and also page 8-83), and half of Alyssa (Stefan, Brent, and Dagmar were

Figure 12 *Breakfast with the Buhacs and the Moores.*

sitting at the head of the table, not in the photograph). We were well served and had many things to discuss. The most important information of the evening, however, was Brian and Caryn's announcement that they, after many years of being best friends, had decided to get married later in the year.

The picture of their wedding on October 29, 2006 at the famous Biltmore Estate in Asheville, NC (see pages 9-8–9 and Figure 10 of Chapter 9) is recorded with the Figure 29 of Chapter 11.

Figure 13 *'151' Birthdays at the Orangerie in 2006.*

Three additional weddings of the younger members of the family were held in 2005, 2007, and 2008 by my niece, nephew, and grand nephew (see page 11-2, lower left column). The nuptials of Julie, Ian, and Roland were in Nebraska, California, and Oregon, respectively. Unfortunately we could only attend Julie's wedding in Omaha, NE. For the others, we had unresolvable scheduling problems. Seemingly, I was still not retired enough. Figure 14 is a picture in the entry of the church in Omaha with most of the Wunderlichs who attended. From right to left are Manfred, Joachim, Dagmar, Laura, Shahla, Heidel, Caryn, Carol, and I, with Eberhard being the photographer (for the family correlation see page 11-2 (all are listed in the left column, except for Caryn, who is also found on page 11-4, center).

Figure 14 *Guests at Julie's wedding in Omaha, 2005.*

After the wedding we drove on to Hastings, NE, our third hometown (see pages 3-36–37). We visited Vati and Mutti's grave and delivered two of the white roses from Julies bridal bouquet (see Figure 39 of Chapter 9 which was taken on this occasion). We had not been in Hastings for a while, and so we could revisit on this occasion many of the places which made Hastings, NE our third hometown. Figure 15 is a montage of these impressions. It starts on the left with the Railway station (Amtrak), where the Zephyr arrives still early in the morning (see pages 6-2–3). The next picture is of our first apartment

Figure 15 *Nostalgic visit of Hastings, NE. Pictures taken in 2005.*

in the US (see page 6-11). The house in which we lived in 1955, at 113 East 4th Street, close to the college, still looks as it did then. Just that the trees were all gone, very much to the detriment of the area (compare to Figure 22 of Chapter 11). Note that I had parked my car just in the same position for the new photo. The next picture is of Joachim's car. He is just as much a car enthusiast as I am, the picture shows his exquisite Porsche. The last photograph of the montage shows the historical building of the Dutton Lainson Company, so important for Hastings and also our coming to the US, as mentioned on page 6-4 (see also another view of the same building in Figure 32 of Chapter 3).

The *travel plans after retirement* were introduced on pages 9-84 and 85. The list of major travels (excluding trips to Europe) are given, as always, in Appendix C (pages C-65–67). Through 2003 these were further discussed in Chapter 9 with occasional photos, trying, as always, to avoid 'postcard and travel-folder views' and concentrating on something special for our life. The next four trips will not be further detailed. They went to the Galapagos Islands (2005), the Adriatic Coast (2006), Andalusia (2008), and from the Atlantic to the Pacific (2009). All were worth the trip, time, and expense. A fifth trip, however, needs attention. All 90 members of the 'expedition' agreed, that this was the 'trip of our lifetime.' This means, I must end with a brief description of it, since I have described two other 'trips of our lifetime' before, starting on page 8-25 and 102, respectively.

This third trip was the 2007 **travel around the world**. It was not a trip "Around the World in 80 Days" as written in Jules Verne's 1873 novel, or seen in the 1956 movie by Michael Todd,[1] based on the novel. We managed to cut the time to 21 days and went westward, instead of eastward. How could we cut the time so short? We went all long distances across the oceans and continents by private jet instead by balloons, steamers, and trains, as was necessary in the 19th century. Figure 16 illustrates our travel route. We started February 24 from Knoxville and returned on March 19. To collect the different passengers from all over the US and to arrange for convenient trips home, extra nights were arranged at the Ritz-Carlton Resort in Orlando before and after the round trip.

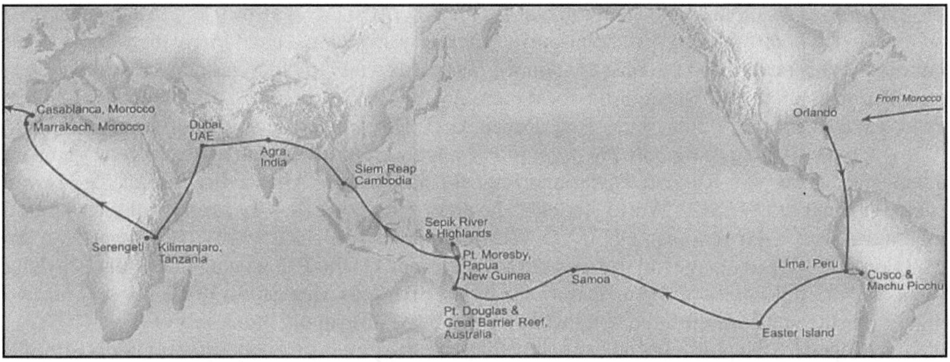

Figure 16 *Travel around the world touching on five continents and crossing two oceans 2/25–3/18.*

[1] Winner of five Academy Awards with David Niven as Phileas Fogg, Mario 'Cantinflas' Moreno as Passepartout, Robert Newton as Fix, a Scotland Yard agent, and Shirley MacLaine as Aouda, a young Parsi princess. Overall, 68,894 persons were photographed in 34,685 special costumes. The film was shot in 252 locations with more than 40 Hollywood stars as 'extras.'

Figure 17 *'Our' private jet parked on Easter Island.*

All was well organized. Everyone was furnished before with identical, wheeled luggage and a back pack; all well marked, so that after checking it at the Knoxville airport, we did not have to handle suitcases until we were back in Knoxville. They were standing in the hotel room wherever we arrived. All sleeping was in the best hotels available, all travel during the daylight hours. Our private jet, a Boeing 757, is shown in Figure 17. On every stop, the plane was parked at the airport, waiting with its crew for our return. The plane was specially outfitted in first-class comfort for the 90 passengers (with one change of seat assignments after half of the trip, for us, from right rear to left front). In addition to the passengers, there was the following personnel on board: (A) Three expedition leaders (a local crew was always arriving a few days before, by commercial transportation and saw that all was ready for our arrival). (B) four lecturers (university-connected experts giving five presentations at the hotels and three during the longer flights, for the latter, using the plane's TV screens and noise-cancelling earphones for easy listening). (C) One physician (from the EU of Massachusetts General Hospital). (D) The flight-crew of 17 (three pilots, two engineers, nine flight-attendants, and a chef and two caterers who prepared the in-flight meals in cooperation with the local hotels).

All passports (with proper visa and health certificates) were collected, and all forms needed to enter and leave the various countries were filled out for us by the crew. We only needed to give the signatures. In many places, we could abbreviate entry and exit formalities using the side doors for 'diplomats.' Rain gear and an alarm clock with a thermometer was supplied to keep everyone on time and in comfort, even in the rainforests. On every arrival in a hotel room, there was a small gift of a local specialty and a couple of stamped post cards ready to report home. An envelope was provided with about $10 worth of the local currency for small expenses, i.e., Nuevo Sol, Chilean Peso, Tala, Australian Dollar, Kina, Rial, Rupee, UAE Dirham, Shilling, or Moroccan Dirham.

On the first leg of the trip, *the flight to Peru*, we were greeted with a rose on each seat, and a welcome snack with Russian Caviar and French Champagne. After a city tour of the historic center of Lima (a UNESCO World Heritage site since 1988), we flew by separate charter jet into the Andes, to Cuzco (another UNESCO World Heritage site since 1983). The next day, we experienced a narrow-gauge rail trip to the 50-mile distant Machu Picchu with the Hiram Bingham train, the 'Orient Express' of Peru. It is owned by the luxurious Monasterio Hotel, our stay for two nights. As the name implies, the hotel was lavishly fitted into an old monastery.

Helped by the physician on board and the customary coca tea, nobody was unduly affected by the altitude (see page 8-35). I had seen Machu Picchu before (Chapter 8, Figures 41–44 and pages 8-34–35). All was as before, just that now the site had not only to be protected from the elements of nature, but also from the tourists. Since 1983 Machu Picchu is a UNESCO World Heritage site. Figure 18 illustrates one of the earth-quake-proof, narrow streets of the Inca city with houses of carefully fitted white granite. All is overshadowed by the towering Huayna Picchu. Extending the focal length of my Nikon S10 (equivalent to 380 mm of a 35 mm camera) yielded

Figure 18 *Street in Machu Picchu, 2007.*

Figure 19 *Construction on the top of Huayna Picchu.*

Figure 19. In Figure 18 the view of the agricultural terraces and the watchtower on the steep mountain are hardly visible. Walking in Machu Picchu, our wireless guide system proved its value. Each of us had earphone receivers and the local guides were fitted with microphone transmitters broadcasting on given frequencies. One could stray up to 100' from the group without missing the comments and, thus, I could spend more time on closer inspection and photography.

This first step of our trip was already one of the highlights. The second was reached on the next major stop. After the return to Cuzco and the second night at the Monasterio Hotel, we flew back to Lima (1.5 h) and took 'our' jet to *the 45 square mile small Easter Island* (Rapa Nui, one of the world's most remote, inhabited islands, belonging to Chile and being on the UNESCO World Heritage list since 1995). The flight took 5 h across the South Pacific to Mataveri Airport (see Figure 17). This airport with its two-mile-long runway serves as an abort site for the US Space Shuttle. The following flight from Rapa Nui to Apia, Samoa, was even longer, 9 h. For many years it had been a puzzle how anyone could reach this small island by traveling the ocean with only a knowledge of the stars, winds, and currents. It helped that Thor Heyerdahl had made such long trips with primitive balsa and papyrus rafts (pages 9-76). By now it is quite certain that the immigrants came between 300 and 400 AD from Polynesia, using large outrigger canoes, still in use today. Hawaii, which we visited in 2000 (see page 9-77–79) was similarly colonized.

Figure 20 *The fifteen giant "moai" of the Ahu Tongariki, backs to the ocean. Restored in 1996.*

Over 1,000 giant statues ("moai"), carved out of volcanic rock, add beauty to Easter Island. They represented the symbol of the native culture. When their society collapsed, starting at ≈1500 AD, the "moai" were, in time, all toppled from their "ahu," (platform). In 1960, a 25' tidal wave scattered the ruins of the Ahu Tongariki, the largest on the Island. The site was well described in its early, toppled state, so that a detailed reconstruction of the destroyed architectural monument was possible, as shown in Figure 20. Tadano, a Japanese company spearheaded the effort by providing funding and modern cranes.

To us, the similarity of Easter Islanders and Samoans was obvious when listening to their traditional songs and dances, although we could not check on the authenticity of the presentations given to us as tourists. The DNA evidence is much more convincing. Figure 21 is a picture from a 'Kava Ceremony' as it was performed by the 'maid of the village' in front of the dignitaries (us?). After carefully washing out her mouth in our presence, she chewed the bits of the peeled kava root and then ejected them into her hands and deposited them into the wooden bowl in front of her. The so concocted, bitter and intoxicating beverage was offered to us from a communal cup. Unfortunately I cannot describe the taste in detail, since I, with several others, gracefully declined the honor.

Figure 21 *Performing the Kava Ceremony (the chewed, extracted roots are on the basin rim).*

This ceremony was performed in Vailima, the former home and 100 acre estate of Robert Lewis Stevenson (1850–94), the Scottish writer of "Treasure Island" (published in 1883). In his constant search for places on earth favorable to his health, he moved in 1890 to Samoa. There he adopted the name "Tusitala" (story writer) and got involved in local politics. He was well respected by the natives and was buried by them on nearby Mount Vaea.

For one night we stayed at 'Aggie Grey's Lagoon Beach Resort and Spa' (March 2–3). Mrs. (Agnes) Grey was the daughter of a British settler who married a Samoan and developed the upscale hotel and resort from a sandwich stand, catering to American servicemen during WW II. She is known to have been one of the several businesswomen who inspired the composite character of 'Bloody Mary' in the musical and movie adapted from James A. Michener's 'Tales of the South Pacific.' After a somewhat rainy day, we experienced a beautiful sunset seen in Figure 22, highlighted by the lagoon and the outlying coral reef. In the evening we were then treated to the well-known 'Grey Dancers' with a "fiafia" performance, remembered in Figure 23.

Next was the flight to Cairns and bus travel to Port Douglas to see the Australian Daintree rainforest (see also page 9-61). It took a surprising 21 h for the six hours in the air. Unlike Phileas

Figure 22 *Sunset from our room at Aggie Grey.*

Figure 23 *Grey's Dancers with Samoan "tatau."*

Fogg mentioned on page 10-13, we were reminded that going west across the international dateline (between Samoa and Australia) we had to set our *calendar* ahead by one day. The *clock* needed to be set back three hours due to the normal westward progress. Fogg went eastwards and regularly set the clock ahead. He never looked at the calendar during his final travel through the USA and by ship to Europe (or asked for the day of the week—an unlikely story). He arrived one day 'earlier' than he thought, and we 'lost' a day. Our hotel was the 'Sea Temple Resort' from which one group directly explored the Great Barrier Reef (see Chapter 9, Figure 65C), while our group spent the day in the rainforest by bus, on foot, and by boat. During the next morning, we were bussed to Kuranda and inspected the rainforest from the Skyrail to Cairns. Figures 24 and 25 give a view of the rainforest from below and above. In the evening, the aborigines presented their dances and songs.

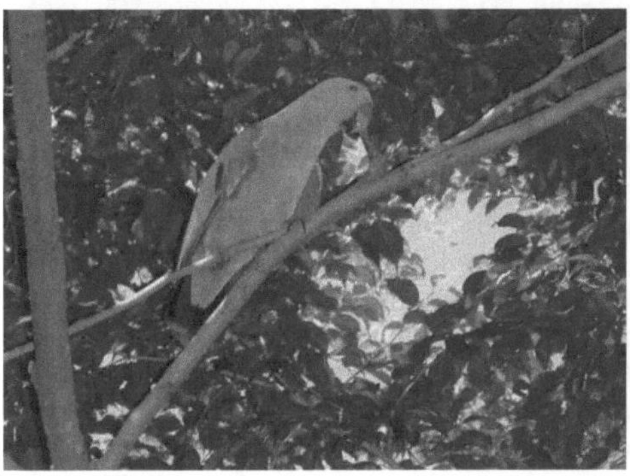

Figure 24 *Parrot under the rainforest canopy.*

Figure 25 *From the Skyrail above the Rainforest.*

In the afternoon of March 6, the third highlight of the trip was approached, the visit to the rainforest highlands of Papua New Guinea, PNG, the eastern half of the Island of New Guinea. The trip started with a short flight of 1.5 h from Cairns to Port Moresby, the capital of PNG. In 2004, Port Moresby was ranked as the worst capital in the world to live in, due to the gangs known as "raskols." We made a visit of the interesting new museum where we were greeted by one of the 'Asaro Mudmen,' shown in Figure 26. The tribal legend has it that their mud cover from the Asaro river in the highlands made them appear as scary spirits to a warring neighbor so that they won the fight.

Figure 26 *Live 'Mudman' protecting the museum.*

After a night at the Airways Hotel, the adventure began with a 1.5 h charter flight in a turboprop to Tari in the highlands. At the primitive airport we were greeted by the Huli tribe as visitors from the sky. This tribe, one of the biggest in PNG, had seen the first white man only some 60 years before, as they said, coming from heaven.

The majority of the PNG agriculture involves the fallow systems by clearing and cutting of forest, cultivation of crops, and abandonment to allow natural processes of regeneration. This agriculture evolved over 6,500 years ago, earlier than in any other agricultural area in the world. Their most elaborate farming techniques apply a system of mounding, terracing, mulching, ditching, draining, and irrigation. Figure 27 gives an areal view of the tribal area of the Hulis, not far from the Kuk UNESCO World Historical site (listed since 2008). There the earliest archeological evidence is preserved.

Figure 27 *View of the highlands approaching Tari.*

At the airport we were picked up by four-wheel-drive transporters and transferred about 50 miles to the Ambua Lodge along the only 'superhighway,' consisting of a barely maneuverable dirt road. The lodge had about 40 huts, as shown in Figure 28, and a big assembly hall as restaurant and lounge. All buildings were thatched. Although these huts look superficially like the native dwellings, they were quite modern inside. The modern bathroom had hot and cold water, and electric blankets were provided for the cold nights. Water and electricity were supplied from a dammed mountain stream with a generator. Besides our trips to the nearby clan center and into the rainforest to observe birds of paradise, natives were invited to the lodge to talk about their life before and since the white man interfered. Interesting for Heidel was the plight of the local women. A native lady's wisdom was summarized by: "Make sure that you do not ever marry a Huli!"

As a special treat, a group of a dozen Huli Wigman were invited to the lodge. They explained how they grew their wigs by treating the hair while growing. After the proper legth and shape is reached, the wig is cut and prepared to be worn. The next wig can then be started. Figure 29 shows three particularly interesting faces prepared for a traditional dance. The wig is adorned with feathers from birds of paradise. The necklaces are from coral, shells, and beads, acquired through trade with coastal Papuans. Facial and body paints are made from red or yellow iron ore (ocher), clays, and natural oils. Branches are to mimic bird tails. We could observe their

Figure 28 *Our hut at the Ambua Lodge in PNG.*

Figure 29 *Huli Tribesmen during their dance.*

preparations which took more than one hour. All was still done with the original materials and tools, except for modern mirrors. When asked, they said it is worth the expense since this is so much easier than using a water surface to properly apply the paint. Similarly, when the women of Tombete Village showed their expertise in weaving traditional rain-gear, it started to rain. Quickly they all opened modern umbrellas. The traditional rain-gear was only to illustrate old skills. We got away with some insight into the traditional culture and the problems caused by 'modernization' which lay ahead and the expected growth of population outrunning the agricultural productivity.

From the South Pacific islands we traveled to South-east Asia (7 h). We landed in Siem Reap, Cambodia, near the Buddhist Angkor Temple Complex. This complex was built by the Khmer Kings in the 9th–13th century, predating the Siamese Ayuttaya Kings. Our attention was drawn to this UNESCO World Heritage Cultural site (since 1992) when we visited Thailand (see page 10-9). There, in the Royal Palace in Bangkok, we were shown the perhaps 20' long model of Angkor Wat, sculpted out of granite. This model was commissioned in modern times by Rama IV, King of Siam (1804–1868). He wanted to show the grandeur within the lands under his rule. At that time, Cambodia was under Thai control until it became part of French Indochina in 1887. We also noted the similarity of the Thai and Cambodian past when watching and listening to the classical folklore dances and songs. They were similar to the ones illustrated with Figure 8.

We stayed in the formerly French 'Grand Hotel d'Anchor' and visited the magnificent, but overgrown "Ta Prohm" (jungle temple), and the well-restored Angkor Wat, a work still in progress. Figures 30 and 31 show two selected photos of Ta Prohm. Angkor Wat offered many of the widely publicized views, so that no additional photos are necessary. Besides the beauty of the sites, I must

Figure 30 *Takeover of a temple by the jungle, 2007.*

Figure 31 *A most serene face we discovered.*

remark that there, I also experienced the most oppressive heat and humidity which just drinking water could not ameliorate. To feel better, I had to replace the lost electrolytes on advice of the expedition physician. I was restored by next morning.

At this next morning, a five-hour flight brought us to Agra, India, a city of a population of ≈1.5 million. It has no convenient airport, but two UNESCO World Heritage sites, the Agra Fort and the Taj Mahal, both were added to the list in 1983. For our quick visit, we flew into a military airfield and had to promise not to take pictures of the facility. With this shortcut, we were in the early afternoon in the Taj Mahal. In 1980, I had seen already this 17th-century Mughal monument, the tribute to Shah Jahan's (favorite) wife. At that time, not many tourists were seen, as documented

Figure 32 *At the Taj Mahal 2007.*

with Figure 92C in Chapter 8. One could even venture into the lower rooms to the marble tombs of both, Shah Jahan (1592–1666) and Empress Mumtaz Mahal. By now, it was not possible to get a picture without the crowds of tourists. Figure 32 shows what a professional photographer could do, to eliminate the many people in front of the Taj Mahal.

A surprise occurred when we entered 'The Oberoi Armavilas,' one of the world's finest hotels. The hotel was only about 650 yards from the Taj Mahal. From every room, one could see the 'Taj' in the distance. Figure 33 is a photo taken with my telescopic lens from our balcony. The gardens surrounding the Taj Mahal cover the crowds beautifully. Next morning, March 12, we had a half-day tour to the second major attraction, the red Agra Fort, enclosed by a 1.5-mile-long, mighty wall. In it, Shah Jahan was under house arrest since 1657. His son, Aurangzeb, had declared him incompetent. The view left to Shah Jahan of his beloved Taj Mahal during the last years of his life is given in Figure 34.

Figure 33 *Taj Mahal from our hotel balcony.*

Figure 34 *Taj Mahal seen from the Fort Agra.*

The way to the airport allowed me to take a few pictures of street scenes in Agra. As seen in Figure 35, they do not display the splendor of the UNESCO World Hertage sites. At noon, we started our trip to Dubai in the United Arab Emirates, UAE. It took a 4 h flight over Pakistan and the Strait of Hormuz. From the airport we drove into the desert with a fleet of 4-wheel-drive SUVs. An alternate transportation in the desert is shown in Figure 36. After a romantic sunset and evening meal in the desert, we drove to our hotel, Al Qasr, in the city.

Figure 35 *A typical street scene in Agra.*

Figure 36 *The alternate transportation in the desert.*

The next morning we could visit the city built into the desert. It looked just as impressive as all the new, fast-growing cities in Asia. We saw the world's largest building taking shape, the ski slope in the desert, and what little could be saved from the 'Old Dubai Heritage,' was displayed in a modern museum just being completed. A boat ride along the Dubai Creek with a traditional, but motorized 'dhow' showed the view of the city and is given in Figure 37. The afternoon was spent at the tea room on the top floor of the Burj al Arab, the famous signature hotel of Dubai, located near our hotel. A beautiful, but hazy view let us imagine the construction on the Palm Islands along the coast of the Persian Gulf. The rest of the day was spent in the Souk Medinat (shopping center) which was located between the two hotels, offering a dazzling array of upscale goods and gold jewelry at affordable prices. This was a whirlwind tour of a city created out of what before might have been called a wasteland.

Figure 37 *View of Dubai from the Dubai Creek.*

On March 14, the next step of our tour led us to the Kilimanjaro International Airport in Tanzania. The peak of Mount Kilimanjaro, unfortunately, hid behind a layer of clouds every time we passed it. A group of tall Masai greeted us at the airport with traditional songs and dance. Next, we boarded a fleet of five small planes for the 1.5 h flight west into <u>*the Serengeti Seronera Airstrip in the middle of the National Park*</u>. On this flight we crossed the Ngorongoro Conservation Area and from above could look into the up to 10,000' high extinct and active volcanoes (without pressurization!).

Figure 38 *Cheetah on top of the expedition car surveying the Serengeti plane for prey in Tanzania.*

As on our earlier trip to Kruger National Park, RSA. (see pages 9-58–59), we had four half-day safaris in 4-wheel drive vehicles to observe the wild-life. The most memorable photo is Figure 38. One of a pair of cheetahs jumped onto the roof of one of our cars. Fortunately it ignored the passengers who had ducked away from the opening between the roof and the body of the car. The roof of these vehicles was raised whenever we wanted to observe and photograph without obstructions. We were in another car, called by radio to see the spectacle. It took about 20 min before the cheetah decided that nothing interesting was to be seen from the roof, and jumped to the ground to unite with his companion. Both walked slowly to the nearest water hole with another chance of impressive photos for us. This was a special experience for us, because all through the trip in Kruger National Park one of the members of our group absolutely wanted to see a cheetah, but despite all efforts, there, it was not to be.

The Serengeti National Park, is a UNESCO World Heritage site since 1981. It is most famous for its annual migration of wildebeest and zebra, first researched in more detail by the Grzimeks (father and son) of the Frankfurt Zoo in the 1950s,[1] well known to us from magazine articles during our time in Frankfurt. In this March, however, the migration expected from the south had not started because of sufficient rain. Instead, we had to be satisfied to watch the (considerable) local wildlife of giraffes, elephants, apes, gazelles, lions, etc., and the large numbers of birds. We quickly learned to stay away from cars carrying enthusiastic bird watchers. They were armed with field glasses and coerced the guides to stand for seemingly unending times at the same spot to watch little birds which, for us, were often barely visible (and of little interest). In the park we stayed at the Sopa Lodge, a much more basic accommodation than the luxurious Al Qasr. From there, the safaris started with stops at the "kopjes" (Afrikaans for a small, isolated, usually rocky hill), the places for predators, often lions, to hide in, and offering opportunities for photography.

Our last major stop was Marrakech, the third largest city in Morocco. As when first visiting South Africa (page 9-57), we again had misjudged the size of the African continent. The flight from east to northwest, from Kilimanjaro to Marrakech, lasted 10 h! This was just as long as the flight from Morocco to Orlando! We were back in an Islamic country and stayed in a luxury hotel (the Sofitel Marrakech). The Figures 39 and 40 give two impressions of the more traditional sights, a typical shopping area (souk) and a Berber village at the foot of the nearby Atlas mountains.

Figure 39 *The Marrakech souk (shopping area).*

Figure 40 *Berber village in the Ourike Valley.*

[1] Documented by Prof. Dr. Bernhard Grzimek with the movie "Serengeti darf nicht sterben" (Serengeti Shall Not Die). It won the Academy Award for a Documentary Feature in 1959.

As in the other cities, interesting sights and cultural items were presented during a tour of the Medina (a UNESCO World Heritage Site since 1985) with the Museum of Marrakech, the Madrasah Ben Youssef (a Saadian school and example of Moroccan architecture), the Qoubba Almoravide (the oldest remaining building in Marrakech), and a walk through the souk (see Figure 39) to the Place Djamaa el Fna (the square and heart of the old city facing the famous mosque). The afternoon was spent with a small group interested in the Berbers[1] by visiting a village (see Figure 40) and having peppermint tea (sweet) in one of the houses which we were permitted to inspect. The evening was reserved for the festive farewell dinner at the famous Yacout Restaurant.

All this was very helpful for the understanding of our trip to Andalusia, Spain, in 2008. Andalusia was for many years under control of the Moroccan Moors. Our return trip home from Marrakech had to be rerouted through Casablanca since the US Transportation Security Administration, TSA, thought that only Casablanca had a secure airport. Our own experience was that Marrakech seemed to be more secure. Late at night we arrived in Casablanca to go to our alternate hotel for the last night. There were a number of expedition members who first wanted to have a drink at Rick's[2] 'Café Américain.' This detour made them arrive in the hotel long after midnight. Others, like us, preferred a good night sleep before the trip to Orlando in the next morning (with a refueling stop on the island of Maria in the Azores).

Back at the Ritz-Carlton Resort in Orlando, we got ready for the flight back to Knoxville. Overall, we had traveled by air for 83.5 h and covered a distance of 32,000 miles (125% of the circumference of the earth, but all without frequent flyer miles!). Besides the sightseeing, we listened to and discussed the following lectures of about one hour each: "The World Wildlife Federation, a Summary of their Projects;" "Climate Changes;" "Conserving Eco-regions;" "Easter Island, Myth and Reality;" "Changing Oceans and Land Bridges of the Pacific;" "Polynesians;" "Religion and Castes in India;" "The Geo-politics of the Middle East in the 20th Century."

After arriving home, it took some time to recover from the many new impressions, answer the hundreds of e-mails that had arrived, complete the sorting, storing and printing of the over 1,000 pictures. Finally, two albums with photos and memorabilia and a DVD needed to be completed.

One final statement and a comment will close the book. The first ten years of the 20th century were recorded on page 1-1 as part of the 'the Good Old Times.' It went down-hill from there with WW I, WW II, and the far reaching Cold War. By now, in many parts of the world the first ten years of the 21st century can hardly be called 'good times.' Hopefully, they will not be considered as such in 50 years. On page 3-38 it was already said that the various weapons of mass destructions (WMDs) of the 20th century are still available and are increasingly in danger of proliferation.

---

[1] The indigenous people of North Africa. Historically, they have been known under various names, for example as Libyans by the ancient Greeks, as Numidians and Mauri by the Romans, and as Moors by medieval and early modern Europeans.

[2] The Café is featured in the movie Casablanca (1942), a romantic drama film directed by Michael Curtiz with: Humphrey Bogart, Ingrid Bergman, Paul Henreid, Claude Rains, Conrad Veidt, Sydney Greenstreet, and Peter Lorre. It is set during the early years of World War II. It won three Academy awards, including 'Best Picture.' By now, it is considered to be one of the greatest movies ever made, and its characters, dialog, music, and the Café have become iconic.

Furthermore, the use of suicide as a means to launch terror attacks to gain religious, political, or military goals was relatively unknown to Western Civilization because of the strong opposition of its dominant religions to suicide. By now, it is a new and dangerous WMD.

This appraisal of the beginning of the 21st century is not too optimistic. Our travels allowed us not only to visit European and Mediterranean antiquities, but also make a sampling of the whole world. We could not only inspect the beauty of our globe, but also had a chance to learn about the successes and failures of a number of its civilizations and societies which arose and declined over the years. We managed to get an impression about the South American Incas (Figures 18and 19 and Chapter 8, Figures 41–44); the Mesoamerican Mayas (Chapter 9, Figures 68–71, 73–74), Olmecs, Teotihuacans, and Aztecs (Chapter 8, Figures 90 and 91); the Northamerican Pueblo and Tlingit Indians (see page 8-95, and Chapter 9, Figure 76, respectively); the Mughal Empire in India and beyond (see Figure 33–34 and Chapter 8, Figure 92C); the South East Asian Khmer in Cambodia (see Figures 30 and 31) and the Ayuttaya Kings of Siam (Thailand, see Figure 8 and page 10-9); the Easter Islanders (see Figure 20), the Hawaian Islanders (see Chapter 9, Figure 90), and the tribes of the New Guinea Highlands (see Figure 26 and 29); and, finally, the varied civilizations in East Asia, as represented by Japan and China (see Chapter 8, Figures 33C and D, 93D and E, and 96B and C, Chapter 9, Figures 56, 58, and 59). It behooves us to apply the knowledge about these and other civilizations to the success of a global unity which we are approaching. This integration needs a peaceful and parallel existence of the multitude of religions, traditions, and societal structures—something presently far from being accomplished. If this alone were not already taxing, our ability to handle environmental aspect is still limited. They alone have often posed unsurmountable problems to prior human societies, as is so well described by Diamond.[1]

The environmental problems bring me back to science. It alone can identify possible paths to shape the future. If we agree, that the environment is the key to long-range survival of mankind, a new 'Golden Age of the World' in the sciences needs to be initiated (see page 8-55). The leadership toward this goal would have to be taken by a world-organization. The center of this development may well drift away from the US to countries which value education in science more and can produce and support the large number of scientists for such an effort. (See pages 9-36–37.) Science has not only constantly improved the knowledge about matter, but also about the origin and development of the universe, the solar system, the earth, and life on earth. It also has produced increasing quantitative information about the climate as it changed over the last 3,000 years.[2]

Again, drawing on my own experience, I was taught in mid-century that there *might be* the danger of a change to a colder climate (documented by extrapolation of temperature measurements over 20 years). It did not happen. The solar radiance fluctuated and reversed the trend. Now, it is public opinion that there *will be* continued global warming, One should tackle reasons of this global warming (and also the not fully forgotten global cooling) using the 'scientific method' as developed over the last few centuries. If scientific problems are attempted to be solved by political discussions and public opinion, resources will be wasted and diverted from identifying possible solutions.

---

[1] "Collapse, How Societies Choose to Fall or Succeed." VIKING, Penguin Group, New York, 2005. By Jared Diamond, scientist, specializing in evolutionary biology and biogeography.

[2] "Environmental Effects of Increased Atmospheric Carbon Dioxide" A. B. Robinson, N. E. Robinson, W. Soon, J. Am. Physicians and Surgeons, 12, 70–90 (2007). A readable review of the present knowledge of climate changes with more than 100 references to the original literature.

CHAPTER-11

A BRIEF GENEALOGY

On pages 11-2–4, the three columns list the more recent members of the family on the left while the prior generations are listed stepwise toward the right. The underlining and bolding should help in finding the connection to the family members mentioned in the text. Marriages are indicated by:

▲
▼

The two branches of grandparents on pages 11-2 and 3 are separated by a single asterisk: ✱ . Additional information about children, siblings, cousins, nephews, nieces, etc., are given in the empty spaces not used by the main family tree. The different groups of relatives are collected under the appropriate titles, separated by a line of asterisks: ... ✱✱✱ .... On pages 11-2 and 11-3 the genealogy starts on the right with the grandparents of Bernhard and Heidel, respectively, and on page 11-4, it continues to our children and grandchildren.

The relations which had frequent contact with us have been featured in more detail. Neither one of us has ever seen our great-grandparents, so the meager information about them known to us is not included.

Throughout the general text, the *German titles* of aunts, uncles, siblings, parents, grandparents and in-laws, are used as they were customary in our families and duplications were avoided as much as possible. The names are listed in the Index of Personal Names on pages 12-1–10 with the page numbers whenever they are involved in major happenings. References to figures are given in *Italics*. To easily locate the *German titles* in the genealogy, their names are written in **bold type**. Less frequently quoted and more removed relations are <u>underlined</u>. Names which are not emphasized in the listing are explained in the text, if mentioned.

As much as available, *pictures* are given on pages 11-5–12. Figure 1 shows the complete nuclear Wunderlich family as they first were reunited in the US in 1957. On pages 11-5 and 6 my direct relatives, connected with the listings on page 11-2 are shown at various ages (Figures 2–8). On pages 11-7 and 8 the connections to Heidel's direct relatives of page 11-3 are highlighted with Figures 9–16. On page 11-9, Figure 17 is of the silver wedding anniversary of Tante Marie and Onkel Rudi (July 1953), and Figure 18 is of our engagement party (Easter, 4/5/1953) and marks the joining our two families.

Finally, the listings on page 11-4 are connected to the figure on pages 11-10–12. Page 11-10 briefly illustrates our immigration to the US (Figures 19–21). Details about these three pictures can be found in Chapters 4, 5, and 6, respectively. On page 11-11 some important happenings are highlighted with Figures 12–17 from our time in Hastings, NE, Chicago, IL, and Ithaca, NY. The specifics of these time periods can be seen in Chapters 6 and 7. With page 11-12, a 25-year jump is made to the 21st century, i.e., to the events mentioned in the Epilogue (Chapter 10). Photographs are shown of our golden wedding anniversary in 2003 in Knoxville, TN, the marriage of Caryn and Brian in 2006, and finally, the graduation of our grandchildren, Stefan and Alyssa, from Ridge View High School in Columbia, SC in 2005 and 2007, respectively.

The Children of Mutti und Vati:

1. Gisela Ruth Margarete,
born: May 30, 1927
died: June 6, 1927.

2. Richard Herrmann **Frank**
born: Juli 18, 1928,
died: Jan. 11, 1992,
married: to **Dagmar**, née Hanf, April, 26, 1951, in Brandenburg; Son: Eberhard (*born* in 1952).

3. **Bernhard** (Otto Klaus)
born: May 28, 1931 in Brandenburg an der Havel;
married: to **Heidel**, née Felix, Dec. 28, 1953, in Frankfurt/Main. Daughter: **Caryn**. Son: **Brent**; for details see page 11-4.

4. **Joachim**
born: Jan. 12, 1944,
married: to **Laura**, née Marvel, June, 7, 1998 in Hastings, NE.

Son: Ian Eric (*born* 1972, when Joachim was *married* to Sharon, née Kailey, 1966).

5. **Manfred** Dieter
born: May 12, 1946,
married: to **Carol** née Groom, Dec. 21, 1968, in Lincoln, NE. Son: David (*born* in 1974), Daughter: Julie (*born* in 1976).

****

Bernhard's Nephews & Niece:
A. Eberhard, *married* to Shahla née Mozzafari in 1978.
1. Son Roland (*born* 1979),
2. Daughter Zeba (*born* 1981).
B. Ian, *married* to Amy, née Gray in Napa Valley in 2007,
C. David (who changed his last name to Andrae), *married* to Amanda née Meinert in 2001.
1. Daughter Mabel (*born* 2002),
2. Son Finn (*born* 2004).
D. Julie, *married* in Omaha in 2005 to Nicholas Bernard.

Vati's Siblings:
1. Elsbeth Anna, née W. (1900–97); *married* to R.A, Pass. **(Tante Else, Onkel Richard II)**. [*Children*: Henry, 1923; Berthold (Bertie), 1927; Klaus-Jürgen, 1935; Konrad, 1943; Anna-Margarete (Annegret), 1945].

2. Margarete (1901–91); **(Tante Grete)**, not *married*, lived last in Stralsund, buried in Roskow by Brandenburg.

****

Bernhard's Parents:

Richard Otto Wunderlich **(Vati)**
born: July 8, 1903
in Paunsdorf, Sachsen,
married: Oct. 10, 1925, in
Brandenburg a.d. Havel.
died: July 29, 1985, in
Hastings, Nebraska.

▲
▼

Johanne Marie Wohlgefahrt **(Mutti)**
born: March 3, 1906 in
Vechelde, Braunschweig,
married: Oct. 10, 1925, in
Brandenburg a.d. Havel,
died: May 30, 1999, in
Hastings, Nebraska.

****

Mutti's Siblings:
1. Wilhelmine **(Tante Minna)** *born:* 1894, *married* P. Schmalz **(Onkel Paul)** *born:* 1895, children: Kurt, Werner, and the twins Irmgard and Rudi (b.1926).
2. Hermann **(Onkel Hermann** *born:* 1899, *married* to **(Tante Frieda)**, children: Heinz, Harald.
3. Elise (1901–2005), *married* to R. Zaroban; **(Tante Lieschen & Onkel Richard)**, children: Richard, Don, Jim, Marilyn.

Bernhard's Grandparents:

Otto Ernst Julius Wunderlich **(Opa)**
born: June 19, 1877,
in Stargard, Pommern;
married: March 24, 1900, in Landsberg/Warthe, now Gorzow Wielkopolski (Poland).
(In 1943 *remarried* to "**Tante Lorchen**");
died: June 7, 1953 in
Brandenburg a.d. Havel.

▲
▼

Anna Berta Meiling **(Oma)**
born: July 26, 1879, Landsberg/ Warthe (Gorzow Wielkopolski);
married: March 24, 1900, in Landsberg, moved ≈1936 from Brandenburg to live with Tante Grete in Stralsund, Pommern, where she *died:* May 30, 1944.

*

Hermann O. K. Wohlgefahrt **(Opa Siedlung)**
born: August 27, 1868 in
Zinthen, Ostpreußen,
married: Sept. 24, 1893 in
Pannawitz, Ostpreußen.
died: May 21, 1938, in
Brandenburg a.d. Havel.

▲
▼

Johanna L. Neumann **(Oma Siedlung)**
born: Dec. 18, 1865 in
Stolzenberg, Ostpreußen,
married: Sept. 24, 1893, in Pannawitz, Kreis Heiligenbeil (now Russia), The Wohlgefahrts moved later to Vechelde by Braunschweig, and in 1914 to Brandenburg, to work in the newly built Walzwerk of the Mitteldeutsche Stahlwerke.,
died: Jan. 24, 1942 in Brdbg.

Mama's Siblings Part B:

4. Luzie **(Tante Luzie)**
born: 12/10, 1910,
married: 1931 to Franz Merkel.

5. Hubert **(Onkel Hubert)**
born: 1917,
married: 1951, to Margareta (née Müller) **(Tante Grete II)**
born: 1920;
died: 2007

****

Child of Mama and Papa:

Adelheid Christa (**Heidel** or Heidi), *born:* March 26, 1930, in Mechtal/Beuthen (now Bytom)# *married* **Bernhard** Wunderlich in 1953 in Frankfurt am Main. Daughter: **Caryn**. Son: **Brent**; for details see page 11-4.

****

Selected Cousins of Heidel:

1. Son of Tante Marie: Herbert Czichos (1931–1983) *married* to Charlotte née Kräner in 1960 (three sons).

2. Daughters of Tante Gertrud: Irmgard Surowski (1928–2002), *married* 1957 to Henning Schuer. Ria Surowski (1940–1998) *married* 1976 to Karl Bűhler.

3. Daughters of Victor Maleika Helga (*born* 1934). Renate (*born* 1936).

4. Children of Tante Luzie: Total of 12, only one listed, Adelheid Erbe, *born:* 1932.

5. Sons of Onkel Hubert: Manfred (*born* 1945) Wolfgang (*born* 1951) Rainer (*born* 1962).

Papa's Siblings:

1 Robert Feliks **(Onkel Robert)** (1897–1980), *married* to Maria (1902–1985); children: Maria (1923), Gertrud (1925), Steffi (1928), Georg (1935), Resi (1940); Onkel Robert lived in Sczedrzik, Oberschlesien.#
2. Anton Felix **(Onkel Anton)** *born:* 1898, *married* to Gertrud, née Felix; (son, *died* as child).

****

Heidel's Parents:

Alois Felix **(Papa)**
born: July 21, 1899 in Sczedrzik, Oberschlesien,#
married: May 7, 1928, in Beuthen, Oberschlesien,#
died: June 7, 1992 in Quadrath-Ichendorf, Nordrhein-Westfalen.

▲▼

Katharina Emma **(Mama)**
born: April 25, 1902, in Schlesiengrube,#
married: May 7, 1928, in Beuthen, Oberschlesien,#
died: Feb. 6, 1979, Bergheim, Nordrhein-Westfalen.

****

Mama's Siblings Part A:

1. Marie (1900–84) **(Tante Marie),** *married* to Rudolf Czichos **(Onkel Rudi)**.
2. Gertrud Antonie (1908–85), **(Tante Gertrud)** *married* 1931 to Joseph Surowski (1907–1988) **(Onkel Sepp)**.
3. Victor (1909–1947) **(Onkel Victor),** *married:* 1931 to Anna (1911–2003) (née Witoschek) **(Tante Anna)**.

Mama's Siblings Part B:
>>See at the upper left<<

Heidel's Grandparents:

Konstantin Felix **(Opa II)**
born: April 10, 1864, in Kreuzthal, Oberschlesien#,
married: ≈1895 in Sczedrzik in Oberschlesien,# (note that Sczedrzik was called Hitlersee from 1934–1945),
died: 1939 in Sczedrzik in Oberschlesien.#

▲▼

Marie Jonietz **(Oma II)**
born: Sept. 10, 1864 in Sczedrzik, Oberschlesien,#
married: ≈1895 in Sczedrzik, Oberschlesien,#
died: 1934 in Sczedrzik, Oberschlesien.#

*

Emanuel Maleika **(Opa III)**
born: Sept. 23, 1873 in Mechtal# by Beuthen, Oberschlesien (now Bytom, Poland),
married: ≈1900 in Mechtal,#
died: 1929, in Mechtal by Beuthen, Oberschlesien.#

▲▼

Marie Froehlich **(Oma III)**
born: June 8, 1871 in Oberschlesien,#
married: 1898 in Mechtal, Oberschlesien,#
died: 1904, in Mechtal by Beuthen, Oberschlesien.#

[Pauline, née Golla **(Stiefoma)**, second wife of Opa
married: 1905;
died: 1945 in Mechtal.#]

Now located in Poland

Our Grandchildren:

Stefan, *born:* August 30, 1987, in Syracuse, NY.

Early interests included karate (black belt), playing the clarinet in the school band. His Boy Scout activities led to the rank of Eagle. In 2005 he became a chemistry student at Clemson University, SC.

Alyssa Marie, *born:* May 1, 1989, in Syracuse, NY.

Early interests included competitive swimming and playing the flute in the high school orchestra. Her other accomplishments included drawings and paintings. In 2007 she became a student in the SC University System.

****

The Kosar Family:

Parents Michael (1922–83) and **Maria** (*born:* 1926), *married* in 1947, then they emigrated from Germany in 1949. In Germany he was in a concentration camp, and she, a forced laborer from the vicinity of L'vov, Ukraine (at that time, part of the USSR).

Their children:

1. Rosalia (Rosie) (*born*: 1948).
2. Daniel (*born*: 1953), *married* to Cherryl (née Stein) in 1990.
3. **Anna** (*born* 1955) *married* to **Brent**, see bottom center.
4. John (*born:* 1958).

The Murphy Family

Parents: James (died 1968), *married:* 1938 to Anna (died 1975), née, Follenus.

Their Children:

1. **Brian** Bernard, emigrated from Ireland in 1965, *married:* to **Caryn** (see above, center)
2. Marie, *married:* to Finbar (Barry) Fennell, residing in Dublin, Ireland.

Our Children:

Caryn Cornelia **Wunderlich**
born: Sept. 20, 1956,
in Evanston IL

Caryn stayed at home through her college education at RPI in Troy and went 1967/68 for ½ year to a German school in Mainz-Kastell (parents sabbatic). The year 1976/77 was spent as an exchange student in Zürich, Switzerland at the ETH . Her stay at home was followed by medical school at the University of Chicago (1978–1982, MD). A series of internships and employments were in Baltimore, MD, San Antonio, TX, and Richmond, VA, before moving to Knoxville, TN in 1992.

married: 2006 at the Biltmore Estate in Asheville, NC to Dr. **Brian** Murphy, Oak Ridge, TN.

**

Brent Bernhard **Wunderlich**
born: Sept. 22, 1959,
in Ithaca, NY.

Brent stayed at home through his college education at RPI in Troy, NY and went for one half year to the German elementary school in Mainz-Kastell in 1967/68 during the sabbatic leave of his parents. In 1981 he accepted employment with Bausch & Lomb in Rochester, in 1983 with Carrier in Syracuse NY where he also received his MBA at Syracuse University. Later employment brought him to Columbia, SC with a ½ year stay in Göttingen. His employer is CSC in Columbia, SC (Computer Sciences Corporation) and he teaches graduate classes at Webster University since 1999.

married: Nov. 24, 1984, in Utica, NY to **Anna** Kosar.

Our Vital Statistics:

Bernhard Wunderlich
born: May 28, 1931,

in Brandenburg an der Havel, Germany. At that time Germany was a new democracy, established in 1919, after WW I as the Weimarer Republik, which in 1933 changed to the "3rd Reich" under the dictatorship of Adolf Hitler. After WW II, in 1945, it became the zone of Soviet Russian occupation (1945–49), and finally, changed to the GDR dictatorship (German Democratic Republic). In 1953 the whole Wunderlich Family fled to Berlin and then to the, Bundesrepublik Germany (FRG).

married: Dec. 28, 1953,

in Frankfurt am Main (FRG). In 1954, the Wunderlich Family emigrated to the US.

▲
▼

Adelheid Christa **Felix**
(**Heidel** or Heidi),
born: March 26, 1930,

in Mechtal by Beuthen, Oberschlesien, which is now Bytom, in Poland. In 1939, she moved with the Felix Family to Brandenburg an der Havel. This was followed by the same moves as for the Wunderlich Family after 1945.

married: Dec. 28, 1953,

in Frankfurt am Main (FRG). In 1954 emigration to the US as part of the Wunderlich Family. Both of us became US citizens on May 2, 1960. In the US we moved in 1955 to Chicago, IL, in 1958 to Ithaca, NY, in 1963 to Troy NY, and finally, in 1988 to Knoxville, TN. In 1967/68 and 1986/87 we had two extensive sabbatic leaves in Germany (at the Universities of Mainz, Freiburg, and Ulm).

Pictures to Page 11-2

Figure 1 *The Wunderlich Family, Christmas 1957 in Hastings Nebraska in front of Vati and Mutti's House, W 12th Street. From left to right are: Dagmar and Frank with Eberhard; Manfred, Mutti and Vati, Joachim; Heidel with Caryn and Bernhard.*

Figure 2 *(Right) The Wunderlich Family, 1949 in Brandenburg an der Havel. From left to right: Bernhard, Joachim, Frank (above), Manfred (below), Mutti, and Vati.*

Figure 3 *(Below) Family get-together in 1937 on the occasion of the visit of Tante Lieschen from the USA. In our 1st-floor apartment in Brandenburg, Wilhelmsdorfer Straße 70. From left to right: Oma and Opa Siedlung, Opa with Tante Lorchen, Tante Lieschen, Mutti, and Vati.*

Figure 4 *(Above) The only picture we have from my grandmother (Oma), Anna née Meiling. Excerpted from a picture taken in Stralsund, ≈1940.*

Figure 5 *The gathering in Oma Siedlung's garden in Brandenburg in 1937. From right to left, top row: Tante Minna's oldest son, Onkel Paul, Tante Frieda (her husband, Onkel Hermann is shown in Fig. 7), Tante Minna (Onkel Paul's wife) and Opa Siedlung. Middle row, from left to right: Heinz (Onkel Hermann's and Tante Frieda's son, died 1940 in Finland), Oma Siedlung and the twins of Tante Minna and Onkel Paul. Bottom row left to right: Jim, the visiting son of Tante Lieschen, and Harald (Onkel Hermann and Tante Frieda's second son).*

Figure 6 *Christmas 1956 in Hastings Nebraska in Vati and Mutti's 'new' house. From left to right: Marilyn (Tante Lieschen and Onkel Richard's daughter), Tante Lieschen, Onkel Richard with Caryn, and Grandmother Zaroban, née Boehl (Onkel Richard's mother).*

Figure 7 *(Left) Oma Siedlung and Onkel Hermann in in Oma's garden in 1937.*

Figure 8 *(Right) A picture of Tante Grete in 1974.*

Pictures to Page 11-3

Figure 9 *(Left) Mama's Communion in 1911. From left to right are the Stiefoma, Opa, and Mama.*

Figure 10 *(Right) The two sisters, Mama, left, and Tante Marie, about 1924.*

Figure 11 *Tante Luzie's Communion in 1921. From left to right: Tante Gertrud, Opa III, Onkel Victor, Stiefoma, Tante Luzie. In front, Onkel Hubert.*

Figure 12 *Wedding picture of Mama and Papa, 1928.*

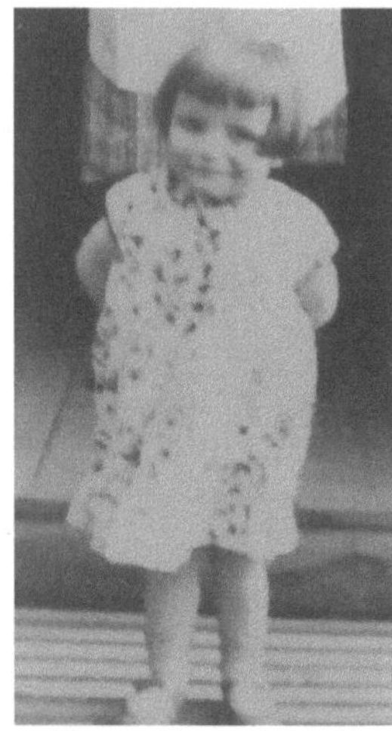

Figure 13 *Heidel, 1933.*

Figure 14 *Heidel, 1936.*

Figure 15 *Mama and Heidel, 1939.*

Figure 16 *Heidel 1946.*

Figure 17 *Silver anniversary of Tante Marie und Onkel Rudi. From left to right bottom row: Christa (first wife of Herbert) and mother, Tante Marie and Onkel Rudi, Mama and Papa, Center Row: Friends of the anniversary couple. Top row: Herbert with a friend.*

Figure 18 *Engagement 1953. From left to right: Mama, Heidel, Bernhard, Mutti, Joachim, Vati, Manfred, Papa.*

Pictures to Page 11–4

Figure 19: *1951, The first portrait made of Heidel and Bernhard, shortly after getting to know each other (on the occasion of the initial OK for emigration of the Wunderlich family from the GDR to the US! Which, however, finally was not permitted and Heidel did not have to be left behind).*

Figure 20 *The last visit from Mama and Papa in Frankfurt-Main in 1954 before we left for the US.*

Figure 21 *Arrival in the 'New World' on the pier in New York City in the middle of all our possessions. The two big wooden crates were made so that they later could serve as boxes for the winter-supply of potatoes. Little did we know that in the US potatoes are not bought in 100-pound quantities and stored for the winter. Left: Joachim and Mutti, Center: Tante Lieschen, Right: Vati in leather coat.*

Figure 22 *Our first car (a used 1949 Chevrolet) and our first apartment in Hastings, NE, 113 East 4th Street (in the background, lower left half of the house) early in 1955.*

Figure 23 *Fall 1955, our apartment and car in Chicago facing Sheridan Road. The windows are on the immediate left of the red light at the street corner.*

Figure 24 *1956, Caryn, our new arrival in Chicago. Born on September 20 in the Hospital of Evanston, IL.*

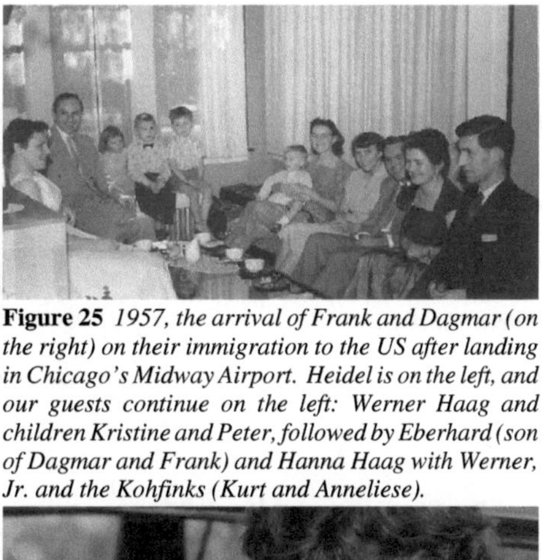

Figure 25 *1957, the arrival of Frank and Dagmar (on the right) on their immigration to the US after landing in Chicago's Midway Airport. Heidel is on the left, and our guests continue on the left: Werner Haag and children Kristine and Peter, followed by Eberhard (son of Dagmar and Frank) and Hanna Haag with Werner, Jr. and the Kohfinks (Kurt and Anneliese).*

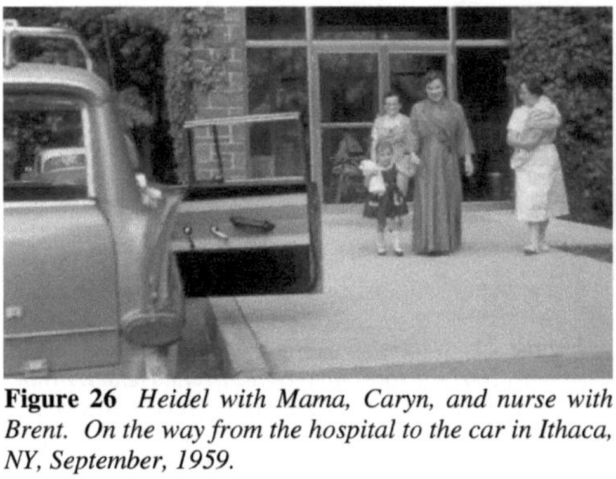

Figure 26 *Heidel with Mama, Caryn, and nurse with Brent. On the way from the hospital to the car in Ithaca, NY, September, 1959.*

Figure 27 *1959, Brent, the new addition to the family.*

Figure 28 *Family-picture on the day of the golden wedding anniversary. Top row from left to right Alyssa, Caryn, Brent, Stefan, and Anna. In the center you see Heidel and Bernhard, December 28, 2003.*

Figure 29 *Family picture on Caryn and Brian's wedding at the Biltmore on October 29, 2006. Left to right: Brian's sister Marie and her husband (Barry Fennell), Heidel and Bernhard, followed by the wedded couple Caryn and Brian, and Brian's daughter (Eileen Murphy)*.

Figure 30 *Ridge View High School, graduation of Stefan in 2005. From left to right: Bernhard, Heidel, Brent, Rosie Kosar (Anna's Sister), Caryn, Stefan, Anna, and Anna's Mother (Maria). Low center: Alyssa and John (Anna's Brother).*

Figure 31 *Ridge View High School, graduation of Alyssa (2007). From left to right: Brent, Heidel, Brian, Alyssa, Caryn, Bernhard, Dagmar (Alyssa and Stefan's aunt), Stefan, and Anna. The graduation ceremony was held in the Coliseum of USC in Columbia.*

Indices

The page numbers are given by the Chapter numbers in **bold** type, followed by the page number, as in: **10-**17. Page numbers in *italics* refer to figures. The Preface is indexed with the letters 'Pref' without the (roman) page numbers. The subjects prominently referred to in specific Chapters are listed by the Chapter and its number. Specific page references point in this case to major items only. The 'Index of Personal Names' starts below; for the 'Subject Index,' see the pages from 12-11–37.

Ranges of the page numbers for the Chapters:

Index of Personal Names

The index entries for the names of the immediate family (Vati, Mutti, and the author, Bernhard, as well as Heidel, Caryn, and Brent) are limited to major events and references to figures (italics). The full identity of relatives can be recognized by references to the genealogy on pages **11**-2–4. The last names of friends listed under their first names can be identified on the text page of the first entry.

A

B

C

D

E

F

G

H

I, J

K

L

M

N

O

P, Q

R

S

T

U, V

W

X, Y, Z

Subject Index

For page numbering, see page 11-1. For unexplained acronyms and abbreviations see the first text entry.

B

C

D

E

G

H

I

L

M

R

S

T

U

V

W

X

Y

Z

Springer Nature Customer Service Center
Europaplatz 3, 69115 Heidelberg, Germany

MIX
Papier aus verantwortungsvollen Quellen
Paper from responsible sources
FSC® C105338

If you have any concerns about our products,
you can contact us on
ProductSafety@springernature.com

In case Publisher is established outside the EU,
the EU authorized representative is:
Springer Nature Customer Service Center GmbH
Europaplatz 3, 69115 Heidelberg, Germany

Printed by Libri Plureos GmbH
in Hamburg, Germany